Fundamental Theories of Physics

Volume 187

The international monograph series “Fundamental Theories of Physics” aims to stretch the boundaries of mainstream physics by clarifying and developing the theoretical and conceptual framework of physics and by applying it to a wide range of interdisciplinary scientific fields. Original contributions in well-established fields such as Quantum Physics, Relativity Theory, Cosmology, Quantum Field Theory, Statistical Mechanics and Nonlinear Dynamics are welcome. The series also provides a forum for non-conventional approaches to these fields. Publications should present new and promising ideas, with prospects for their further development, and carefully show how they connect to conventional views of the topic. Although the aim of this series is to go beyond established mainstream physics, a high profile and open-minded Editorial Board will evaluate all contributions carefully to ensure a high scientific standard.

More information about this series at http://www.springer.com/series/6001

Jasjeet Singh Bagla · Sunu Engineer
Editors

Gravity and the Quantum

Pedagogical Essays on Cosmology, Astrophysics, and Quantum Gravity

Subtle is the Malice of the Lord

Springer

Editors
Jasjeet Singh Bagla
Department of Physical Sciences
IISER Mohali
Mohali, Punjab
India

Sunu Engineer
Ryussi Technologies Pvt. Ltd.
Pune
India

ISSN 0168-1222 ISSN 2365-6425 (electronic)
Fundamental Theories of Physics
ISBN 978-3-319-51699-8 ISBN 978-3-319-51700-1 (eBook)
DOI 10.1007/978-3-319-51700-1

Library of Congress Control Number: 2016963293

Printed on acid-free paper

This Springer imprint is published by Springer Nature
The registered company is Springer International Publishing AG
The registered company address is: Gewerbestrasse 11, 6330 Cham, Switzerland

Preface

Professor T. Padmanabhan (aka Paddy) from the Inter-University Centre for Astronomy and Astrophysics in Pune, India, is well known as an astronomer and cosmologist. He has contributed immensely to many areas though the core theme has been different aspects of gravitation. He has written a number of books on physics, gravitational theories, and cosmology. He has taught and mentored a large number of students. In developing the many themes that constitute the body of his research, he has collaborated with numerous scientists across the planet.

This volume of collected papers from various collaborators, colleagues, and students across the ages has been compiled to commemorate his contributions to physics on the day of his sixtieth birthday, March 10, 2017. In many ways, it is an expression of appreciation of Paddy as a scientist, expositor, teacher, and mentor. The spectrum of topics that these papers touch upon is quite variegated though connected via a gravitational core as the title indicates. We are confident that you will enjoy them.

We thank all the contributors for putting in effort to prepare articles that take a critical look at fundamental issues. We are greatly appreciative of their contributions that have made this volume a reality. We also thank our patient editor at Springer Ms. Angela Lahee, who has borne our delays with fortitude.

Mohali, India
Pune, India
November 2016

Jasjeet Singh Bagla
Sunu Engineer

Contents

Prof. Padmanabhan: A Personal and Professional History 1
Jasjeet Singh Bagla and Sunu Engineer

Measuring Baryon Acoustic Oscillations with Angular Two-Point Correlation Function 11
Jailson S. Alcaniz, Gabriela C. Carvalho, Armando Bernui, Joel C. Carvalho and Micol Benetti
1 Introduction 11
2 The Angular Two-Point Correlation Function 12
3 Cosmological Constraints 16
4 Conclusions 18
References 19

Resonant Disruption of Binary Stars by a Catalytic Black Hole 21
C.M. Boily
1 Introduction 21
2 Anisotropic Motion 22
3 Binary Stars 25
4 Numerical Examples 26
5 Outlook 28
References 28

Mechanics of Apparent Horizon in Two Dimensional Dilaton Gravity 31
Rong-Gen Cai and Li-Ming Cao
1 Introduction 31
2 General Dilaton Gravity Theory in Two Dimensions 33
3 Some Solutions of the Theory 34
4 Apparent Horizon 35
5 The Mechanics of the Apparent Horizon 37
References 41

Boundary Terms of the Einstein–Hilbert Action 43
Sumanta Chakraborty
1 Introduction 43
2 Reconciling Boundary Terms for the Einstein–Hilbert Action 45
3 (1+3) Decomposition, Time Derivatives and Canonical Momenta 51
4 Null Surfaces: Completing the Circle 55
5 Concluding Remarks 58
References 59

Decay of the Cosmic Vacuum Energy 61
Timothy Clifton and John D. Barrow
1 Introduction 61
2 Background Evolution 63
3 Energy Exchange in the Presence of a Non-interacting Fluid 64
4 Perturbations Generated During Inflation 66
5 Discussion 70
References 71

Understanding General Relativity After 100 Years: A Matter of Perspective 73
Naresh Dadhich
1 Prologue 73
2 Introduction 74
3 At the Very Beginning 76
4 Self Interaction and Vacuum Energy 79
5 In Higher Dimension 82
6 Outlook and Perspective 84
References 86

Piecewise Conserved Quantities 89
Tevian Dray
1 Introduction 89
2 Piecewise Smooth Tensors 90
3 Piecewise Conserved Quantities 90
4 The Patchwork Divergence Theorem 91
5 Shells of Matter 92
6 Signature Change 94
7 Lorentzian Vector Calculus 95
8 Summary 97
References 97

Units of the Nonlinear Universe 99
Sunu Engineer
1 Introduction 99
2 General Features of Nonlinear Evolution 100

3 Correlation Functions, Density Profiles and Stable Clustering 103
4 Self-similar Evolution 107
5 Units of the Nonlinear Universe 109
References 116

Self-similarity and Criticality in Gravitational Collapse 117
Pankaj S. Joshi
References 126

Notes on Semiclassical Weyl Gravity 127
Claus Kiefer and Branislav Nikolić
1 Notes on Semiclassical Einstein Gravity 127
2 Quantization of Conformal (Weyl) Gravity 131
3 Semiclassical Weyl Gravity and the Recovery of Time 139
4 Outlook 142
References 143

Accelerated Observers, Thermal Entropy, and Spacetime Curvature 145
Dawood Kothawala
1 Gravity and Thermodynamics 145
2 Gravity $\Longleftrightarrow$ Accelerated Frames $\Longleftrightarrow$ Thermodynamics 147
3 Thermal Entropy and Spacetime Curvature 153
4 Concluding Remarks 157
References 166

100 Years of the Cosmological Constant: Past, Present and Future 169
Ofer Lahav
1 Introduction 169
2 Background 170
3 The Dark Energy Survey 171
4 The Globalization of Research: Pros and Cons 173
References 174

Pedagogical and Real Physics 175
Malcolm S. Longair
1 Introduction 175
2 A Personal Prelude 175
3 Galileo and the Nature of the Physical Sciences 177
4 Maxwell, Reduced Momenta, Differential Gears and Electromagnetism 179
5 Einstein and Statistics 186
6 Cosmological Conundrums—Space–time Diagrams for the Standard World Models 196

7 Epilogue 206
References 207

A Local Stress Tensor for Gravity Fields 209
D. Lynden-Bell
1 Introduction 209
2 Gravity Field's Energy Density for Spheres 212
3 Gravity's Energy Density Discussed 215
4 Potentials Do Not Yield Energy Densities 215
5 Gravitational Field Stresses 215
6 A Flat 3-Space for Every Static Space 217
7 Thanks Paddy 218
References 219

Nonlocal Infrared Modifications of Gravity. A Review 221
Michele Maggiore
1 Introduction 221
2 Nonlocality and Quantum Effective Actions 223
3 The Anomaly-Induced Effective Action 227
4 Nonlocality and Mass Terms 234
5 How Not to Deal with Effective Nonlocal Theories 239
6 Localization of Nonlocal Theories 243
7 Cosmological Consequences 251
8 Toward a Fundamental Understanding 276
References 278

Emergence of Gravity and RG Flow 283
Ayan Mukhopadhyay
1 A Route to Explore the Holographic Origin of Gravity 283
2 Reformulating Gravity as a *Highly Efficient RG Flow* 286
3 The Field Theory Perspective and the Hydrodynamic Limit 295
4 Outlook 299
References 300

Modelling Non-paradoxical Loss of Information in Black Hole Evaporation 303
Sujoy K. Modak and Daniel Sudarsky
1 Introduction 303
2 Measurement Problem and Models of Wave-Function Collapse 306
3 Callan–Giddings–Harvey–Strominger (CGHS) Model 308
4 Gravitational Induced Collapse of Wave-Function and Loss of Information 310
5 Discussions 315
References 315

Relativistic Paths: A Feynman Problem 317
Jayant V. Narlikar
1 Preamble 317
2 Introduction 317
3 Motion in One Space + One-Time Dimensions 319
4 Motion in 3 + 1 Dimensions 324
5 The Non-relativistic Approximation 326
6 Conclusion 332
References 337

Einstein Equations from/as Thermodynamics of Spacetime 339
Krishnamohan Parattu
1 Introduction 340
2 Comparing the Two Results in GNC 341
3 Discussion 349
References 351

Thoughts on 50 Years in Astrophysics and Cosmology and on What Comes Next 353
Martin J. Rees
1 Introduction 353
2 The Resurgence of Relativity: Some History 354
3 Probing the Early Universe 356
4 The Geometry and the Expansion 357
5 Beyond the Horizon 359
6 Concluding Comments 361
References 361

Area Theorem: General Relativity and Beyond 363
Sudipta Sarkar
References 373

What Are the Atoms of the Space Time? 375
S. Shankaranarayanan
1 Introduction 375
2 Black Hole Thermodynamics 377
3 Fluid Gravity Correspondence in Arbitrary Dimensions 378
4 Horizon Fluid Energy 381
5 Modeling the Horizon Fluid 384
6 Fluctuations of the Horizon Fluid 389
7 What Are the Atoms of Space Time? 393
References 395

From Quantum to Classical in the Sky 397
Suprit Singh
References 409

Classical and Quantum: A Conflict of Interest 411
T.P. Singh
1 Some Limitations of Quantum Theory 411
2 The Quantum Measurement Problem 413
3 The Problem of Time in Quantum Theory 418
4 Quantum Non-locality and Space-Time Structure 420
5 Implications for a Quantum Theory of Gravity? 422
References 423

Four Principles for Quantum Gravity 427
Lee Smolin
1 Introduction 427
2 Four Principles for Quantum Gravity 430
3 Recovery of the Einstein Equations 436
4 AdS/CFT as an Example of the Quantum Equivalence Principle 439
5 The Origin of MOND from the Quantum Equivalence Principle 441
6 Conclusions 446
References 447

What Do Detectors Detect? 451
L. Sriramkumar
1 Introduction 451
2 Response of the Unruh–DeWitt Detector in Flat Spacetime 453
3 'Inversion of Statistics' in Odd Dimensions 461
4 Detecting Planck Scale Effects 465
5 Finite Time Detectors 474
6 Summary 475
References 476

Stability Longevity and All That: False Vacua and Topological Defects 479
Urjit A. Yajnik
1 Cuisine and a Canine 479
2 Introduction 480
3 Topological Solutions and Fractionalised Fermion Number 481
4 Assignment of Fermion Number 483
5 Energetics and Dynamics of the Thin, False String 485
6 Instantons and the Bulge 487
7 Conclusion 491
References 492

Prof. Padmanabhan: A Personal and Professional History

Jasjeet Singh Bagla and Sunu Engineer

Prof. T. Padmanabhan (known to friends and students as Paddy) was born on 10 March 1957 in a lower middle class family at Trivandrum, Kerala, India. His mother, Lakshmi, was a home-maker. His father, Thanu Iyer, had a genius for mathematics but had to abandon his academic pursuits because of family circumstances and take up a job in the Forest department of the Government of Kerala. However, his father, as well as several other family members of his father's generation, had a great passion for all of mathematics, especially geometry. Two strong inspirations in Paddy's early life, which influenced him to take up academic pursuits, were his father and another senior member of the family, Neelakanta Sarma. Both of them had a high level of personal integrity and passion for knowledge — two qualities which Paddy inherited. Paddy recalls that the code of life emphasised in the family circles in which he grew up was simple: "*Excellence is not negotiable!*"

Given this background, it was no surprise that Paddy acquired a high level of expertise in mathematics — well ahead of what was taught in his school, the Government Karamana High School, Trivandrum, where he did his schooling in the vernacular, Malayalam medium — and developed a strong interest in geometry. Other than his mathematical abilities, Paddy was not a child prodigy of any kind; while he was within the top three students in his class all along, he was not even a consistent class topper in his school. (His major problem was the Hindi language, which was compulsory; he regrets that he still hasn't learnt it!) Another passion during his school days was chess, which also he learnt from Neelakanta Sarma. Unfortunately, even state level competitive chess needed devoting so many hours which he could ill afford, and, at some stage, he decided to pursue academics rather than chess — a decision about which he has occasional regrets even today!

After ten years of schooling (1963–1972), he joined the Government Arts College in Trivandrum for two years of the Pre-degree (1973–1974) as it was called then. Three major events occurred during this period, which, sort of, decided his future lifeline.

J.S. Bagla (✉) · S. Engineer
Department of Physical Sciences, IISER, Mohali, Punjab, India
e-mail: jasjeet@iisermohali.ac.in

J.S. Bagla and S. Engineer (eds.), *Gravity and the Quantum*,
Fundamental Theories of Physics 187, DOI 10.1007/978-3-319-51700-1_1

First, he came across the *Feynman Lectures in Physics*, and found Physics to be more fascinating than pure mathematics, which was the original career he was planning to pursue. “It appeared to me”, says Paddy, “that theoretical physics beautifully combines the best of objective science and the elegance of pure mathematics.” Though the Feynman lectures did influence his decision to change his mind as regards his career, Paddy is hardly a fan of Feynman as a person! In fact, other than influencing his career decision, the Feynman lectures did not figure in his physics education directly; he learnt the first round of physics from the 5 volumes of the *Berkeley Physics Course*, and later on from the 10 volumes of Landau and Lifshitz’s *Course of Theoretical Physics*.

Second, he came across a wonderful organization — and later became an active member of it — called the Trivandrum Science Society. “Thinking back, I find it hard to believe that such an organization existed and flourished, influencing a handful of us so strongly. It was a transient phenomenon, which lasted for just about a decade,” says Paddy. This was an organization entirely run by students in Trivandrum colleges, financed by membership fees and donations from well-wishers. Here, the members devoted themselves to the pursuit of science, unshackled by curricula and examinations. Paddy and a few others had an informal self-study group which concentrated on theoretical physics, and in a span of about 3 years, Paddy managed to master the volumes of the *Course of Theoretical Physics* by Landau and Lifshitz.

Third, Paddy took the National Science Talent Search (NSTS) examination organized by the NCERT, Government of India, which was probably one of the greatest stimuli that the government provided to students who wanted to pursue pure science. Success in this examination guaranteed a handsome scholarship for the rest of one’s scientific career, as long as one pursued pure science. The money was very important to people like Paddy, whose family’s financial position was never too good. In addition, NSTS scholars could participate in one-month summer camps at leading institutes in the country, allowing them to interact with researchers even while they were pursuing their college education. The Trivandrum Science Society also used to run “classes” to prepare students for the NSTS exam. These classes were run essentially by senior students and sometimes by one’s own contemporaries, but it was really a wonderful procedure for entrapment: “Whether one got through the NSTS exam or not,” says Paddy, “the classes made you realize that the pursuit of pure science is the best thing one can do. The indoctrination was done rather subtly but very effectively!”

After his pre-degree, Paddy joined the University College, Trivandrum for his Bachelor’s Degree (B.Sc, 1974–1977) in Physics. His final year of pre-degree and the first two years of B.Sc were again noteworthy in two respects. First, this was the time when he worked through the Landau-Lifshitz volumes and various other books in theoretical physics, spending sometimes 14 hours a day in this pursuit. He also acquired a fascination for anything related to gravity and was strongly influenced by the epic book *Gravitation* by Misner, Thorne and Wheeler (Freeman and Co.). “This book was an eye opener for me; I am probably one of the few people who have worked through every problem in this book. I still have hundreds of pages of handwritten notes I took from this book, since I could not afford to buy it — and xeroxing was unheard of in the seventies in Trivandrum!" says Paddy. The second key event, which has nothing to do directly with his academic pursuits but shaped his entire attitude towards life, was his exposure to Upanishads, Zen philosophy, meditation techniques, etc. (This is not directly relevant here, since this article is mainly about Paddy's academics; his homepage contains two articles with a clear description of these aspects.)

The mastery of theoretical physics, especially General Relativity (GR), helped him to publish his first technical paper in GR in 1977 when he was still a B.Sc student. He started working seriously on several ideas in GR and quantum field theory around this time. Paddy was a Gold Medalist for topping the B.Sc exam in Kerala University, and joined for his Master's in Physics (M.Sc) in the same University college. Given the fact that he already knew all the standard stuff which was taught in the M.Sc course in Physics, he had sufficient spare time for his research work. The interaction with the other members of the Science Society was very helpful in these academic activities.

Needless to say, Paddy was again a Gold Medalist for topping the M.Sc. in Kerala university in 1979. By now, he had caught the attention of several leading scientists both in India and abroad. The NSTS summer camps (at IIT, Kharagpur and the Raman Research Institute, Bangalore) as well as an Einstein's Centenary Symposium (1979) at PRL, Ahmedabad which he attended, helped significantly in this regard. Pursuing a Ph.D in the US or the UK (like some of his close friends in the Trivandrum Science society did) would have been a logical course to follow, but his family circumstances prevented him from doing so. As a result, he decided to join what was then probably the best research institute in the country (and an internationally acclaimed one), viz., the Tata Institute of Fundamental Research (TIFR) for his Ph.D. He joined TIFR in August 1979 and became its tenured faculty member (called Research Associate, which was the entry-level faculty position in those days) in February 1980, while still working towards his Ph.D. His thesis work was in Quantum Cosmology (done under the supervision of J.V. Narlikar) and he got his degree in 1983. This work developed a particular formalism of quantum cosmology which had the potential to solve the cosmological singularity problem - an idea which echoes in many of the currently fashionable quantum gravity models. His thesis also contained the notion of the wave function of the universe, which was being developed independently by Hartle and Hawking around the same time, from a different perspective.

During his Ph.D years, he met and fell in love with Vasanthi (who was also a research scholar, one year junior to him, in TIFR). They got married in March, 1983 when he had just completed his Ph.D and Vasanthi was still pursuing hers. Vasanthi's entry into his life had a strong influence in - amongst other things - his academic pursuits. She was working under the supervision of Ramnath Cowsik on the nature and distribution of dark matter in the universe. Paddy found this area fascinating and started collaborating with her in this subject. This broadened his interest into several aspects of astrophysics - an interest which has continued since then - and resulted in his entering the area of cosmology, in which he later made his mark.

Prompted by this new found passion, Paddy decided to take up a research associate position at the Institute of Astronomy, Cambridge for one year (1986–1987) rather than go for post-doctoral work in his thesis area. He was strongly influenced during this time by Donald Lynden-Bell, whose scholarship and breadth of scientific interests resonated well with his own way of doing science. He found the subject of the Statistical Mechanics of Gravitating Systems particularly fascinating, and spent a fair amount of time working on different aspects of it. His own contributions to this area are highlighted in the first single-authored review he wrote for *Physics Reports* in 1990 and the later lecture notes of the prestigious *Les Houches Schools* in 2002 and 2008. His authority in this subject is well recognized not only by the astrophysics community, but also by the condensed matter community interested in the statistical mechanics of long range systems.

Interestingly enough, he started supervising two students (T.R. Seshadri and T.P. Singh were the first two) just after finishing his own thesis. The steady flow of students continued — in spite of him being rather selective — and he has so far supervised the thesis work of sixteen students. It is commendable that ten of them hold faculty positions in different institutes in India and are guiding their own students. At present his academic family tree has forty grand-students! "I can take no credit for the achievements of my students, except to say that I did not spoil them", says Paddy, "I was fortunate because good students always wanted to work with me. In the early years, when my age gap with the students was moderate, Vasanthi and I maintained very close personal contact with them and they were like members of our family. It was wonderful." In fact, almost all the young ($\lesssim$ 45 years) cosmologists working in various institutes/universities in India today have been associated with Paddy and mentored by him in the Ph.D/post-doctoral stage of their career, in one way or another.

Another facet of Paddy's career, which also took root during his TIFR days and flourished in the years to come, is his public outreach involvement. In addition to the numerous popular lectures he gave, Paddy became a regular contributor to two science magazines of India (the *Science Today* and the *Science Age*) which existed at that time. He ran several regular columns in these magazines, like *Playthemes* (on recreational mathematics), *Let us think it over* (on everyday physics applications), and *Milestones in Science* (on the history of science). His strong interest in the history of physics prompted him to present the *Story of Physics* as a comic strip serial in the *Science Age*. This was extremely popular, and, later on, was published as a book, translated into several Indian regional languages and made available to the school children at an affordable price. This was made possible, in large part, because Paddy does *not* take any royalties from this work. More recently, he ran a 24 part serial in the science journal *Resonance*, called the *Dawn of Science*, dealing with the history of all sciences, from pre-history to the 17th century. A popular science book which he wrote, *After the first three minutes* [2000; Cambridge University Press (CUP)] has also been very well received and was translated into Portuguese, Chinese and Polish. Paddy remains strongly committed to the responsibilities of scientists towards the society, and continues to be very active in public outreach programmes.

Around 1990, Paddy started working on his first single-authored book, Structure Formation in the Universe (CUP 1993). He was persuaded to write this book by Martin Rees, who introduced him to Rufus Neal, the CUP commissioning editor.

This book was extremely well received and made Paddy well known among the astrophysics and cosmology community. Since then, Paddy has taken to writing books like a duck to water, and has published 9 more — with a few more in the offing! "Recently," Paddy recalls, "Martin Rees commented that he had triggered a run-away process when he persuaded me to write that first book!"

People often ask him how he manages to write so many high-quality books while keeping up with his research, and maintaining an average of more than 8 research publications per year. "Well, you need two things", says Paddy, "first, you need the discipline to work on it 2 hours each day – in which you can write 6 pages, if you have everything ready in your head. So, a 600 page book will just take 100 days. But, of course, you won't have everything in your head, so it will take about 5–10 times more time; so you can turn out one book every 2–3 years. The second thing you need is Vasanthi. She worked with me in all the books, taking care of much of the typing, latexing and back-end processing!" Every one of his books acknowledges Vasanthi's contribution.

In 1992 he shifted from TIFR to the Inter-University Centre for Astronomy and Astrophysics (IUCAA). During the early part of his career at IUCAA, he concentrated on various aspects of structure formation in the universe. Along with his Ph.D student, Jasjeet Bagla, and later on with Sunu Engineer, he developed a code which describes the dynamics of a large number of (of the order of a few million) astrophysical particles, which is known as an N-body simulation code. *This code was the first of its kind in India at that time*. "What took nearly 3 weeks to compute with the best computers available to us in those days, is being done in half an hour with a laptop today; but it was fun developing the code from scratch, bringing in as much innovation as possible," says Paddy.

Coming from a theoretical physics background, Padmanabhan's perspective on astrophysics was rather different from that of many other people whose initial training itself was in astrophysics. In particular, he noticed that there was no comprehensive treatise covering all of astrophysics, like, for example, the Landau–Lifshitz course for theoretical physics. He was lamenting about this to Jerry Ostriker, during his visit to Princeton in 1996, when Jerry asked him "Why don't *you* write it?" During the next few weeks, Jerry was very supportive and helpful in making concrete the structure of a 3-volume *Course of Theoretical Astrophysics* which Paddy came up with. These 3 volumes were published by CUP during 2000–2002 and reviewers have called them magnificent achievements. With these, Paddy became well known to a very large community of astrophysicists and his breadth of scholarship was recognized all around. More recently, he wrote two graduate level textbooks, on *Gravitation* (CUP 2010) and on *Quantum Field Theory* (Springer 2016) with which he has covered almost all the frontiers of theoretical physics and astrophysics.

One reason he could write so many books is his passion for innovative teaching. Paddy strongly believes that research and teaching should go hand-in-hand. He has taught virtually every aspect of theoretical physics and astrophysics at the graduate schools in TIFR and IUCAA, in addition to occasional courses at the Pune University and IISER, Pune. He is considered a fantastic teacher and his lectures on even routine topics are punctuated by creative and original approaches — something which echoes in his books, which elaborate on his classroom teaching. At present, he has started on designing a course for senior undergraduates which will teach them *all of* theoretical physics in about 150 lectures. This course will eventually appear as a four-volume text book.

While completely at home with any aspect of theoretical physics, Paddy's real passion is for quantum aspects of gravity. He never let go of this, having tasted blood in the early years of his career. From the early part of 2000, he decided to spend more time in this area, with his astrophysical interests taking a back seat. Given the importance and potential of this work, a detailed description will be appropriate:

The two most important conceptual advances in theoretical physics, made during the twentieth century, were General Relativity and Quantum Theory. However, all attempts to put together the principles of these two disciplines have repeatedly failed, often after a lot of hope and hype which accompanied each attempt. The research work of Paddy, over the last decade or so, suggests that this is because we have misunderstood the nature of space-time structure, and are applying the principles of quantum theory to the wrong physical entity.

An analogy will make this clear. A fluid, or a solid, is described by certain mathematical equations in classical physics. Applying the principles of quantum theory directly to these equations will allow you to discover what are known as 'phonons' – the microscopic quanta of the vibrations of a solid. But this approach will never get you to the atoms, which are the true basic constituents of the solid. To obtain the correct description, you first need to recognize that matter is made of atoms, and then apply the quantum principles to these atoms.

The key new insight provided by Paddy's research shows that the status of space-time, in General Relativity, is completely analogous to the status of a fluid in classical physics. There is a sufficient amount of *internal* evidence — uncovered by his work — to indicate that space-time, itself, is made of more fundamental and microscopic degrees of freedom, which are analogous to the atoms in a solid. This means that applying the principles of quantum theory directly to Einstein's field equations — which is what almost all models of quantum gravity attempt to do — will inevitably lead to failure. Paddy's research allows one to identify, and *actually count* these degrees of freedom (which is similar to counting the number of molecules in, say, one litre of a fluid). It is then possible to apply well defined principles of statistical mechanics to describe the dynamics of these degrees of freedom, and show that the result, in the appropriate limit, leads precisely to Einstein's theory. Further, his research allows one to extend these results towards a broad class of theories far more general than Einstein's theory, showing that Einstein's theory is just one special case of a much deeper paradigm.

Every good paradigm shift should allow us to recognize the key theoretical problems from a deeper, more fundamental perspective and thereby solve them. Paddy's approach is no exception, and it gives fresh insights into solving the following problems:

- Observations tell us that the present-day universe is composed not only of the normal atoms we see around us, but also an exotic type of matter called dark matter, and another component termed as dark energy. It is very likely that this dark energy is equivalent to what is known as the cosmological constant – which was a term that Einstein introduced into his equations. The numerical value of this constant was a mystery, and was considered to be *the* key problem in present-day theoretical physics. Paddy's approach shows how the value of the cosmological constant can be naturally related to the amount of *information* which an observer in the universe can access, and predicts a mathematical formula to determine its value. The numerical value predicted by Paddy's work is in perfect agreement with cosmological observations!

This work also makes a *falsifiable prediction* — which is more than any other approach to quantum gravity has done — about the very early, inflationary phase of the universe. This prediction is also borne out by all present-day cosmological observations and future observations can test this with greater precision.

[Incidentally, part of this work was done in collaboration with his daughter, Hamsa Padmanabhan. Born in 1989, to two parents with Ph.Ds in astrophysics, she grew up in the academic atmosphere of the campuses of research institutes. While the parents encouraged her to pursue whatever she wanted — and she had a taste of arts and literature — she could not escape the charming world of science and ended up as the third Ph.D in astrophysics in the family!]

- Paddy's approach allows one to understand several peculiar features of gravity, including the thermodynamic behaviour of black holes, from a natural and broader perspective. That is, the microscopic picture based on the atoms of space-time allows us to understand almost all the features of macroscopic, classical gravity in a simple and intuitive manner. Again, this has wide-ranging consequences going well beyond the realm of Einstein's theory, which no other model has achieved so far.
- A second key problem in theoretical physics concerns the origin of the universe, as well as the ultimate fate of the collapsing matter which forms a black hole. While Paddy's approach has not yet completely solved these deep issues (together known as 'the singularity problem'), it sheds light on the way forward towards the final resolution of these problems. With the development of a novel mathematical framework to address the properties of the microscopic nature of space-time, Paddy's approach has the potential to address the singularity problem in a direct and elegant manner. Thus, it has made further advances than any other approach in the literature, towards the resolution of the deepest problems in theoretical physics.

It is fair to say that our understanding of gravity underwent a fundamental paradigm shift in 1915, with the advent of Einstein's General Theory of Relativity. One hundred years down the line, we are poised on the brink of another breakthrough, with the novel paradigm shift led by Paddy's research.

Another feature of Paddy's career – which again sets him apart from many other scientists in his peer group – is his willingness and capability to provide scientific leadership in various ways. He has served in several key committees and has taken a leading role in the development of astronomy in India. Here are a few examples from the recent years: (a) The Department of Science and Technology has appointed him as the Convener of the Advisory Group (2008–2010) to facilitate India's entry into one of the international collaborations building the next generation Giant Segmented Mirror Telescopes. He has played a key role in taking this initiative and developing a consensus in the Indian astronomy community in this task, which has now led India into joining the TMT. (b) He served as the Chairman (2006–2009) of the Time Allocation Committee of the Giant Meterwave Radio Telescope (GMRT), has introduced many innovative aspects into its working and been instrumental in streamlining several aspects of the GMRT. (c) He was the Chairman (2008–2011) of the Indian National Science Academy's National Committee which interfaces with the activities of the International Astronomical Union. In addition to advising the Government on policy issues, this also required him to coordinate the International Year of Astronomy 2009 activities in the country. In the international arena, he was the President of the Commission 47 on Cosmology of the International Astronomical Union (2009 – 2012), and the Chairman of the Commission 19 (Astrophysics) of the International Union of Pure and Applied Physics (2011 – 2014).

Paddy has received numerous awards and distinctions in India and abroad for his contributions. He is an elected Fellow of all the three Science Academies of India as well as of the Third World Academy of Sciences. The national and international awards received by him include the Padma Shri (2007), the J.C.Bose Fellowship (2008-), the Inaugural Infosys Prize in Physical Sciences (2009), the Third World Academy of Sciences Prize in Physics (2011), the Millennium Medal (2000), the Shanti Swarup Bhatnagar Award (1996), the INSA Vainu-Bappu Medal (2007), the Al-Khwarizmi International Award (2002), the Sackler Distinguished Astronomer of the Institute of Astronomy, Cambridge (2002), the Homi Bhabha Fellowship (2003), the G.D.Birla Award for Scientific Research (2003), the Miegunah Award of the Melbourne University (2004), the Goyal Prize in Physical Sciences (2012–2013), the Birla Science Prize (1991) and the INSA Young Scientist Award (1984). His research work has won prizes from the Gravity Research Foundation, USA seven times, including the First Prize in the Gravity Essay Contest in 2008.

Paddy is unique in his breadth of scholarship and his passion for knowledge. Few scientists in the world – and no one in India – is so competent and knowledgeable in a wide spectrum of areas ranging from the numerical analysis of astrophysical data to the realm of quantum gravity! While reviewing his book on Gravitation, one reviewer actually chose to comment on the author by saying, ".... *There is immense erudition, and mastery of both formal tools and calculational details; it is really impressive that one individual can understand so much, so deeply ....*", a feeling echoed by many who have closely interacted with Paddy.

Personally, Paddy feels that theoretical physicists are a fortunate lot. In a preface to a recent book, *Sleeping Beauties in Theoretical Physics* (Springer 2015), he makes his view quite clear: "Theoretical physics *is* fun. Most of us indulge in it for the same reason a painter paints or a dancer dances — the *process* itself is so enjoyable! Occasionally, there are additional benefits like fame and glory and even practical uses; but most good theoretical physicists will agree that these are not the primary reasons why they are doing it. The fun in figuring out the solutions to Nature's brain teasers is a reward in itself."

Measuring Baryon Acoustic Oscillations with Angular Two-Point Correlation Function

Jailson S. Alcaniz, Gabriela C. Carvalho, Armando Bernui, Joel C. Carvalho and Micol Benetti

Abstract The Baryon Acoustic Oscillations (BAO) imprinted a characteristic correlation length in the large-scale structure of the universe that can be used as a standard ruler for mapping out the cosmic expansion history. Here, we discuss the application of the angular two-point correlation function, $w(\theta)$, to a sample of luminous red galaxies of the Sloan Digital Sky Survey (SDSS) and derive two new measurements of the BAO angular scale at $z = 0.235$ and $z = 0.365$. Since noise and systematics may hinder the identification of the BAO signature in the $w - \theta$ plane, we also introduce a potential new method to localize the acoustic bump in a model-independent way. We use these new measurements along with previous data to constrain cosmological parameters of dark energy models and to derive a new estimate of the acoustic scale r_s.

1 Introduction

Along with measurements of the luminosity of distant type Ia supernovae (SNe Ia) and the anisotropies of the cosmic microwave background (CMB), data of the large-scale distribution of galaxies have become one of the most important tools to probe the late-time evolution of the universe. This kind of measurement encodes not only information of the cosmic expansion history but also of the growth of structure, a fundamental aspect to probe different mechanisms of cosmic acceleration as well as to distinguish between competing gravity theories. In particular, recent measurements of a tiny excess of probability to find pairs of galaxies separated by a characteristic scale r_s – the comoving acoustic radius at the drag epoch – was revealed in the two-point spatial correlation function (2PCF) of large galaxy catalogs.

This BAO signature arise from competing effects of radiation pressure and gravity in the primordial plasma which is well described by the Einstein-Boltzmann equations in the linear regime [1–5]. The first detections of the BAO scale in the galaxy

J.S. Alcaniz (✉) · G.C. Carvalho · A. Bernui · J.C. Carvalho · M. Benetti
Coordenação de Astronomia & Astrofísica, Observatório Nacional,
20921-400 Rio de Janeiro – RJ, Brazil
e-mail: alcaniz@on.br

J.S. Bagla and S. Engineer (eds.), *Gravity and the Quantum*,
Fundamental Theories of Physics 187, DOI 10.1007/978-3-319-51700-1_2

distribution were obtained only in the past decade from galaxy clustering analysis of the Two Degree Field Galaxy Survey (2dFGRS) [6] and from the Luminous Red Galaxies (LRGs) data of the Sloan Digital Sky Survey (SDSS) [7]. More recently, higher-z measurements at percent-level precision were also obtained using deeper and larger galaxy surveys [8–12] (see [13] for a recent review).

The BAO signature defines a statistical standard ruler and provides independent estimates of the angular diameter distance $D_A(z)$ and the Hubble parameter $H(z)$ through the transversal ($dr_\perp = (1+z)D_A\theta_{BAO}$) and radial ($dr_\parallel = c\delta z/H(z)$) BAO modes, respectively. However, it is worth mentioning that the detection of the BAO signal through the 2PCF, i.e., using the 3D positions of galaxies, makes necessary the assumption of a fiducial cosmology in order to transform the measured angular positions and redshifts into comoving distances. Such conversion may bias the parameter constraints, as discussed in Refs. [7, 14] (see also [15]).

On the other hand, the calculation of the angular 2-point correlation function (2PACF), $w(\theta)$, involves only the angular separation θ between pairs, yielding information of $D_A(z)$ almost model-independently, provided that the comoving acoustic scale is known. In order to extract information using 2PACF, the galaxy sample is divided into redshift shells whose width has to be quite narrow ($\delta z \leq 10^{-2}$) to avoid large projection effects from the radial BAO signal. Another important issue in this kind of analysis is how to identify the actual BAO bump once the 2PACF is noisy and usually exhibits more than one single bump due to systematic effects present in the sample (we refer the reader to [16] for a detailed discussion on this point). In what follows, we discuss the application of the 2PACF to large galaxy samples and introduce a potential new method to identify the BAO signature in a model-independent way. We exemplify the method with a sample of 105,831 LRGs from the seventh data release of the Sloan Digital Sky Survey (SDSS) and obtain two new measurements of θ_{BAO} at $z = 0.235$ and $z = 0.365$.

2 The Angular Two-Point Correlation Function

2.1 Theory

In the cosmological context, the two-point correlation function, $\xi(s)$, is defined as the excess probability of finding two pairs of galaxies at a given distance s. This function is obtained by comparing the real catalog to random catalogs that follow the geometry of the survey [17, 18]. The most commonly used estimator of the 2PCF is the one proposed in Ref. [19]:

$$\xi(s) = \frac{DD(s) - 2DR(s) + RR(s)}{RR(s)}, \tag{1}$$

where $DD(s)$ and $RR(s)$ correspond to the number of galaxy pairs with separation s in real-real and random-random catalogs, respectively, whereas $DR(s)$ stands for the number of pairs with comoving separation s calculated between a real-galaxy and a random-galaxy.

Assuming a flat universe, as indicated by recent CMB data [20, 21], the comoving distance s between a pair of galaxies at redshifts z_1 and z_2 is given by

$$s = \sqrt{r^2(z_1) + r^2(z_2) - 2r(z_1)r(z_2)\cos\theta_{12}}\ , \tag{2}$$

where θ_{12} is the angular distance between such pair of galaxies, and the radial distance between the observer and a galaxy at redshift z_i, $r(z_i) = c\int_0^{z_i} dz/H(z, \mathbf{p})$, depends on the parameters $\mathbf{p}$ of the cosmological model adopted in the analysis.

Similarly to the 2PCF, the 2PACF is defined as the excess joint probability that two point sources are found in two solid angle elements $d\Omega_1$ and $d\Omega_2$ with angular separation θ compared to a homogeneous Poisson distribution [17]. As mentioned earlier, this function can be used model-independently, considering only angular separations in narrow redshift shells of small δz in order to avoid contributions from the BAO mode along the line of sight. The function $w(\theta)$ is calculated analogously to Eq. (1) with s being replaced by θ. The expected 2PACF, w_E, is given by [22]

$$w_E(\theta, \bar{z}) = \int_0^\infty dz_1\,\phi(z_1)\int_0^\infty dz_2\,\phi(z_2)\,\xi_E(s, \bar{z})\,, \tag{3}$$

where $\bar{z} \equiv (z_1 + z_2)/2$, with $z_2 = z_1 + \delta z$, and $\phi(z_i)$ is the normalised galaxy selection function at redshift z_i. Note that, for narrow bin shells, $\delta z \approx 0$, then $z_1 \approx z_2$ and $\xi_E(s, z_1) \simeq \xi_E(s, z_2)$. Therefore, one can safely consider that $\xi_E(s, \bar{z})$ depends only on the constant parameter $\bar{z}$, instead of on the variable z. The function $\xi_E(s, z)$ is given by [23]

$$\xi_E(s, z) = \int_0^\infty \frac{dk}{2\pi^2}\,k^2\,j_0(ks)\,b^2\,P_m(k, z)\,, \tag{4}$$

where j_0 is the zeroth order Bessel function, $P_m(k, z)$ is the matter power spectrum and b is the bias factor. For shells of arbitrary δz, we refer the reader to [24].

2.2 Application to the Data

As illustrated in Fig. 1, the 2PACF, derived from Eq. (1), usually exhibits more than one single bump which, in general, is due to systematic effects present in the galaxy samples. In order to identify, among all bumps, which one corresponds to the real BAO scale the usual procedure in the literature (see, e.g., [22]) is to compare the bump scales observed in the 2PACF with a cosmological model prediction obtained from Eqs. (3) and (4). Here, since we want to perform an analysis as model-independent as possible, we adopt the following criterium [16]: if the BAO bump is present in the sample and is robust, then it will survive to changes in the galaxies angular

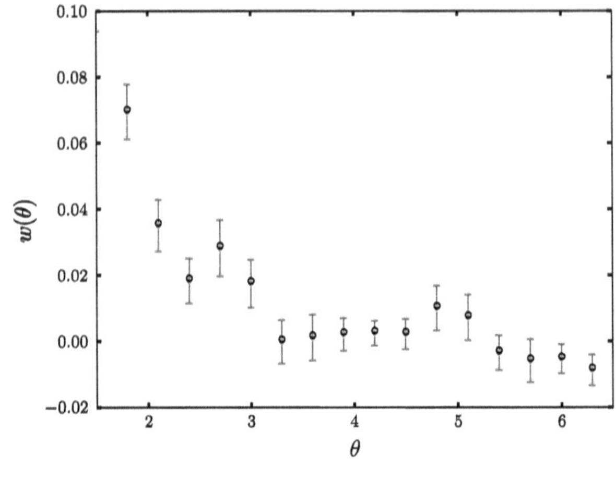

Fig. 1 An example of the 2PACF. As mentioned in the text, noise and systematics may give rise to bumps at different scales which hinder the identification of the BAO signature imprinted on the data. Clearly, two possible BAO bumps are observed at $\theta = 2.5° - 3.0°$ and $\theta = 4.8° - 5.2°$

coordinates by small and random amount whereas the bumps produced by systematic effects will not.

We apply this procedure to a sample of 105,831 LRGs of the seventh public data release of the Sloan Digital Sky Survey distributed in the redshift interval $z = [0.16 - 0.47]$ [25]. The transversal signatures as a function of redshift are obtained by dividing the data into two shells of redshift: $z = [0.20 - 0.27]$ and $z = [0.34 - 0.39]$, containing 19,764 and 24,879 LRGs, whose mean redshifts are $\bar{z} = 0.235$ and $\bar{z} = 0.365$, respectively. The 2PACF as a function of the angular separation θ for this distribution of galaxies is shown in Fig. 2a and b, where several bumps at different angular scales are observed. However, using the criterium mentioned earlier, we calculate a number of 2PACF performing random displacements in the angular position of the galaxies, i.e., following Gaussian distributions with $\sigma = 0.25$, 0.5, and 1.0 [16]. Without assuming any fiducial cosmology, we find that only the bumps localised at $\theta_{\rm FIT} = 7.75°$ and $\theta_{\rm FIT} = 5.73°$ remain.[1]

Another potential method for identifying the real BAO scale can be achieved by counting the number of neighbours of the sub-sample of galaxies contained in each shell. To make this explicit, we first select all the pairs contributing to the BAO bump at $\theta_{\rm FIT}(\bar{z} = 0.235) = 7.75°$ in the 2PACF, i.e., pairs of galaxies that have an angular separation distance in the interval $[7.6°, 7.8°]$. Within this set, we analyse the repetition rate: given a galaxy, we compute how many galaxies (that is, neighbours) are apart by an angular distance between 7.6° to 7.8° from that galaxy. The same procedure is applied to the galaxies in the interval $[5.6°, 5.8°]$, where a bump is also observed at $\simeq 5.7°$ Fig. 2a. The result is shown in Fig. 3, where we find that, while the mean number of neighbours in the latter interval is $N_n = 16 \pm 7$, the value of N_n in the former (the BAO bump) is 20 ± 9 Fig. 3a. By similar proceeding, we also find this same characteristic in the sub-sample of galaxies contained in the shell

[1] Only for comparison, the predictions of a flat ΛCDM cosmology, assuming $\Omega_m = 0.27$ and $r_s = 100$ Mpc/h, are $\theta_{\rm BAO}(0.235) = 8.56°$ and $\theta_{\rm BAO}(0.365) = 5.68°$.

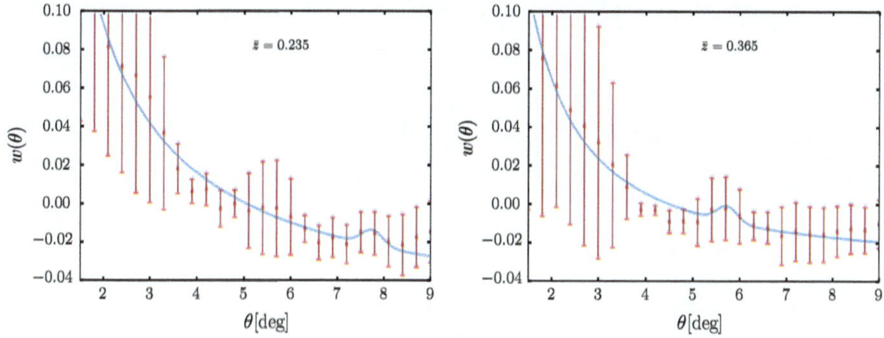

Fig. 2 The 2PACF for two *redshift* intervals obtained from the LRGs sample of the SDSS-DR7 (*red bullets*). The continuous line is derived from the 2PACF parameterisation proposed in Ref. [23]

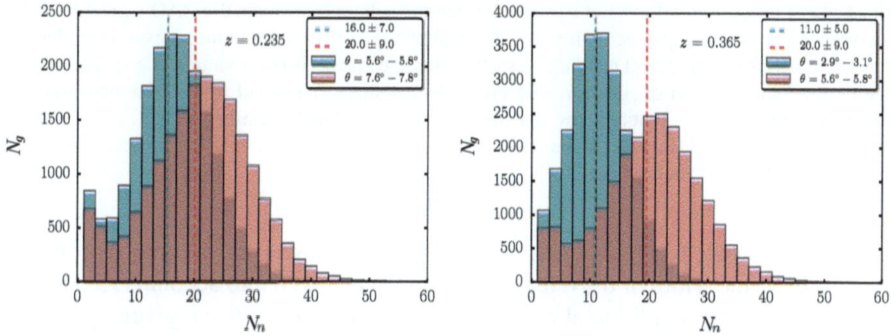

Fig. 3 *Left:* Histograms of the number of neighbours for galaxy pairs in the intervals [7.6°, 7.8°] and [5.6°, 5.8°] ($\bar{z} = 0.235$). The horizontal axis, N_n, represents the number of neighbours of each galaxy whereas vertical *dashed lines* indicate the mean value of each distribution. *Right:* The same as in the previous panel for galaxy pairs in the intervals [5.6°, 5.8°] and [2.9°, 3.1°] ($\bar{z} = 0.365$)

$z = [0.34 - 0.39]$, as shown in Fig. 3b. It is worth mentioning that similar results are also found for the samples of LRGs of the tenth and eleventh data release of the SDSS. In other words, it seems that galaxies around the BAO bumps have a much larger number of neighbours than those around non-BAO bumps. If confirmed in other galaxy samples, this property could also be used to identify the real BAO scales in a model-independent way.

After finding the real BAO signature, we obtain the angular BAO scale using the method of Ref. [23], which parameterises the 2PACF as a sum of a power law, describing the continuum, and a Gaussian peak, which describes the BAO bump, i.e.,

$$w_{FIT}(\theta) = A + B\theta^{\nu} + Ce^{-\frac{(\theta-\theta_{FIT})^2}{2\sigma_{FIT}^2}}, \tag{5}$$

where A, B, C, ν, and σ_{FIT} are free parameters, θ_{FIT} defines the position of the acoustic scale and σ_{FIT} gives a measure of the width of the bump. Note that, if $\delta z = 0$,

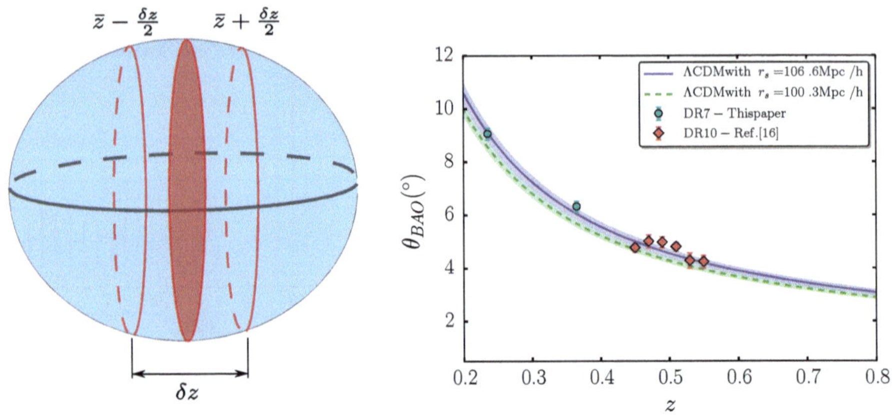

Fig. 4 *Left:* Projection effect: Let us consider that the sphere represents the BAO signature. When the rings at $\bar{z} \pm \delta z/2$ are projected into the the shaded region ($\bar{z}$), the signature appears smaller than true one. Therefore, this effect produces a shift of the BAO peak to lower values. *Right:* The angular BAO scale as a function of *redshift*. The *green* data points correspond to the two measurements obtained in this analysis whereas the *red* ones are the data points obtained in Ref. [16] using the LRGs of the SDSS-DR10. The curves stand for the ΛCDM prediction with the acoustic scale fixed at the WMAP9 and Planck values

the true BAO scale θ_{BAO} and θ_{FIT} would coincide. However, for $\delta z \neq 0$ projection effects due to the width of the redshift shells must be taken into account (see Fig. 4a). Therefore, assuming a fiducial cosmology, the function $w_E(\theta, \bar{z})$, given by Eqs. (3) and (4), has to be calculated for both $\delta z = 0$ and $\delta z \neq 0$ in order to compare the position of the peak in the two cases. This allows to find a correction factor α that, given the value of θ_{FIT} estimated from Eq. (5), provides the true value for θ_{BAO}. To perform the calculation of α, we assume the standard ΛCDM cosmology[2] and also consider the correction due to the photometric error of the sample by following Ref. [22]. After all these corrections, the two new measurements of the BAO angular scales at $\theta_{\rm BAO} = [9.06 \pm 0.23]°$ ($\bar{z} = 0.235$) and $\theta_{\rm BAO} = [6.33 \pm 0.22]°$ ($\bar{z} = 0.365$) are obtained.

3 Cosmological Constraints

We present cosmological parameter fits to the BAO data derived in the previous section along with the data set obtained in Ref. [16] (see Fig. 4). The angular scale θ_{BAO} is related to the acoustic radius r_s and the angular diameter distance $D_A(z)$ through

---

[2]For narrow redshift shells, such as the ones considered in this analysis ($\delta z \sim 10^{-2}$), it can be shown that the correction factor depends weakly on the cosmological model adopted (see Fig. 3 of [23]).

$$\theta_{BAO}(z) = \frac{r_s}{(1+z)D_A(z)} , \tag{6}$$

where

$$D_A(z) = \frac{c}{(1+z)} \int_0^z \frac{dz'}{H(z, \mathbf{p})} . \tag{7}$$

In what follow, we explore three classes of cosmological models, starting from the minimal ΛCDM cosmology. We also consider a varying dark energy model whose equation-of-state parameter evolves as $w = w_0 + w_a[1 - a/(2a^2 - 2a + 1)]$ [26] and a particular case when $w_0 < 0$ and $w_a = 0$ (wCDM) (for a discussion on theoretical models of dark energy, see [27]). In our analysis, we use the WMAP9 final estimate of the comoving acoustic radius at the drag epoch, $r_s = 106.61 \pm 3.47h^{-1}$ Mpc [20]. Plots of the resulting cosmological constraints are shown in Fig. 5. Although the $\theta_{\rm BAO}$ data alone (gray contours) are consistent with a wide interval of w_0 and w_a values, their combination with the CMB data limits considerably the range of w, favouring values close to the cosmological constant limit $w = -1.0$. This can be seen when we combine the BAO data points with CMB measurements of the shift parameter (red contours), defined as $\mathscr{R} = \sqrt{\Omega_m} \int_0^{z_{ls}} H_0/H(z)dz$, where z_{ls} is the redshift of the last scattering surface. In order to avoid double counting of information with the r_s value from WMAP9 used in the BAO analysis, we use $\mathscr{R} = 1.7407 \pm 0.0094$, as given by the latest results from the Planck Collaboration [21]. The joint results (brown contours) improve significantly the cosmological constraints, providing the values shown in Table 1.

Finally, it is important to observe that, given a cosmological model, estimates of the acoustic scale r_s can be obtained directly from Eq. (6), i.e., independently of CMB data. Using the data set shown in Fig. 4 and assuming the ΛCDM scenario, we find $r_s = 103.6 \pm 4.1\ h^{-1}$Mpc, which is in good agreement with both the WMAP9 and Planck estimates as well as with the value obtained in [28].

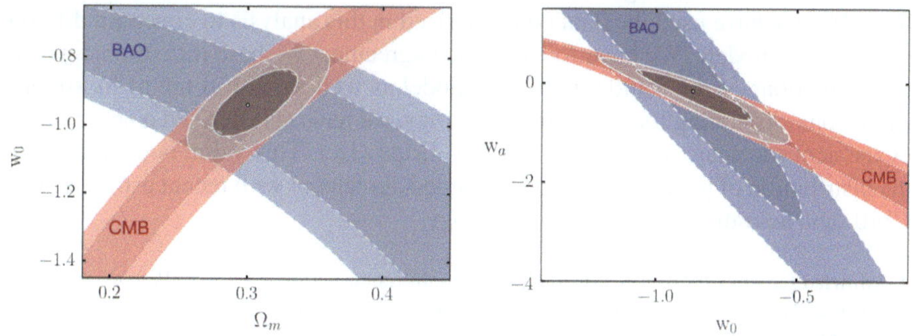

Fig. 5 *Left:* Confidence regions in the $\Omega_m - w_0$ plane. The *grey* contours correspond to the region allowed by the current $\theta_{\rm BAO}$ data (SDSS - DR7/DR10) whereas the *red* contours are obtained from CMB data. The combination between $\theta_{\rm BAO}$ and CMB limits considerably the allowed interval of the cosmological parameters. *Right:* The same as in the previous panel for the $w_0 - w_a$ plane

Table 1 Constraints on model parameters. The error bars correspond to 1σ

Model	Ω_m	w_0	w_a
ΛCDM	0.27 ± 0.04	-1	–
wCDM	0.30 ± 0.02	-0.94 ± 0.06	–
$w(z)$CDM	0.29 ± 0.03	-0.87 ± 0.13	-0.16 ± 0.32

4 Conclusions

Sourced by the initial density fluctuations, the primordial photon-baryon plasma supports the propagation of acoustic waves until the decoupling of photons and baryons. These oscillations imprinted a preferred clustering scale in the large scale structure of the universe which have been detected either as a peak in the real space correlation function or as a series of peaks in the power spectrum. In this paper, we have discussed the application of the angular two-point correlation function to a sample of 105,831 LRGs of the seventh public data release of the Sloan Digital Sky Survey (SDSS) distributed in the redshift interval $z = [0.16 - 0.47]$ [25]. Differently from analysis that use the spatial correlation function, $\xi(s)$, where the assumption of a fiducial cosmology is necessary in order to transform the measured angular positions and redshifts into comoving distances, the calculation of the 2PACF, $w(\theta)$, involves only the angular separation θ between pairs, yielding measurements of the BAO signal almost model-independently.

After identifying the BAO peaks using the method of Ref. [16] and introducing a new potential method based on the mean number of neighbours of the galaxies contained in the redshift shells, we have derived two new measurements of the BAO angular scale: $\theta_{\rm BAO} = [9.06 \pm 0.23]°$ ($\bar{z} = 0.235$) and $\theta_{\rm BAO} = [6.33 \pm 0.22]°$ ($\bar{z} = 0.365$). As shown in Fig. 4b, these low-z measurements are important to fix the initial scale in the $\theta_{\rm BAO} - z$ plane, which improves the constraints on cosmological parameters. Along with six measurements of $\theta_{\rm BAO}(z)$ recently obtained in Ref. [16], we have used the data points derived in this analysis to constrain different dark energy models. We have found a good agreement of these measurements with the predictions of the standard ΛCDM model as well as with some of its simplest extensions. Assuming the standard cosmology, we have also derived a new estimate of the acoustic scale, $r_s = 103.6 \pm 4.1\ h^{-1}$Mpc ($1\sigma$). This value is obtained from the distribution of galaxies only and is in good agreement with recent estimates from CMB data assuming the ΛCDM cosmology.

Acknowledgements Jailson S. Alcaniz dedicates this contribution to Prof. T. Padmanabhan with affection and profound admiration. May he continue to inspire us with new ideas about the Universe. Happy 60th birthday, Paddy!

The authors acknowledge support from the Conselho Nacional de Desenvolvimento Científico e Tecnológico (CNPq) e Fundação Carlos Chagas de Amparo à Pesquisa do Estado do Rio de Janeiro (FAPERJ). Joel C. Carvalho is also supported by the DTI-PCI/Observatório Nacional program of the Brazilian Ministry of Science, Technology and Innovation (MCTI).

References

1. P.J.E. Peebles, J.T. Yu, Astrophys. J. **162**, 815 (1970)
2. R.A. Sunyaev, Y.B. Zeldovich, Astrophys. Space Sci. **7**, 3 (1970)
3. J.R. Bond, G. Efstathiou, Mon. Not. Roy. Astron. Soc. **226**, 655 (1987)
4. W. Hu, S. Dodelson, Annu. Rev. Astron. Astrophys. **40**, 171 (2002)
5. C. Blake, K. Glazebrook, Astrophys. J. **594**, 665 (2003)
6. S. Cole et al., 2dFGRS Collaboration. Mon. Not. Roy. Astron. Soc. **362**, 505 (2005)
7. D.J. Eisenstein, (SDSS Collaboration) et al. Astrophys. J. **633**, 560 (2005)
8. W.J. Percival, B.A. Reid, D.J. Eisenstein et al., Mon. Not. Roy. Astron. Soc. **401**, 2148 (2010)
9. N. Padmanabhan, X. Xu, D.J. Eisenstein, R. Scalzo, A.J. Cuesta, K.T. Mehta, E. Kazin, Mon. Not. Roy. Astron. Soc. **427**(3), 2132 (2012)
10. F. Beutler, C. Blake, M. Colless et al., Mon. Not. Roy. Astron. Soc. **416**, 3017 (2011)
11. C. Blake, E.A. Kazin, F. Beutler et al., Mon. Not. Roy. Astron. Soc. **418**, 1707 (2011)
12. A. Sánchez et al., Mon. Not. Roy. Astron. Soc. **425**, 415 (2012)
13. D.H. Weinberg, M.J. Mortonson, D.J. Eisenstein, C. Hirata, A.G. Riess, E. Rozo, Phys. Rept. **530**, 87 (2013)
14. A.G. Sánchez, M. Crocce, A. Cabré, C.M. Baugh, E. Gaztañaga, Mon. Not. Roy. Astron. Soc. **400**, 1643 (2009)
15. S. Salazar-Albornoz, et al., (BOSS Collaboration), arXiv:1607.03144 astro-ph.CO
16. G.C. Carvalho, A. Bernui, M. Benetti, J.C. Carvalho, J.S. Alcaniz, Phys. Rev. D **93**, 023530 (2016)
17. P.J.E. Peebles, *The Large-Scale Structure of the Universe* (Princeton University press, New Jersey, 1980)
18. T. Padmanabhan, *Structure Formation in the Universe* (Cambridge University press, Cambridge, 1993)
19. S.D. Landy, A.S. Szalay, ApJ **412**, 64 (1993)
20. G. Hinshaw et al., Astrophys. J. Suppl. **208**, 19 (2013)
21. P.A.R. Ade et al., Astron. Astrophys. **571**, A16 (2014)
22. A. Carnero et al., Mon. Not. Roy. Astron. Soc. **419**, 1689 (2012)
23. E. Sánchez et al., Mon. Not. Roy. Astron. Soc. **411**, 277 (2011)
24. T. Matsubara, A.S. Szalay, A.C. Pope, ApJ **606**, 1 (2004)
25. E.A. Kazin et al., Astrophys. J. **710**, 1444 (2010)
26. E.M. Barboza Jr., J.S. Alcaniz, Phys. Lett. B **666**, 415 (2008)
27. T. Padmanabhan, Phys. Rept. **380**, 235 (2003)
28. A. Heavens, R. Jimenez, L. Verde, Phys. Rev. Lett. **113**, 241302 (2014)

Resonant Disruption of Binary Stars by a Catalytic Black Hole

C.M. Boily

Abstract The motion of a black hole (BH) about the centre of gravity of its host galaxy induces a strong response from the surrounding stars. We revisit the case of a harmonic potential and argue that half of the stars on circular orbits in that potential shift to an orbit of lower energy, while the other half receives a positive boost. The black hole itself remains on an orbit of fixed amplitude and merely acts as a catalyst for the evolution of the stellar energy distribution function $f(E)$. We then consider the response of binary stars to the motion of a central BH, and find that they are selectively heated up to disruption according to the binary's total mass and semi-major axis a. This enhanced depletion of binaries (compared with the case when the BH is fixed) might hold an important key to a more complete history of BH dynamics at the heart of the Milky Way.

1 Introduction

Black hole ($\equiv$ BH) dynamics in galactic nuclei has attracted much attention for many years (e.g., [1] for a survey of the field). The influence of a BH on its surrounding stars is felt first through the large velocity dispersion and rapid orbital motion of the inner-most cluster stars ($\sigma \sim v_{1d} \lesssim 10^3$ km/s). This sets a scale $\gtrsim GM_{bh}/\sigma^2$ ($\simeq 0.015 - 0.019$ pc for the Milky Way, henceforth MW; see [2]) within which large-angle scattering and stellar stripping and disruption take place. For the MW, large-angle scattering star-BH encounters are likely given the high density of $\rho \sim 10^7 M_\odot/\mathrm{pc}^3$ within a radius of ≈ 10 pc (see e.g. [3, 4]). Star-BH scattering leads to the formation of a Bahcall–Wolf stellar cusp of density $\rho_\star \sim r^{-\gamma}$ where γ falls in the range 3/2 to 7/4 [5–7]. Genzel et al. [8] modeled the kinematics of the inner few parsecs about Sgr A$\star$ with a mass profile $\rho_\star \sim r^{-1.4}$. This was later improved upon [9] with a double power-law fit to the data, where the power index $\simeq 1.2$ inside a break radius $r_{br} \simeq 0.2$ pc, and $\simeq 7/4$ outside it. More recently it was found that the

C.M. Boily (✉)
Observatoire Astronomique de Strasbourg & CNRS UMR 7550,
11 Rue de l'Université, 67000 Strasbourg, France
e-mail: christian.boily@astro.unistra.fr

J.S. Bagla and S. Engineer (eds.), *Gravity and the Quantum*,
Fundamental Theories of Physics 187, DOI 10.1007/978-3-319-51700-1_3

central density profile of *old type* stars is not inconsistent with a flat core, while on the whole the nuclear cluster would have a cuspy density profile $\rho \propto r^{-2}$ [10, 11]. It is puzzling that the mass profile should vary so much depending on the chosen tracers. Here we will discuss a mechanism which operates selectively on binary stars according to their mass and separation.

2 Anisotropic Motion

Most, if not all, studies of galactic nuclei dynamics assume a fixed BH (or BH binary) at the centre of coordinates. It has been argued [12] that the (not so) rapid migration of the BH to the centre would induce anisotropy in the stellar velocity field, on a scale $\sim 3\times$ the BH radius of influence ($\equiv$ volume enclosing the same mass in stars as the BH mass, M_{bh}). The catalytic role of the BH in the process may in fact slow down its return to the centre of coordinates [13]. Reid and collaborators [14] used maser emission maps to compute the mean velocity of 15 SiO emitters relatively to Sgr A$\star$. They compute a mean (three-dimensional) velocity of up to 45 km/s, a result obtained from sampling a volume of $\simeq$ 1 pc about the centre.[1] Spectroscopic surveys found a significant signature for rotation in their $f(E, L_z)$ modelling of spectroscopic central parsec stars (see [2, 9] for an update). It thus appears that stars within the central stellar cusp experience significant streaming motion with respect to Sgr A$\star$.

2.1 *BH Motion as Catalyst of Anisotropy*

A rough calculation helps to get some orientation in the problem: this one follows closely [12]. Let a point mass fall from rest from a radius R_o in the background potential of the MW stellar cusp. Then let the radial mass profile of the cusp $\rho_\star(r) \propto r^{-3/2}$, a rough fit to MW data. If we set the BH radius of influence $\simeq 1$ pc for a BH mass $M_{bh} \simeq 4.0 \pm 0.2 \times 10^6 M_\odot$ [2], then R_o may be expressed in terms of the maximum BH speed in the MW potential as $[\max\{v\}/100\,\mathrm{km/s}]^{\frac{4}{3}} = R_o/1\,\mathrm{pc}$. For a maximum velocity anisotropy in the range 10 to 40 km/s, this yields $R_o \simeq 0.3 - 0.5$ pc, which is the same *fraction* of its radius of influence.[2] To see what impact the BH motion will have on the velocity d.f. of the stars, we outline a basic argument set out in [12]. Let us focus on a circular stellar orbit outside R_o in the combined potential of the BH and an axisymmetric galaxy. When the BH is at rest at the centre of coordinates, the star draws a closed circular orbit of radius r and constant velocity

[1] Statistical root-n noise $\sim 25\%$ remains large owing to the small number of sources but is inconsequential to the argument being developed here.

[2] These figures are robust to details of the stellar cusp mass profile, so for instance a steeper density profile ($3/2 \geq \gamma \leq 7/4 < 2$) would yield R_o in the range 0.2 to 0.3 pc; and $\gamma = 0$, $R_o \simeq 0.6$ pc.

v. We now set the BH on a radial path of amplitude R_o parallel to the horizontal x-axis. We write the angular frequency of the stellar orbit be $\omega_\star$, and that of the BH $\omega \geq \omega_\star$. The ratio $\omega/\omega_\star \geq 1$ in general, but is otherwise unbounded. The net force $\boldsymbol{F}$ acting on the star can always be expressed as the sum of a radial component $\boldsymbol{F}_r$ and a force parallel to the x-axis which we take to be of the form $F_x \cos(\omega t + \varphi)$; clearly the constant $F_x = 0$ when $R_o = 0$. The net mechanical work done on the star by the BH as the star completes one orbit is

$$\delta W = \int \boldsymbol{F} \cdot v \mathrm{d}t = \int F_x v \sin(\omega_\star t) \cos(\omega t + \varphi) \, \mathrm{d}t \tag{1}$$

where φ is the relative phase between the stellar and BH orbits. The result of integrating (1) is set in terms of the variable $\varpi \equiv \omega/\omega_\star$ as

$$\begin{aligned} \frac{2\delta W}{v F_x} \omega_\star &= \frac{1}{\varpi + 1} \left[\cos(2\pi \varpi + \varphi) - \cos(\varphi) \right] \\ &+ \frac{1}{\varpi - 1} \left[\cos(2\pi \varpi - \varphi) - \cos(\varphi) \right] \end{aligned} \tag{2}$$

when $\varpi > 1$, and $2\delta W / v F_x \omega_\star = 2\pi \sin(\varphi)$ when $\varpi = 1$. Equation (2) encapsulates the essential physics, which is that δW changes sign when the phase φ shifts to $\varphi + \pi$. Thus whenever the stellar phase-space density is well sampled and all values of $\varphi : [0, 2\pi]$ are realised with equal probability, half the stars receive mechanical energy ($\delta W > 0$) and half give off energy ($\delta W < 0$). By construction, the BH neither receives nor loses energy but merely acts as a *catalyst* for the redistribution of mechanical energy between the stars.

An illustration of how stars may be stirred by increased kinetic energy is shown on Fig. 1 for the case of a logarithmic potential for the host galaxy (see below) and for stellar orbits in co-planar motion with the BH. Our approach does not integrate the full response of the stars to their own density enhancements, when they adjust their orbit to the BH perturbation. These could become self-bound structures which would alter the global dynamics of the galaxy. To inspect whether this could have an influence over the evolution of the velocity field, we compute the Toomre Q_J ratio defined as

$$Q_J \equiv \frac{\sigma \Omega}{G \Sigma} = \frac{\sigma^2}{G \Sigma \mathrm{d}l}$$

on a mesh of 30×30 points in real space. We computed the dispersion σ by first *subtracting* the dispersion that arises from the equilibrium velocity d.f. (with BH at the centre of coordinates); hence $\sigma = 0$ when the BH is at rest. Stars are stable against self-gravitating local modes of fragmentation when $Q_J \gtrsim 1$. When that condition is satisfied, the BH contributes through its orbital motion more than 58% of the *total* square velocity dispersion required to prevent local fragmentation instability [7].

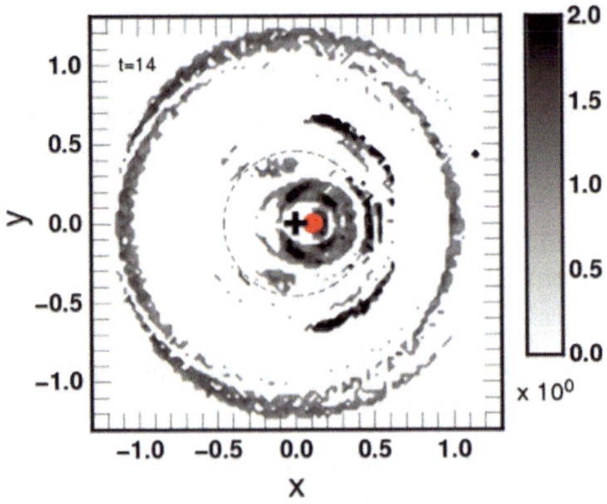

Fig. 1 Map of the Toomre number Q_J for a calculation with BH parameters $\tilde{m}_{bh} = R_o/R_c = 0.3$ shown at time $t = 14$ in computational units. A set of 80,000 orbits were integrated with an Bulirsch–Stoer integrator with compact kernel [12]; all stars are singles. The *dark colours* indicate a high value $Q_J > 1$ which shuts off self-gravitation of density enhancements. The *dash circle* of radius 0.5 marks the influence radius of the BH, shown as a *light red dot*. The cross is the origin of the coordinates. The ring-like feature of radius $\simeq 1.1$ indicates large velocity dispersion at the $\varpi = 5:2$ resonance

Given this background, we are now in a position to ask what / how would a population of binary stars responds to the motion of a BH. The combined galactic- and BH tidal field will tend to disrupt any binary, well before they reach the (stellar) disruption radius $r_d \simeq 0.015$ pc. To setup the calculation, we first recall the properties of the host galaxy's potential in the next subsection before turning to binary stars in Sect. 3.

2.2 *Galaxy Logartihmic Potential*

We cast our problem in the framework of the logarithmic potential, which we write as

$$\Phi_g(\boldsymbol{r}) = -\frac{1}{2}v_o^2 \ln\left|\frac{R^2}{R_c^2} + 1\right| \tag{3}$$

with v_o the constant circular velocity at large distances, and the radius R_c defines a core length inside of which the density is nearly constant. Thus when $r \ll R_c$ all bodies follow harmonic motion of angular frequency $\omega = v_o/R_c$. If we define $u \equiv r/R_c$, the integrated mass $M_g(u)$ reads

$$M_g(< u) = \frac{v_o^2 R_c}{G}\frac{u^3}{u^2+1}\,. \tag{4}$$

The mass $M_g(u \gg 1) \propto u$ diverges at large distances, however this is not a serious flaw since we consider only the region where $u \sim 1$. The mass $M_g(u = 1) = v_o^2 R_c/2G$ fixes a scale against which to compare the BH mass M_{bh}. For the case when the BH orbits inside the harmonic core, we set

$$M_{bh} \equiv \tilde{m}_{bh} \frac{v_o^2 R_c}{2G} = \tilde{m}_{bh} \frac{M_c}{2} \tag{5}$$

with $0 < \tilde{m}_{bh} \leq 1$, and the definition $M_c = v_o^2 R_c/G$ allows for simplifications (the 'core mass' $M_c = M_g(1)/2$ from [4]). For computational purposes one sets $G = 1$ and uses R_c as unit of length. The physical dimensions of the system are set in Sect. 3.

3 Binary Stars

We may now look into the stability of a binary star, and focus on the tidal heating by the combined galaxy potential $\Phi_g(\mathbf{r})$ of Eq. (3) and BH potential $-GM_{bh}/||\mathbf{R} - \mathbf{r}||$, where $\mathbf{R}$ denotes the position of the BH, and $\mathbf{r}$ the barycenter of the binary. Let $\mathbf{R} = 0$ for the moment. We first write out the condition for the binary to sit well inside its Roche radius as (cf. [7])

$$\bar{\rho}_{bin} \gtrsim \frac{1}{3} \left(\rho_\Phi + \bar{\rho}_{bh} \right) \tag{6}$$

where $\bar{\rho}_{bh} = M_{bh}/(4\pi r^3/3)$ is the contribution of the BH to the tidal field at $\mathbf{r}$; in the same way we compute $\bar{\rho}_{bin} = M_{bin}/(4\pi a^3/3)$ with a the semi-major axis of the binary. The use of Poisson's equation with (3) and the definitions (5) allows to write (with $u = r/R_c$)

$$\bar{\rho}_{bin} \gtrsim \frac{1}{3} \left(\frac{5}{4\pi} \left[\frac{u^2 + 3}{(u^2 + 1)^2} \right] + \frac{3}{8\pi} u^{-3} \right) \tilde{m}_{bh} \, M_c R_c^{-3} \tag{7}$$

which we can rearrange once we fix $\tilde{m}_{bh} = 0.2$ to replace the core mass M_c by the BH mass. To be more specific we fix the scales of the binary- and BH masses and lengths. The distribution of binary separations in the field is well adjusted with a Gaussian profile which peaks at $a \simeq 50\,\text{AU}$ [15]. To simplify the calculations, we will express the binary separation in units of 100 AU, the BH mass in units of $10^6 M_\odot$, and the core length R_c of the logarithmic potential in units of 1 pc $\simeq 2 \times 10^5$ AU. With the notation $R_c = \mathscr{R}_c \times 1\,\text{pc}$, $M_{bh} = \mathscr{M}_{bh6} \times 10^6 M_\odot$, and $a = a_{100} \times 100\,\text{AU}$, the condition for stability becomes

$$8 \times 10^4 \frac{\mathscr{M}_{b10}/\mathscr{M}_{bh6}}{(a_{100}/\mathscr{R}_c)^3} \gtrsim \frac{1}{3} \left(\frac{5}{4\pi} \left[\frac{u^2 + 3}{(u^2 + 1)^2} \right] + \frac{3}{8\pi} u^{-3} \right). \tag{8}$$

In this last relation, we have expressed the binary mass $M_{bin} = \mathcal{M}_{b10} \times 10 M_{\odot}$ because most binary stars of interest will be more massive than Solar (bright components, high binding energy).

For numerical examples of Eq. (8), we chose reference values such that the BH's radius of influence is $= 1$ pc, with a core length $R_c = 2$ pc. (In that way the amplitude of BH motion $R_o = 0.2 < 1$ pc is expressed as a fraction of the BH's influence radius.) Clearly for a binary on an orbit such that $u = 1$ with a total mass of $\mathcal{M}_{b10} = 0.1 (M_{bin} = 1 M_{\odot}$, two half-Suns, say) then (8) is satisfied for all separations a up to 30,000 AU. If, on the other hand, the binary orbits at a radius such that $u = 10^{-2}$, then only robust binaries with $a < 100$ AU may survive.

The situation remains (basically) the same when the BH is set in motion, but with the important difference that now orbital energy may be transferred to / taken from the BH. The situation when the BH motion is commensurate with the binary's internal period will then be more effective in heating up the binary (or cooling it down). If a significant fraction of all the stars are binaries, then presumably the binaries with low-mass companions will be destroyed more efficiently by the wandering BH, while at the same time little or no net orbital energy will be transferred to the BH (same argument as in Sect. 2).

The internal period of a binary star of total mass M_{bin} and semi-major axis a is $P_{bin} = 2\pi \sqrt{a^3/GM_{bin}}$; the characteristic orbital period of a BH orbiting in the core of the galaxy is $P_{BH} \simeq 1/\sqrt{G\rho_{\Phi}}$. The dimensionless ratio of these two quantities reads

$$\frac{P_{bin}}{P_{BH}} = \frac{\sqrt{2\pi}}{40} \left(\frac{a_{100}}{\mathcal{R}_c}\right)^{3/2} \left(\frac{\mathcal{M}_{bh6}}{\mathcal{M}_{b10}}\right)^{1/2} \left[\frac{u^2+3}{(u^2+1)^2}\right]^{1/2} . \tag{9}$$

This ratio $\to 0$ when $u \gg 1$ and reaches a constant to second order in u near $u = 0$. It is clear from (9) that many high $m : n$, $m < n$ commensurate period ratios will be achieved if the binary components have low total mass, or, if the semi-major axis a is large. Thus clearly tight and massive binary stars stand a much better chance of survival under those conditions.

4 Numerical Examples

We can explore the difference brought by BH motion if we pick a relatively stable binary star in the background potential of the BH + galaxy with $\mathbf{R} = 0$. To make things more concrete, we have integrated a number of binary stars in the BH + logarithmic potential for the same galactic circular orbit at a radius $r = 0.5$ pc, or half the BH radius of influence $R_{BH} \simeq 1$ pc. We reduce the problem to an exploration of the effect of changing the period ratio of the binary / BH orbit, by fixing the initial semi-major axis $a = 400$ AU and varying only the total mass of the binary: we chose $M = 4M_{\odot}, 2M_{\odot}$ and $\sqrt{2}\, M_{\odot}$. The period ratio derived from (9) is then close to $4:3, 9:5$ and $2:1$, respectively, in each case. The integrations were performed

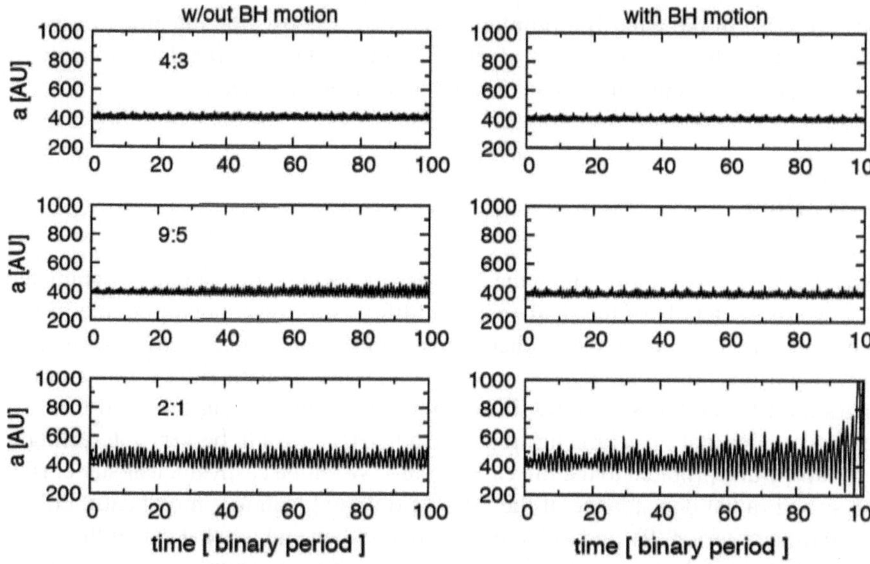

Fig. 2 The semi-major axis a (in AU) of binary stars as function of time. Relative variations of a and binding energy correspond one-to-one. All binaries were set on the same galactic orbit with $r/R_c = 1/4$ which is half the BH influence radius. The *left*-hand side columns are the solutions for a fixed BH at the origin; the *right*-hand side are for a radial BH orbit of amplitude $R_{BH}/R_c = 1/10$ (or, $0.2\times$ its influence radius). From *top* to *bottom*, each row is for an equal-component binary respectively of total mass $= 4, 2$ and $\sqrt{2}$; this leads to the period ratios derived from (9) indicated on the *left*-hand panels. Note that $a = 400$ AU initially in each case

with a fixed galactic potential but self-consistent BH - binary interactions. Since the mass ratio star / BH $\sim 10^{-6}$ is small, we made used of a code with regularised equations of motion kindly provided by S. Mikkola. Details of the regularisation method are in [16, 17]; see [18] for a review. The orbits were started aligned with the x-axis with velocity vetor parallel to the y-axis (planar motion with orthogonal initial conditions). All orbits maintained a relative energy precision of 1×10^{-8}; checks with a higher accuracy criterion did not affect the outcome significantly.

Figure 2 graphs the semi-major axis a as function of time given in units of the binary period. Clearly when the BH is fixed, all orbits displayed give relatively mild evolution - the secular nature of which is clear from the rapid fluctuations on the scale of the binary period (= unity on the horizontal [time] axis). That is due to the fact that the orthogonal initial conditions did not factor in the external tidal field of the BH + galactic potential, which leads to periodic oscillations of low-amplitude. Not much difference takes place even when the BH is set in motion, except for the case of the 2:1 resonance, or when to total binary mass $= \sqrt{2}$. That case is displayed on the bottom-right panel on the figure. The beating-mode of period ~ 20 binary periods leads to a rapid increase of a after some 80 binary revolution; integration beyond $t = 100$ periods shows that the binary splits shortly thereafter, as the separation quickly reaches the size of the galactic orbit (and becomes even larger). That situation just does not arise in the fixed BH problem, even if one boosts the integration time by a large factor. The beat-frequency identified with (9) leads to rapid energy transfer and the selective slicing of binaries through enhanced tidal heating.

5 Outlook

It has not been possible to explore the response of a wide range of binary parameters and give more details of the mathematics. Still, it was very much possible to find other unstable orbits with ever higher ratios than the unstable 2:1 case shown here. Thus at constant amplitude $R_{BH}/R_c = 0.2$, all binaries with the right combination of separation and total mass will be sliced up more efficiently than expected from the standard static tidal field (arising from a fixed BH). It would be very interesting to try an apply this problem to the full case of the MW with a cuspy nuclear star cluster. The strong radial dependance of the tidal field would lead to a modification of the period ratio criterion (9), and a clearer picture of when and to what extent the MW's BH wandered off its current position in the past. What the simple model developed here shows is that we can expect the binary stellar population to bear witness to the past dynamics of the BH.

As a final example, let us mention that the stable 9:5 case displayed on Fig. 2 in fact becomes rapidly unstable once the initial amplitude of the BH's orbit shifts from $R_{BH}/R_c = 1/10$ to $3/20$. When that is the case, the 9:5 resonance (or, a binary with $2 \times 1M_\odot$ stars) is split in the same way as the 2:1 case displayed ($2 \times 1/\sqrt{2}M_\odot$ binary). A displacement of the BH will take place in any significant merger event (e.g., the accretion of a giant molecular cloud). Hence the demographics of binary stars, and their dependence on radius, may shed light on the recent history of BH dynamics in the MW.

Acknowledgements It is a pleasure to thank S. Mikkola for generously forwarding me a tailor-made version of his code AR4bb with which the calculations outlined here were performed. I would also like to thank J.S. Bagla for his invitation to revisit the problem of black hole dynamics, and T. Padmanabhan for sharing his insight into galactic dynamics over the years.

References

1. D. Merritt, *Dynamics and Evolution of Galactic Nuclei* (Princeton University Press, Princeton, 2013)
2. S. Chatzopoulos, T.K. Fritz, O. Gerhard, S. Gillessen, C. Wegg, R. Genzel, O. Pfuhl, MNRAS **447**, 948 (2015). doi:10.1093/mnras/stu2452
3. Q. Yu, S. Tremaine, ApJ **599**, 1129 (2003). doi:10.1086/379546
4. R.M. O'Leary, A. Loeb, MNRAS **383**, 86 (2008). doi:10.1111/j.1365-2966.2007.12531.x
5. J.N. Bahcall, R.A. Wolf, ApJ **216**, 883 (1977). doi:10.1086/155534
6. M. Preto, D. Merritt, R. Spurzem, ApJ Lett. **613**, L109 (2004). doi:10.1086/425139

7. J. Binney, S. Tremaine, *Galactic Dynamics*, 2nd edn. (Princeton University Press, Princeton, 2008)
8. R. Genzel, R. Schödel, T. Ott, F. Eisenhauer, R. Hofmann, M. Lehnert, A. Eckart, T. Alexander, A. Sternberg, R. Lenzen, Y. Clénet, F. Lacombe, D. Rouan, A. Renzini, L.E. Tacconi-Garman, ApJ **594**, 812 (2003). doi:10.1086/377127
9. R. Schödel, A. Eckart, T. Alexander, D. Merritt, R. Genzel, A. Sternberg, L. Meyer, F. Kul, J. Moultaka, T. Ott, C. Straubmeier, A&A **469**, 125 (2007). doi:10.1051/0004-6361:20065089
10. T.K. Fritz, S. Chatzopoulos, O. Gerhard, S. Gillessen, R. Genzel, O. Pfuhl, S. Tacchella, F. Eisenhauer, T. Ott, ApJ **821**, 44 (2016). doi:10.3847/0004-637X/821/1/44
11. R. Schödel, A. Feldmeier, D. Kunneriath, S. Stolovy, N. Neumayer, P. Amaro-Seoane, S. Nishiyama, A&A **566**, A47 (2014). doi:10.1051/0004-6361/201423481
12. C.M. Boily, T. Padmanabhan, A. Paiement, MNRAS **383**, 1619 (2008). doi:10.1111/j.1365-2966.2007.12682.x
13. S. Inoue, MNRAS **416**, 1181 (2011). doi:10.1111/j.1365-2966.2011.19122.x
14. M.J. Reid, K.M. Menten, S. Trippe, T. Ott, R. Genzel, ApJ **659**, 378 (2007). doi:10.1086/511744
15. D. Raghavan, H.A. McAlister, T.J. Henry, D.W. Latham, G.W. Marcy, B.D. Mason, D.R. Gies, R.J. White, T.A. ten Brummelaar, ApJ Suppl. **190**, 1 (2010). doi:10.1088/0067-0049/190/1/1
16. S. Mikkola, S.J. Aarseth, Celest. Mech. Dyn. Astron. **47**, 375 (1990)
17. S. Mikkola, S.J. Aarseth, Celest. Mech. Dyn. Astron. **57**, 439 (1993). doi:10.1007/BF00695714
18. S. Mikkola, IAU symposium, in *Dynamical Evolution of Dense Stellar Systems, IAU Symposium*, ed. by E. Vesperini, M. Giersz, A. Sills (2008), pp. 218–227. doi:10.1017/S1743921308015639

Mechanics of Apparent Horizon in Two Dimensional Dilaton Gravity

Rong-Gen Cai and Li-Ming Cao

Abstract In this article, we give a definition of apparent horizon in a two dimensional general dilaton gravity theory. With this definition, we construct the mechanics of the apparent horizon by introducing a quasi-local energy of the theory. Our discussion generalizes the apparent horizons mechanics in general spherically symmetric spacetimes in four or higher dimensions to the two dimensional dilaton gravity case.

1 Introduction

Quantum theory together with general relativity predicts that black hole behaves like a black body, emitting thermal radiation, with a temperature proportional to the surface gravity of the black hole and with an entropy proportional to the area of the cross section of the event horizon [1, 2]. The Hawking temperature and Bekenstein-Hawking entropy together with the black hole mass obey the first law of thermodynamics [3]. The first law of thermodynamics of black hole has two different versions—phase space version or passive version and physical process or active version [4]. In these two versions of discussion, the stationary of the black hole is essential, and the discussion is focused on the event horizon of these stationary spacetimes. However, this kind of horizon strongly depends on the global structure of the spacetime and there exist some practical issues which can not be easily solved [5]. It is very interesting to note that gravitational field equation on the black hole horizon can be expressed into a first law form of thermodynamics [6–8]. The usual approach to black hole thermodynamics is to start with the dynamics of gravity and ends with

R.-G. Cai (✉)
Institute of Theoretical Physics, Chinese Academy of Sciences,
P.O. Box 2735, Beijing 100190, China
e-mail: cairg@itp.ac.cn

L.-M. Cao
Interdisciplinary Center for Theoretical Study, University of Science
and Technology of China, Hefei 230026, Anhui, China
e-mail: caolm@ustc.edu.cn

J.S. Bagla and S. Engineer (eds.), *Gravity and the Quantum*,
Fundamental Theories of Physics 187, DOI 10.1007/978-3-319-51700-1_4

the thermodynamic of black hole spacetimes. Can we turn the logic around and get the dynamics of gravity from some thermodynamic considerations? This can not be fulfilled in this traditional approach because of the dependence of the global spacetime information of the event horizon. Apparent horizon defined by Hawking does not rely on the causal structure of the spacetime. However, it still depend on some global information of the spacetime—one has to select a slicing of the spacetime in advance. Furthermore, it is not clear how to establish thermodynamics on general apparent horizons. To reveal the relation between the spacetime dynamics and thermodynamics, probably local or quasilocal defined horizons are necessary.

A local or quasi-local definition of horizon is based on the local geometry of the spacetime. So it has a potential possibility to provide us more hints to study the relation between some fundamental thermodynamics and the gravitational equations. Along this way, fruitful results have been obtained. In fact, based on local Rindler horizon, Jacobson et al. [9, 10] was able to derive gravity field equation from the fundamental Clausius relation. With the assumption of FRW spacetime, Cai and Kim have obtained the Friedmann equations from the fundamental thermodynamical relation $\mathrm{d}E = T\mathrm{d}S$ on the apparent horizon of the spacetimes [11]. A simple summary on the relation between the spacetime dynamics and thermodynamical first law can be found in Ref. [12, 13], while the further understandings of gravitational dynamics from thermodynamical aspects can be seen in Ref. [14].

On the other hand, for general dynamical black holes, Hayward has proposed a new horizon, trapping horizon, to study associated thermodynamics in 4-dimensional Einstein theory without the stationary assumption [15]. In this theory, for general spherically symmetric spacetimes, Einstein equations can be rewritten in a form called "unified first law". Projecting this unified first law along trapping horizon, one gets the first law of thermodynamics for dynamical black holes. This trapping horizon can be null, spacelike, and timelike, and has no direct relation to the causal structure of the spacetime. Inspired by this quasilocal definition of horizon, Ashtekar et al. have proposed two types of horizons, i.e., isolated horizon and dynamical horizon. The former is null, while the later is spacelike [5]. The mechanics of these horizons also has been constructed. In some sense, the trapping horizon is a generalization of the Hawking's apparent horizon. However, the slicing of the spacetimes is not necessary to define this horizon. Nevertheless, in this paper, we still use the terminology of apparent horizon. Of course, it has the same meanings as the trapping horizon, and can be understood as a generalized apparent horizon.

Two dimensional dilation gravity theory has been widely studied over the past twenty years. One can get this kind of gravity from the spherically symmetric reduction of Einstein gravity theory in higher dimensions. To eliminate Weyl anomaly on string world sheet, one also has such a kind of gravity theory, for example, the famous CGHS model and others, for a nice review see [16]. In these theories, a lot of black hole solutions and cosmological solutions have been found. When some matter fields are included, in general the situations become complicated, and usually we have to study general dynamical solutions. In these two dimensional dilaton gravity theories, the apparent horizon has been used for a long time. However, what is the meaning of the apparent horizon in a two dimensional theory? Obviously, apparent horizon can not be defined in the usual way because we can not define any expansion scalar of a null congruence in two dimensional spacetimes. In other words, the codimension-2 surface shrinks to a point in a two dimensional spacetime, and intuitively, the size of the point does not change along a light-like geodesic, so the expansion does not make any sense. In this paper, we will propose a definition of apparent horizon in these two dimensional general dilaton gravity theories. With this definition in hand, we can construct the mechanics of the apparent horizon by introducing a quasilocal energy. This energy is similar to the Misner-Sharp energy in four or higher dimensional Einstein gravity theory. Actually, it can be found that this energy reduces to the usual Misner-Sharp energy for a special kind of dilaton gravity theory coming from the spherical reduction of higher dimensional Einstein gravity theory.

2 General Dilaton Gravity Theory in Two Dimensions

For a general two dimensional dilaton gravity theory, its action can be written into a following form [16]

$$I = \int d^2x\sqrt{-h}\left[\Phi R + U(\Phi)D^a\Phi D_a\Phi + V(\Phi) + \mathscr{L}_m\right], \tag{1}$$

where Φ is the so-called dilaton field, and R is the two dimensional Ricci scalar. The matter Lagrangian is represented by $\mathscr{L}_m$ which may contain tachyon (and others). The matter is denoted by ψ for simplicity. So, in general, the matter Lagrangian can be expressed as

$$\mathscr{L}_m = \mathscr{L}_m(\psi, D_a\psi\,, \ldots\,, \Phi\,, D_a\Phi\,, \ldots)\,.$$

The equation of motion for the dilaton field Φ can be written as

$$R - U'(\Phi)D^a\Phi D_a\Phi + V'(\Phi) - 2U(\Phi)D^a D_a\Phi + \mathscr{T}_m = 0\,, \tag{2}$$

where the prime stands for the derivative with respect to Φ: $d/d\Phi$, while the scalar $\mathscr{T}_m$ is defined as

$$\mathscr{T}_m = \frac{\partial\mathscr{L}_m}{\partial\Phi} - D_a\frac{\partial\mathscr{L}_m}{\partial D_a\Phi} + \cdots\,.$$

Of course, if the dilation field does not couple to the matter field, this term vanishes. The Euler–Lagrangian equation for the matter field ψ can be obtained in a similar way. The equations of motion for the metric h_{ab} can be put into a form

$$\begin{aligned} U(\Phi)D_a\Phi D_b\Phi - \frac{1}{2}U(\Phi)D^c\Phi D_c\Phi h_{ab} - D_a D_b\Phi \\ + D^c D_c\Phi h_{ab} - \frac{1}{2}V(\Phi)h_{ab} = T_{ab}\,, \end{aligned} \tag{3}$$

where T_{ab} is the energy-momentum tensor of the matter field. Straightforward calculation shows the covariant divergence of this energy-momentum tensor is given by

$$D^a T_{ab} = -\frac{1}{2}\big[R - U'(\Phi) D^c\Phi D_c\Phi + V'(\Phi) - 2U(\Phi) D^c D_c\Phi\big] D_b\Phi\,, \quad (4)$$

and this suggests that

$$D^a T_{ab} = \frac{1}{2}\mathscr{T}_m D_b\Phi\,. \quad (5)$$

Thus we see that the dilation provides an external force to the matter field when the coupling between the matter field and dilaton is present.

3 Some Solutions of the Theory

When the matter field is absent, we have $T_{ab} = 0$ and $\mathscr{T}_m = 0$. In Eddington–Finkelstein gauge, the general solution of Eq. (3) has a simple form [16]

$$h = e^{Q}\left[2\mathrm{d}v\mathrm{d}\Phi - (w - 2m)\mathrm{d}v^2\right]. \quad (6)$$

Here, two functions $Q(\Phi)$ and $w(\Phi)$ have been introduced, and they are defined by

$$U = -Q'\,, \quad V = e^{-Q}w' \quad (7)$$

up to some constants. Note that $e^{Q}\mathrm{d}\Phi$ is a closed form, from Poincare lemma, there exists a function r satisfying $\mathrm{d}r = e^{Q}\mathrm{d}\Phi$. This means the general solution of the metric can be transformed into a familar form

$$h = \left[2\mathrm{d}v\mathrm{d}r - e^{Q}(w - 2m)\mathrm{d}v^2\right]. \quad (8)$$

One can replace the constant m by a function of v, i.e., $m(v)$, and then construct a typical dynamical spacetime, i.e., Vaidya-like spacetime. In that case, the energy-momentum tensor of matter field will no longer vanish, and has a nontrivial component T_{vv} satisfying

$$\frac{\mathrm{d}m(v)}{\mathrm{d}v} = T_{vv}\,. \quad (9)$$

Naively one can read off the apparent horizon of this spacetime in this Eddington–Finkelstein gauge—it is given by equation

$$w - 2m(v) = 0 = e^{Q} D_a\Phi D^a\Phi\,. \quad (10)$$

However, we may ask a question here — What is the apparent horizon in this two dimensional spacetime? In the above discussion, we have read off naively the location of apparent horizon from the metric in Eddington–Finkelstein gauge. But in what sense it is an apparent horizon? Usually, the definition of apparent horizon depends on the extrinsic geometry of codimenstion-2 spacelike surface, i.e., the expansion scalars of the surface. Now we are considering two dimensional spacetime, the codimension-2 surface shrinks to a point, and the expansion scalars can not be defined. In the next section, we will focus on this question, and explain why the location of $D_a \Phi D^a \Phi = 0$ can be viewed as the apparent horizon in the two dimensional case.

4 Apparent Horizon

In this section, we define the apparent horizons in the two dimensional spacetime of the dilaton gravity theory. Assume $\{\ell^a, n^a\}$ is a null frame in the spacetime, and the metric can be expressed as

$$h_{ab} = -\ell_a n_b - n_a \ell_b \,, \tag{11}$$

where ℓ^a and n^a are two null vector fields which are globally defined on the spacetime and satisfy $\ell_a n^a = -1$. We assume ℓ^a and n^a are both future pointing, and furthermore, ℓ^a and n^a are outer pointing and inner pointing respectively. On the spacetime, there is a natural vector field $\phi^a = D^a \Phi$. Obviously, the causality of the vector field is determined by the signature of $\phi_a \phi^a$. According to the causality of this vector, the spacetime can be divided into several parts, and in each part the vector field ϕ^a either spacelike or timelike. ϕ^a is null on the boundary of any part, and this boundary can be defined as a kind of horizon. It is easy to find that on this horizon we have

$$\phi_a \phi^a = D_a \Phi D^a \Phi = -2 \mathscr{L}_\ell \Phi \mathscr{L}_n \Phi = 0 \,.$$

So on these horizons we have $\mathscr{L}_\ell \Phi = 0$ or $\mathscr{L}_n \Phi = 0$. We can further classify the horizons as follows. The horizon is called future if $\mathscr{L}_\ell \Phi = 0$ and $\mathscr{L}_n \Phi < 0$. In this case, if $\mathscr{L}_n \mathscr{L}_\ell \Phi < 0$, we call the horizon is outer. The future horizon with $\mathscr{L}_n \mathscr{L}_\ell \Phi > 0$ is called inner. The past horizon is defined by $\mathscr{L}_n \Phi = 0$, and $\mathscr{L}_\ell \Phi > 0$. Similarly, the past horizon with $\mathscr{L}_\ell \mathscr{L}_n \Phi > 0$ is called outer, and the case with $\mathscr{L}_\ell \mathscr{L}_n \Phi < 0$ is called inner. Mimicking the cases in higher dimensions, the region with $\mathscr{L}_\ell \Phi < 0$ and $\mathscr{L}_n \Phi < 0$ (or $\phi_a \phi^a < 0$) can be called trapped region of the spacetime [15].

At the first sight, these definitions have nothing to do with the geometry of the spacetime. How do these definitions realize the description of the spacetime region where light can not escape? To answer this question, we have to investigate the detailed structure of the definition and the equations of motion of the dilaton theory. Some calculation shows

$$\begin{aligned}
\mathscr{L}_n\mathscr{L}_\ell\Phi &= -\kappa_{(n)}(\mathscr{L}_\ell\Phi) - (1/2)\Box\Phi\,,\\
\mathscr{L}_\ell\mathscr{L}_n\Phi &= -\kappa_{(\ell)}(\mathscr{L}_n\Phi) - (1/2)\Box\Phi\,,\\
\mathscr{L}_\ell\mathscr{L}_\ell\Phi &= \kappa_{(\ell)}(\mathscr{L}_\ell\Phi) + U(\Phi)(\mathscr{L}_\ell\Phi)^2 - T_{ab}\ell^a\ell^b\,,\\
\mathscr{L}_n\mathscr{L}_n\Phi &= \kappa_{(n)}(\mathscr{L}_n\Phi) + U(\Phi)(\mathscr{L}_n\Phi)^2 - T_{ab}n^a n^b\,,
\end{aligned} \tag{12}$$

where we have used the Eq. (3) and introduced two scalars $\kappa_{(\ell)} = -n_a\ell^b D_b\ell^a$, $\kappa_{(n)} = -\ell_a n^b D_b n^a$. From these equations, it is easy to find that on the future outer horizon, we have

$$\Box\Phi > 0\,. \tag{13}$$

Similarly, on the past outer horizon, we have $\Box\Phi < 0$. In the following discussion, we will focus on the future outer horizon.

Here, we give some explanation why we can define the horizon in this way. From Eq. (12), we have

$$\mathscr{L}_k(\phi_a\phi^a) = \alpha\Big[\Box\Phi(\mathscr{L}_\ell\Phi) + 2(T_{ab}\ell^a\ell^b)(\mathscr{L}_n\Phi) - 2U(\Phi)(\mathscr{L}_n\Phi)(\mathscr{L}_\ell\Phi)^2\Big], \tag{14}$$

where $k^a = \alpha\ell^a$ is some null vector field and α is a positive function such that the parameter of k^a is affine. We consider the region very near the future outer horizon where $\mathscr{L}_\ell\Phi = 0$ and $\mathscr{L}_n\Phi < 0$. In this small neighbourhood, from continuity, $\Box\Phi$ should be positive and $\mathscr{L}_\ell\Phi$ is very small. Now let us consider the part of the neighbourhood inside the trapped region of the spacetime (where $\mathscr{L}_\ell\Phi$ is a small negative quantity and $\mathscr{L}_n\Phi$ is a finite negative quantity). In this case, we have

$$\mathscr{L}_k\|\phi\| < 0\,, \qquad \|\phi\| = \sqrt{|\phi_a\phi^a|}\,, \tag{15}$$

unless the null energy condition is broken, i.e., $T_{ab}\ell^a\ell^b < 0$. This mathematical relation suggests the light with wave vector k^a can approach the line with $\|\phi\| = 0$ (inside the trapped region) only when the null energy condition of the matter field is broken. This can not happen for usual classical matter field. So the outward propagating light do not exist near the future outer horizon. Similarly, we have

$$\mathscr{L}_k(\phi_a\phi^a) = \alpha\Big[\Box\Phi(\mathscr{L}_n\Phi) + 2(T_{ab}n^a n^b)(\mathscr{L}_\ell\Phi) - 2U(\Phi)(\mathscr{L}_n\Phi)^2(\mathscr{L}_\ell\Phi)\Big], \tag{16}$$

where α is still a positive function. Obviously, inside the trapped region and near the future outer horizon, we have $\mathscr{L}_k\|\phi\| > 0$. This means that inward propagating light is always allowed whatever the energy condition is satisfied or not.

Now, let us consider the neighborhood of the horizon inside the region where $\phi_a\phi^a > 0$. In the region, $\mathscr{L}_\ell\Phi$ is a small positive quantity. From Eq. (14), it is easy to find that it is possible to get $\mathscr{L}_k\|\phi\| > 0$ in this case especially when matter field is absent. This means the light has possibility to escape from this region to the region with large value of $\|\phi\|$. For the inner pointing light, Eq. (16) suggests it can cross the horizon and can reach the trapped region.

In a word, light cannot escape from the trapped region we have defined. So the horizon we have defined has the same properties as the apparent horizon in higher dimensions which, roughly speaking, can be viewed as the boundary of trapped region. So in this paper we still use the terminology of apparent horizon to describe this kind of horizon. In the next section, by introducing a quasilocal energy in this dilaton gravity theory, the mechanics of the apparent horizon will be established.

5 The Mechanics of the Apparent Horizon

To study the mechanics of the apparent horizon, we have to define the quasilocal energy inside the horizon. Generally, this is not an easy task. However, in the dilaton gravity we are considering, there is a well defined quasilocal energy. This can be found as follows. From the energy-mometum tensor of the matter field, we can define two useful quantities, i.e., a scalar called generalized pressure

$$P = -\frac{1}{2}T^a{}_a\,, \tag{17}$$

and a vector called energy-supply,

$$\Psi_a = T_a{}^b e^Q D_b\Phi + Pe^Q D_a\Phi\,. \tag{18}$$

It is easy to find

$$\Psi_a = e^Q\left[\frac{1}{2}U(\Phi)D^c\Phi D_c\Phi D_a\Phi - D_a D^b\Phi D_b\Phi + \frac{1}{2}D^c D_c\Phi D_a\Phi\right]. \tag{19}$$

Thus we have

$$\begin{aligned}\Psi_a + Pe^Q D_a\Phi = \frac{1}{2}e^Q\big[&U(\Phi)(D^c\Phi D_c\Phi)D_a\Phi\\ &-D_a(D^c\Phi D_c\Phi) + V(\Phi)D_a\Phi\big]\,.\end{aligned} \tag{20}$$

It is not hard to find that the right hand side of the Eq. (20) can be written as

$$\frac{1}{2}D_a\left[w\left(1 - \frac{e^Q}{w}D^c\Phi D_c\Phi\right)\right]. \tag{21}$$

In the above equation, Q and w are the same as those given in Eq. (7). Comparing with the unified first law in higher dimensional spherical symmetric spacetime [15], we can define a similar quasi-local energy

$$E = \frac{1}{2}\left[w\left(1 - \frac{e^Q}{w} D^c\Phi D_c\Phi\right)\right] . \tag{22}$$

Then the equations of motion for the metric, i.e., Eq. (3) can be put into the form

$$D_a E = \Psi_a + P e^Q D_a \Phi . \tag{23}$$

Since $e^Q D_a\Phi$ is a closed one form in the spacetime, at least locally, it can be expressed as $e^Q D_a\Phi = D_a\mathscr{V}$ for some function $\mathscr{V}$. This suggests the above relation can be transformed into a simple form, i.e.,

$$\mathrm{d}E = \Psi + P\mathrm{d}\mathscr{V} . \tag{24}$$

This energy E generalizes the Misner-Sharp energy in higher dimensional theory to the two dimensional dilaton gravity theory, and at the same time, the above equation establishes the unified first law in this two dimensional gravity theory.

Now let us consider the special case where the matter field is absent, i.e., T_{ab} is vanishing, so do P and Ψ. From the above unified first law (24), we have $\mathrm{d}E = 0$. This means that E is a constant which can be denoted by $\mathfrak{m}$. By this consideration, from the energy form (22), we have Eq. (10) with $\mathfrak{m}(v)$. When the energy-momentum tensor is given by some radiation matter, the general solution is just the Vaidya-like spacetime mentioned in the previous section. In this case, it is easy to find that E is nothing but $\mathfrak{m}(v)$ and the first law (24) reduces to the Bondi's energy balance Eq. (9).

Assume on the apparent horizon, i.e., on the spacetime points which satisfy $D^c\Phi D_c\Phi = 0$, that the dilaton field Φ takes value Φ_A, then, on the apparent horizon, the quasi-local energy becomes

$$E = \frac{1}{2} w(\Phi_A) . \tag{25}$$

In general, this energy is not constant because Φ_A may depend on the coordinates. For example, for the Vaidya-like spacetime, the total energy inside the apparent horizon is $\frac{1}{2}w(\Phi_A) = \mathfrak{m}(v)$. For the static case without matter, the apparent horizon coincides with the event horizon (if it can be defined), this energy becomes $\frac{1}{2}w(\Phi_A) = \frac{1}{2}w(\Phi_+) = \mathfrak{m}$, where Φ_+ is the value of dilation on the event horizon.

On the apparent horizon, the energy-supply becomes

$$\Psi_a = \frac{1}{2} e^Q \left[-D_a(D^c\Phi D_c\Phi) + \Box\Phi D_a\Phi\right] . \tag{26}$$

Let ξ be the vector tangent to the apparent horizon. Since $D^c\Phi D_c\Phi$ is a constant on the apparent horizon, $\xi^a D_a(D^c\Phi D_c\Phi) = 0$, then we find

$$\xi^a \Psi_a = \frac{1}{2} e^Q (\Box\Phi) \mathscr{L}_\xi \Phi , \tag{27}$$

where $\mathscr{L}_\xi$ is Lie derivative along the vector ξ. In higher dimensions, the surface gravity of an apparent horizon is defined by the Kodama vector field $K^b D_{[b} K_{a]} = \kappa K_a$ [17]. Here we can also introduce a Kodama-like vector field as $K^a = -e^Q \epsilon^{ab} D_b \Phi$. Some calculation shows

$$K^b D_{[b} K_{a]} = \frac{1}{2}\Big[e^Q \Box\Phi K_a - U(e^Q D_b\Phi D^b\Phi) K_a\Big]. \tag{28}$$

From the definition of the surface gravity, it is easy to find the surface gravity of the apparent horizon can be expressed as

$$\kappa = \frac{1}{2} e^Q (\Box\Phi).$$

On the future outer apparent horizon, from Eq. (13), we see this surface gravity is always positive. Therefore, we find that the energy-supply projecting onto the apparent horizon gives

$$\xi^a \Psi_a = \kappa \, \mathscr{L}_\xi \Phi. \tag{29}$$

As a result, on the apparent horizon, we have a relation

$$\mathscr{L}_\xi E = \frac{\kappa}{2\pi} \mathscr{L}_\xi S + P \mathscr{L}_\xi \mathscr{V}, \tag{30}$$

where $S = 2\pi\Phi$. This relation is the same as the first law of thermodynamics if we identify $T = \kappa/2\pi$ and regard S as entropy. Actually, the "entropy" of the future outer apparent horizon can also be written as

$$S = 2\pi \Phi_A. \tag{31}$$

In general, this entropy is not a constant, and it might change with some coordinate. In the static case, the future outer apparent horizon coincides with the event horizon, this entropy becomes $S = 2\pi\Phi_+$, this result has been found in many static black holes in the two dimensional dilaton gravity theories [16].

The function $\mathscr{V}$ can be viewed as a kind of "volume" of the system. It comes from the divergence free of the Kodama vector K^a, i.e., $D_a K^a = 0$, and can be viewed as a conserved quantity of the theory. Besides the function $\mathscr{V}$, the energy E can also be viewed as a conserved quantity. Actually, from Eqs. (3) and (5), one can prove that $J^a = T^a{}_b K^b$ is conserved, i.e., we have $D_a J^a = 0$. This suggests that the Hodge dual of J_a is a closed one form, and locally it is the exterior derivative of a function. This function is nothing but the energy E. Such kind of discussion can also be found in [18] where some special matter Lagrangian has been considered.

For the case of the energy-momentum tensor given by conformal matter, for example, tachyon, the trace of the energy-momentum tensor vanishes. In this case, from the trace part of Eq. (3), the surface gravity becomes $\kappa = -\frac{1}{2}w'(\Phi_A)$. This is very similar to the static case where the surface gravity is just given by $-\frac{1}{2}w'(\Phi_+)$.

The work term vanishes due to the traceless of the energy-momentum tensor, so the first law on the apparent horizon becomes

$$\mathcal{L}_\xi E = \frac{\kappa}{2\pi} \mathcal{L}_\xi S \,. \tag{32}$$

This means for the system with conformal matter, there is no external work term. The above result shows that, in the case with conformal matter, all the thermodynamic quantities of the apparent horizon can be obtained from the static case through the replacement of Φ_+ by Φ_A.

For the two dimensional dilation gravity coming from the spherically symmetric reduction of n-dimensional Einstein gravity, the potential U and V are given respectively by

$$U(\Phi) = \frac{n-3}{n-2}\Phi^{-1} \,, \quad V(\Phi) = (n-2)(n-3)\lambda^2 \Phi^{\frac{n-4}{n-2}} \,, \tag{33}$$

where λ is a constant with dimension of mass square, and $\Phi = (\lambda r)^{n-2}$. If the function U and V have the forms (33), the quasi-local energy (22) is just the n-dimensional Misner-Sharp energy

$$E_{MS} = \frac{1}{2}(n-2)r^{n-3}\left(1 - D^a r D_a r\right) \,. \tag{34}$$

It is straightforward to see that the entropy of apparent horizon in the two dimensional dilaton gravity is given by the area of the horizon sphere in n-dimensions. This can be obtained by replacing Φ_A in Eq. (31) by $(\lambda r_A)^{n-2}$. It should be noted here that our discussion is not restricted to the future outer apparent horizon. Actually, the same discussions can be applied to other types of apparent horizons. For example, in an FRW universe, the entropy of the apparent horizon is given by the one quarter of the area of the apparent horizon, and the radius of the apparent horizon can be expressed by Hubble parameter as

$$\frac{1}{\tilde{r}_A^2} = H^2 + \frac{k}{a^2} \,,$$

where $\tilde{r}_A = a r_A$, and a is the scale factor in the FRW universe, see for example [19]. Thus the apparent horizon associated with the FRW universe also applies here.

Acknowledgements This work is dedicated to celebrate the 60th birthday of Prof. T. Padmanabhan. The work was supported in part by the National Natural Science Foundation of China with grants No.11622543, No.11235010, No.11375247 and No.11435006.

References

1. S.W. Hawking, Commun. Math. Phys. **43**, 199 (1975)
2. J.D. Bekenstein, Phys. Rev. D **7**, 2333 (1973)
3. J.M. Bardeen, B. Carter, S.W. Hawking, Commun. Math. Phys. **31**, 161 (1973)
4. R.M. Wald, Living Rev. Rel. **4**, 6 (2001)
5. A. Ashtekar, Living Rev. Rel. **7**, 10 (2004)
6. T. Padmanabhan, Class. Quant. Grav. **19**, 5387 (2002) gr-qc/0204019
7. T. Padmanabhan, arXiv:0910.0839 [gr-qc]
8. R.G. Cai, N. Ohta, Phys. Rev. D **81**, 084061 (2010). arXiv:0910.2307 [hep-th]
9. T. Jacobson, Phys. Rev. Lett. **75**, 1260 (1995). arXiv:gr-qc/9504004
10. C. Eling, R. Guedens, T. Jacobson, Phys. Rev. Lett. **96**, 121301 (2006)
11. R.G. Cai, S.P. Kim, JHEP **0502**, 050 (2005). hep-th/0501055
12. R.G. Cai, J. Phys. Conf. Ser. **484**, 012003 (2014)
13. R.G. Cai, Prog. Theor. Phys. Suppl. **172**, 100 (2008). arXiv:0712.2142 [hep-th]
14. T. Padmanabhan, Rept. Prog. Phys. **73**, 046901 (2010). arXiv:0911.5004 [gr-qc]
15. S.A. Hayward, Class. Quant. Grav. **15**, 3147 (1998). arXiv:gr-qc/9710089
16. D. Grumiller, W. Kummer, D.V. Vassilevich, Phys. Rept. **369**, 327 (2002). hep-th/0204253
17. H. Maeda, M. Nozawa, Phys. Rev. D **77**, 064031 (2008). arXiv:0709.1199 [hep-th]
18. R.B. Mann, Phys. Rev. D **47**, 4438 (1993). hep-th/9206044
19. R.G. Cai, L.M. Cao, Phys. Rev. D **75**, 064008 (2007). arXiv:gr-qc/0611071

Boundary Terms of the Einstein–Hilbert Action

Sumanta Chakraborty

Abstract The Einstein–Hilbert action for general relativity is not well posed in terms of the metric g_{ab} as a dynamical variable. There have been many proposals to obtain an well posed action principle for general relativity, e.g., addition of the Gibbons–Hawking–York boundary term to the Einstein–Hilbert action. These boundary terms are dependent on what one fixes on the boundary and in particular on spacetime dimensions as well. Following recent works of Padmanabhan we will introduce two new variables to describe general relativity and the action principle with these new dynamical variables will turn out to be well posed. Then we will connect these dynamical variables and boundary term obtained thereof to existing literature and shall comment on a few properties of Einstein–Hilbert action which might have been unnoticed earlier in the literature. Before concluding with future prospects and discussions, we will perform a general analysis of the boundary term of Einstein–Hilbert action for null surfaces as well.

1 Introduction

Action principle is the starting point of any field theory. Along with the action functional one need to fix the spacetime volume, its boundary and what variable should be fixed on the boundary. When the boundary conditions imposed on an action are compatible with the derived field equation(s), we refer that action principle as well posed. It turns out that the widely used action principle for general relativity, the Einstein–Hilbert action is *not* well posed. To be more precise, with Ricci scalar as the gravitational Lagrangian, derivation of Einstein's equations requires fixing both metric and its first derivative on the boundary — inconsistent with Einstein's equations.

S. Chakraborty (✉)
IUCAA, Post Bag 4, Ganeshkhind, Pune University Campus, Pune 411 007, India
e-mail: sumanta@iucaa.in

S. Chakraborty
Department of Theoretical Physics, Indian Association for the Cultivation of Science, Kolkata 700 032, India
e-mail: sumantac.physics@gmail.com

J.S. Bagla and S. Engineer (eds.), *Gravity and the Quantum*,
Fundamental Theories of Physics 187, DOI 10.1007/978-3-319-51700-1_5

This feature arises, since the action principle for general relativity is peculiar. It contains second derivatives of the dynamical variables, the metric g_{ab}, unlike any other existing Lagrangians. At first glance it seemed quite exotic, since the field equations derived from an action which has second derivatives of the dynamical variable are supposed to have third order derivatives, leading to existence of ghost fields. However it is again the structure of the action principle for general relativity that comes to rescue. The Ricci scalar can be separated into a bulk term and a surface term. The bulk term has the structure Γ^2, where Γ^a_{bc} are the connection coefficients and along with being quadratic it contains only first derivatives of the metric. In any action principle the surface terms do not contribute to the derivation of field equations, so Einstein's equations also have second derivatives of the metric. However all the second derivatives of the metric hides in the surface term and it is the surface term that leads to boundary contribution. Hence quite naturally, in the case of Einstein–Hilbert action one ends up fixing both the metric and its derivative on the boundary.

The above arguments pose the problem but also solves it — it suffices to remove the surface term and consider a new action functional for general relativity, namely, $L = R - L_{\rm sur}$, as proposed by Einstein in 1916 [1]. Then one obtains Einstein's equations without worrying about the boundary terms. But the problem with the above approach is that, the action is not invariant under diffeomorphism, while we want every action to have the symmetries that the underlying system has. Fortunately, the boundary term that one need to add to the Einstein–Hilbert action is by no means unique. Any boundary term that kills all the normal derivatives of the metric on the boundary surface is good enough for our purpose and there could be infinitely many of them as demonstrated by Charap and Nelson in [2]. The most popular boundary term that keeps the action invariant under diffeomorphism and also makes it well posed is the Gibbons–Hawking–York term [3–5]. The Gibbons–Hawking–York term depends on the extrinsic curvature K of the boundary surface and is given by $2K\sqrt{|h|}$, where h stands for the determinant of the induced metric on the boundary surface. Note that even though the Gibbons–Hawking–York boundary term is invariant under diffeomorphism, is not covariant in a strict sense, because of its dependence on the foliation. Further, the Gibbons–Hawking–York term was guessed and then shown to yield a well posed variational principle without a first principle derivation. This gap was filled by providing a direct derivation of the Gibbons–Hawking–York boundary term from the action itself in [6] while another important issue, the boundary term for null boundaries has been tackled recently in [7]. Even then the structure of the boundary term can change depending on what one needs to fix on the boundary, the induced metric or the conjugate momentum and it also changes depending on the spacetime dimensions. In this work we will try to provide a broad overview on the possible boundary term structures of the Einstein–Hilbert action along with what one needs to fix on the boundary surfaces. This will be performed for both null and non-null cases, besides discussing some other important features of the Einstein–Hilbert action.

The paper is organized as follows, in Sect. 2 we will present various boundary terms used in various dimensions for an well-posed action of general relativity and their possible connections. Then in Sect. 3 we will explicitly demonstrate some common notions in the context of general relativity starting from the well known $(1+3)$ decomposition. Finally we comment on the nature of the boundary terms in the context of null surfaces in Sect. 4 before concluding with a brief discussion.

Notation: We will work in D spacetime dimensions in Sect. 2, while the rest of the analysis will be performed in four spacetime dimensions following the mostly positive signature $(-,+,+,+,\ldots)$. The fundamental constants c, G and $\hbar$ have been set to unity.

2 Reconciling Boundary Terms for the Einstein–Hilbert Action

The origin of boundary value problem for general relativity is due to the fact that Einstein–Hilbert action contains second derivatives of the metric — as a consequence one needs to fix both the metric and its derivatives on the boundary rendering the action ill posed. The above problem arises for using the metric as a fundamental variable and hence to obtain a well posed variational principle we have to add boundary terms to the Einstein–Hilbert action. However, it is possible to rewrite the Einstein–Hilbert action in the momentum space and the resulting variational principle becomes well posed. The momentum space representation of the Einstein–Hilbert action can be obtained by introducing two new variables [8] (see [9] for a generalization to Lanczos–Lovelock gravity),

$$f^{ab} = \sqrt{-g}g^{ab}; \qquad N^a_{bc} = Q^{ad}_{be}\Gamma^e_{cd} + Q^{ad}_{ce}\Gamma^e_{bd} = -\Gamma^a_{bc} + \frac{1}{2}\left(\Gamma^d_{bd}\delta^a_c + \Gamma^d_{cd}\delta^a_b\right) , \tag{1}$$

where f^{ab} is a tensor density and N^a_{bc} stands for a linear combination of the connections. Note that the above relation holds for any number of spacetime dimensions as $Q^{ab}_{cd} = (1/2)(\delta^a_c\delta^b_d - \delta^a_d\delta^b_c)$ is independent of spacetime dimensions. However the inverse relation connecting Γ^a_{bc} in terms of N^a_{bc} depends on the spacetime dimensions and reads in general,

$$\Gamma^c_{ab} = -N^c_{ab} + \frac{1}{D-1}\left(N^d_{ad}\delta^c_b + N^d_{bd}\delta^c_a\right) , \tag{2}$$

which reduces to the expression in [8] for $D = 4$. Then the expressions for various curvature components are also modified. For example, the Ricci tensor can be expressed in terms of N^c_{ab} such that,

$$R_{ab} = -\left(\partial_c N^c_{ab} + N^c_{ad}N^d_{bc} - \frac{1}{D-1}N^c_{ac}N^d_{bd}\right) , \tag{3}$$

reducing to the one given in [8] for four spacetime dimensions. These variables can be used in the action principle as well, in which case the Einstein–Hilbert Lagrangian density becomes $\sqrt{-g}R = f^{ab}R_{ab}$, where R_{ab} can be written in terms of N^c_{ab} following Eq. (3). This leads to momentum space representation of the Einstein–Hilbert action, which follows from the result that $N^c_{ab} = \partial(\sqrt{-g}R)/\partial(\partial_c f^{ab})$ and hence the set (f^{ab}, N^c_{ab}) acts as a set of canonically conjugate variables. Further Einstein–Hilbert action when varied reads in terms of variations of these canonically conjugate variables as,

$$\delta\left(\int_{\mathcal{V}} d^D x\sqrt{-g}R\right) = \int_{\mathcal{V}} d^D x R_{ab}\delta f^{ab} - \int_{\mathcal{V}} d^D x f^{ab}\nabla_c \delta N^c_{ab} \tag{4}$$

$$= \int_{\mathcal{V}} d^D x R_{ab}\delta f^{ab} - \int_{\partial\mathcal{V}} d^{D-1}x \bar{n}_c f^{ab}\delta N^c_{ab} , \tag{5}$$

where $\mathcal{V}$ stands for the spacetime volume under interest with boundary being denoted by $\partial\mathcal{V}$. The last term has been obtained through the use of the following relation $f^{ab}\nabla_c\delta N^c_{ab} = \partial_c\left(\sqrt{-g}g^{ab}\delta N^c_{ab}\right)$. Also $\bar{n}_c$ in the final expression is the unnormalized normal. If the surface $\partial\mathcal{V}$ is some $\phi =$ constant surface, then $\bar{n}_c = \delta^\phi_c$. With suitable normalization one obtains, $\bar{n}_c = \varepsilon(1/N)n_c$, where n_c is the normalized normal, $\varepsilon = \pm 1$ depending on the normal being spacelike or timelike and N is $\sqrt{|g^{\phi\phi}|}$. Thus note that one can obtain the Einstein's equations provided N^c_{ab} is fixed at the boundary, leading to an well posed action principle for general relativity, since N^c_{ab} and f^{ab} are treated as independent variables.

On the other hand, it is also well known that the variation of the Einstein–Hilbert action leads to $\delta(2K\sqrt{h})$, where K is the extrinsic curvature of the boundary surface and h is the determinant of the induced metric on that surface, along with variations of the induced metric with proper coefficients as the boundary term [6]. Thus for being consistent one must have the $f^{ab}\delta N^c_{ab}$ to yield $\delta(2K\sqrt{h})$ along with variations of the induced metric. It is not at all clear a priori, how this can be achieved. In order to fill this gap we would like to connect the boundary term obtained above in Eq. (5) with the standard literature. As a first step towards the connection, we will present a simplified analysis and shall subsequently provide a general derivation.

2.1 A Warm-Up Example: Analysis in Synchronous Frame

Before jumping into the formal derivation let us consider an explicit example as a warm-up. Let us use all the gauge degrees of freedom due to diffeomorphism to eliminate four degrees of freedom from the metric and reduce it to synchronous form, in which the line element reads,

$$ds^2 = -d\tau^2 + h_{\alpha\beta}(\tau, x^\mu)dx^\alpha dx^\beta . \tag{6}$$

As explicitly demonstrated in [10], any metric can be written in the synchronous coordinate system. The boundary $\partial\mathcal{V}$ of the full spacetime volume can be taken to be $\tau =$ constant hypersurface in this coordinate system, such that the unnormalized normal becomes $\bar{n}_c = \delta_c^\tau$ and hence the surface term reads,

$$\bar{n}_c f^{ab}\delta N^c_{ab} = f^{ab}\delta N^0_{ab} = -\sqrt{h}\delta N^0_{00} + \sqrt{h}h^{\alpha\beta}\delta N^0_{\alpha\beta} , \tag{7}$$

where in obtaining the last line we have used the synchronous frame metric as in Eq. (6). From the definition of N^a_{bc} in terms of connections as in Eq. (1) and the metric in Eq. (6) it follows that,

$$N^0_{00} = \Gamma^\alpha_{0\alpha} = -K; \qquad N^0_{\alpha\beta} = -\Gamma^0_{\alpha\beta} = K_{\alpha\beta} . \tag{8}$$

Thus one can substitute both N^0_{00} and $N^0_{\alpha\beta}$ in the boundary term which finally leads to,

$$\begin{aligned} \bar{n}_c f^{ab}\delta N^c_{ab} &= \sqrt{h}\delta K + \sqrt{h}h^{\alpha\beta}\delta K_{\alpha\beta} \\ &= \delta\left(2K\sqrt{h}\right) + \sqrt{h}\left(K^{\alpha\beta} - Kh^{\alpha\beta}\right)\delta h_{\alpha\beta} . \end{aligned} \tag{9}$$

This shows the equivalence of the boundary term with (f^{ab}, N^c_{ab}) as the dynamical variables with the standard boundary term. The above expression explicitly shows that one needs to add $2K\sqrt{h}$ as the boundary term to the Einstein–Hilbert action and as a consequence one needs to fix only the spatial part of the metric $h_{\alpha\beta}$ on the boundary $\partial\mathcal{V}$, i.e., on $\tau =$ constant surfaces.

However the above derivation is a special case and more importantly the boundary term even though is independent of coordinate choices depends heavily on foliation, thus it is not clear from the above result whether the same conclusion should hold for arbitrary foliation as well. This is precisely what we will prove next.

2.2 Boundary Terms: A General Analysis

As explained above the demonstration in synchronous frame is a specific one among many possible foliations and one needs to provide a general analysis for an arbitrary foliation to grasp the complete structure. To proceed with the general analysis, we will start with the boundary term and shall write N^c_{ab} in terms of the connections. Using the fact that variations of the connections are tensors one can ultimately write down the boundary term in terms of the normal and variations in the metric tensor,

$$\int_{\partial\mathcal{V}} d^{D-1}x\bar{n}_c f^{ab}\delta N^c_{ab} = -\int_{\partial\mathcal{V}} d^{D-1}x\bar{n}_c\nabla_d\left(-\delta g^{cd} + g^{cd}g_{ik}\delta g^{ik}\right) , \tag{10}$$

where the following algebraic identity, $-g^{ab}\delta N^c_{ab} = \nabla_d \left(-\delta g^{cd} + g^{cd} g_{ik}\delta g^{ik}\right)$ have been used in order to arrive at the final result. Given the above Eq. (10) we can immediately incorporate the normal inside the covariant derivative and the above expression reads,

$$-\int_{\partial\mathcal{V}} d^{D-1}x\bar{n}_c f^{ab}\delta N^c_{ab} = \int_{\partial\mathcal{V}} d^{D-1}x\varepsilon\sqrt{h}\Big\{\nabla_d\left(-n_c\delta g^{cd} + n^d g_{ik}\delta g^{ik}\right) - \nabla_d n_c\left(-\delta g^{cd} + g^{cd} g_{ik}\delta g^{ik}\right)\Big\}, \tag{11}$$

where $\varepsilon = -1$ for spacelike hypersurfaces and is $+1$ for timelike hypersurfaces respectively. The variations of the metric can be divided into two pieces, variations in the induced metric h_{ij} and variations in the normal n^i. Using the contractions properly and the fact that $\delta(n_i n^i) = 0$, we immediately obtain the following expression for the boundary term of the Einstein–Hilbert action,

$$\begin{aligned}-\int_{\partial\mathcal{V}} d^{D-1}x\bar{n}_c f^{ab}\delta N^c_{ab} &= \int_{\partial\mathcal{V}} d^{D-1}x\varepsilon\delta\left(2K\sqrt{h}\right) \\ &\quad - \int_{\partial\mathcal{V}} d^{D-1}x\varepsilon\sqrt{h}\left(K_{ab} - Kh_{ab}\right)\delta h^{ab} \\ &\quad + \int d^{D-1}x\varepsilon\sqrt{h}D_i\left(-n_c h^i_b\delta g^{bc} + 2n_k h^i_l\delta g^{kl}\right).\end{aligned} \tag{12}$$

The last term is again a surface term and would contribute only on the two surface and hence is neglected. It is useful and instructive to define the momentum conjugate to the induced metric h_{ab} on the hypersurface $\partial\mathcal{V}$ as,

$$\Pi_{ab} = \sqrt{h}\left(K_{ab} - Kh_{ab}\right). \tag{13}$$

Note that $n_a\Pi^{ab} = 0$. Thus finally using the expression for Π_{ab} and neglecting the surface term, we obtain the simplified version of the boundary term from Eq. (12) in the most general case as,

$$-\int_{\partial\mathcal{V}} d^{D-1}x\bar{n}_c f^{ab}\delta N^c_{ab} = \int_{\partial\mathcal{V}} d^3x\varepsilon\delta\left(2K\sqrt{h}\right) - \int_{\partial\mathcal{V}} d^{D-1}x\varepsilon\Pi_{ab}\delta h^{ab}. \tag{14}$$

The result in the synchronous frame can be derived immediately from the above relation by substituting $\varepsilon = -1$, since $\tau =$ constant surfaces are spacelike. However note that the two-dimensional surface terms identically vanishes in the synchronous frame. The above result suggests that if we add the boundary term $-2\varepsilon K\sqrt{h}$ to the Einstein–Hilbert action the normal derivatives of the metric will be removed from the boundary and one needs to fix only the induced metric h^{ab}. It is important to emphasis at this stage that fixing h^{ab} is different from fixing h_{ab}. Since by construction we have $n_a \propto \nabla_a\phi$, and $n_a h^{ab} = 0$, this suggests $h^{ab} = h^{\alpha\beta}$, where α, β are spacetime indices excluding ϕ, while h_{ab} has all the metric components. Due to the momentum and Hamiltonian constraints of general relativity one cannot fix all the metric components on the hypersurfaces and hence the correct variational principle would be the one which fixes only h^{ab}, i.e., $h^{\alpha\beta}$ on the boundary $\partial\mathcal{V}$.

Let us now illustrate the fact that $2\varepsilon K\sqrt{h}$ is not the only boundary term that can lead to a well-posed action principle for general relativity, there are infinitely many. However for our illustration we will pick two of them. Since we are working in a D dimensional spacetime we have the following identity, $\Pi_{ab}h^{ab} = -(D-2)K\sqrt{h}$. We can use the above identity to convert the original result in Eq. (14) to two different results,

$$\begin{aligned}
-\int_{\partial\mathcal{V}} d^{D-1}x\bar{n}_c f^{ab}\delta N^c_{ab} &= \int_{\partial\mathcal{V}} d^{D-1}x\varepsilon\delta\left(2K\sqrt{h}\right) - \int_{\partial\mathcal{V}} d^{D-1}x\varepsilon\delta\left(\Pi_{ab}h^{ab}\right) \\
&+ \int_{\partial\mathcal{V}} d^{D-1}x\varepsilon h^{ab}\delta\Pi_{ab} \\
&= \int_{\partial\mathcal{V}} d^{D-1}x\varepsilon\delta\left(DK\sqrt{h}\right) \\
&+ \int_{\partial\mathcal{V}} d^{D-1}x\varepsilon h^{ab}\delta\Pi_{ab}\,. \qquad (15)
\end{aligned}$$

The above result depicts that one can also add $-D\varepsilon K\sqrt{h}$ as the boundary term to the Einstein–Hilbert action and hence obtain an well-posed variational principle if Π_{ab} is fixed at the boundary. Note that as we have argued earlier, the only non-zero components of h^{ab} are $h^{\alpha\beta}$ and hence one need to fix only $\Pi_{\alpha\beta}$ at the boundary $\partial\mathcal{V}$. This result can also be casted in a different form, for that we need to use the identity, $\Pi_{ab}\delta h^{ab} = -\Pi^{ab}\delta h_{ab}$. Use of which enables one to write Eq. (14) in the following form

$$\begin{aligned}
-\int_{\partial\mathcal{V}} d^{D-1}x\bar{n}_c f^{ab}\delta N^c_{ab} &= \int_{\partial\mathcal{V}} d^{D-1}x\varepsilon\delta\left(2K\sqrt{h}\right) + \int_{\partial\mathcal{V}} d^{D-1}x\varepsilon\delta\left(\Pi^{ab}h_{ab}\right) \\
&- \int_{\partial\mathcal{V}} d^{D-1}x\varepsilon h_{ab}\delta\Pi^{ab} \\
&= \int_{\partial\mathcal{V}} d^{D-1}x\varepsilon\delta\left[(4-D)\,K\sqrt{h}\right] \\
&- \int_{\partial\mathcal{V}} d^{D-1}x\varepsilon h_{ab}\delta\Pi^{ab}\,. \qquad (16)
\end{aligned}$$

This is another form of the boundary contribution recently discussed in [11] which essentially follows from the original boundary term in terms of the canonically conjugate variables (f^{ab}, N^c_{ab}). In this case the boundary term one has to add to the Einstein–Hilbert action corresponds to, $(4-D)\varepsilon K\sqrt{h}$, with the peculiarity that at $D = 4$ this term identically vanishes. While in this case one need to fix Π^{ab} at the boundary $\partial\mathcal{V}$. Hence the original boundary term from whichall possible versions of the boundary terms including the well-known $2\varepsilon K\sqrt{h}$ can be derived is the $f^{ab}\delta N^c_{ab}$ combination. Further we have shown two explicit examples in which one can add different boundary term at the expense of fixing either Π^{ab} or Π_{ab} at the boundary (Table 1). Even though it is tempting to assume $h_{ab}\delta\Pi^{ab} = -h^{ab}\delta\Pi_{ab}$, this relation is actually not correct. This can be seen from the following algebraic manipulation straightforwardly,

$$\begin{aligned} h^{ab}\delta\Pi_{ab} &= h^{ab}\delta\left(h_{ac}h_{bd}\Pi^{cd}\right) = h_{cd}\delta\Pi^{cd} + 2\Pi^{ac}\delta h_{ac} \\ &= h_{cd}\delta\Pi^{cd} - 2\Pi_{ac}\delta h^{ac} = -h_{cd}\delta\Pi^{cd} + \delta\left[(4-2D)\,K\sqrt{h}\right], \end{aligned} \tag{17}$$

reconciling the two results presented in Eqs. (15) and (16) respectively. Through this exercise we have achieved two important goals, which are,

- By introducing the canonically conjugate variables (f^{ab}, N^c_{ab}), one obtains the Einstein's equations from variations of f^{ab}, while variations of N^c_{ab} leads to the boundary term. Hence the Einstein–Hilbert action becomes action in the momentum space such that one need to fix the momentum N^c_{ab} at the boundary. However there were no clear consensus how this boundary term is related to the existing ones, e.g., the Gibbons–Hawking–York boundary term. In this section we have explicitly demonstrated the connection, by deriving the Gibbons–Hawking–York counter term starting from the boundary term consisting of $f^{ab}\delta N^c_{ab}$.
- Secondly, in most of the literatures people always take the Gibbons–Hawking–York boundary term to be the only boundary term possible. In the last part of this section we have explicitly demonstrated two more boundary terms. Our result clearly shows that the structure of the boundary term depends crucially on what one fixes at the boundary. If one fixes the induced metric h^{ab}, then Gibbons–Hawking–York term is the only option. But if one fixes the conjugate momentum, then depending on whether one fixes Π^{ab} or Π_{ab}, one arrives to different boundary terms. In particular when Π^{ab} is fixed one need not add any boundary term in four dimensions, which is a peculiar feature of general relativity.

Thus we have reconciled the possible boundary terms that one can add to the Einstein–Hilbert action. Their non-uniqueness and derivation from a first principle starting from Einstein–Hilbert action in momentum space has also been presented. We will now turn to the $(1+3)$ decomposition of the Einstein–Hilbert action and related comments.

Table 1 A comparison of various boundary terms of Einstein–Hilbert action

Bulk term	Surface term	Boundary term[a]	What to fix on boundary	Well-posed action
$R_{ab}\delta f^{ab}$	$-\bar{n}_c f^{ab}\delta N^c_{ab}$	None	N^c_{ab}	$\sqrt{-g}R$
$G_{ab}\delta g^{ab}$	$\varepsilon\delta(2K\sqrt{h})$ $-\varepsilon\Pi_{ab}\delta h^{ab}$	$\varepsilon\delta(2K\sqrt{h})$	h^{ab}	$\sqrt{-g}R -$ $\varepsilon\delta(2K\sqrt{h})$
$G_{ab}\delta g^{ab}$	$\varepsilon\delta(DK\sqrt{h})$ $\varepsilon h^{ab}\delta\Pi_{ab}$	$\varepsilon\delta(DK\sqrt{h})$	Π_{ab}	$\sqrt{-g}R -$ $\varepsilon\delta(DK\sqrt{h})$
$G_{ab}\delta g^{ab}$	$\varepsilon\delta[(4-D)K\sqrt{h}]$ $-\varepsilon h_{ab}\delta\Pi^{ab}$	$\varepsilon\delta[(4-D)K\sqrt{h}]$	Π^{ab}	$\sqrt{-g}R$ $-\varepsilon\delta[(4-$ $D)K\sqrt{h}]$

[a]Note that in the last case for $D = 4$ no boundary term is needed and Einstein–Hilbert action is well posed, with Π^{ab} fixed on the boundary (see also [9])

3 (1+3) Decomposition, Time Derivatives and Canonical Momenta

In general relativity space and time are treated on an equal footing. However for many application, e.g., canonical quantization schemes, one need the notion of time and hence the splitting of four dimensional spacetime into one time and three spatial coordinates becomes immediate. This has been performed successfully by Arnowitt, Deser and Misner (henceforth referred to as ADM) in a seminal work [12], in which the ten independent metric components are split into three pieces — $h_{\alpha\beta}$, N^α and N, such that, the line element becomes

$$ds^2 = -N^2dt^2 + h_{\alpha\beta}\left(dx^\alpha + N^\alpha dt\right)\left(dx^\beta + N^\beta dt\right) . \tag{18}$$

Thus note that the spatial metric $g_{\alpha\beta}$ is just $h_{\alpha\beta}$, the off-diagonal entries are $N_\alpha \equiv h_{\alpha\beta}N^\beta$, while the temporal component of the metric becomes, $g_{00} = -N^2 + h_{\alpha\beta}N^\alpha N^\beta$. For the inverse metric the temporal component is simple but not the spatial components such that,

$$g^{tt} = -\frac{1}{N^2}, \quad g^{t\alpha} = \frac{N^\alpha}{N^2}, \quad g^{\alpha\beta} = \left(h^{\alpha\beta} - \frac{N^\alpha N^\beta}{N^2}\right) . \tag{19}$$

The next point one can address from the ADM splitting corresponds to the $(1+3)$ decomposition of the Einstein–Hilbert action. This would require projection of the Riemann tensor components on the spacelike hypersurface, leading to ${}^{(3)}R$, the Ricci scalar of the spacelike hypersurface and invariants like $K_{ab}K^{ab}$, K^2 constructed out of the extrinsic curvature components [12, 13]

$$\begin{aligned}\sqrt{-g}R &= \sqrt{-g}\left[{}^{(3)}R + K_{ab}K^{ab} - K^2 - 2\nabla_i\left(Kn^i + a^i\right)\right] \\ &= \sqrt{-g}L_{\text{ADM}} - 2\sqrt{-g}\nabla_i\left(Kn^i + a^i\right) ,\end{aligned} \tag{20}$$

where n_i is the normal to the spacelike hypersurface and a^i is the corresponding acceleration. Thus the Einstein–Hilbert Lagrangian can be written in terms of the ADM Lagrangian and an additional boundary term which coincides with the Gibbons–Hawking–York counter term since $n_i a^i = 0$. It is well known that the ADM Lagrangian does not contain time derivatives of N and N^α and hence their conjugate momentums vanish. Thus these variables are non-dynamical. However we have just witnessed that boundary terms are not unique, one can in principle add any boundary term that cancels the normal derivative. Then a natural question arises — are the time derivatives of N and N^α zero for for any possible boundary term? If not can they be dynamical? These questions get firm ground as the following example is considered.

Dynamical or Non-dynamical?

Let us consider a cosmological spacetime. Being homogeneous and isotropic it is described by a single function, the scale factor $a(t)$. The line element for cosmological spacetime by imposition of these symmetry conditions become,

$$ds^2 = -dt^2 + a^2(t)\left[dr^2 + r^2 d\Omega^2\right], \tag{21}$$

where the spatial section has been assumed to be flat for simplicity. The above metric is manifestly in ADM form, with $N = 1$, $N^\alpha = 0$ and $h_{\alpha\beta} = a^2(t)\delta_{\alpha\beta}$ respectively. Thus it is evident that N and N^α are not dynamical, all the dynamics comes from the scale factor $a(t)$ as expected. One can now introduce a new coordinate r, such that $R = a(t)r$ and write the metric in the (t, R, θ, ϕ) coordinate system such that,

$$ds^2 = -\left(1 - H^2R^2\right)dt^2 - 2HRdtdR + dR^2 + R^2 d\Omega^2 . \tag{22}$$

Surprisingly, now the metric is again in ADM form but with a completely different structure. This time the spatial metric is flat, i.e., $h_{\alpha\beta} = \delta_{\alpha\beta}$ and hence cannot have any dynamics. On the other hand, one obtains $N = 1$ and $N^\alpha = HR\delta^\alpha_R$ and would conclude that cosmological spacetime is non-dynamical! This explicitly shows that the standard argument for ADM variables N and N^α to be non-dynamical based on their time derivatives is misleading.

To resolve the dilemma we will explicitly illustrate, depending on the boundary term, Einstein–Hilbert action do contains time derivatives of N and N^α but they are *not* dynamical. For this purpose we make use of the following decomposition of the Einstein–Hilbert action,

$$\sqrt{-g}R = \sqrt{-g}g^{ab}\left(\Gamma^i_{ja}\Gamma^j_{ib} - \Gamma^i_{ab}\Gamma^j_{ij}\right) + \partial_c\left\{\sqrt{-g}\left(g^{ik}\Gamma^c_{ik} - g^{ck}\Gamma^m_{km}\right)\right\} . \tag{23}$$

Here the first term is quadratic in the connection and is known as the Γ^2 Lagrangian, while the second term is the boundary term and contains normal derivatives of the metric as elaborated in [13]. Thus an alternative to Gibbons–Hawking–York boundary term is the total divergence term introduced above and hence a possible well-posed Lagrangian corresponds to the Γ^2 Lagrangian. We will show that this Lagrangian depends on time derivatives of N and N^α. To achieve this we shall expand out the Γ^2 Lagrangian in terms of the ADM variables and separate out the time derivatives of N and N^α. Any term X which contains time derivatives of N and N^α will be denoted by $[X]_{t.d}$. By Expressing all the connections in terms of the ADM variables we find that only Γ^t_{tt} and Γ^α_{tt} depends on time derivatives of N and N^α. Hence the time derivative part for the full Γ^2 Lagrangian reads,

$$\left[\sqrt{-g}L_{\rm quad}\right]_{t.d} = \frac{\sqrt{h}}{N^2}\partial_t N \partial_\alpha N^\alpha - \sqrt{h}\frac{\partial_t N^\alpha \partial_\alpha N}{N^2} + \frac{\partial_t N^\alpha}{N}\partial_\alpha \sqrt{h}\,. \tag{24}$$

Hence we have explicitly demonstrated, that the Γ^2 Lagrangian contains time derivatives of N and N^α. Then one question naturally arises, how is that the ADM Lagrangian does not contain these time derivative terms, as evident from the expression for $L_{\rm ADM}$? The answer to this question is hiding in the boundary terms, since they are not identical. Thus in order to understand this, we will have to compare the two boundary terms, the surface term in Eq. (23) and the Gibbons–Hawking–York boundary term, that separate Γ^2 Lagrangian and ADM Lagrangian, respectively, from the Einstein–Hilbert Lagrangian $\sqrt{-g}R$.

Let us now evaluate the Einstein and the Gibbons–Hawking–York boundary terms using the ADM variables. We shall not evaluate the integrands of the surface integrals, but the corresponding divergence terms present in the bulk Lagrangians given by Eqs. (20) and (23) respectively. One can again use the Christoffel symbols to calculate $Kn^i + a^i$ required for evaluating the Gibbons–Hawking–York term in divergence form. Performing the same, terms in the Gibbons–Hawking–York boundary contribution containing time derivatives of N and N^α has the expression

$$\begin{aligned}\left[-2\partial_i\left\{\sqrt{-g}\left(Kn^i + a^i\right)\right\}\right]_{t.d} &= \sqrt{h}\frac{\partial_t N \partial_\alpha N^\alpha}{N^2} - 2\sqrt{h}\frac{\partial_t \partial_\alpha N^\alpha}{N} - 2\frac{\partial_t \sqrt{h}\partial_t N}{N^2}\\ &\quad + 2\frac{\partial_t N}{N^2}N^\alpha \partial_\alpha \sqrt{h} - 2\frac{\partial_t N^\alpha \partial_\alpha \sqrt{h}}{N}\,.\end{aligned} \tag{25}$$

Having derived the relevant expressions related to Gibbons–Hawking–York boundary term, let us next concentrate on the boundary term in the Einstein–Hilbert action given in Eq. (23), which has the expression $\partial_i(\sqrt{-g}V^i)$, where $V^i = g^{ab}\Gamma^i_{ab} - g^{im}\Gamma^k_{mk}$. Computation of each individual components of the boundary term which contains time derivatives of N and N^α are thus given by

$$\left[\partial_i\left(\sqrt{-g}V^i\right)\right]_{t.d} = -\frac{2}{N^2}\partial_t N\partial_t\sqrt{h} + \frac{\sqrt{h}}{N^2}\partial_t N\partial_\alpha N^\alpha - \frac{\sqrt{h}}{N}\partial_t\partial_\alpha N^\alpha + \frac{2}{N^2}\partial_t N N^\alpha\partial_\alpha\sqrt{h}$$
$$-\frac{2}{N}\partial_t N^\alpha\partial_\alpha\sqrt{h} + \frac{\sqrt{h}}{N^2}\partial_\alpha N\partial_t N^\alpha - \frac{\sqrt{h}}{N}\partial_\alpha\partial_t N^\alpha - \frac{\partial_t N^\alpha\partial_\alpha\sqrt{h}}{N}\,. \tag{26}$$

Hence, from Eqs. (25) and (26), we finally arrive at the total contribution from the boundary terms

$$\left[\partial_c\left(\sqrt{-g}V^c\right) + 2\partial_i\left\{\sqrt{-g}\left(Kn^i + a^i\right)\right\}\right]_{t.d}$$
$$= -\frac{\sqrt{h}}{N^2}\partial_t N\partial_\alpha N^\alpha + \sqrt{h}\frac{\partial_t N^\alpha\partial_\alpha N}{N^2} - \frac{\partial_t N^\alpha\partial_\alpha\sqrt{h}}{N}\,. \tag{27}$$

Thus, we observe that the surface terms in Einstein–Hilbert action in Einstein's original decomposition and ADM decomposition are different. The difference contains time derivatives of N^α and N. These time derivatives should exactly match the time derivatives in Γ^2 Lagrangian as we know that the ADM Lagrangian does not have time derivatives of N and N^α. Evaluating time derivatives in ADM Lagrangian using Eqs. (24) and (27), we obtain

$$\begin{aligned}\left[\sqrt{-g}L_{\mathrm{ADM}}\right]_{t.d} &= \left[\sqrt{-g}R + 2\partial_i\left\{\sqrt{-g}\left(Kn^i + a^i\right)\right\}\right]_{t.d}\\ &= \left[\sqrt{-g}L_{\mathrm{quad}} + \partial_c\left(\sqrt{-g}V^c\right) + 2\partial_i\left\{\sqrt{-g}\left(Kn^i + a^i\right)\right\}\right]_{t.d}\\ &= 0\,,\end{aligned} \tag{28}$$

which confirms the ADM Lagrangian does not contain any time derivatives of N and N^α and demonstrates that the time derivatives of N and N^α in the Γ^2 action arise because of the difference in surface terms.

Since the Γ^2 Lagrangian contains time derivatives of N and N^α, it is pertinent to ask what are the conjugate momenta corresponding to N and N^α. From Eq. (24), the conjugate momenta for N and N^α turn out to be

$$p_{(N)} = \frac{\partial\left(\sqrt{-g}\Gamma^2\right)}{\partial\left(\partial_t N\right)} = \frac{\sqrt{h}}{N^2}\partial_\alpha N^\alpha \tag{29}$$

$$p_{\alpha\ (N^\alpha)} = \frac{\partial\left(\sqrt{-g}\Gamma^2\right)}{\partial\left(\partial_t N^\alpha\right)} = -\sqrt{h}\frac{\partial_\alpha N}{N^2} + \frac{1}{N}\partial_\alpha\sqrt{h}\,. \tag{30}$$

Note that the conjugate momenta to N and N^α do not depend on time derivatives of N and N^α respectively. Hence, these relations cannot be inverted to obtain $\partial_t N$ and $\partial_t N^\alpha$ in terms of $p_{(N)}$ and $p_{\alpha\ (N^\alpha)}$. Returning back to our example of cosmological spacetime, this means that H is indeed non-dynamical and that is clear since in terms of Hubble parameter, the Einstein's equations contain only single time derivative of H. Thus we conclude:

Even though the ADM Lagrangian does not contain time derivatives of N and N^{α}, the quadratic Lagrangian $L_{\rm quad}$ differing from the ADM Lagrangian by total derivative do contains time derivatives of N and N^{α}. However, the corresponding canonical momentums are non-invertible, i.e., one cannot obtain time derivatives of N and N^{α} in terms of their canonical momentum. Hence follows their non-dynamical nature.

This explicitly demonstrates standard statements, showing truth in non-dynamical behavior of N and N^{α} but also demonstrating existence of time derivatives of non-dynamical variables.

4 Null Surfaces: Completing the Circle

The boundary terms and ADM decomposition discussed earlier depends crucially on the timelike (or spacelike) nature of the boundary surface. However, the most ubiquitous surfaces in general relativity are the null surfaces, e.g., in a black hole spacetime the standard boundary would consist of the surface at infinity and the event horizon, which is a null surface. The limit of non-null surfaces to null surfaces is not at all straightforward, since many quantities including the extrinsic curvature, induced metric can either blow up or vanish on the null surface if proper care is not taken. Thus it is important to consider the boundary term from a first principle in connection to null hypersurfaces. The first step towards this direction was taken in [6] by constructing a general formalism and its explicit implementation was carried out in [7]. There it was argued that for a null vector ℓ_a (i.e., $\ell^a\ell_a = 0$) the boundary term one should add corresponds to $2\sqrt{q}(\Theta + \kappa)$, where q stands for the determinant of the induced metric on the null surface, Θ stands for the expansion of the null geodesics and κ is the non-affinity parameter. Since null surfaces are intrinsically two-dimensional, use of a single vector field ℓ_a is not sufficient. One need to introduce another auxiliary vector field k_a, satisfying $k_ak^a = 0$ and $\ell_ak^a = -1$. In the above derivations it has been assumed that the null surface is preserved under variations, i.e., the following three conditions hold: $\delta(\ell_a\ell^a) = 0, \delta(\ell_ak^a) = 0$ and finally $\delta(k_ak^a) = 0$. In this work we will relax all these assumptions and shall investigate the effect of these constraints on the boundary term and degrees of freedom on the boundary. We will start with the general expression for boundary term of Einstein–Hilbert action having the form [14]

$$\begin{aligned}\sqrt{-g}Q[\ell_c] &= \sqrt{-g}\nabla_c\left(\delta u^c\right) - 2\delta\left(\sqrt{-g}\nabla_a\ell^a\right) + \sqrt{-g}\left[\nabla_a\ell_b - g_{ab}\left(\nabla_c\ell^c\right)\right]\delta g^{ab} \\ &= Q_1 + Q_2 + Q_3\,, \end{aligned} \tag{31}$$

where, $\delta u^a = \delta\ell^a + g^{ab}\delta\ell_b$. We have separated the boundary term in three natural combinations, one is a divergence term, Q_1, second one corresponds to total variation Q_2 and finally the degrees of freedom term Q_3 respectively. We will explore each of these terms and subsequently shall evaluate the boundary term on the null surface following the convention, if some relation holds *only* on the null surface it will be denoted by $A := 0$. As explained above we will assume the following conditions on the null surface only, $\ell_a\ell^a := 0$, $\ell_a k^a := 0$ and $k_a k^a := 0$ respectively, but we would not assume anything about off the null surface relations, i.e., variations can be arbitrary. Then one can introduce the *partial* projector P^a_b through the vectors ℓ^a and k^a as, $P^a_b = \delta^a_b + k^a\ell_b$ and can write the first divergence term Q_1 in Eq. (31) as:

$$\begin{aligned} Q_1 := &\ \partial_\alpha\left(\sqrt{-g}P^\alpha_d\delta u^d\right) - \delta\left(\sqrt{-g}k^c\partial_c\ell^2\right) + \left(k^c\partial_c\ell^2\right)\delta\sqrt{-g} \\ &+ \sqrt{-g}\delta k^c\partial_c\ell^2 - \partial_c\left(\sqrt{-g}k^c\right)\delta\ell^2 \,, \end{aligned} \tag{32}$$

while the second term can also be expressed using the partial projector P^a_b and then the complete boundary term on using the variation of $\sqrt{-g}$, takes the following form

$$\begin{aligned} \sqrt{-g}Q\left[\ell_c\right] := &\ \partial_\alpha\left(\sqrt{-g}P^\alpha_d\delta u^d\right) - 2\delta\left(\sqrt{-g}P^a_b\nabla_a\ell^b\right) + \sqrt{-g}\delta k^c\partial_c\ell^2 \\ &- \partial_c\left(\sqrt{-g}k^c\right)\delta\ell^2 + \sqrt{-g}\left(\nabla_a\ell_b - g_{ab}\left\{P^c_d\nabla_c\ell^d\right\}\right)\delta g^{ab} \,. \end{aligned} \tag{33}$$

Note that the first term is a pure surface term — it has no component along the normal ℓ_a. Then we can decompose the metric in terms of the induced metric q_{ab} and the null vectors ℓ^a and k^a as: $g_{ab} = q_{ab} - \ell_a k_b - \ell_b k_a$. Thus variations of the metric now gets transformed to variations of the induced metric and the null vectors. One important point to keep in mind is the fact that $\delta\ell^a = g^{ab}\delta\ell_b + \ell_b\delta g^{ab}$ but *not* $g^{ab}\delta\ell_b$. Using the properties of the null vectors outside variation and decomposition of $\nabla_a\ell_b$ in terms of the extrinsic curvature ultimately lands us into the following expression for the boundary term

$$\begin{aligned} \sqrt{-g}Q := &\ \partial_\alpha\left(\sqrt{-g}P^\alpha_a\delta u^a\right) - 2\delta\left(\sqrt{-g}P^a_b\nabla_a\ell^b\right) \\ &+ \sqrt{-g}\left[\Theta_{ab} - \left(P^c_d\nabla_c\ell^d\right)q_{ab}\right]\delta q^{ab} \\ &- \sqrt{-g}\left\{k^m\nabla_m\ell_a + k^n\nabla_a\ell_n + \left(k^m k^n\nabla_m\ell_n\right)\ell_a - 2\left(P^c_d\nabla_c\ell^d\right)k_a\right\}\delta\ell^a \\ &+ \sqrt{-g}\left\{k^m\nabla_m\ell^a + k^n\nabla^a\ell_n + \left(k^m k^n\nabla_m\ell_n\right)\ell^a - 2\left(P^m_n\nabla_m\ell^n\right)k^a\right\}\delta\ell_a \\ &+ \sqrt{-g}\left\{\partial_c\ell^2\right\}\delta k^c - \partial_c\left(\sqrt{-g}k^c\right)\delta\ell^2 \,. \end{aligned} \tag{34}$$

Before commenting on the structure of the boundary term let us quickly check one possible limit we have derived in our earlier works [7]. This corresponds to the situation in which $\ell_a = \nabla_a\phi$, implying $\delta\ell_a = 0$ and also $\delta\ell^2 = 0 = \delta\left(\ell^a k_a\right)$, such that we have $P^a_b\nabla_a\ell^b = \Theta + \kappa$. Under imposition of these conditions, the boundary term reduces to:

$$\begin{aligned}\sqrt{-g}Q\left[\nabla_c\phi\right] := {} & \partial_\alpha\left(\sqrt{-g}P^\alpha_a\delta u^a\right) - 2\delta\left[\sqrt{-g}\left(\Theta+\kappa\right)\right] \\ & + \sqrt{-g}\left[\Theta_{ab} - \left(\Theta+\kappa\right)q_{ab}\right]\delta q^{ab} \\ & - 2\sqrt{-g}\left\{k^m\nabla_m\ell_a - \left(\Theta+\kappa\right)k_a\right\}\delta\ell^a\,. \end{aligned} \tag{35}$$

This is exactly what we had derived by various other routes in [7]. Having checked the consistency with earlier derived results we now concentrate on the physical implications of Eq. (34). The first term as emphasized earlier corresponds to another boundary term[1] and contributes only on the two surface without much significance. The second term is the boundary term that one should add (negative of that term, to be precise) to the Einstein–Hilbert action as evaluated with volume encompassing null boundaries. The rest of the terms are related to degrees of freedom and what one should fix on the null surface. Among them fixing induced metric is expected, with its conjugate momentum being $\pi_{ab} = \sqrt{-g}\left[\Theta_{ab} - \left(P^m_n\nabla_m\ell^n\right)q_{ab}\right]$. In this case as well one can write the last term as a total divergence leading to a different boundary term and conjugate momentum to fix on the boundary. Unlike the cases of timelike or spacelike surfaces the situation is not so simple for null surfaces, since even after fixing the induced metric one needs to fix the components of the null vectors as well. But one can improve on that. Since the normalization of the null vector is arbitrary one can always choose ℓ_a to be a pure gradient such that $\delta\ell_a = 0$. Further since the choice of k^a is arbitrary one might chose it such that its expansion vanishes and further with $\delta(k^a\ell_a) = 0$. As these seemingly natural conditions are being satisfied the boundary term simplifies a lot, ultimately leading to,

$$\begin{aligned}\sqrt{-g}Q := {} & \partial_\alpha\left(\sqrt{-g}P^\alpha_a\delta u^a\right) - 2\delta\left(\sqrt{-g}P^a_b\nabla_a\ell^b\right) \\ & + \sqrt{-g}\left[\Theta_{ab} - \left(P^c_d\nabla_c\ell^d\right)q_{ab}\right]\delta q^{ab} \\ & - \sqrt{-g}\left\{k^m\nabla_m\ell_a + k^n\nabla_a\ell_n + \left(k^m k^n\nabla_m\ell_n\right)\ell_a - 2\left(P^c_d\nabla_c\ell^d\right)k_a\right\}\delta\ell^a\,. \end{aligned} \tag{36}$$

Hence along with q_{ab} one need to fix the components of the null vector ℓ^a. One more point should be noted, since $\delta\ell_a = 0$, one obtains $\delta(\ell_a\ell^a) = \ell_a\delta\ell^a$ and hence any contribution from $\delta\ell^2$ can be dumped into the contribution from $\delta\ell^a$. Hence this suggests that on the null surface one need to fix the induced metric q^{ab} as well as ℓ^a. This has interesting consequences for degrees of freedom on the null surfaces à la degrees of freedom on spacelike or timelike surfaces. One interesting consequence could be, as the diffeomorphisms are gauged away one can eliminate the four degrees of freedom in $\delta\ell^c$, keeping the true (physical) degrees of freedom in the two metric

[1]This kind of terms are also present in the the calculation for spacelike (or timelike) surfaces, see for example the last term of Eq. (12).

q_{ab} of the null surface. This can have interesting implications for black hole entropy, which we will pursue elsewhere.

5 Concluding Remarks

The peculiarity of the Einstein–Hilbert action can be traced back to its boundary terms. In the standard treatments it is often overlooked that Einstein–Hilbert action is not well posed, one has to add boundary terms to get an well posed action for gravity. There have been parallel results on this issue, one is the well-known Gibbons–Hawking–York boundary term, while the other is recent and more promising from a thermodynamic hindsight which invokes two new variables f^{ab} and N^c_{ab} to describe gravity, with $f^{ab}\delta N^c_{ab}$ as the boundary term. In this work we have explicitly derived the equivalence between these two formalisms in any spacetime dimensions. Further we have also demonstrated the argument that "boundary terms are not unique" by constructing two more boundary terms starting from the Gibbons–Hawking–York term. To our surprise these boundary terms depends strongly on the spacetime dimensions and even can vanish in $D = 4$. Then we have elaborated the meaning of another statement made often in the literature, "the ADM variables N and N^α are not dynamical". The standard argument goes by saying that the ADM Lagrangian does not depend on time derivatives of N and N^α. We have shown that one can add boundary terms to the ADM Lagrangian leading to a new Lagrangian which contain time derivatives of N and N^α, (so it might appear they can be made dynamical by adding boundary terms) but still they are non-dynamical as conjugate momentums cannot be inverted. This finishes our discussion on spacelike or timelike surfaces and we turn to the case of null surfaces. In earlier works regarding boundary term on null surfaces, various assumption about variations of the null vectors were imposed, here we have derived the structure of the boundary term for most general variation. Imposing some minimal restrictions we could show that besides the induced metric, the null vector ℓ^a contains additional degrees of freedom. If they can be removed by diffeomorphism (as [7] suggests) then the induced metric might contain all the physical degrees of freedom associated with null surfaces, having greater implications for emergent paradigms of gravity [15–18].

Acknowledgements Research of S.C. is funded by a SPM fellowship from CSIR, Government of India. He thanks T. Padmanabhan for suggesting this problem and for comments on an earlier version of the manuscript. He also thanks J. Bagla and S. Engineer for invitation to write this article along with K. Parattu, S. Singh and K. Lochan for interesting discussions and comments. A part of this work was completed during a visit to ICTP, Italy and I thank ICTP for warm hospitality.

Dedication

One of the great quality of Prof. Padmanabhan is his ability to ask the correct question. This work stems from such questions asked by him during our discussions: "What is the connection between various action principles for general relativity? Why N and N^{α} are non-dynamical? What one should fix on a null surface?". I have tried to answer them in this work and I respectfully dedicate it to Prof. Padmanabhan on the occasion of his 60th birthday.

References

1. A. Einstein, Hamilton's principle and the general theory of relativity. Sitzungsber. Preuss. Akad. Wiss. Berlin (Math. Phys.) **1916**, 1111–1116 (1916)
2. J. Charap, J. Nelson, Surface integrals and the gravitational action. J. Phys. A: Math. General **16**, 1661 (1983)
3. J. York, W. James, Role of conformal three geometry in the dynamics of gravitation. Phys. Rev. Lett. **28**, 1082–1085 (1972)
4. G. Gibbons, S. Hawking, Action integrals and partition functions in quantum gravity. Phys. Rev. D **15**, 2752–2756 (1977)
5. J.W. York, Boundary terms in the action principles of general relativity. Found. Phys. **16**(3), 249–257 (1986). doi:10.1007/BF01889475
6. T. Padmanabhan, A short note on the boundary term for the Hilbert action. Mod. Phys. Lett. A **29**, 1450037 (2014)
7. K. Parattu, S. Chakraborty, B.R. Majhi, T. Padmanabhan, Null surfaces: counter-term for the action principle and the characterization of the gravitational degrees of freedom. arXiv:1501.01053 [gr-qc]
8. K. Parattu, B.R. Majhi, T. Padmanabhan, Structure of the gravitational action and its relation with horizon thermodynamics and emergent gravity paradigm. Phys. Rev. D **87**, 124011 (2013). doi:10.1103/PhysRevD.87.124011. arXiv:gr-qc/1303.1535 [gr-qc]
9. S. Chakraborty, T. Padmanabhan, Geometrical variables with direct thermodynamic significance in Lanczos-Lovelock gravity. Phys. Rev. D **90**(8), 084021 (2014). arXiv:1408.4791 [gr-qc]
10. L.D. Landau, E.M. Lifshitz, *The Classical Theory of Fields*, vol. 2, 4th edn., Course of Theoretical Physics Series (Butterworth-Heinemann, 1980)
11. C. Krishnan, A. Raju, A Neumann boundary term for gravity. arXiv:1605.01603 [hep-th]
12. R.L. Arnowitt, S. Deser, C.W. Misner, The dynamics of general relativity. arXiv:gr-qc/0405109 [gr-qc]
13. T. Padmanabhan, *Gravitation: Foundations and Frontiers* (Cambridge University Press, Cambridge, 2010)
14. K. Parattu, S. Chakraborty, T. Padmanabhan, Variational principle for gravity with null and non-null boundaries: a unified boundary counter-term. Eur. Phys. J. C **76**(3), 129 (2016). arXiv:1602.07546 [gr-qc]
15. T. Padmanabhan, Gravity and the thermodynamics of horizons. Phys. Rep. **406**, 49–125 (2005). arXiv:gr-qc/0311036 [gr-qc]
16. T. Padmanabhan, Thermodynamical aspects of gravity: new insights. Rep. Prog. Phys. **73**, 046901 (2010). arXiv:0911.5004 [gr-qc]
17. T. Padmanabhan, Emergent gravity paradigm: recent progress. Mod. Phys. Lett. A **30**(03n04), 1540007 (2015). arXiv:1410.6285 [gr-qc]
18. S. Chakraborty, T. Padmanabhan, Thermodynamical interpretation of the geometrical variables associated with null surfaces. Phys. Rev. D **92**(10), 104011 (2015). arXiv:1508.04060 [gr-qc]

Decay of the Cosmic Vacuum Energy

Timothy Clifton and John D. Barrow

Abstract In his 2005 review, *Gravity and the Thermodynamics of Horizons*, Paddy suggested that a vacuum in thermal equilibrium with a bath of radiation should have a gradually diminishing energy. We work through the consequences of this scenario, and find that a coupling between the vacuum and a bath of black-body radiation at the temperature of the horizon requires the Hubble rate, H, to approach the same type of evolution as in the "intermediate inflation" scenario, with $H \propto t^{-1/3}$, rather than as a constant. We show that such behaviour does not conflict with observations when the vacuum energy is described by a slowly-rolling scalar field, and when the fluctuations in the scalar field are treated as in the "warm inflation" scenario. It does, however, change the asymptotic states of the universe. We find that the existence of the radiation introduces a curvature singularity at early times, where the energy densities in both the radiation and the vacuum diverge. Furthermore, we show that the introduction of an additional non-interacting perfect fluid into the space-time reveals that radiation can dominate over dust at late times, in contrast to what occurs in the standard cosmological model. Such a coupling can also lead to a negative vacuum energy becoming positive.

1 Introduction

Inflationary cosmology is based on the hypothesis of a period of accelerated expansion in the very early history of the universe. This surge in the expansion solves the horizon problem, and is generically expected to drive the observable curvature of space, and any expansion or curvature anisotropies, to unobservably small values

T. Clifton (✉)
School of Physics and Astronomy, Queen Mary University of London,
Mile End Road, London E1 4NS, UK
e-mail: t.clifton@qmul.ac.uk

J.D. Barrow
DAMTP, Centre for Mathematical Sciences, University of Cambridge,
Wilberforce Road, Cambridge CB3 0WA, UK
e-mail: j.d.barrow@damtp.cam.ac.uk

J.S. Bagla and S. Engineer (eds.), *Gravity and the Quantum*,
Fundamental Theories of Physics 187, DOI 10.1007/978-3-319-51700-1_6

today. In addition, inflationary cosmology provides a natural mechanism for creating the seeds of structure formation, from tiny quantum mechanical fluctuations. Such fluctuations are a manifestation of the thermal nature of quantum fields in curved spaces. But the existence of a thermal space also implies the existence of a bath of radiation [1–3]. In this paper we will consider the gravitational consequences of this radiation on the large-scale expansion of the universe, as well as on the observables that emerge from a period of inflation, as suggested by Paddy in his review [4]. We will work in Planck units throughout, such that $G = c = \hbar = k_B = 1$.

The energy density of radiation with a black-body spectrum, at temperature T, is given in Planck units as

$$\rho_r = 4\sigma T^4 , \tag{1}$$

where $\sigma = g_* \pi^2/120$ is the Stefan–Boltzmann constant, and g_* is the effective number of relativistic degrees of freedom. In a Friedmann–Lemaître–Robertson–Walker (FLRW) geometry, the energy-conservation equation for this radiation fluid is given by

$$\dot{\rho}_r + 4H\rho_r = Q , \tag{2}$$

where $Q = Q(t)$ is an energy exchange term that is required in order for Eq. (1) to be satisfied at all times, and overdots denote derivatives with respect to the comoving proper time, t. The Q term parameterizes the energy flow into the radiation field, and thermalization is assumed to be instantaneous.

Energy-momentum conservation now requires that $T^{\mu\nu}{}_{;\nu} = 0$, where $T_{\mu\nu}$ is the total energy-momentum tensor of all matter fields in the space-time. If we are considering a space-time that contains effectively just radiation (r) and vacuum (v) energy, with $T_{\mu\nu} = T^{r}_{\mu\nu} + T^{v}_{\mu\nu}$, then we must therefore also have

$$\dot{\rho}_v = -Q , \tag{3}$$

where $\rho_v = -p_v$ is the energy density of the vacuum, and p_v is its pressure. This equation shows that the vacuum energy density must be decaying, and is the cosmological counterpart of the requirement that radiating black holes must reduce in mass, in order for the total energy-momentum in the space-time to be conserved [5].

The equations above, together with the Friedmann equation, $H^2 = \frac{8\pi}{3}(\rho_r + \rho_v)$, can be used to write

$$\dot{H} + \frac{g_*}{90\pi}(2\pi T)^4 = 0 . \tag{4}$$

If we can find an expression for T as a function of H, then we have a differential equation that can be solved to find the rate of expansion.

In this study we will assume that the radiation is in thermal equilibrium with the vacuum, so that T in Eq. (4) is given by the usual semi-classical expression for the temperature of space:

$$T = \frac{|\kappa|}{2\pi} = \left| \frac{H}{2\pi}\left(1 + \frac{\dot{H}}{2H^2}\right) \right| , \tag{5}$$

where κ is the surface gravity of the apparent horizon. The expression after the second equality is found by evaluating the surface gravity of the horizon in a spatially flat FLRW geometry [6].

2 Background Evolution

It can immediately be seen from Eqs. (4) and (5) that $\dot{H} \leq 0$, and that $\dot{H} = 0$ if and only if $H = 0$. This shows that H is always decreasing, and that (in an initially expanding universe) it is bounded from below by zero. We can therefore use H as a proxy for time. Figure 1 shows the evolution of the energy density in both the radiation and vacuum fields as a function of H, as the Universe expands. We have chosen to display this information in terms of the density parameters, defined as $\Omega_i \equiv 8\pi\rho_i/3H^2$.

At early times, when $H \rightarrow \infty$, it can be seen that $\dot{H} \rightarrow -2H^2$. Using $H = \dot{a}/a$ this can be shown to correspond to a scale factor of the form

$$a(t) \propto t^{\frac{1}{2}}\,. \tag{6}$$

This is the same as that which occurs in a standard radiation-dominated FLRW cosmology. It can be seen from Eq. (4) that in this limit $T \propto H^{\frac{1}{2}} \propto 1/a$, and from Eq. (1) that $\rho_r \propto H^2 \propto 1/a^4$. These are again exactly the forms of these expressions that one would expect from a radiation filled universe without energy exchange. However, while $\Omega_v \rightarrow 0$ at early times, it can be seen that $\rho_v \rightarrow \infty$. That is, we find that a coupling to radiation can reduce the magnitude of the vacuum energy, even if it is initially very large.

At late-times, on the other hand, we have $H \rightarrow 0$, and hence $\dot{H}/H^2 \rightarrow 0$. In this limit the leading-order contribution to the temperature takes the form of the Gibbons–Hawking value, $T = H/2\pi$ [7], and Eq. (4) becomes $\dot{H} + g_* H^4/90\pi = 0$. This leads to a scale factor that evolves as

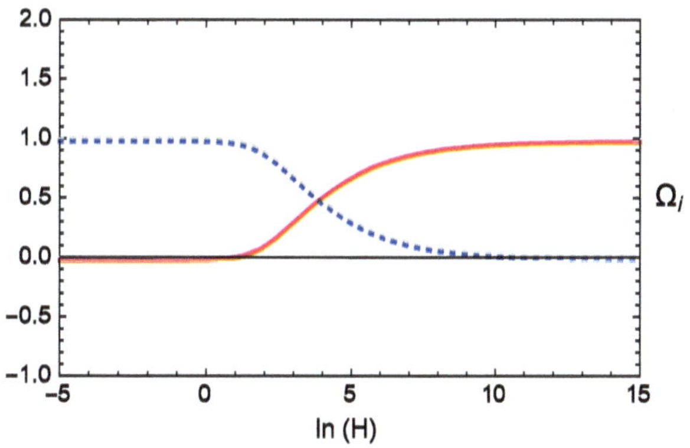

Fig. 1 The evolution of the density parameters, $\Omega_i = 8\pi\rho_i/3H^2$, for the radiation field (*red solid line*) and the vacuum field (*blue dotted line*). We have chosen $g_* = 2$ in order to produce this plot

$$a(t) \propto \exp\left\{\frac{3}{2}\left(\frac{30\pi}{g_*}\right)^{\frac{1}{3}}(t-t_0)^{\frac{2}{3}}\right\}, \tag{7}$$

where $t_0 =$ constant. This type of expansion is of a type known as "intermediate inflation" [8], which generally has $a(t) \propto \exp\{At^n\}$, where $A > 0$ and $0 < n < 1$ are constants. Intermediate inflation has been studied by several authors [9–11], and is known to arise in rainbow gravity theories [12].

The particular form of intermediate inflation with $a(t) \propto \exp\{\lambda t^{2/3}\}$ is special. When generated from a minimally coupled scalar field in a suitably chosen potential, it is the only form of intermediate inflation (other than perfect de Sitter) that gives an exact Harrison–Zeldovich spectrum of first-order density perturbations. Unlike standard slow-roll scenarios, however, it is also known that this type of intermediate inflation can produce large amounts of gravitational radiation [8, 9, 13, 14].

Equation (7) is a significant departure from the usual exponential expansion, and occurs even though the energy density of radiation may be small. This can be attributed to the dual requirements of an almost constant density of radiation, as well as the exponential dilution of that radiation with inflationary expansion. Therefore, the vacuum energy must constantly replace the quickly dissipating radiation, and even though the amount of radiation required at any given time may be small, it must effectively be replenished at every moment of time.

3 Energy Exchange in the Presence of a Non-interacting Fluid

If we also include in the universe a non-interacting perfect fluid, with equation of state

$$p = (\gamma - 1)\rho,$$

then the Friedmann equation becomes

$$H^2 = \frac{8\pi}{3}\left(\rho_r + \rho_v + \rho\right), \tag{8}$$

while the energy conservation equations for ρ_r and ρ_v can again be written as in Eqs. (2) and (3). The energy density in the non-interacting field, which we take to be separately conserved, is given by

$$\dot{\rho} + 3\gamma H\rho = 0. \tag{9}$$

At this point it is convenient to use the number of e-foldings, $N \equiv \ln a$, as a replacement for the time variable. We can then integrate Eq. (9) to find

$$\rho = \rho_0 e^{-3\gamma N}, \tag{10}$$

where ρ_0 is a constant. The corresponding energy densities in the radiation and the vacuum can be found from Eqs. (1), (5) and (8) to be

$$\rho_r = \frac{g_* H^4}{480\pi^2}\left(1+\frac{H'}{2H}\right)^4 \tag{11}$$

and

$$\rho_v = \frac{3H^2}{8\pi} - \frac{g_* H^4}{480\pi^2}\left(1+\frac{H'}{2H}\right)^4 - \rho_0 e^{-3\gamma N}\,, \tag{12}$$

where we have used a prime to denote a derivative with respect to N. Finally, differentiating Eq. (8), and using the conservation equations for ρ_r, ρ_v and ρ we obtain

$$HH' = -\frac{4\pi}{3}\left[\frac{g_* H^4}{120\pi^2}\left(1+\frac{H'}{2H}\right)^4 + 3\gamma\rho_0 e^{-3\gamma N}\right]. \tag{13}$$

This is a first-order ODE for H, as a function of N. Once we have $H = H(N)$, then the equations above give us ρ_r, ρ_v and ρ as functions of N, too.

An example evolution of the density parameters of the radiation field, the vacuum field, and non-interacting fluid are shown in Fig. 2. To produce this plot we have taken the non-interacting fluid to be dust, so that $\gamma = 1$. We have also chosen to consider the case $g_* = 2$, and have set initial conditions so that $H = 0.1$ and $\rho = 10^{-4}$ at $\ln a = 0$. For a finite period of time the energy density in the non-interacting dust dominates over the radiation and vacuum energies, and we have $a(t) \sim t^{2/3}$. After this we have a period of intermediate inflation occurring, of the type given in Eq. (7). Before it we have radiation domination, as described in Eq. (6). During this early period of radiation domination, the vacuum energy in fact becomes negative, and starts to diverge. This change of sign does not appear in the absence of the non-interacting dust.

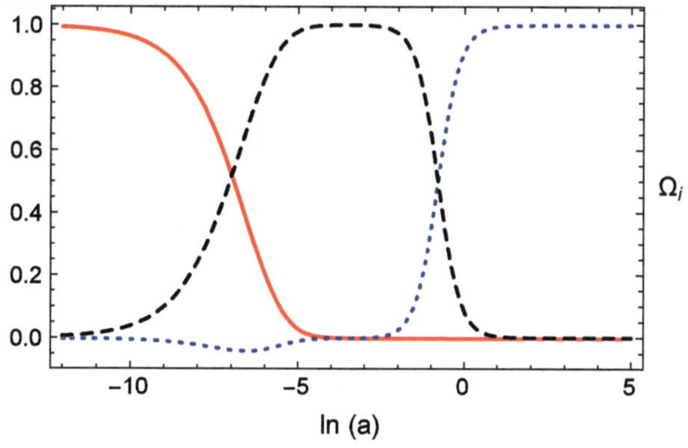

Fig. 2 The evolution of the density parameters for the radiation field (*red solid line*), the vacuum energy (*blue dotted line*) and the non-interacting fluid (*black dashed line*). To produce this plot we have taken the non-interacting fluid to be a pressure-less fluid of dust, such that $\gamma = 1$. We have also taken $g_* = 2$, and set $H = 0.1$ and $\rho = 10^{-4}$ at $\ln a = 0$

At late times the energy density in radiation generically dominates over matter, in contrast to the usual case in cosmological models with non-interacting radiation and dust [15]. In Fig. 3 we plot the energy density in dust as a fraction of the energy density in radiation, again for $g_* = 2$ and with $H = 0.1$ and $\rho = 10^{-4}$ at $\ln a = 0$. It can be seen that there is a transient period when the dust dominates over the radiation, but that the radiation dominates over the dust both before and after this. The opposite result is true for $\gamma \leqslant 0$, in which case the non-interacting fluid dominates over radiation at late times, while being sub-dominant beforehand. If we had introduced an effective perfect 'fluid' with $\gamma = 2/3$, to mimic the presence of negative spatial curvature in the Friedmann equation, then the same general evolution occurs and the curvature 'fluid' does not dominate at late times. This shows that flatness is approached at late times, just as in standard inflation.

We should point out at this stage that the black-body radiation that is created by the assumed thermal equilibrium with the vacuum is not the only radiation that one should expect in a realistic cosmology. For example, there is also a bath of radiation in the late Universe, which is very close to being a black-body, and that is at a much higher temperature than that of the apparent horizon. The synthesis of the light elements occurs during the epoch in which this fluid dominates over all other matter, and the energy that creates this additional radiation comes (originally) from the reheating process that occurs after inflation. A first approximation to the cosmological consequences of including an additional radiation field of this type, at a different temperature to the vacuum, could be studied by adding a non-interacting fluid, as described above. In this case the primordial synthesis of light elements should be expected to occur in exactly the same way that it usually does, as the energy density of the vacuum is tiny compared to that of radiation during nucleosynthesis.

In reality, of course, one would expect any additional fluid to also interact with the vacuum in some way, and perhaps stimulate an increase or decrease in the vacuum energy density by some small amount. A calculation to determine how this proceeds would require a knowledge of the non-equilibrium thermodynamics of the interaction. If this were known, and all of the carriers of entropy could be identified, then it would also be possible to investigate the stability of the assumed thermal equilibrium. We will leave a detailed study of these points for future work.

4 Perturbations Generated During Inflation

It is natural to consider the effects of the interaction introduced in Sect. 1 on the observables that result from thermal fluctuations during inflation. To do this, we model the vacuum energy as a scalar field with a self-interaction potential, $V(\phi)$, such that

$$\rho_v = \frac{1}{2}\dot{\phi}^2 + V(\phi) \quad \text{and} \quad p_v = \frac{1}{2}\dot{\phi}^2 - V(\phi)\,. \tag{14}$$

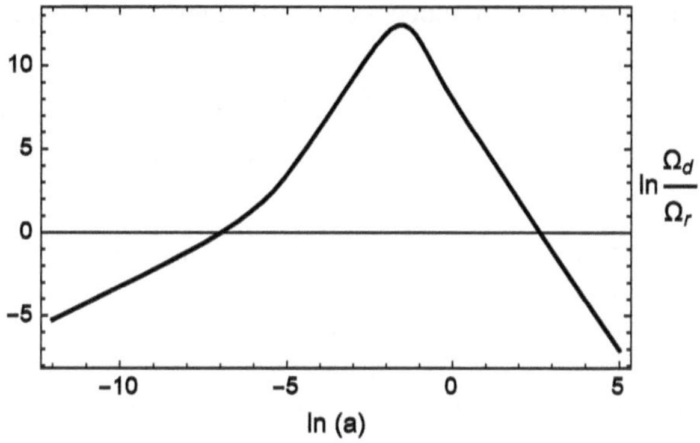

Fig. 3 The logarithm of the ratio in energy densities of dust and radiation, when $g_* = 2$, and when $H = 0.1$ and $\rho = 10^{-4}$ at $\ln a = 0$. The radiation field dominates over the dust at both late and early times

The evolution we get in this case will be different to that obtained in the previous section, where we considered a fluid with equation of state $p_v = -\rho_v$, but will reduce to it in the appropriate limits.

In terms of these new variables, the Friedmann and conservation equations can be manipulated into the form

$$\dot{H} - \frac{4g_*\pi^3}{45}T^4 + 3H^2 - 8\pi V = 0 \tag{15}$$

$$\dot{\phi}^2 + 2V - \frac{3H^2}{4\pi} + \frac{g_*\pi^2}{15}T^4 = 0\,, \tag{16}$$

with $T = T(H, \phi)$ given by the solution of $H T - g_*\pi^2 T^4/45 = 2V - H^2/4\pi$, and where we have taken $H > 0$ and $\dot{H} + 2H^2 > 0$. This simple system of equations represents a 2-dimensional dynamical system, the solutions of which can be found only after the form of the potential $V(\phi)$ is specified. In the absence of the terms involving g_*, these equations reduce to the usual ones for a scalar field-filled Friedmann model.

Unlike in normal inflation, the existence of radiation should be expected to have an effect on the power spectrum of comoving curvature perturbations. This is because the thermal radiation should drive the evolution of the perturbations in ϕ through the existence of both noise and dissipation terms. Such scenarios have been considered in the literature under the title "warm inflation". In this case the origin of the fluctuations in the CMB are not only from a quantum origin, but also from the thermal nature of the radiation.

If we differentiate Eq. (16), and make use of Eq. (15), then we can write the Klein–Gordon equation for ϕ as

$$\ddot{\phi} + 3H\,(1 + \Gamma)\,\dot{\phi} + \frac{dV}{d\phi} = 0\,, \tag{17}$$

where we have included a dissipation term, with coefficient

$$\Gamma = -\frac{8g_* \pi^3 T^4}{45H} \frac{\left(H + \frac{\dot{T}}{T}\right)}{\left(\dot{H} + \frac{8g_*\pi^3}{45} T^4\right)}, \tag{18}$$

and where T can be given in terms of H using Eq. (5). The effect of Γ on the spectrum of comoving curvature perturbations, $\mathscr{P}_{\mathscr{R}}$, has been studied in a series of papers by Berera and collaborators [16–22].

If the noise source that the radiation creates is taken to be Markovian, then the spectrum of perturbations is found to be [20–22]

$$\mathscr{P}_{\mathscr{R}} = \frac{H^4}{(2\pi)^2 \dot{\phi}^2} \left[1 + \left(\frac{2\pi T}{H}\right) \frac{\Gamma}{\sqrt{1 + \frac{4\pi}{3}\Gamma}} \right]. \tag{19}$$

This expression reduces to the usual one when $\Gamma \to 0$. Generalised to include non-trivial distributions of inflaton particles [20], and noise sources for the radiation [21], have also recently been derived in the literature.

The spectral index of the primordial curvature perturbations, n_s, is given in this case by the usual expression:

$$n_s - 1 \equiv \frac{d \ln \mathscr{P}_{\mathscr{R}}}{d \ln k}, \tag{20}$$

where the right-hand side is to be evaluated at horizon crossing, when $k \simeq aH$. This spectral index can be calculated explicitly, to find a lengthy expression, using the equations above. Below we will calculate the leading-order contributions that occur during slow-roll inflation.

Likewise, the spectrum of tensor perturbations is found to be given by [19]

$$\mathscr{P}_{\text{grav}} = \frac{4H^2}{\pi}, \tag{21}$$

which can be used to write the spectral index of tensor perturbations as

$$n_T - 1 \equiv \frac{d \ln \mathscr{P}_{\text{grav}}}{d \ln k} = \frac{2\dot{H}}{(\dot{H} + H^2)}, \tag{22}$$

where the derivative of $\ln \mathscr{P}_{\text{grav}}$ has been evaluated at horizon crossing, to get the simple expression after the last equality. The spectral index of tensor perturbations can be seen to be unchanged by the presence of radiation.

Lastly, Eqs. (19) and (21) can also be used to define the tensor-to-scalar ratio:

$$r \equiv \frac{A_T}{A_S} = 4\frac{\mathscr{P}_{\rm grav}}{\mathscr{P}_{\mathscr{R}}} = -\frac{16\left(\frac{\dot{H}}{H^2} + \frac{g_* H^2}{90\pi}\right)}{\left[1 + \left(\frac{2\pi T}{H}\right)\frac{\Gamma}{\sqrt{1+\frac{4\pi}{3}\Gamma}}\right]}, \tag{23}$$

where all roots are taken to be positive, and where we have used Eqs. (15) and (16) to write this expression in terms of H and its derivatives only. Here we have taken the amplitude of tensor perturbations to be $A_T = 4\mathscr{P}_{\rm grav}$, and the amplitude of scalar perturbations to be $A_S = \mathscr{P}_{\mathscr{R}}$. Once more, this expression reduces to the usual one when $g_* \to 0$.

Recent observations imply that the scalar spectral index generated during inflation is given by [23]

$$n_S - 1 = -0.0365 \pm 0.0094, \tag{24}$$

while the amplitude of curvature perturbations is inferred to be

$$A_S = 2.19^{+0.12}_{-0.14} \times 10^{-9}, \tag{25}$$

and the tensor-to-scalar ratio is constrained by [24]

$$r \lesssim 0.2\,. \tag{26}$$

If we apply the latter two of these together, then Eq. (21) can be seen to imply

$$|H| \lesssim 10^{-5}\,. \tag{27}$$

This severely limits any effect that the radiation can have on the scalar spectral index, and the amplitude of scalar fluctuations.

In fact, if we define slow-roll parameters by

$$\varepsilon_H \equiv \frac{3\dot{\phi}^2}{2V + \dot{\phi}^2} = -\frac{(\dot{H} + \frac{8g_*\pi^3}{45}T^4)}{(H^2 - \frac{4g_*\pi^3}{45}T^4)} \tag{28}$$

$$\eta_H \equiv -\frac{\ddot{\phi}}{H\dot{\phi}} = -\frac{(\ddot{H} + \frac{32g_*\pi^3}{45}T^3\dot{T})}{2H(\dot{H} + \frac{8g_*\pi^3}{45}T^4)}, \tag{29}$$

then we can write our expressions for A_S and n_S as functions of ε_H, η_H and H only. If we expand these first in H (this quantity having been found to be small already), and then in ε_H and η_H, we find that the leading-order parts A_S and n_S are given by

$$n_S - 1 \simeq -4\varepsilon_H + 2\eta_H - \frac{g_* H^2}{45\pi}\frac{(4\varepsilon_H - \eta_H)}{\varepsilon_H} \tag{30}$$

and

$$A_S \simeq -\frac{H^2}{\pi \varepsilon_H}\left(1+\frac{g_* H^2}{90\pi \varepsilon_H}\right). \tag{31}$$

The observational constraints from Eqs. (24) and (25) then imply

$$\varepsilon_H \lesssim 0.1 \qquad \text{and} \qquad \eta_H \lesssim 0.1\,, \tag{32}$$

as long as $g_* \ll 10^7$. The observational constraints on ε_H and η_H are therefore unchanged from their usual values. The expressions for the spectral indices of scalar and tensor fluctuations, as well as their amplitudes, are effectively given by the usual expressions in terms of the slow-roll parameters, with leading-order corrections as given in Eqs. (30) and (31).

Given these constraints, one might naively expect that the level of non-gaussianity should be the same as that in standard slow-roll inflationary models, where $f_{NL} \sim O(\varepsilon_H \text{ and } \eta_H) \sim 10^{-2}$, [25]. Detailed calculations, however, show that the amplitude of non-gaussianity are strongly dependant on the parameters involved in the interactions of the warm inflationary model, and hence on the microscopic physics and dynamics [26]. For models with weak interactions, as suggested by the observational constraints above, the shape of the bispectrum is found to be close to equilateral.

5 Discussion

In this paper, we have studied the cosmological consequences of the vacuum being in thermal equilibrium with a bath of black-body radiation, as suggested by Paddy in [4]. In this situation, energy is exchanged between the vacuum and the radiation. In the absence of other matter fields, the assumption of a vacuum equation of state $p_v = -\rho_v$, and a temperature corresponding to the surface gravity of the cosmological horizon, results at late times in intermediate inflation with $H \propto t^{-1/3}$, and the introduction of an initial curvature singularity.

We also calculated the evolution of such a universe when it contains a non-interacting barotropic perfect fluid, in additional to the interacting radiation and vacuum energy. We found that, as long as the non-interacting fluid has an equation of state $p > -\rho$, it is dominated at both late and early times by the radiation. We also find that it is possible for the non-interacting fluid to dominate for a finite period at intermediate times, and that during this time the energy density in the vacuum can change sign from negative to positive.

We then proceeded to study the observational consequences of this energy exchange when the vacuum is treated as a slowly-rolling minimally-coupled scalar field with a self-interaction potential. We find that observational constraints on the amplitude of scalar and tensor perturbations, and the spectral index of primordial curvature perturbations, result in expressions that are very close to the usual ones, written in terms of the slow-roll parameters. There are therefore no strong observational constraints to distinguish this scenario from a standard slow-roll inflation.

While the generic end-state of these models is intermediate inflation driven by the vacuum energy, we find that the generic initial state is a radiation dominated universe in which all energy densities diverges. The occurrence of an early universe with a large negative vacuum energy, that can evolve into one a late universe with positive vacuum energy, is an intriguing consequence of this scenario, and would appear to be consistent with the picture of the negative Planck-sized vacuum energy that is generically expected to result from the lowest-order super-gravity terms in string and M-theories [27]. We leave further study of this feature for future work.

Acknowledgements We are grateful to J. Lidsey, A. Linde and D. Mulryne for helpful discussions. TC and JDB are both supported by the STFC.

References

1. L.P. Grishchuk, Lett. Nuovo Cim. **12**, 60 (1975)
2. J. Haro, E. Elizalde, J. Phys. A **41**, 372003 (2008)
3. L. Parker, D. Toms, *Quantum Field Theory in Curved Spacetime* (Cambridge University Press, Cambridge, 2009)
4. T. Padmanabhan, Phys. Rep. **406**, 49 (2005)
5. J.D. Barrow, E.J. Copeland, A.R. Liddle, Mon. Not. R. Astron. Soc. **253**, 675 (1991)
6. R.-G. Cai, S.P. Kim, JHEP **0502**, 050 (2005)
7. G.W. Gibbons, S.W. Hawking, Phys. Rev. D **15**, 2738 (1977)
8. J.D. Barrow, Phys. Lett. B **235**, 40 (1990)
9. J.D. Barrow, A.R. Liddle, C. Pahud, Phys. Rev. D **74**, 127305 (2006)
10. A.G. Muslimov, Class. Quantum Gravity **7**, 231 (1990)
11. A.A. Starobinsky, JETP Lett. **82**, 169 (2005)
12. J.D. Barrow, J. Magueijo, Phys. Rev. D **88**, 103525 (2013); J.D. Barrow, M. Lagos, J. Magueijo, Phys. Rev. D **89**, 083525 (2014)
13. J.D. Barrow, P. Saich, Phys. Lett. B **249**, 406 (1990)
14. J.D. Barrow, A.R. Liddle, Phys. Rev. D **47**, 5219(R) (1993)
15. L.M. Krauss, R.J. Scherrer, Phys. Rev. D **75**, 083524 (2007)
16. A. Berera, M. Gleiser, R.O. Ramos, Phys. Rev. D **58**, 123508 (1998)
17. A. Berera, Nucl. Phys. B **585**, 666 (2000)
18. L.M.H. Hall, I.G. Moss, A. Berera, Phys. Rev. D **69**, 083525 (2004)
19. M. Bastero-Gil, A. Berera, Int. J. Mod. Phys. A **24**, 2207 (2009)
20. S. Bartrum et al., Phys. Lett. B **732**, 116 (2014)
21. R.O. Ramos, L.A. da Silva, JCAP **03**, 032 (2013)
22. M. Bastero-Gil et al., JCAP **05**, 004 (2014)
23. P.A.R. Ade et al., (Planck Collaboration). Astron. Astrophys. **571**, A22 (2014)
24. P.A.R. Ade et al., (BICEP-2 Collaboration). Phys. Rev. Lett. **112**, 241101 (2014)
25. X. Cheng, Adv. Astron. (special issue) **2010**, 638979 (2010)
26. M. Bastero-Gil, A. Berera, I.G. Moss, R.O. Ramos, JCAP **12**, 008 (2014)
27. J. Maldacena, C. Nunez, Int. J. Mod. Phys. A **16**, 822 (2001); B. de Wit, D.J. Smit, N.D. Hari Dass, Nucl. Phys. B **283**, 165 (1987)

Understanding General Relativity After 100 Years: A Matter of Perspective

Naresh Dadhich

Abstract This is the centenary year of general relativity, it is therefore natural to reflect on what perspective we have evolved in 100 years. I wish to share here a novel perspective, and the insights and directions that ensue from it.

1 Prologue

It is the contradiction between observed phenomena and the prevalent theory that drives search for a new theory. Then it takes few to several years of intense activity of tentative guesses, effective and workable proposals, and slowly a new understanding evolves inch by inch that finally leads to new theory. This is indeed the case for all physical theories. Take the case of special relativity (SR). Electromagnetic theory when it was ultimately completed by Maxwell by synthesizing Coulomb, Ampere and Faraday, and introducing the ingenious displacement current in 1875, it predicted an invariant velocity of light. The famous Michelson–Morley experiment verified the prediction with great accuracy leaving no room for any suspicion or doubt. The universally constant velocity obviously conflicted with the Newtonian mechanics. Then the search began which culminated in 1905 after 30 years in Einstein's discovery of special relativity (SR).

In the journey to special relativity there were contributions from severe people, most notably of Poincare and Lorentz who had it all but for one bold statement that velocity of light is universally constant. Then came a young man of 26, and simply did what Poincare and Lorentz hesitated to do, and walked away with the credit of discovering one of the fundamental theories of physics. He said that the velocity of light was constant and its incorporation in mechanics naturally led to special relativity. After 30 years of probing, the atmosphere was sufficiently charged for someone to take the crucial bold step to pick up lowly hanging SR. It was strange

N. Dadhich (✉)
Center for Theoretical Physics, Jamia Millia Islamia, New Delhi 110 025, India
e-mail: nkd@iucaa.in

N. Dadhich
Inter-University Center for Astronomy and Astrophysics, Post Bag 4, Pune 411 007, India

J.S. Bagla and S. Engineer (eds.), *Gravity and the Quantum*,
Fundamental Theories of Physics 187, DOI 10.1007/978-3-319-51700-1_7

that Poincare and Lorentz, who had explored various properties of SR, failed to take the critical step. It was perhaps their great scientific reputation that came in the way. If it did not turn out right, their hard earned reputation over a life time would go down the drain and the whole world would laugh at them. It was this hesitation that costed them dearly in losing their rightful claim on discovery of SR. On the other hand Einstein had neither much reputation nor even an academic job to worry about. He had nothing at stake as he was a clerk in a patent office. If it did not come out right, nothing much would have been lost.

Like all other discoveries it was a situational discovery. If it were not him, someone else would have done it in a year or two. Had he discovered only SR, he would have been one among many great scientists, but not in a different league altogether. For that he had to do something very special which none else could have done. By that I mean that when atmosphere is sufficiently charged and ripe for a new theory to sprout, it is a matter of chance, who happened to take the critical last step. So far as SR was concerned, Einstein was really lucky.

Following SR, the real action at that time was in understanding atomic structure by building a new theory of quantum mechanics. He did make a pretty interesting little contribution in that which was good enough to win him the Nobel prize. But then he totally withdrew from the action and devoted 10 long years for completion of the principle of relativity, from special to general. In the process, he arrived at a new relativistic theory of gravity – general relativity (GR). True, there was no observation or experiment that asked for anything beyond the Newtonian theory at that time. He was therefore not driven by contradiction with experiment but was entirely propelled by the principle. That is why GR was born as a whole and also much ahead of its time. This is something none else could have discovered.

To put it all in perspective, had it not been for him, nobody would have asked for a new theory of gravity until quasars were discovered in mid 1960s. This is what puts him in a class of his own. And so is GR as well because it was born out of a principle without any bearing on observation and experiment, whatsoever. More importantly it makes demand on spacetime, which no other force makes, that it has to curve to describe its dynamics. Above all, not only it still stand tall and firm after 100 years, but its centenary is in fact being celebrated with the detection of gravitational waves which were also predicted by Einstein 100 years back! This is indeed a discovery of the same proportion as that of the electromagnetic waves, and hence it is one of the greatest of all times. It is time to salute with utmost reverence and admiration both the theory as well as its creator.

2 Introduction

General relativity is in many ways unique and different from all other physical theories. The first and foremost among them is the fact that, unlike all other forces, relativistic gravitational law is not prescribed but instead it is dictated by spacetime geometry itself. It naturally arises from inhomogeneity of spacetime and that is why it is universal – links to everything that physically exists. Presence of any force makes spacetime inhomogeneous for particles to which the force links but not for others. For instance, presence of electric field makes spacetime inhomogeneous for electrically charged particles while for neutral particles it remains homogeneous. By universal force we mean a force that links to everything that physically exists irrespective of particle parameters like mass, charge and spin. Since relativistic gravity is universal and hence it can only be described by spacetime geometry. Thus unlike Newton, Einstein had no freedom to prescribe a relativistic gravitational law because it is entirely governed by spacetime itself which does not obey anyone's dictate or prescription. Since relativistic gravity encompasses Newtonian gravity, it is remarkable that now Newton's inverse square law simply follows from spacetime geometry without any external prescription.

Note that spacetime is a universal entity as it is the same for all and equally shared by all and so is the universal force. Hence the two respond to each-other leaving no room for any external intervention. By simply appealing to inhomogeneity of spacetime curvature, we will derive an equation of motion for universal force which would be nothing other than Einsteinian gravity. It is remarkable that we make no reference to gravity at all yet spacetime curvature yields gravitational equation. This happens because both *spacetime and Einstein gravity are universal* [1]. A general principle that emerges is that *all universal things respond to each other and they must therefore be related.*

The equation of motion that emerges from Riemann curvature is non-linear involving square of first derivative of metric. It indicates that gravity is self interactive. As a matter of fact it is the universal character that demands energy in any form must gravitate. Since gravitational field like any other field has energy, it must hence also gravitate – self interact. Isn't it wonderful that spacetime curvature automatically incorporates this feature through nonlinearity inherent in Riemann tensor? The important aspect of self interaction is that it gravitates without changing the Newtonian inverse square law. This is rather strange because self interaction would, in the classical framework, have asked for $\nabla^2\Phi = 1/2\Phi'^2$ which would have disturbed the inverse square law. The situation is exactly as it is for photon (light) to feel gravity without having to change its velocity. Within classical framework it is impossible to accommodate these contradictory demands.

The answer could however be in the enlargement of framework in which gravity curves space and photon freely floats on it without having to change its velocity. What should curve space and the obvious answer is gravitational field energy which is not supposed to contribute to acceleration, $\nabla\Phi$. Thus gravity self interacts via space curvature and that also facilitates photon's interaction with gravity [2]. These are the two new aspects of Einstein gravity which wonderfully take care of each-other leaving Newtonian inverse square law intact. Einstein is therefore Newton with space curved [2]. Since space and time are bound together in spacetime through universal light velocity, and hence spacetime must be curved. This is how spacetime curvature enters in description of the universal force – Einstein gravity. In this way the self interaction gets automatically incorporated in Riemann curvature and is reflected through occurrence of square of first derivative of metric.

General relativity (GR) is undoubtedly the most elegant and beautiful theory and it is for nothing that Paul Dirac termed it as the greatest feat of human thought!

In what follows we would further explore its elegance and richness of structure and form in relation to what new insights and understanding we have gained in past 100 years, and marvel on new questions and directions that ensue.

3 At the Very Beginning

Let us begin by characterizing free state of space and time in absence of all forces. Space is homogeneous and isotropic and time is homogeneous. As space is homogeneous, one can freely interchange x and y. Since time is also homogeneous which means both space and time are homogeneous, and hence x and t should also be similarly interchangeable. But they are not of the same dimension. Never mind, homogeneity of space and time is a general property which must always be respected and it demands their interchange. One has therefore to bring x and t to the same dimension by demanding existence of a universal invariant velocity c so that x and ct could be interchanged [1]. Thus it is homogeneity of space and time that demands existence of a universally constant velocity without reference to anything else. It is identified through Maxwell's electrodynamics with velocity of light. Thus space and time get bound together into spacetime through the universal velocity of light. It thus arises in a natural way as a constant of spacetime structure independent of anything else.

Spacetime free of all dynamics and forces is therefore homogeneous (space and time being homogeneous and space being isotropic). The next question that arises is, what is its geometry? As spacetime is homogeneous so should be its geometry. Geometry is defined by Riemann curvature and so it should be homogeneous which means it should be covariantly constant; i.e. $R_{abcd;e} = 0$. The Riemann curvature should therefore be written in terms of something which is constant for covariant derivative. That is the metric tensor and hence we write

$$R_{abcd} = \Lambda(g_{ac}g_{bd} - g_{ad}g_{bc}). \tag{1}$$

A homogeneous spacetime free of all forces is thus a spacetime of constant curvature, Λ and not necessarily Minkowskian of zero curvature. It is a maximally symmetric spacetime and that is what is required for absence of all dynamics. The important point to note is that Minkowski is not dictated by homogeneity of spacetime but it is rather an external imposition by setting $\Lambda = 0$. Of course one is free to choose Λ zero but then it has to be justified on physical grounds. In classical physics, force free state is characterized by constant potential or constant velocity while for the Einstein gravity it is done by constant curvature. The important point to realize is that constant curvature means no dynamics – it is on the same footing as constant potential in classical physics. This is because it is Riemann curvature which is the basic element for description of the Einstein gravity and hence it is this, like potential for classical physics, that should have freedom of addition of a constant. Force free homogeneous spacetime is in general described by maximally symmetric dS/AdS and not necessarily by flat Minkowski.

The point we wish to emphasize is that Λ arises naturally as a constant of spacetime structure on the same footing as c without reference to any physical force or phenomenon. These two are pure constants of spacetime structure itself and arise as the characteristics of *force free* state. They are therefore the most fundamental constants of Nature. No other constant can claim this degree of fundamentalness simply because none else arises directly from spacetime structure itself.

The next question that arises is, what happens when spacetime is not homogeneous? Obviously it should indicate presence of force which makes spacetime inhomogeneous for all particles irrespective of their mass, charge or any other attributes. This force should therefore be universal meaning it links to everything that physically exists. Everything that physically exists must have energy-momentum – a universal attribute/charge, and hence this force must link to the universal charge – energy-momentum. How do we then determine its dynamics? Since presence of this universal force makes spacetime curvature inhomogeneous, hence its dynamics cannot be prescribed but has to follow from the curvature itself. What is it that we can do to Riemann curvature to get to an equation of motion for the universal force? The Riemann curvature satisfies the Bianchi differential identity, vanishing of the Bianchi derivative, $R_{ab[cd;e]} = 0$. Let's take its trace which leads to

$$G^{ab}{}_{;b} = 0 \tag{2}$$

where

$$G_{ab} = R_{ab} - \frac{1}{2} R g_{ab}, \tag{3}$$

is the second rank symmetric tensor with vanishing divergence. Then we can write

$$G_{ab} = -\kappa T_{ab} - \Lambda g_{ab} \tag{4}$$

with $T^{ab}{}_{;b} = 0$ and the second term on the right is constant relative to covariant derivative. Could this be an equation of motion for the universal force responsible for inhomogeneity of spacetime curvature? On the left is a second rank symmetric tensor derived from Riemann tensor involving second order derivative of the metric and hence it is a second order differential operator like ∇^2 operating on the metric potential g_{ab}. If we identify the new tensor T_{ab} with energy-momentum distribution, which is universal, as source, then the above equation becomes equation of motion for the Einstein gravity. Thus emerges GR from spacetime curvature all by itself.

The principle of Equivalence had played very important role in discovery of GR but we made no reference to it in our derivation of the Einstein equation. This is simply because gravitational law being described by curved spacetime which admits a tangent plane at every point. This property of curved space automatically incorporates the Principle of equivalence. Since Einsteinian gravity is universal, this is why its equation of motion is geometric and so is motion under gravity – a geodesic with no reference to any particle parameter. It is purely a geometric statement. The point to be noted is that Newton's second law does not apply to relativistic gravity because it is universal. For a geodesic motion there is no inertial and gravitational mass, what we need to experimentally verify is that how accurately particles follow the geodesic. Thus the question why should inertial and gravitational mass be equal becomes impertinent.

Note that we began by characterizing force or dynamics free state of spacetime and it is defined by (homogeneous) constant curvature. What happens when curvature is inhomogeneous, the Einstein gravity naturally arises even though we had not asked for it. Like c and Λ characterize homogeneous spacetime, similarly the Einstein gravity characterizes inhomogeneous spacetime. In other words, gravity is inherent in inhomogeneity of spacetime curvature. This is different from rest of physics where a force law like Newton's gravity is always prescribed from outside. Thus the Einstein gravitational law cannot be prescribed instead it is dictated by inhomogeneity of spacetime. This is so simply because it is universal – links to everything that physically exists, the unique distinguishing feature of the Einstein gravity.

Further note that Λ enters into the equation on the same footing as energy-momentum tensor T_{ab}. It is therefore as solid a piece in the equation as energy-momentum, and hence it should not be subjected to one's whims and fancy without due physical justification. When $T_{ab} = 0$, we are back to homogeneous spacetime of constant curvature.

> Had Einstein followed this line of reasoning to get to his equation, he would have certainly not treated Λ as a blunder instead would have respectfully recognized it as a true constant of spacetime structure alongside the velocity of light. This would have saved us all from this monumental confusion that has gone over a century and yet no sign of diminishing. Further, perhaps he would have made the greatest prediction of all times that the Universe would experience accelerated expansion some time in the future. Had that been the case it would have been the most remarkable and truly Einstein like. Alas that didn't happen.

The picture thus emerges is that homogeneity, characterizing force-free state, of spacetime requires two invariants, a velocity that binds space and time into spacetime and a length that gives a constant curvature to it. Introduction of matter/forces makes spacetime inhomogeneous and so emerges the Einsteinian gravitational dynamics. In the conventional picture, for absence of matter spacetime is taken to be flat and then matter makes zero curvature to non-zero. There is a discontinuity and break from flat to non-flat while in our picture there is continuous transition from homogeneity to inhomogeneity. There is therefore a paradigm shift, all dynamics free spacetime is thus not flat but is (homogeneous) of constant curvature and introduction of matter makes it inhomogeneous. As invariant velocity is needed to bind space and time into spacetime so as to provide a relativistic platform for physical phenomena, exactly with the same force of argument and spirit, Λ is needed to provide an appropriate curved spacetime platform for the relativistic gravitational phenomenon – the Einstein gravity to unfold [1].

What it tells is that like constant potential is irrelevant for classical physics, similarly constant curvature is irrelevant for gravitational dynamics. It is because the constant curvature spacetime is maximally symmetric characterizing the 'force free' state of spacetime. On the other hand, it turns out that constant potential for radially symmetric field in the usual Schwarzschild coordinates is indeed, unlike the Newtonian gravity, non-trivial for the Einstein gravity because it produces inhomogeneous spacetime of non-zero curvature [2, 3].

4 Self Interaction and Vacuum Energy

The driving force for GR is to universalize gravity which meant all forms of energy distribution including its own self energy as well as zero mass particles must participate in gravitational interaction [4]. The only way zero mass particle can, since its velocity cannot change, be brought in the fold is that gravity must curve space and zero mass particle simply floats freely on it. Since space is already bound with time by the invariant velocity, gravity thus curves spacetime. We have seen above how Einstein gravity naturally follows purely from differential geometric property of Riemann curvature of spacetime. This is all very fine, but how is self interaction taken care of; i.e. how does gravitational field energy gravitate? Does it do through a stress tensor like any other matter field? No, we write no stress tensor on the right of the Einstein equation given above. As a matter of fact gravitational potential in the Schwarzschild solution describing field of a mass point is the same Newtonian going as $1/r$ indicating that it is solution of the good old Laplace equation. There is no self interaction contribution in it, and that is why the inverse square law remains intact.

If the self interaction were to be incorporated in the equation, then it should have been modified to $\nabla^2 \Phi = 1/2\Phi'^2$. This would have of course been relative to flat spacetime, and it would have modified the inverse square law. The latter is, as we all know, the cornerstone of classical physics as it ensures conservation of flux and thereby of charge. Hence that should not be tampered. The self interaction should therefore have to be accommodated without modifying the gravitational law.

Also note that the only way photon can respond to gravity is that gravity curves space. Could it be that the self interaction is responsible for curving space while the matter/energy distribution produces the inverse square law? This is exactly what happens in GR, and so we can say Einstein is Newton with space curved [2]. It is most remarkable that the two new aspects of GR, self interaction and photon feeling gravity, take care of each other so beautifully that the former curves space and that is precisely what is required for the latter. This is indeed the mark of sheer elegance and profoundity.

Why the self interaction is not visible in the Schwarzshild solution because it has been automatically absorbed in the space curvature when we write $g_{rr} = (1 + 2\Phi)^{-1}$ while the Newtonian acceleration is accounted for by $g_{tt} = 1 + 2\Phi$. For solving the Einstein vacuum equation, when we write $R^t_t = R^r_r$ which demands $g_{tt}g_{rr} = -1$ and then $R^t_t = \nabla^2\Phi = 0$ leading to the inverse square law. This is how the equation requires space to be curved and the self interaction gets absorbed in that. If space were flat, $g_{rr} = -1$, then $R^t_t = 0$ implies $\nabla^2\Phi \; \Phi'^2$, clearly showing the self interaction. When $g_{tt} = -g_{rr}^{-1} = 1 + 2\Phi$, it gets beautifully absorbed in the space curvature leaving the Newtonian law intact [2].

There is yet another subtlety that the Einstein potential can be zero only at infinity and nowhere else – it is determined absolutely. This follows from $R^\theta_\theta = 0$ which determines $\Phi = -M/r$ exactly without a possibility of addition of a constant. This happens because gravitational field energy can vanish only at infinity, and hence so should the space curvature. That means potential can only vanish at infinity.

The important lesson that follows is that gravitational field energy gravitates not through a stress tensor as a source on the right hand side of the equation but instead by enlarging the framework from flat to curved space. Why does this happen, what is it that is different for gravitational field energy? The answer is, that it is a secondary source which is produced by the primary source, matter-energy. It has no independent existence of its own – it is matter fields that produce gravitational field. It is therefore natural that a secondary source produced by the primary source should not sit alongside in the equation. It can therefore only be incorporated by enlarging the framework [1, 6].

This suggests a general principle that anything that doesn't have independent existence of its own is a secondary source and hence *must not* gravitate via a stress tensor but instead by enlarging the framework. The point in question is that of vacuum energy produced by quantum fluctuations of vacuum by the matter fields. It is exactly on the same footing as gravitational field energy. Never mind one is able to compute its stress tensor relative to flat spacetime, which has exactly the same form as Λg_{ab}, it must not sit alongside the matter fields, T_{ab}. This is precisely the reason for its association with Λ and then its incredible mismatch, 10^{120} with the Planck length. This is the root cause of the confusion which arises from making vacuum energy gravitate through a stress tensor. This defies and violates the above general principle just enunciated.

If we adhere to the principle, there is no relation between Λ and vacuum energy. It is then free to have any value that observations determine. Recall that both c and Λ arose purely from the symmetries of (homogeneous) spacetime as constants of its structure. Then c got identified with velocity of light and Λ remained dangling until the 1997 supernova observations of accelerating Universe [5]. Thus Λ simply represents acceleration of the Universe as all observations are wonderfully consistent with it. There is no need for any kind of dark energy involving exotic matter or outlandish modifications of gravitation theory.

Of course the moot question remains, how to enlarge the framework to make vacuum energy gravitate? Vacuum energy is a quantum creature and hence it would be difficult to guess the enlargement of framework until there emerges a quantum theory of gravity. If we had asked the same question for gravitational field energy in 1912, say, before the advent of GR, it would have been hard to guess that enlargement is in curving space. This means we won't know exactly without quantum gravity how to enlarge framework for making vacuum energy gravitate [1, 6].

There could however be some informed guesses based on the lessons learnt from gravitational field energy. In GR, the real question was how to make light feel gravity which required space to be curved. Since space and time were already bound into spacetime by the velocity of light, it meant curving of spacetime. That is how gravity can only be described by spacetime curvature which automatically incorporated gravitational self interaction. Since the Newtonian inverse square law remains intact, self interaction can only curve space. This suggests that framework should be so enlarged that keeps GR intact. The real question therefore is to identify some phenomenon which has so far remained aloof, like light in the case of GR, and that has to be brought into the gravitational fold. Answering this question would require framework enlargement which would automatically incorporate gravitational interaction of vacuum energy. This is what will perhaps show the road to quantum gravity. Unfortunately we have not yet been able to clearly identify this critical question. That is the problem.

Spacetime curves or bends like a material object, it should therefore have physical structure – a micro-structure as is the case for any material object. That means space should have some micro building blocks – "atoms of space". Such a micro structure is also required for vacuum to quantum fluctuate giving rise to vacuum energy. Thus micro structure of space is intimately related to vacuum energy and hence incorporation of the former would perhaps automatically, as anticipated, take care of gravitational interaction of the latter. The key question is then how to bring in atoms of space into the fray. Loop quantum gravity seems to follow this route but has not been able go far enough.

Another possible avenue could be that vacuum energy may gravitate via higher dimension [7–9] leaving GR intact in the four dimensional spacetime. It is conceivable that at very high energy gravity may not entirely remain confined to four dimension, it may leak into higher dimension [10]. The basic variable for gravity is the Riemann curvature tensor, for high energy exploration, we should include its higher powers in the action. Yet we want the equation of motion to retain its second order character, then this requirement uniquely identifies Lovelock Lagrangian. Even though Lovelock action is a homogeneous polynomial of degree N in the Riemann curvature, it has remarkable unique property that the resulting equation is always second order. Note that Lovelock gravity includes GR for $N = 1$, and $N = 2$ is the quadratic Gauss-Bonnet (GB) gravity, and then cubic and so on. But the higher order terms make non-zero contribution in the equation only in dimensions higher than four. If we want to explore high energy sector of gravity, which should indeed be the case for quantum gravity, we have to go to higher dimensions [11]. This is purely a classical motivation for higher dimensions. What it suggests is that the road to quantum gravity may go via higher dimensions notwithstanding the fact that higher dimensions are natural playground for string theory. Though string theory is a very popular approach to quantum gravity, yet it has also not gone far enough.

What it all suggests is that like quantum gravity, gravitational interaction of vacuum energy is an open question, and the solution of the latter is perhaps inseparable from the former. In the absence this, incorporation of it through a stress tensor is simply a tentative attempt similar to inclusion of gravitational self interaction by writing $\nabla^2 \Phi = 1/2\Phi'^2$. This was not borne out by the correct theory of gravity – GR. So would be the case for vacuum energy when quantum gravity emerges.

5 In Higher Dimension

In the previous section, we have hinted that consideration of gravity in higher dimensions may not be entirely outrageous. Then the question arises, what should be the equation in there? Could it very well be the Einstein equation which is valid in all dimensions larger than two? Yes, that could be the case. However how did we land in four rather than three dimension? This is because in three dimension, gravity is kinematic which means Riemann is entirely determined by Ricci tensor and hence there exists no non-trivial vacuum solution. This translates into the fact that there are no free degrees of freedom for free propagation of gravitational field. This is how we come to four dimension where Riemann has 20 while Ricci has 10 components allowing for non-trivial vacuum black hole solutions. Could this feature be universalized for all odd dimensions in a new theory which reduces to Einstein gravity for dimension, $d \leq 4$? It would be nice to incorporate this feature in higher dimension.

Another desirable feature that one can ask for is existence of bound orbit around a static object like a black hole. It is easy to see that in GR, bound orbits can exist only in $d = 4$ and in none else. This is because gravitational potential goes as $1/r^{d-3}$ which becomes sharper and sharper with dimension while centrifugal potential always falls off as $1/r^2$ and hence the two can balance only in four dimension to give bound orbits.

If we take these two as the guiding features for gravitational equation in higher dimension, then pure Lovelock gravity is uniquely singled out [12–15]. In Lovelock gravity, Lagrangian is $\sum \alpha_N \mathscr{L}^N$ where each α_i is a dimensionful coupling constant. Note that $\alpha_0 = \Lambda$ is the cosmological constant and $\alpha_1 = G$, $\mathscr{L} = \mathscr{R}$ are respectively the Newtonian gravitational constant and the Einstein–Hilbert Lagrangian. By pure Lovelock we mean that Lagrangian has the only one Nth order term without sum over lower orders, and consequently the equation also has only one term. For pure Lovelock, gravitational potential goes as $1/r^{(d-2N-1)/N}$ for $d \geq 2N + 1$ [16]. For existence of bound orbits, what is required is $(d - 2N - 1)/N < 2$ which is always true for $N \geq 1$, and it means $d < 4N + 1$. Further $d > 2N + 1$ else gravitational potential becomes constant, and hence we have the dimensional range, $2N + 1 < d < 4N + 1$ for existence of bound orbits. For the linear $N = 1$ Einstein, it is $3 < d < 5$ (and hence only $d = 4$) while for the quadratic $N = 2$ GB, $5 < d < 9$.

In pure Lovelock gravity, potential becomes constant in all critical odd $d = 2N + 1$ dimensions and hence gravity must be kinematic. For $N = 1$ Einstein gravity, potential is constant in $d = 3$, and gravity is kinematic in the sense that Riemann is given in terms of Ricci tensor. This means that it should be possible to define an Nth order Lovelock Riemann tensor which is then given in terms of the corresponding Ricci in all $d = 2N + 1$ dimensions. This is indeed the case [17–19]. The Lovelock Riemann, which is a homogeneous polynomial in Riemann, is defined by the property that vanishing of trace of its Bianchi derivative gives a divergence free second rank symmetric tensor – Lovelock analogue of Einstein tensor, and it is exactly the same as what one obtains by varying Nth order Lovelock action [17]. Then the pure Lovelock gravitational equation reads as follows:

$$^{(N)}E^a_b \equiv -\frac{1}{2^{N+1}}\delta^{ac_1d_1...c_Nd_N}_{ba_1b_1...a_Nb_N} R^{a_1b_1}_{c_1d_1} \dots R^{a_Nb_N}_{c_Nd_N} = -8\pi T^a_b\,. \tag{5}$$

It is then shown that Nth order Lovelock Riemann can be entirely written in terms of Nth order Einstein tensor, $^{(N)}E^a_b$ [19]. The pure Lovelock gravity is kinematic in all critical odd $d = 2N + 1$ dimensions.

We have identified the two critical properties of Einstein gravity, kinematicity in odd three dimension and existence of bound orbits around a static source, which we would like to carry over to higher dimensions. It is the universalization of these properties that leads to pure Lovelock equation uniquely. This is the right equation in higher dimensions [13, 15]. For a given N, existence of bound orbits prescribes the dimensional range, $2N + 1 < d < 4N + 1$. On the other hand stability of static black hole requires $d \geq 3N + 1$ [20] and hence the range gets further refined to $3N + 1 \leq d < 4N + 1$. For $N = 1$, it admits only one $d = 4$ while for $N = 2$, there are two $d = 6, 7$, and in general number of allowed dimensions are equal to Lovelock order N. It is interesting that stability threshold is though included but not the entire range $d \geq 3N + 1$. That is, bound orbits exist for unstable black hole for $2N + 1 < d < 3N + 1$ while for $d \geq 4N + 1$, black hole is stable without bound orbits around it.

Further pure Lovelock gravity gives rise to an interesting situation that $1/r$ potential on which whole of astrophysics and cosmology rest could occur not only in four but higher dimensions as well [20]. This is because potential goes as $1/r^{(d-2N-1)/N}$ which will be $1/r$ in all dimensions, $d = 3N + 1$. Static black holes are thus indistinguishable in this entire dimensional spectrum. In particular four dimensional Schwarzschild black hole is indistinguishable from its pure GB seven dimensional counterpart. Not only that cosmology is also the same as FRW expanding Universe evolves with the same scale factor [20].

So far as gravity is concerned, the situation is indistinguishable in dimensional spectrum $d = 3N + 1$ for all astrophysical and cosmological observations, except for gravitational degrees of freedom determining number of polarizations of gravitational wave. Number of degrees of freedom are given by $d(d - 3)/2$ [21] which is two in four and 14 in seven dimension. The Hulse–Taylor pulsar observations do verify two polarizations for emitted gravitational wave. But for that it would not be possible to decide whether it is Einstein gravity in four or Nth order pure Lovelock in $(3N + 1)$ dimension. It is an interesting feature of pure Lovelock gravity.

6 Outlook and Perspective

GR is purely a principle and concept driven theory and it is therefore born as a whole complete theory. There was no observation or phenomenon driving it. That is why it was not developed as step by step but it emerged as a complete full theory. One can envisage the driving principle as inclusion of zero mass particle in mechanics and gravitational interaction. This meant universalization of mechanics and gravity for all particles including zero mass particles – photons/light [4]. The former leads to relativistic mechanics known as special relativity while the latter to relativistic theory of gravitation – general relativity. Of course in the former case there was the compelling phenomenological demand – velocity of light was observed to be constant for all observers while for the latter there was no such phenomena asking for it. As a matter of fact, the first serious challenge to the Newtonian gravity only came as late as in mid 1960s in the form of observation of highly energetic quasi-stellar objects – quasars.

It was a theory at least 50 years ahead of its times. This was because it was principle rather than experiment or observation driven. The same situation still holds as there is no strong observational challenge to it as yet. The accelerating Universe observation did pose some concern, and it did generate enormous amount of activity in building models of dark energy which were rather too many for comfort, and involved exotic matter fields and outlandish modifications of the theory. However it has all settled down to Λ successfully accounting for the observations. It is the symmetry of homogeneous spacetime that gives rise to Λ as a true constant of spacetime structure on the same footing as velocity of light, and the accelerating Universe determines its value [5].

Had Einstein followed the natural and straightforward geometric path to arrive at GR, he could have in fact realized the true significance of Λ and would have predicted that the Universe would experience accelerated expansion some time in the future. That would have been the greatest prediction of all times. Then there won't have been any reason for questioning Λ but instead one would have questioned how does vacuum energy stress tensor has the same form as Λ? Neither there would have been that much thrust for dark energy models nor would there have been acrobatics for getting Λ as a constant of integration via trace-free or unimodular gravity [22, 23]. Though not much material difference but it would have been a different and perhaps the right perception.

With this backdrop, the viewpoint, that vacuum energy cannot gravitate via a stress tensor but instead it would require an enlargement of framework, would have perhaps been appreciated with a positive disposition. It is then a principle that dictates that secondary gravitational sources like self interaction and vacuum energy caused by primary matter source do not gravitate via a stress tensor but instead they do by enlargement of the framework as is the case for the former – by curving space [1, 6]. For inclusion of vacuum energy in gravitational interaction, it is the principle that directs us to go beyond GR. It is therefore not only GR was principle driven but a journey beyond it is as well. Since vacuum energy is a quantum effect, for its inclusion we need a quantum theory of gravity. As and when it comes about, what is expected is that vacuum energy would be automatically included. The gravitational interaction of vacuum energy cannot be accommodated in GR itself and would remain an open question until quantum gravity is discovered.

Another point to note is that the vacuum energy has equation of state, $\rho + p = 0$, which defines inertial density for fluid equation of motion. What happens when inertial mass of a particle is zero, it cannot be accommodated in the existing framework. It asks for a new framework of relativistic mechanics – SR. Thus vanishing of inertial quantity is a serious matter, it indicates that it cannot be accommodated in the existing theory, and a new theory would be required for its incorporation.

Once Λ is liberated from vacuum energy, it can hence have any value that the observations determine. Then there is no embarrassing discrepancy of 10^{120} orders with the Planck length. What this number then indicates is simply the fact that the Universe measures this much in units of the Planck area [6]!

The important point to be noted is that universal velocity and universal force cannot be described by Newton's second law [24] instead they could only be described by spacetime geometry. Motion under gravity is free of particle mass, it simply follows geometry of spacetime – geodesic. The response to gravity is therefore not through gravitational mass and hence passive gravitational mass is not defined. The question of equality of inertial and passive gravitational mass is therefore not admitted for the relativistic gravitational force. Hence there is no need to pose the question of its equality with inertial mass, and so the formidable question of explaining their equality doesn't arise. It is simply the reflection of the fact that gravity is no longer an external force, it is synthesized in spacetime geometry and hence Newton's second law is inapplicable.

Let me point out that light cannot bend, what bends is space. We measure bending of space by light because it freely floats on space [25]. The question is how does space bend like a wire? Wire bends because it is made of small discrete units like atoms and molecules. This means discrete micro structure is necessary for anything to bend [1]. At deep down Space should therefore also have discrete micro structure – some fundamental entities as its building blocks. Given that, the question arises what is the natural geometric structure of such a system? Should it be flat with zero curvature or be of constant curvature. There are efforts afoot on building spacetime from an evolving system of causal sets. It turns out that constant curvature is more probable than zero curvature though the latter cannot be ruled out [26].

For a classical force which is produced by a charge, when all charges in the Universe are summed over, the total charge must vanish. This is quite clear for electric charge because positive and negative charges are created by pulling out a positive or negative charge from a neutral entity, then what remains behind has opposite polarity. This must also be true for gravity. Energy-momentum is charge for gravity which is unipolar. How could it be balanced to yield total charge zero. The only possible way is that gravitational field must have charge of opposite polarity [4]. This implies three things: gravity is self-interactive, negative gravitational charge is non-localizable and spread over whole space and gravity is always attractive.

There is yet another general principle we would like to invoke is that all universal concepts must be related [4] and that relation must also be universal – the same for all. By universality we mean a concept or phenomenon which is the same for all and equally shared by all. Space and time are universal entities and they must be related by a universal relation – universal velocity. It binds space and time into spacetime. This then leads to the special relativity. Next gravity is a universal force, and hence its dynamics should be described by spacetime curvature – general relativity [27]. Is there anything else which is universal that could be bound to spacetime structure? Like gravity, the primary quantum uncertainty principle is also universal and hence it must also be related to spacetime. This is what has not been achieved and until that happens quantum theory remains incomplete. As and when that happens, it would give rise to a quantum theory of spacetime as well as of gravity. This is perhaps the deepest question of all times, probing the building blocks of spacetime itself. It is therefore the most formidable problem and it is not for nothing that it has so far defied all attempts by the best of the minds for over half a century.

Of late there has been lot of work on gravity as an emergent force. It began with the seminal work of Ted Jacobson who could deduce the first law of thermodynamics from the Einstein vacuum equation [28]. The activity picked up considerably in past decade or say, and among others, Paddy along with his coworkers is one of the leading players [29]. It raises the question, if it is emergent like thermodynamical laws, it is not a fundamental force. It is a bulk property of some underlying kinetic structure of "atoms of space". It is indeed a very deep question, is gravity fundamental or emergent? I believe that the question is similar to asking, is photon a wave or particle? Both gravity and photon are self-dual, meaning their dual is contained in themselves. Gravity is different from all other forces in several ways, and perhaps the most remarkable of them all is that it is both fundamental and emergent or neither.

Notwithstanding all this, probing and understanding of quantum structure of building blocks of spacetime is very pertinent and is the most challenging question of the day.

Finally I conclude it with cheering Paddy warmly and affectionately on turning 60, which is simply a number like any other, and it doesn't matter at all if one doesn't mind it. It is all anyway a matter of mind.

References

1. N. Dadhich, Int. J. Mod. Phys. D **20**, 2739 (2011). arXiv:1105.3396
2. N. Dadhich, Curr. Sci. **109**, 260 (2015). arXiv:12060635
3. M. Barriola, A. Vilenkin, Phys. Rev. Lett. **63**, 341 (1989)
4. N. Dadhich, Universalization as a physical guiding principle. arXiv:gr-qc/0311028
5. S. Perlmutter et al., Astrophys. J. **483**, 566 (1997). Nature **391**, 51 (1998)
6. N. Dadhich, Λ is a constant of spcetime structure and has nothing to do with vacuum energy, The talk given in the workshop on questioning fundamental physical principles, CERN, 6–9 May 2014 and in MG14, Rome (2015)
7. C.P. Burgess, Extra dimensions and the cosmological constant problem. arXiv:0708.0911

8. S. Sengupta, Class. Quantum Gravity **32**, 195005 (2015). arXiv:1501.0079
9. V. Soni, N. Dadhich, R. Adhikari, The cosmological constant from the zero point energy of compact dimensions. arXiv:1509.02156
10. L. Randall, R. Sundrum, Phys. Rev. Lett. **83**, 4690 (1999)
11. N. Dadhich, Universalization as a physical guiding principle. arXiv:gr-qc/0311028; Universal features of gravity and higher dimensions. arXiv:1105.0988; Universality, gravity, the enigmatic Λ and beyond, in *Proceedings of 12th Regional Conference on Mathematical Physics*, eds. by M.J. Islam, F. Hussain, A. Qadir, H. Saleem (World Scientific, 2006). arXiv:hep-th/0509126
12. N. Dadhich, S.G. Ghosh, S. Jhingan, Phys. Lett. B **711**, 196 (2012). arXiv:1202.4575
13. N. Dadhich, The gravitational equation in higher dimensions, in *Relativity and Gravitation: 100 Years After Einstein in Prague*, ed. by J. Bicak, T. Ledvinka (Springer, New York, 2013). arXiv:1210.3022
14. N. Dadhich, S.G. Ghosh, S. Jhingan, Phys. Rev. D **88**, 124040 (2013). arXiv:1308.4770
15. N. Dadhich, Eur. Phys. J. D **78**, 1 (2016). arXiv:1506.08764
16. N. Dadhich, J.M. Pons, K. Prabhu, General Relat. Gravit. **45**, 1131 (2013). arXiv:1201.4994
17. N. Dadhich, Pramana **74**, 875 (2010). arXiv:0802.3034
18. D. Kastor, Class. Quantum Gravity **29**, 155007 (2012). arXiv:1202.5287
19. X.O. Camanho, N. Dadhich, On Lovelock analogues of the Riemann tensor. arXiv:1503.02889
20. S. Chakraborty, N Dadhich, Do we really live in four or higher dimension. arXiv:1605.01961
21. N. Dadhich, R. Durka, N. Merino, O. Miskpvic, Phys. Rev. D **93**, 064009 (2016). arXiv:1511.02541
22. G. Ellis, H. van Elst, J. Murgan, J.-P. Uzan, Class. Quantum Gravity **28**, 225007 (2011). arXiv:1008.1196
23. T. Padmanabhan, H. Padmanabhan, Int. J. Mod. Phys. D **23**, 1430011 (2014). arXiv:1404.2284
24. N. Dadhich, Phys. News **39**, 20 (2009). arXiv:1003.2359
25. N. Dadhich, Subtle is the gravity. arXiv:gr-qc/0102009
26. M. Ahmed, D. Rideout, Phys. Rev. D **81**, 083528 (2010). arXiv:0909.4771
27. N. Dadhich, Why Einstein (Had I been born in 1844!)? arXiv:physics/0505090
28. T. Jacobson, Phys. Rev. Lett. **75**, 1260 (1995)
29. T. Padmanabhan, Exploring the nature of gravity. arXiv:1602.01474; S. Chakraborty, T. Padmanabhan, Phys. Rev. D **92**, 104011 (2015). arXiv:1508.04060; T. Padmanabhan, S. Chakraborty, D. Kothawala, General Relativ. Gravity **48**, 55 (2016). arXiv:1507.05669

Piecewise Conserved Quantities

Tevian Dray

Abstract We review the treatment of conservation laws in spacetimes that are glued together in various ways, thus adding a boundary term to the usual conservation laws. Several examples of such spacetimes will be described, including the joining of Schwarzschild spacetimes of different masses, and the possibility of joining regions of different signatures. The opportunity will also be taken to explore some of the less obvious properties of Lorentzian vector calculus.

1 Introduction

In 1987, my wife (Corinne Manogue) and I found ourselves in India for 6 months, where we were both Indo-American Fellows. The highlight of our visit was the 3 months we spent at TIFR, working with Paddy and others in the Theoretical Astrophysics Group. During this visit, Paddy and I wrote a paper on piecewise Killing vectors [1], bringing additional mathematical clarity to the intriguing results I had previously obtained with 't Hooft [2] on shells of matter in Schwarzschild spacetimes. Little did I know at the time that this theme would recur in my research in a variety of contexts (and more than a dozen papers) over the next 20 years.

As I congratulate Paddy on the occasion of this Festschrift, it is with great pleasure that I look back on this period near the beginning of our careers. It seems only fitting that I use this opportunity to summarize my own journey through several quite different applications of piecewise conserved quantities.

In Sects. 2 and 3, I lay out the framework for analyzing piecewise structures, then in Sect. 4 provide the main mathematical result, the Patchwork Divergence Theorem, generalizing my work with Paddy [1]. The two basic applications are considered next, namely shells of matter in Sect. 5, and signature change in Sect. 6. In Sect. 7, I point out how the underlying framework of the Patchwork Divergence Theorem also provides insight into vector calculus in Lorentzian geometry, providing access to these ideas for non-experts, including undergraduates. Finally, a very brief summary is given in Sect. 8.

T. Dray (✉)
Department of Mathematics, Oregon State University, Corvallis, OR 97331, USA
e-mail: tevian@math.oregonstate.edu

J.S. Bagla and S. Engineer (eds.), *Gravity and the Quantum*,
Fundamental Theories of Physics 187, DOI 10.1007/978-3-319-51700-1_8

2 Piecewise Smooth Tensors

The long history of the distributional curvature due to piecewise smooth metric tensors is summarized in [3]. As discussed there, the basic setup is two smooth manifolds $M^{\pm}$ joined along a (possibly null) hypersurface Σ, with smooth metric tensors $g^{\pm}_{ab}$ on either side. Introducing a step function Θ which is 0 on M^{-} and 1 on M^{+}, the metric on $M = M^{-} \cup M^{+}$ is given by

$$g_{ab} = (1 - \Theta)\, g^{-}_{ab} + \Theta\, g^{+}_{ab}. \tag{1}$$

One traditionally assumes that the metric is continuous at Σ,

$$[g_{ab}] := g^{+}_{ab}\big|_{\Sigma} - g^{-}_{ab}\big|_{\Sigma} = 0, \tag{2}$$

in which case the connection $\Gamma^{c}{}_{ab}$ is at worst discontinuous and, as discussed briefly in [2], it is straightforward to compute distributional curvature tensors. For example, the distributional Ricci tensor is given by

$$\begin{aligned} R_{ab} &= (1 - \Theta)\, R^{-}_{ab} + \Theta\, R^{+}_{ab} + \delta_c[\Gamma^{c}{}_{ab}] - \delta_b[\Gamma^{c}{}_{ac}] \\ &= (1 - \Theta)\, R^{-}_{ab} + \Theta\, R^{+}_{ab} + \delta \rho_{ab}, \end{aligned} \tag{3}$$

where

$$\delta_c = \delta n_c = \nabla_c \Theta, \tag{4}$$

so that n_c is normal to Σ, and the distribution δ can be thought of as a Dirac delta function. However, as shown in [3], it is enough for the *pullbacks* of $g^{\pm}_{ab}$ to agree on Σ in order to have a well-defined tangent space on all of M, which we will exploit in Sect. 6.

More generally, we can consider other piecewise smooth tensors on M, such as a vector field of the form

$$\xi^{a} = (1 - \Theta)\, \xi^{a}_{-} + \Theta\, \xi^{a}_{+} \tag{5}$$

and its discontinuity $[\xi^{a}]$, defined in analogy with (2).

3 Piecewise Conserved Quantities

As is well-known, a *Killing vector* ξ^{a} can be contracted with the stress-energy tensor T_{ab} to yield a conserved quantity

$$X^{a} = T^{ab} \xi_{b} \tag{6}$$

satisfying

$$\oint X^a N_a dS = 0 \tag{7}$$

where S is any closed, piecewise smooth hypersurface (assumed for the moment to be nowhere null) with unit normal vector N^a. The vanishing of this integral is a consequence of the Divergence Theorem, since

$$\nabla_a X^a = \nabla_a (T^{ab}\xi_b) = (\nabla_a T^{ab})\xi_b + T^{ab}\nabla_a \xi_b = 0; \tag{8}$$

the first term vanishes by energy conservation, and the second by Killing's equation.

As defined in [1], a *piecewise Killing vector* is a piecewise smooth vector field ξ^a on M of the form (5), such that $\xi^a_\pm$ are Killing vectors on $M^\pm$. However, piecewise Killing vectors are *not* in general Killing vectors on M, since

$$\nabla_{(a}\xi_{b)} = \left[\xi_{(a}\right]\delta_{b)} \tag{9}$$

which is nonzero if ξ^a is discontinuous. Referring to (8), we see that

$$\nabla_a (T^{ab}\xi_b) = \left[T^{ab}\xi_a\right]\delta_b. \tag{10}$$

If Σ is a spacelike hypersurface, and if the Darmois junction conditions (continuity of both the intrinsic metric and the extrinsic curvature) are satisfied there, then the stress-energy tensor is continuous at Σ ($[T^{ab}] = 0$), and (10) reduces to

$$\nabla_a (T^{ab}\xi_b) = T^{ab}\left[\xi_a\right]\delta_b. \tag{11}$$

A natural condition on ξ^a is for its tangential components to agree on Σ, in which case

$$[\xi_a] = \Xi n_a \tag{12}$$

for some function Ξ defined on Σ. Given (12), and using (4), we will obtain a conserved quantity so long as

$$T^{ab} n_a n_b = 0 \tag{13}$$

on Σ. If Σ is spacelike, (13) asserts that the energy density at Σ seen by an observer orthogonal to Σ must vanish.

4 The Patchwork Divergence Theorem

In the language of differential forms, the *divergence* of a vector field X is defined in terms of the volume element ω as

$$\operatorname{div}(X)\,\omega := \pounds_X \omega \tag{14}$$

where £ denotes the Lie derivative. Using Stokes' Theorem in the form

$$\oint_{\partial W} \alpha = \int_W d\alpha \tag{15}$$

and the identity

$$£_X \alpha = d(i_X \alpha) + i_X(d\alpha), \tag{16}$$

where i denotes the interior product, one obtains the Divergence Theorem in the form

$$\int_W \operatorname{div}(X)\,\omega = \oint_S i_X \omega, \tag{17}$$

where $S = \partial W$. Any 1-form m orthogonal to S determines a unique volume element σ on S through the requirement that

$$m \wedge \sigma = \omega; \tag{18}$$

σ is compatible with the induced orientation on W precisely when m is outward pointing. Using the properties of the interior product, the Divergence Theorem becomes

$$\int_W \operatorname{div}(X)\,\omega = \oint_S m(X)\,\sigma. \tag{19}$$

So long as S is not null, the right-hand side of (19) is the same as the integral in (7) after obvious identifications.

For piecewise smooth tensors, we can apply (19) separately on $M^{\pm}$ and then add the results. Given a region $W = W^+ \cup W^-$ overlapping Σ, we let $S = \partial W$ and $S_0 = W \cap \Sigma$. We can extend m to outward-pointing 1-forms $m_{\pm}$ orthogonal to $S^{\pm} = \partial W^{\pm}$; on Σ we have $m_- = -m_+ =: m_0$. The *Patchwork Divergence Theorem* [4] then takes the form

$$\int_W \operatorname{div}(X)\,\omega = \oint_S m(X)\,\sigma - \int_{S^0} m_0([X])\,\sigma^0. \tag{20}$$

Our convention is that m_0 is the *1-form* pointing from M_- to M_+; which way the physically equivalent vector field points depends on whether Σ is spacelike, timelike, or null.

5 Shells of Matter

A simple model for matter falling into a black hole consists of spherical shells of massless matter. Remarkably, as shown originally by 't Hooft [2], this situation can be described by an exact solution of the Einstein field equation, at least in the context of piecewise smooth tensors. The special case of a single massless particle sitting at the horizon of a Schwarzschild black hole [5] remains the only explicitly known exact solution in general relativity that describes a test particle moving in the field of another object, and is in this sense the only known solution to the relativistic two-body problem. These models have been generalized to charged black holes [6], to colliding shells [2, 7, 8], and, more recently, to shells of negative energy [9].

In [1], we considered two Schwarzschild spacetimes with different masses joined along a null cylinder $\Sigma = \{u = \alpha\}$ representing a spherical shell of massless dust. The corresponding metric is

$$ds^2 = \begin{cases} -\frac{32m^3}{r} e^{-r/2m} du\, dv + r^2\, d\Omega^2 & (u \le \alpha) \\ -\frac{32m^3}{r} e^{-r/2M} dU\, dV + r^2\, d\Omega^2 & (u \ge \alpha) \end{cases} \tag{21}$$

where U and V are functions (only) of u and v, respectively, and

$$uv = -\left(\tfrac{r}{2m} - 1\right) e^{r/2m} \qquad (u \le \alpha), \tag{22}$$

$$UV = -\left(\tfrac{r}{2M} - 1\right) e^{r/2M} \qquad (u \ge \alpha). \tag{23}$$

Continuity of the metric requires that on Σ we have

$$\frac{\alpha}{m} = \frac{U(\alpha)}{MU'(\alpha)} =: \gamma, \tag{24}$$

which implies that

$$\frac{u\partial_u}{m} = \frac{U\partial_U}{M} \tag{25}$$

on Σ. The only nonzero component of the stress-energy tensor is

$$T_{uu} = \frac{\delta}{\gamma \pi r^2}\,(M - m), \tag{26}$$

and we have the piecewise Killing vector

$$\xi = (1 - \Theta)\,\frac{v\partial_v - u\partial_u}{4m} + \Theta\,\frac{V\partial_V - U\partial_U}{4M}. \tag{27}$$

Thus, $[\xi]$ is proportional to ∂_V, satisfying condition (12), while (13) is satisfied by virtue of the double-null form of the stress-energy tensor.

We therefore obtain an integral conservation law of the form (7). Since the support of T_{ab} is on Σ, we obtain a conserved quantity Q by evaluating this integral over any hypersurface S intersecting Σ only once. We choose

$$S = \begin{cases} \{t = \text{const}\} & (u \le \alpha) \\ \{T = \text{const}\} & (u \ge \alpha) \end{cases} \tag{28}$$

where t and T denote Schwarzschild time in the regions $u \leq \alpha$ and $u \geq \alpha$, respectively, and where the constants are chosen so that Σ is continuous. Putting this all together, we have

$$-Q = \int_S \left((1-\Theta)\, T^t{}_t + \Theta\, T^T{}_T\right) dS \tag{29}$$

which appears to involve the distributional product $\delta\Theta$. However, since

$$\frac{\partial u}{\partial r} = \frac{u}{4m}\frac{1}{1-r/2m}, \tag{30}$$

it turns out that

$$T^t{}_t = T^T{}_T = -\frac{\delta(r-r_0)}{4\pi r^2}(M-m) \tag{31}$$

where r_0 is the radius of the shell where S intersects Σ. Thus, there is no actual step function present in the integrand in (29). Finally, evaluating the integral leads to

$$Q = M - m \tag{32}$$

which shows that the energy of the shell is precisely the difference of the two Schwarzschild masses, as expected.

Thus, the results of [1] can be regarded as an application of the Patchwork Divergence Theorem in this setting.

6 Signature Change

"Spacetimes" combining both Lorentzian and Euclidean regions were proposed independently by George Ellis's group in the context of early universe cosmology [10, 11] and by our group in the context of quantum field theory in curved space [12–14]. Subsequent work by both groups addressed tensor distributions [15, 16] and the distributional field equations [17, 18], in the process realizing that conservation laws would take a different form at a change of signature [19], ultimately leading to the Patchwork Divergence Theorem [4].

The key point is that, even though the metric is clearly discontinuous at a change of signature,[1] the same need not be true for the volume element. The easiest way to see this surprising fact is to construct orthonormal frames on both $M^\pm$, and compare them along Σ. We assume that Σ is spacelike as seen from both sides, in which case an orthonormal frame on Σ can be (separately) extended to orthonormal frames on $M^\pm$ by adding the appropriate normal vector, which is spacelike in one case but timelike in the other. However, this discontinuity lies in the metric; the resulting normal vectors, taken together, form a continuous vector field. Since the volume

[1] We assume the metrics $g^\pm_{ab}$ are non-degenerate on Σ, the only other possibility.

element is just the (wedge) product of the (dual) frame elements, it, too, must be continuous. As mentioned above (and discussed in more detail in [3]), it is enough for the pullbacks of the metric from $M^{\pm}$ to Σ to agree in order for there to be a well-defined tangent space on M, a condition which is satisfied by this construction.

As discussed in [19], Israel's results [20] relating the stress-energy tensor to the intrinsic and extrinsic curvature of the boundary layer Σ must be modified in the presence of signature change. For example, the "energy" density on $\Sigma \subset M^{\pm}$ is now given by

$$\rho := G_{ab}n^a n^b = \frac{1}{2}\left((K^c{}_c)^2 - K_{ab}K^{ab} - \varepsilon R\right) \tag{33}$$

where K_{ab} is the extrinsic curvature of Σ, R is the scalar curvature of Σ, and $\varepsilon = n_a n^a = \pm 1$. Imposing Darmois boundary conditions, the curvatures themselves are continuous—but ε is not. Thus, rather than the Israel condition $[\rho] = 0$, we obtain

$$[\rho] = \left[G_{ab}n^a n^b\right] = -R. \tag{34}$$

Furthermore, we can independently recover the extrinsic curvature term from

$$\left[G^a{}_b n^b l_a\right] = (K^c{}_c)^2 - K_{ab}K^{ab} \tag{35}$$

where l_a is the dual vector satisfying $l_a n^a = 1$ (on both sides), as using l_a instead of n_a is equivalent to adding a factor of ε inside the square brackets. How to interpret "energy" inside a spacelike region is, of course, an open question.

7 Lorentzian Vector Calculus

A differential geometer regards vector calculus as "really" being about differential forms. For the last 20 years, in addition to my traditional research in relativity, I have had the pleasure of attempting to implement an approach to the teaching of vector calculus which can be described as "differential forms without differential forms" [21, 22]. Key to this description is the use of the infinitesimal displacement vector $d\vec{r}$, which is really a vector-valued differential form.[2]

When moving from Euclidean geometry to Riemannian geometry to Lorentzian geometry, vector calculus as expressed in terms of differential forms is virtually unchanged. However, the conversion to traditional vector language does depend on the signature. It's easy to rewrite the dot product in Lorentzian signature; relativists do this routinely when working with Lorentzian metrics. However, it's less obvious whether there is a cross product, or what it looks like. But nowhere is this dependence on signature more apparent than in the statement of the Divergence Theorem. What is an "outward-pointing" vector field, anyway?

---

[2]My recent textbook on general relativity [23] also uses this language.

In order to make the analogy to vector calculus more apparent, let's work in $2+1$-dimensional Minkowski space, with orthonormal basis vectors $\hat{x}$, $\hat{y}$, and $\hat{t}$. This basis satisfies

$$\hat{x}\cdot\hat{x} = 1 = \hat{y}\cdot\hat{y}, \qquad \hat{t}\cdot\hat{t} = -1, \tag{36}$$

with all cross terms vanishing. So consider a vector field

$$\vec{F} = F^x\,\hat{x} + F^y\,\hat{y} + F^t\,\hat{t}. \tag{37}$$

In Minkowski space, the divergence of $\vec{F}$ is

$$\vec{\nabla}\cdot\vec{F} = \frac{\partial F^x}{\partial x} + \frac{\partial F^y}{\partial y} + \frac{\partial F^t}{\partial t}; \tag{38}$$

there are no minus signs. As in the component-based proof of the (ordinary) Divergence Theorem, we can integrate the divergence over a rectangular box W. The first term yields

$$\iiint_W \frac{\partial F^x}{\partial x}\,dx\,dy\,dt = \iint \Delta F^x\,\,dy\,dt = \int_{S_x} \vec{F}\cdot\hat{n}\,dA \tag{39}$$

where S_x consists of the two faces of the box with outward-pointing normal vectors $\hat{n} = \pm\hat{x}$. A similar expression holds for the y-component, but the last equality fails for the t component, since the dot product has the wrong sign. To fix this problem, we must instead choose $\hat{n}$ to be the *inward*-pointing normal vector on S_t.

Why this asymmetry? Stokes' Theorem (15) is really about differential forms, and the 1-form physically equivalent to $\hat{n}$ (with components n_a) is outward-pointing on all of S. The "asymmetry" arises due to the metric when converting from 1-forms to vectors.

Why haven't we noticed this asymmetry in relativity? In practice, the Divergence Theorem is not applied to closed regions W, but rather to infinite "sandwiches", the region between two spacelike hypersurfaces. The integrals over the timelike sides of the box are replaced by falloff conditions at spatial infinity, leaving only the "S_t" contribution in the above argument. The relative sign difference on spacelike and timelike boundaries becomes an overall sign, which can be—and is—safely ignored.

The computations in this section are straightforward, but at first sight the conclusion may be uncomfortable to some readers. The Lorentzian Divergence Theorem does *not*, in general, involve the outward-pointing normal vector (but rather the outward-pointing 1-form). It is precisely this sort of confrontation between expectation and reality that leads students to an enhanced understanding of the underlying mathematics even in the traditional setting.

The investigation of the Divergence Theorem for regions in Minkowski space whose boundaries contain null pieces is left as an exercise for the reader.

8 Summary

We have briefly summarized two quite different bodies of work, the analysis of shells of matter in black-hole spacetimes, and of signature-changing spacetimes, emphasizing the common thread provided by the Patchwork Divergence Theorem, which also sheds new insight on topics in vector calculus. One never knows where the journey will lead.

Thank you, Paddy, for your contributions to my journey.

Acknowledgements It is a pleasure to thank the many collaborators who contributed to my work on piecewise smooth manifolds, including especially Chris Clarke, George Ellis, Charles Hellaby, Gerard 't Hooft, Corinne Manogue, and Robin Tucker.

References

1. Tevian Dray, T. Padmanabhan, Conserved quantities from piecewise Killing vectors. Gen. Rel. Grav. **21**, 741–745 (1989)
2. Tevian Dray, Gerard 't Hooft, The effect of spherical shells of matter on the Schwarzschild black hole. Commun. Math. Phys. **99**, 613–625 (1985)
3. C.J.S. Clarke, Tevian Dray, Junction conditions for null hypersurfaces. Class. Quantum Grav. **4**, 265–275 (1987)
4. Tevian Dray, Charles Hellaby, The patchwork divergence theorem. J. Math. Phys. **35**, 5922–5929 (1994)
5. Tevian Dray, Gerard 't Hooft, The gravitational shock wave of a massless particle. Nucl. Phys. B **253**, 173–188 (1985)
6. Tevian Dray, Pankaj Joshi, Glueing Reissner-Nordstrøm spacetimes together along charged shells of matter. Class. Quantum Grav. **7**, 41–49 (1990)
7. Tevian Dray, Gerard 't Hooft, The gravitational effect of colliding planar shells of matter. Class. Quantum Grav. **3**, 825–840 (1986)
8. Tevian Dray, Bouncing shells. Class. Quantum Grav. **7**, L131–L134 (1990)
9. Jeffrey S. Hazboun, Tevian Dray, The effect of negative-energy shells on the Schwarzschild black hole. Gen. Rel. Grav. **42**, 1457–1467 (2010)
10. G.F.R. Ellis, Covariant change of signature in classical relativity. Gen. Rel. Grav. **24**, 1047–1068 (1992)
11. G. Ellis, A. Sumeruk, D. Coule, C. Hellaby, Change of signature in classical relativity. Class. Quantum Grav. **9**, 1535–1554 (1992)
12. Tevian Dray, Corinne A. Manogue, Robin W. Tucker, Particle production from signature change. Gen. Rel. Grav. **23**, 967–971 (1991)
13. Tevian Dray, Corinne A. Manogue, Robin W. Tucker, The scalar field equation in the presence of signature change. Phys. Rev. D **48**, 2587–2590 (1993)
14. Tevian Dray, Corinne A. Manogue, Robin W. Tucker, Boundary conditions for the scalar field in the presence of signature change. Class. Quantum Grav. **12**, 2767–2777 (1995)
15. Tevian Dray, Tensor distributions in the presence of degenerate metrics. Int. J. Mod. Phys. D **6**, 717–740 (1997)
16. David Hartley, Robin W. Tucker, Philip Tuckey, Tevian Dray, Tensor distributions on signature-changing space-times. Gen. Rel. Grav. **32**, 491–504 (2000)
17. Tevian Dray, Einstein's equations in the presence of signature change. J. Math. Phys. **37**, 5627–5636 (1996)

18. Tevian Dray, George Ellis, Charles Hellaby, Corinne A. Manogue, Gravity and signature change. Gen. Rel. Grav. **29**, 591–597 (1997)
19. Charles Hellaby, Tevian Dray, Failure of standard conservation laws at a classical change of signature. Phys. Rev. D **49**, 5096–5104 (1994)
20. W. Israel, Singular hypersurfaces and thin shells in general relativity. Nuov. Cim. **44B**, 1–14 (1966)
21. Tevian Dray, Corinne A. Manogue, Bridging the Vector Calculus Gap (website) (2000–2016). http://math.oregonstate.edu/bridge. Cited 24 June 2016
22. Tevian Dray, Corinne A. Manogue, The Geometry of Vector Calculus (2009–2016). http://math.oregonstate.edu/BridgeBook. Cited 24 June 2016
23. Tevian Dray, *Differential Forms and the Geometry of General Relativity* (A K Peters/CRC Press, Boca Raton, 2014)

Units of the Nonlinear Universe

Sunu Engineer

Abstract The *late time* evolution of the gravitational clustering in an expanding universe is described based on the nonlinear scaling relations (NSR) which connect the nonlinear and linear two point correlation functions at different length scales. The existence of critical indices for the NSR suggests that the evolution may proceed towards a universal profile which does not change its shape at late times. If the evolution should lead to a halo profile which preserves the shape at late times, then the correlation function should grow as a^2 (in a $\Omega = 1$ universe) even at nonlinear scales. We prove that such *exact* solutions do not exist; however, there exists a class of solutions ("psuedo-linear profiles", PLPs for short) which evolve as a^2 to a good approximation related to halo profiles of isothermal spheres. They are also configurations of mass in which the nonlinear effects of gravitational clustering is a minimum and hence can act as building blocks of the nonlinear universe.

1 Introduction

The evolution of large number of particles under their mutual gravitational influence is a well-defined mathematical problem. In the presence of an expanding background universe characterised by an expansion factor $a(t)$, expansion tends to keep particles apart thereby exerting a civilising influence against newtonian attraction. The average density of particles contribute to the expansion of the background universe and the deviations from uniformity lead to clustering. Particles evaporating from a local overdense cluster cannot escape to "large distances" but necessarily will encounter other deep potential wells. Naively, one would expect the local overdense regions to eventually form gravitationally bound objects, with a hotter distribution of particles hovering uniformly all over. As the background expands, the velocity dispersion of the second component will keep decreasing and they will be captured by the deeper potential wells. Meanwhile, the clustered component will also evolve dynamically and participate in, e.g. mergers. If the background expansion and the initial conditions

S. Engineer (✉)
Ryussi Technologies Pvt Ltd., C 804, Pune IT Park, Bhau Patil Road, Pune 411 020, India
e-mail: sunu@ryussi.com

J.S. Bagla and S. Engineer (eds.), *Gravity and the Quantum*,
Fundamental Theories of Physics 187, DOI 10.1007/978-3-319-51700-1_9

have no length scale, then the clustering will continue in a hierarchical manner *ad infinitum*.

Can one make any general statements about the very late stage evolution of the clustering? For example, does the power spectrum at late times 'remember' the initial power spectrum or does it possess some universal characteristics which are reasonably independent of initial conditions? (This question is closely related to the issue of whether gravitational clustering leads to density profiles which are universal. [1–5]).

This work provides partial answers to these questions based on a simple paradigm. The key assumption is this: Let ratio between mean relative pair velocity $v(a, x)$ and the negative Hubble velocity $(-\dot{a}x)$ be denoted by $h(a, x)$ and let $\bar{\xi}(a, x)$ be the mean correlation function averaged over a sphere of radius x. *We shall assume that $h(a, x)$ depends on a and x only through $\bar{\xi}(a, x)$; that is, $h(a, x) = h[\bar{\xi}(a, x)]$.* This minimal assumption leads to some deep insights into the nature of late time clustering. Such an assumption was originally introduced — in a different form — by Hamilton [6]. The present form, as well as its theoretical implications were discussed in [7], and a theoretical model for the scaling was attempted by Padmanabhan [8]. It must be noted that simulations indicate a dependence of the relation $h(a, x) = h[\bar{\xi}(a, x)]$ on the initial spectrum and also on cosmological parameters [9–12].

2 General Features of Nonlinear Evolution

Consider the evolution of the system starting from a gaussian initial fluctuations with an initial power spectrum, $P_{in}(k)$. The fourier transform of the power spectrum defines the correlation function $\xi(a, x)$ where $a \propto t^{2/3}$ is the expansion factor in a universe with $\Omega = 1$. It is more convenient to work with the average correlation function inside a sphere of radius x, defined by

$$\bar{\xi}(a, x) \equiv \frac{3}{x^3} \int_0^x \xi(a, y)\, y^2 dy \tag{1}$$

This quantity is related to the power spectrum $P(a, k)$ by

$$\bar{\xi}(x, a) = \frac{3}{2\pi^2 x^3} \int_0^\infty \frac{dk}{k} P(a, k)\,[\sin(k\,x) - k\,x\,\cos(k\,x)] \tag{2}$$

with the inverse relation

$$P(a, k) = \frac{4\pi}{3k} \int_0^\infty dx\, x\, \bar{\xi}(a, x)\,[\sin(k\,x) - k\,x\,\cos(k\,x)] \tag{3}$$

In the linear regime we have $\bar{\xi}_L(a, x) \propto a^2 \bar{\xi}_{in}(a_i, x)$.

The conservation of pairs of particles gives an exact equation satisfied by the correlation function [13]:

$$\frac{\partial \xi}{\partial t} + \frac{1}{ax^2}\frac{\partial}{\partial x}[x^2(1+\xi)v] = 0 \tag{4}$$

where $v(a, x)$ denotes the mean relative velocity of pairs at separation x and epoch a. Using the mean correlation function $\bar{\xi}$ and a dimensionless pair velocity $h(a, x) \equiv -(v/\dot{a}x)$, Eq. (4) can be written as

$$\left(\frac{\partial}{\partial \ln a} - h\frac{\partial}{\partial \ln x}\right)\ (1+\bar{\xi}) = 3h(1+\bar{\xi}) \tag{5}$$

Given the key assumption, viz. that h depends on (a, x) only through $\bar{\xi}$, several results follow.

2.1 Formal Solution

Given that $h = h[\bar{\xi}(a, x)]$, one can easily integrate the Eq. (5) to find the general solution (see [7]). The characteristics of this Eq. (5) satisfy the condition

$$x^3(1+\bar{\xi}) = l^3 \tag{6}$$

where l is another length scale. When the evolution is linear at all the relevant scales, $\bar{\xi} \ll 1$ and $l \approx x$. As clustering develops, $\bar{\xi}$ increases and x becomes considerably smaller than l. The behaviour of clustering at some scale x is then determined by the original *linear* power spectrum at the scale l through the "flow of information" along the characteristics. This suggests that *the true correlation function* $\bar{\xi}(a, x)$ *can be expressed in terms of the linear correlation function* $\bar{\xi}_L(a, l)$ *evaluated at a different point*. This is indeed true and the general solution can be expressed as a nonlinear scaling relation (NSR, for short) between $\bar{\xi}_L(a, l)$ and $\bar{\xi}(a, x)$ with l and x related by Eq. (6). This solution can be expressed in terms of two functions $\mathscr{V}(z)$ and $\mathscr{U}(z)$ where $\mathscr{V}(z)$ is related to the function $h(z)$ by

$$\mathscr{V}(z) = \exp\left(\frac{2}{3}\int^z \frac{dz}{h(z)\,(1+z)}\right) \tag{7}$$

and $\mathscr{U}(z)$ is the inverse function of $\mathscr{V}(z)$. Then the solution to the Eq. (5) can be written in either of two equivalent forms as:

$$\bar{\xi}(a, x) = \mathscr{U}\left[\bar{\xi}_L(a, l)\right]; \qquad \bar{\xi}_L(a, l) = \mathscr{V}\left[\bar{\xi}(a, x)\right] \tag{8}$$

where $l^3 = x^3(1+\bar{\xi})$ [7]. Given the form of $h(\bar{\xi})$ this allows one to relate the non-linear correlation function to the linear one.

From general theoretical considerations (see [8]) it can be shown that $\mathscr{V}(z)$ has the form:

$$\mathscr{V}(z) = \begin{cases} 1 & (z \ll 1) \\ z^{1/3} & (1 \lesssim z \lesssim 200) \\ z^{2/3} & (200 \ll z) \end{cases} \tag{9}$$

In these three regions $h(z) \approx [(2z/3), 2, 1]$ respectively. We shall call these regimes, linear, intermediate and nonlinear respectively. More exact fitting functions to $\mathscr{V}(z)$ and $\mathscr{U}(z)$ have been suggested in literature. (see [6, 9, 11]). This paper uses the one given in [6]:

$$\mathscr{V}(z) = z\left(\frac{1+0.0158\,z^2+0.000115\,z^3}{1+0.926\,z^2-0.0743\,z^3+0.0156\,z^4}\right)^{1/3} \tag{10}$$

$$\mathscr{U}(z) = \frac{z+0.358\,z^3+0.0236\,z^6}{1+0.0134\,z^3+0.0020\,z^{9/2}} \tag{11}$$

Equations (8) and (10), (11) implicitly determine $\bar{\xi}(a,x)$ in terms of $\bar{\xi}_L(a,x)$.

2.2 *Critical Indices*

These NSRs already allow one to obtain some general conclusions regarding the evolution. A local index for rate of clustering is defined by

$$n_a(a,x) \equiv \frac{\partial \ln \bar{\xi}(a,x)}{\partial \ln a} \tag{12}$$

which measures how fast $\bar{\xi}(a,x)$ is growing. When $\bar{\xi}(a,x) \ll 1$, then $n_a = 2$ irrespective of the spatial variation of $\bar{\xi}(a,x)$ and the evolution preserves the shape of $\bar{\xi}(a,x)$. However, as clustering develops, the growth rate will depend on the spatial variation of $\bar{\xi}(a,x)$. Defining the effective spatial slope by

$$-[n_{eff}(a,x)+3] \equiv \frac{\partial \ln \bar{\xi}(a,x)}{\partial \ln x} \tag{13}$$

one can rewrite the Eq. (5) as

$$n_a = h\left(\frac{3}{\bar{\xi}(a,x)} - n_{eff}\right) \tag{14}$$

At any given scale of nonlinearity, decided by $\bar{\xi}(a,x)$, there exists a critical spatial slope n_c such that $n_a > 2$ for $n_{eff} < n_c$ (implying rate of growth is faster than predicted by linear theory) and $n_a < 2$ for $n_{eff} > n_c$ (with the rate of growth being slower than predicted by linear theory). The critical index is fixed by setting $n_a = 2$ in Eq. (14) at any instant. This feature will tend to "straighten out" correlation functions towards the critical slope. (It is assumed that $\bar{\xi}(a,x)$ has a slope that is decreasing with scale, which is true for any physically interesting case). From the fitting function it is easy to see that in the range $1 \lesssim \bar{\xi} \lesssim 200$, the critical index is $n_c \approx -1$ and for $200 \lesssim \bar{\xi}$, the critical index is $n_c \approx -2$ [14]. This clearly suggests that the local effect of evolution is to drive the correlation function to have a shape with $(1/x)$ behaviour at nonlinear regime and $(1/x^2)$ in the intermediate regime. Such a correlation function will have $n_a \approx 2$ and hence will grow at a rate close to a^2.

3 Correlation Functions, Density Profiles and Stable Clustering

A nonlinear scaling relation giving $\bar{\xi}(a,x)$ in terms of $\bar{\xi}_L(a,l)$ leads to the question: How does $\bar{\xi}(a,x)$ behave at highly nonlinear scales or, equivalently, at any given scale at large a?

To begin with, it is easy to see that $v = -\dot{a}x$ or $h = 1$ for sufficiently large $\bar{\xi}(a,x)$, that the evolution gets frozen in proper coordinates at highly nonlinear scales (assumed). Integrating Eq. (5) with $h = 1$, leads to $\bar{\xi}(a,x) = a^3F(ax)$; This is "stable clustering". There are two points which need to be emphasised about stable clustering:

(1) At present, there exists some evidence from simulations [10] that stable clustering does *not* occur in a $\Omega = 1$ model. In a *formal* sense, numerical simulations cannot disprove (or even prove, strictly speaking) the occurrence of stable clustering, because of the finite dynamic range of any simulation.

(2) Theoretically speaking, the "naturalness" of stable clustering is often overstated. The usual argument is based on the assumption that at very small scales — corresponding to high nonlinearities — the structures are "expected to be" frozen at the proper coordinates. However, this argument does not take into account the fact that mergers are not negligible at *any scale* in an $\Omega = 1$ universe. In fact, stable clustering is more likely to be valid in models with $\Omega < 1$ — a claim which seems to be again supported by simulations [10].

If stable clustering *is* valid, then the late time behaviour of $\bar{\xi}(a,x)$ *cannot* be independent of initial conditions. In other words the two requirements: (i) validity of stable clustering at highly nonlinear scales and (ii) the independence of late time behaviour from initial conditions, are mutually exclusive. This is most easily seen for initial power spectra which are scale-free. If $P_{in}(k) \propto k^n$ so that $\bar{\xi}_L(a,x) \propto a^2x^{-(n+3)}$, then it is easy to show that $\bar{\xi}(a,x)$ at small scales will vary as

$$\bar{\xi}(a,x) \propto a^{\frac{6}{n+5}} x^{-\frac{3(n+3)}{n+5}}; \qquad (\bar{\xi} \gg 200) \tag{15}$$

if stable clustering is true. Clearly, the power law index in the nonlinear regime "remembers" the initial index. The same result holds for more general initial conditions.

What does this result imply for the profiles of individual halos? To answer this question, start with the simple assumption that the density field $\rho(a, \mathbf{x})$ at late stages can be expressed as a superposition of several halos, each with some density profile; that is

$$\rho(a, \mathbf{x}) = \sum_i f(\mathbf{x} - \mathbf{x}_i, a) \tag{16}$$

where the i-th halo is centered at $\mathbf{x}_i$ and contributes an amount $f(\mathbf{x} - \mathbf{x}_i, a)$ at the location $\mathbf{x}_i$ (This equation is easily generalised to the situation in which there are halos with different properties, like core radius, mass etc. by summing over the number density of objects with particular properties). The power spectrum for the density contrast, $\delta(a, \mathbf{x}) = (\rho/\rho_b - 1)$, corresponding to the $\rho(a, \mathbf{x})$ in (16) can be expressed as

$$P(\mathbf{k}, a) \propto \left(a^3 |f(\mathbf{k}, a)|\right)^2 \left|\sum_i \exp -i\mathbf{k} \cdot \mathbf{x}_i(a)\right|^2 \tag{17}$$

$$\propto \left(a^3 |f(\mathbf{k}, a)|\right)^2 P_{\text{cent}}(\mathbf{k}, a) \tag{18}$$

where $P_{\text{cent}}(\mathbf{k}, a)$ denotes the power spectrum of the distribution of centers of the halos.

If stable clustering is valid, then the density profiles of halos are frozen in proper coordinates giving $f(\mathbf{x} - \mathbf{x}_i, a) = f(a\,(\mathbf{x} - \mathbf{x}_i))$; hence the fourier transform will have the form $f(\mathbf{k}, a) = f(\mathbf{k}/a)$. On the other hand, the power spectrum at scales which participate in stable clustering must satisfy $P(\mathbf{k}, a) = P(\mathbf{k}/a)$ (This is merely the requirement $\bar{\xi}(a, x) = a^3 F(ax)$ re-expressed in fourier space). From Eq. (18) it follows that $P_{\text{cent}}(\mathbf{k}, a) =$ constant independent of $\mathbf{k}$ and a at small length scales. This can arise in the special case of random distribution of centers or — more importantly — because the equation is essentially describing the interior of a single halo at sufficiently small scales. The halo profile can be related to the correlation function using (18). In particular, if the halo profile is a power law with $f \propto r^{-\epsilon}$, it follows that the $\bar{\xi}(a, x)$ scales as $x^{-\gamma}$ (see also [15, 16]) where

$$\gamma = 2\epsilon - 3 \tag{19}$$

Now if the *correlation function* scales as $[-3(n+3)/(n+5)]$, then the halo density profiles should be related to the initial power law index through the relation

$$\epsilon = \frac{3(n+4)}{n+5} \tag{20}$$

So clearly, the halos of highly virialised systems still "remember" the initial power spectrum.

Alternatively, one can try to "reason out" the profiles of the individual halos and use it to obtain the scaling relation for correlation functions. One of the favourite arguments used by cosmologists to obtain such a "reasonable" halo profile is based on spherical, scale invariant, collapse. It turns out that one can provide a series of arguments, based on spherical collapse, to show that — under certain circumstances — the *density profiles* at the nonlinear end scale as $[-3(n+3)/(n+5)]$. The simplest variant of this argument runs as follows: If an initial density profile is $r^{-\alpha}$, then scale invariant spherical collapse will lead to a profile which goes as $r^{-\beta}$ with $\beta = 3\alpha/(1+\alpha)$ (see eg., [8, 17] and references cited therein) [18].

Taking the initial slope as $\alpha = (n+3)/2$ will immediately give $\beta = 3(n+3)/(n+5)$. (Our definition of the stable clustering in the last section is based on the scaling of the correlation function and gave the slope of $[-3(n+3)/(n+5)]$ for the *correlation* function. The spherical collapse gives the same slope for *halo profiles*.) In this case, when the halos have the slope of $\varepsilon = 3(n+3)/(n+5)$, then the correlation function should have slope

$$\gamma = \frac{3(n+1)}{n+5} \tag{21}$$

Once again, the final state "remembers" the initial index n.

The argument for correlation function to scale as $[-3(n+3)/(n+5)]$ is based on the assumption of $h = 1$ asymptotically, which may not be true. The argument, leading to density profiles scaling as $[-3(n+3)/(n+5)]$, is based on scale invariant spherical collapse which does not do justice to nonradial motions.

There are two possibilities where independence from initial conditions can be achieved.

(i) A first example is when the slope of the correlation function is universal and obtain the slope of halos in the nonlinear limit using relation (19). Such an interesting situation can develop *if is assumed that h reaches a constant value asymptotically which is not necessarily unity*. In that case, the Eq. (5) can be integrated to get $\bar{\xi}(a, x) = a^{3h} F[a^h x]$ where h now denotes the constant asymptotic value of of the function. For an initial spectrum which is scale-free power law with index n, this result translates to

$$\bar{\xi}(a, x) \propto a^{\frac{2\gamma}{n+3}} x^{-\gamma} \tag{22}$$

where γ is given by

$$\gamma = \frac{3h(n+3)}{2+h(n+3)} \tag{23}$$

This obtains a γ which is independent of initial power law index provided h satisfies the condition $h(n+3) = c$, a constant. In this case, the nonlinear correlation function will be given by

$$\bar{\xi}(a, x) \propto a^{\frac{6c}{(2+c)(n+3)}} x^{-\frac{3c}{2+c}} \tag{24}$$

The halo index will be independent of n and will be given by

$$\varepsilon = 3\left(\frac{c+1}{c+2}\right) \tag{25}$$

This requires that the asymptotic value of h to *explicitly depend* on the initial conditions though the *spatial* dependence of $\bar{\xi}(a, x)$ does not. In other words, the velocity distribution — which is related to h — still "remembers" the initial conditions. This is indirectly reflected in the fact that the growth of $\bar{\xi}(a, x)$ — represented by $a^{6c/((2+c)(n+3))}$ — does depend on the index n.

As an example of the power of such a — seemingly simple — analysis note the following: Since $c \geq 0$, it follows that $\varepsilon > (3/2)$; invariant profiles with shallower indices (for e.g. with $\varepsilon = 1$) are not consistent with the evolution described above.

(ii) Second example: Make an ansatz for the halo profile and use it to determine the correlation function. It is assumed, based on small scale dynamics, that the density profiles of individual halos should resemble that of isothermal spheres, with $\varepsilon = 2$, irrespective of initial conditions. Converting this halo profile to correlation function in the *nonlinear* regime is straightforward and is based on Eq. (19): If $\varepsilon = 2$, then $\gamma = 2\varepsilon - 3 = 1$ at small scales; that is $\bar{\xi}(a, x) \propto x^{-1}$ at the nonlinear regime. Note that this corresponds to the critical index at the nonlinear end, $n_{eff} = n_c = -2$ for which the growth rate is a^2 — same as in linear theory. (This is, however, possible for initial power law spectra, only if $\varepsilon = 1$, i.e. $h(n + 3) = 1$ at very nonlinear scales. Testing the conjecture that $h(n + 3)$ is a constant is probably a little easier than looking for invariant profiles in the simulations but the results are still uncertain).

The corresponding analysis for the intermediate regime, with $1 \lesssim \bar{\xi}(a, x) \lesssim 200$, is more involved. This is clearly seen in Eq. (18) which shows that the power spectrum (and hence the correlation function) depends *both* on the fourier transform of the halo profiles as well as the power spectrum of the distribution of halo centres. In general, both quantities will evolve with time and we cannot ignore the effect of $P_{\rm cent}(k, a)$ and relate $P(k, a)$ to $f(k, a)$. The density profile around a *local maxima* will scale approximately as $\rho \propto \xi$ while the density profile around a *randomly* chosen point will scale as $\rho \propto \xi^{1/2}$. (The relation $\gamma = 2\varepsilon - 3$ expresses the latter scaling of $\xi \propto \rho^2$). There is, however, reason to believe that the intermediate regime (with $1 \lesssim \bar{\xi} \lesssim 200$) is dominated by the collapse of high peaks [8]. If so the correlation function and the density profile will have the same slope in the intermediate regime with $\bar{\xi}(a, x) \propto (1/x^2)$. Remarkably enough, this corresponds to the critical index $n_{eff} = n_c = -1$ for the intermediate regime for which the growth is proportional to a^2 Thus if: (i) the individual halos are isothermal spheres with $(1/x^2)$ profile and (ii) if $\xi \propto \rho$ in the intermediate regime and $\xi \propto \rho^2$ in the nonlinear regime, it will result in a correlation function *which grows as a^2 at all scales.* Such an evolution, of course, preserves the shape and is a good candidate for the late stage evolution of the clustering.

While the above arguments are suggestive, they are far from conclusive. It is, however, clear from the above analysis and it is not easy to provide *unique* theoretical reasoning regarding the shapes of the halos. The situation gets more complicated if the fact that all halos will not all have the same mass, core radius etc. and we have to modify our equations by integrating over the abundance of halos with a given value of mass, core radius etc. is included. This brings in more ambiguities and depending on the assumptions for each of these components (e.g, abundance for halos of a particular mass could be based on Press-Schecter or Peaks formalism), and the final results have no real significance. It is, therefore easier to address the question based on the evolution equation for the correlation function rather than from "physical" arguments for density profiles.

4 Self-similar Evolution

The above discussion motivates a search for correlation functions of the form $\bar{\xi}(a, x) = a^2 L(x)$, starting with a more general question: Does Eq. (5) possess self-similar solutions of the form

$$\bar{\xi}(a, x) = a^{\beta} F\left(\frac{x}{a^{\alpha}}\right) = a^{\beta} F(q) \tag{26}$$

where $q \equiv x a^{-\alpha}$?. Defining $Q = \ln q = X - \alpha A$ (Defn: $A = \ln a$ and $X = \ln x$)and changing independent variables to from (A, X) to (A, Q) we can transform our Eq. (5) to the form:

$$\left(\frac{\partial \bar{\xi}}{\partial A}\right)_Q - (h + \alpha)\left(\frac{\partial \bar{\xi}}{\partial Q}\right)_A = 3(1 + \bar{\xi})\, h(\bar{\xi}) \tag{27}$$

Using the relations $(\partial\bar{\xi}/\partial A)_Q = \beta\bar{\xi}$, $(\partial\bar{\xi}/\partial Q)_A = (\bar{\xi}/F)(dF/dQ)$ we can rewrite this equation as

$$\frac{\beta\bar{\xi} - 3(1 + \bar{\xi})h(\bar{\xi})}{\left[\alpha + h(\bar{\xi})\right]\bar{\xi}} = \frac{1}{F}\frac{dF}{dQ} \equiv K(Q) \tag{28}$$

The right hand side of this equation depends only on Q and hence will vanish if differentiated with respect to A at constant Q. Imposing this condition on the left hand side and noticing that it is a function of $\bar{\xi}(a, x)$ leads to

$$\left(\frac{\partial \bar{\xi}}{\partial A}\right)_Q \frac{d}{d\bar{\xi}}(\text{Left Hand Side}) = 0 \tag{29}$$

To satisfy this condition either (i) $(\partial\bar{\xi}/\partial A)_Q = \beta\bar{\xi} = 0$ implying $\beta = 0$ or (ii) the left hand side must be a constant.

Considering the two cases separately.

(i) The simpler case corresponds to $\beta = 0$ which implies that $\bar{\xi}(a, x) = F(Q)$. Setting $\beta = 0$ in Eq. (28) gives

$$\left(\frac{d\bar{\xi}}{dQ}\right) = -\frac{3(1+\bar{\xi})h(\bar{\xi})}{\left[\alpha + h(\bar{\xi})\right]} \tag{30}$$

which can be integrated in a straightforward manner to give a relation between $q = \exp Q$ and $\bar{\xi}$:

$$\begin{aligned} q &= q_0(1+\bar{\xi})^{-1/3} \exp\left(-\frac{\alpha}{3}\int \frac{d\bar{\xi}}{(1+\bar{\xi})h(\bar{\xi})}\right) \\ &= q_0(1+\bar{\xi})^{-1/3}\mathscr{V}(\bar{\xi})^{-\alpha/2} \end{aligned}$$

Given the form of $h[\bar{\xi}(a,x)]$, this equation can be in principle inverted to determine $\bar{\xi}$ as a function of $q = xa^{-\alpha}$.

To understand when such a solution will exist, the limit of $\bar{\xi} \ll 1$ is to be looked at. In this limit, when linear theory is valid, $h \approx (2/3)\bar{\xi}$ (see [13]). Using this in Eq. (31) the solution becomes $\ln \bar{\xi} = -(2/\alpha)\ln q$ or

$$\bar{\xi} \propto q^{-\frac{2}{\alpha}} \propto x^{-\frac{2}{\alpha}}a^2 \propto a^2 x^{-(n+3)} \tag{31}$$

with the definition $\alpha \equiv 2/(n+3)$. This clearly shows that the solution is valid, *if and only if* the linear correlation function is a scale-free power law. In this case, of course, it is well known that solutions of the type $\bar{\xi}(a,x) = F(q)$ with $q = xa^{-\frac{2}{(n+3)}}$ exists. (Equation (31) gives the explicit form of the function $F(q)$.) This result shows that this is the *only* possibility. It should be noted that, even though there is no explicit length scale in the problem, the function $\bar{\xi}(q)$ — determined by the above equation — does exhibit different behaviour at different scales of nonlinearity. Roughly speaking, the three regimes in Eq. (9) translates into nonlinear density contrasts in the ranges $\delta < 1$, $1 < \delta < 200$ and $\delta > 200$ and the function $\bar{\xi}(q)$ has different characteristics in these three regimes. This shows that gravity can intrinsically select out a density contrast of $\delta \approx 200$ which, of course, is well-known from the study of spherical tophat collapse.

(ii) Considering the second possibility, *viz.* that the left hand side of Eq. (28) is a constant. If the constant is denoted by μ, then $F = F_0\, q^{\mu}$ and

$$\beta\,\bar{\xi} - 3\,(1+\bar{\xi})\,h(\bar{\xi}) = \mu\,\alpha\,\bar{\xi} + \mu\,h\,\bar{\xi} \tag{32}$$

which can be rearranged to give

$$h = \frac{(\beta - \alpha\mu)\bar{\xi}}{3 + (\mu+3)\bar{\xi}} \tag{33}$$

This relation shows that solutions of the form $\bar{\xi}(a,x) = a^{\beta}\,F(x/a^{\alpha})$ with $\beta \neq 0$ is possible only if $h[\bar{\xi}(a,x)]$ has a *very specific* form given by (33). In this form, h is a monotonically increasing function of $\bar{\xi}(a,x)$. There is, however, firm theoretical and numerical evidence [6, 8] to suggest that h increases with $\bar{\xi}(a,x)$ first, reaches a maximum and then decreases. In other words, the h for actual gravitational clustering is *not* in the form suggested by Eq. (33). *Leading to the conclusion that solutions of the form in Eq.* (26) *with* $\beta \neq 0$ *cannot exist in gravitational clustering.*

By a similar analysis, a stronger result can be proved: There are no solutions of the form $\bar{\xi}(a, x) = \bar{\xi}(x/F(a))$ except when $F(a) \propto a^{\alpha}$. So self-similar evolution in clustering is a very special situation.

This result, incidentally, has an important implication. It shows that power-law initial conditions are very special in gravitational clustering and may not represent generic behaviour. This is because, for power laws, there is a strong constraint that the correlations etc. can only depend on $q = xa^{-2/(n+3)}$. For more realistic — non-power law — initial conditions the shape can be distorted in a generic way during evolution.

All the discussion so far was related to finding *exact* scaling solutions. It is however possible to find *approximate* scaling solutions which are of practical interest. Note that normally constants like α, β, μ etc. are expected to be of order unity while $\bar{\xi}(a, x)$ can take arbitrarily large values. If $\bar{\xi}(a, x) \gg 1$ then Eq. (33) shows that h is approximately a constant with $h = (\beta - \alpha\mu)/(\mu + 3)$. In this case

$$\bar{\xi}(a, x) = a^{\beta} F(q) \propto a^{\beta} q^{\mu} \propto a^{(\beta-\alpha\mu)} x^{\mu} \propto a^{h(\mu+3)} x^{\mu} \tag{34}$$

which has the form $\bar{\xi}(a, x) = a^{3h} F(a^h x)$ which was obtained earlier by directly integrating Eq. (5) with constant h.

5 Units of the Nonlinear Universe

Having reached the conclusion that *exact* solutions of the form $\bar{\xi}(a, x) = a^2 G(x)$ are not possible, the logical question is: Are there such *approximate* solutions? And if so, how do they look like? Such profiles — called "pseudo-linear profiles"— that evolve very close to the the above form can indeed be shown to exist. In order to obtain such a solution and check its validity, it is better to use the results of Sect. 2.1 and proceed as follows:

An approximate solution of the form $\bar{\xi}(a, x) = a^2 G(x)$ to Eq. (5) is what is sought. Since the linear correlation function $\bar{\xi}_L(a, x)$ does grow as a^2 at fixed x, continuity demands that $\bar{\xi}(a, x) = \bar{\xi}_L(a, x)$ for all a and x. (This can be proved more formally as follows: Let $\bar{\xi} = a^2 G(x)$ and $\bar{\xi}_L = a^2 G_1(x)$ for some range $x_1 < x < x_2$. Consider a sufficiently early epoch $a = a_i$ at which all the scales in the range (x_1, x_2) are described by linear theory so that $\bar{\xi}(a_i, x) = \bar{\xi}_L(a_i, x)$. It follows that $G_1(x) = G(x)$ for all $x_1 < x < x_2$. Hence $\bar{\xi}(a, x) = \bar{\xi}_L(a, x)$ for all a in $x_1 < x < x_2$. By choosing a_i sufficiently small, any range (x_1, x_2) can be covered. So $\bar{\xi} = \bar{\xi}_L$ for any arbitrary range. *QED*). Since there exists a formal relation (8) between nonlinear and linear correlation functions, the form of $G(x)$ can be determined.

To do this the form of the linear correlation function is inverted $\bar{\xi}_L(a, l) = a^2 G(l)$ and write $l = G^{-1}(a^{-2}\bar{\xi}_L) \equiv F(a^{-2}\bar{\xi}_L)$ where F is the inverse function of G. The linear correlation function $\bar{\xi}_L(a, l)$ at scale l can be expressed as $\mathscr{V}[\bar{\xi}(a, x)]$ in terms of the true correlation function $\bar{\xi}(a, x)$ at scale x where

$$l = x(1 + \bar{\xi}(a, x))^{1/3} \tag{35}$$

Therefore

$$l = F\left[\frac{\bar{\xi}_L(a, l)}{a^2}\right] = F\left[\frac{\mathscr{V}[\bar{\xi}(x, a)]}{a^2}\right] \tag{36}$$

But x can be expressed as $x = F[\bar{\xi}_L(a, x)/a^2]$; Substituting this in (35) leads to

$$l = F\left[\frac{\bar{\xi}_L(a, x)}{a^2}\right][1 + \bar{\xi}]^{1/3} \tag{37}$$

From our assumption $\bar{\xi}_L(a, x) = \bar{\xi}(a, x)$; therefore this relation can also be written as

$$l = F\left[\frac{\bar{\xi}(a, x)}{a^2}\right](1 + \bar{\xi})^{1/3} \tag{38}$$

Equating the expressions for l in (36) and (38) we get an implicit functional equation for F:

$$F\left[\frac{\mathscr{V}[\bar{\xi}]}{a^2}\right] = F\left[\frac{\bar{\xi}}{a^2}\right](1 + \bar{\xi})^{1/3} \tag{39}$$

which can be rewritten as

$$\frac{F\left[\mathscr{V}(\bar{\xi})/a^2\right]}{F\left[\bar{\xi}/a^2\right]} = (1 + \bar{\xi})^{1/3} \tag{40}$$

This equation should be satisfied by the function F if the relation $\bar{\xi}(a, x) = \bar{\xi}_L(a, x)$ is to be maintained.

To see what this implies, note that the left hand side should not vary with a at fixed $\bar{\xi}$. This is possible only if F is a power law:

$$F(\bar{\xi}) = A\bar{\xi}^m \tag{41}$$

which in turn constrains the form of $\mathscr{V}(\bar{\xi})$ to be

$$\mathscr{V}(\bar{\xi}) = \bar{\xi}\,(1 + \bar{\xi})^{1/3m} \tag{42}$$

Knowing the particular form for $\mathscr{V}$ the corresponding $h(\bar{\xi})$ can be computed from the relation

$$\frac{d\ln\mathscr{V}}{d\bar{\xi}} = \frac{2}{3}\,\frac{1}{(1 + \bar{\xi})\,h(\bar{\xi})} \tag{43}$$

The $\mathscr{V}(\bar{\xi})$ considered in Eq. (42) gives

$$h = \frac{2\bar{\xi}}{3 + (3 + 1/m)\bar{\xi}} \tag{44}$$

which is the same result obtained by putting $\beta = 2$, $\alpha = 0$ in Eq. (26). This vindicates the earlier result that *exact* solutions of the form $\bar{\xi}(a, x) = \bar{\xi}_L(a, x) = a^2\, G(x)$ are *not* possible because the correct $\mathscr{V}(\bar{\xi})$ and $h(\bar{\xi})$ do not have the forms in Eqs. (42) and (44) respectively. But, as in the last section, there are approximate solutions.

From Eq. (42) for $\bar{\xi} \gg 1$, we have

$$\mathscr{V}(\bar{\xi}) = \bar{\xi}^{(1+1/3m)}; \qquad F(\bar{\xi}) \propto \bar{\xi}^m; \qquad G(\bar{\xi}) \propto \bar{\xi}^{1/m} \tag{45}$$

This can be rewritten as

$$\mathscr{V}(\bar{\xi}) = \bar{\xi}^\nu; \qquad F(\bar{\xi}) \propto \bar{\xi}^{1/3(\nu-1)}; \qquad G(\bar{\xi}) \propto \bar{\xi}^{3/(\nu-1)} \tag{46}$$

In other words if $\mathscr{V}(\bar{\xi})$ can be approximated as $\bar{\xi}^\nu$, we have an approximate solution of the form

$$\bar{\xi}(a, x) = a^2\, G(x) = a^2\, x^{3(\nu-1)} \tag{47}$$

Since the $\mathscr{V}$ in Eq. (10) is well approximated by the power laws in (9) so that

$$\mathscr{V}(\bar{\xi}) \propto \bar{\xi}^{1/3} \quad (1 \lesssim \bar{\xi} \lesssim 200) \tag{48}$$

$$\propto \bar{\xi}^{2/3} \quad (200 \lesssim \bar{\xi}) \tag{49}$$

giving $\nu = 1/3$ in the intermediate regime and $\nu = 2/3$ in the nonlinear regime. It follows from (46) that the approximate solution should have the form

$$F(\bar{\xi}) \propto \frac{1}{\sqrt{\bar{\xi}}} \quad (1 \lesssim \bar{\xi} \lesssim 200) \tag{50}$$

$$\propto \frac{1}{\bar{\xi}} \quad (200 \lesssim \bar{\xi}) \tag{51}$$

This gives the approximate form of a pseudo-linear profile which will grow as a^2 at all scales.

There is another way of looking at this solution which is probably more physical and throws light on the scalings of pseudo-linear profiles. Recalling that, in the study of finite gravitating systems made of point particles and interacting via newtonian gravity, isothermal spheres play an important role. They can be shown to be the local maxima of entropy (see [19]) and hence dynamical evolution drives the system towards an $(1/x^2)$ profile. Since one expects similar considerations to hold at small scales, during the late stages of evolution of the universe, it may be that isothermal spheres with $(1/x^2)$ profile may still play a role in the late stages of evolution of clustering in an expanding background. However, while converting the profile to correlation, all the issues noted in Sect. 2 is to be taken into account. In the intermediate regime, dominated by scale invariant radial collapse [8], the density will scale as the correlation function and $\bar{\xi} \propto (1/x^2)$. On the other hand, in the nonlinear end, $\gamma = 2\varepsilon - 3$ (see Eq. (19)) which gives $\bar{\xi} \propto (1/x)$ for $\varepsilon = 2$. Thus, if isothermal spheres are the generic contributors, then the correlation function will vary as $(1/x)$ and nonlinear scales, steepening to $(1/x^2)$ at intermediate scales. Further, since isothermal spheres are local maxima of entropy, a configuration like this should remain undistorted for a long duration. This argument suggests that a $\bar{\xi}$ which goes as $(1/x)$ at small scales and $(1/x^2)$ at intermediate scales is likely to be a candidate for pseudo-linear profile. This is indeed the case.

To go from the scalings in two limits given by Eq. (50) to an actual profile, an interpolating fitting function can be computed. By making the fitting function sufficiently complicated, the pseudo-linear profile can be made more exact. Choosing the simplest interpolation between the two limits as the ansatz:

$$F(z) = \frac{A}{\sqrt{z}\,(\sqrt{z} + B)} \tag{52}$$

where A and B are constants. Using the original definition $l = F[\bar{\xi}_L/a^2]$ and the condition that $\bar{\xi} = \bar{\xi}_L$, we get

$$\frac{A}{\sqrt{\bar{\xi}/a^2}\,(\sqrt{\bar{\xi}/a^2} + B)} = l \tag{53}$$

This relation implicitly fixes our pseudo-linear profile. Solving for $\bar{\xi}$, we get

$$\bar{\xi}(a,x) = \left(\frac{Ba}{2}\left(\sqrt{1 + \frac{L}{x}} - 1\right)\right)^2 \tag{54}$$

with $L = 4A/B^2$. Since this profile is not a pure power law, this will satisfy the Eq. (40) only approximately. Choose B such that the relation

$$F\left(\frac{\mathscr{V}(\bar{\xi})}{a^2}\right) = F\left(\frac{\bar{\xi}}{a^2}\right)(1 + \bar{\xi})^{1/3} \tag{55}$$

is satisfied to greatest accuracy at $a = 1$.

This approximate profile works reasonably well. Figures 1 and 2 show this result. In Fig. 1 is plotted the ratio $F(\mathscr{V}(\bar{\xi})/a^2)/F(\bar{\xi}/a^2)$ on the x-axis and the function $(1 + \bar{\xi})^{1/3}$ on the y-axis. If the function in (54) satisfies Eq. (40) exactly, a 45-degree line in the figure should be obtained shown by a dashed line. The fact that the curve is pretty close to this line shows that the ansatz in (54) satisfies Eq. (40) fairly well. The optimum value of B chosen for this figure is $B = 38.6$. When a is varied from 1 to 10^3, the percentage of error between the 45-degree line and the curve is less than

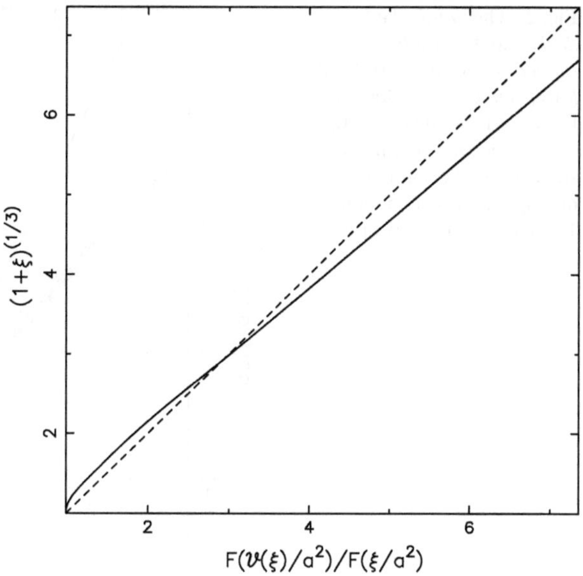

Fig. 1 The approximate solution to the functional equation determining the pseudo-linear profile is plotted. See text for discussion

about 20 percent in the worst case. It is clear that the profile in (54) satisfies Eq. (55) quite well for a dynamic range of 10^6 in a^2.

Figure 2 shows this result more directly. The pseudo linear profile is evolved from $a^2 = 1$ to $a^2 \approx 1000$ using the NSR, and plot $[\bar{\xi}(a,x)/a^2]$ against x. The dot-dashed, dashed and two solid curves (upper one for $a^2 = 100$ and lower one for $a^2 = 900$) are for $a^2 = 1, 9, 100$ and 900 respectively. The overlap of the curves show that the profile does grow approximately as a^2. Also shown are lines of slope -1 (dotted) and -2 (solid); clearly $\bar{\xi} \propto x^{-1}$ for small x and $\bar{\xi} \propto x^{-2}$ in the intermediate regime.

In Eq. (54) the simplest kind of ansatz combining the two regimes is chosen and only two parameters A and B are used. It is quite possible to come up with more elaborate fitting functions which will solve the functional equation far more accurately but it has not been done for two reasons: (i) Firstly, the fitting functions in Eq. (9) for $\mathcal{V}(z)$ itself is approximate and is probably accurate only at 10–20 percent level. There has also been repeated claims in literature that these functions have weaker dependence on n which have been ignored for simplicity in this paper. (ii) Secondly, one must remember that only those $\bar{\xi}$ which correspond to positive definite $P(k)$ are physically meaningful. This happens to be the case for this choice (which can be verified by explicit numerical integration with a cutoff at large x) but this may not be true for arbitrarily complicated fitting functions. Incidentally, another simple fitting function for the pseudo-linear profile is

$$\bar{\xi}(a,x) = a^2 \frac{A'}{(x/L')[(x/L')+1]} \tag{56}$$

with $A' = B^2$ and $L' = L/4$.

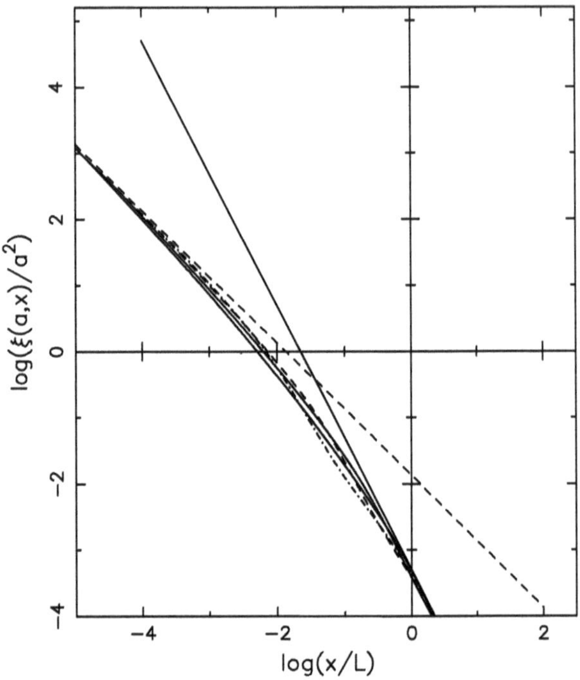

Fig. 2 The *dot-dashed*, *dashed* and *two solid curves* (*upper one* for $a^2 = 100$ and *lower one* for $a^2 = 900$) are for $a^2 = 1, 9, 100$ and 900. The *dotted straight line* is of slope-1 and the solid one is of slope-2 showing both the $1/x$ and $1/x^2$ regions of the profile

If a more accurate fitting is required, one can obtain it more directly from Eq. (14). Setting $n_a = 2$ in that equation predicts the instantaneous spatial slope of $\bar{\xi}(a, x)$ to be

$$\frac{\partial \ln \bar{\xi}(a, x)}{\partial \ln x} = \frac{2}{h[\bar{\xi}(a, x)]} - 3(1 + \frac{1}{\bar{\xi}(a, x)}) \tag{57}$$

which can be integrated to give

$$\ln \frac{x}{L} = \int_{\bar{\xi}[L]}^{\bar{\xi}[x]} \frac{h d\bar{\xi}}{\bar{\xi}(2 - 3h) - 3h} \tag{58}$$

at $a = 1$ with L being an arbitrary integration constant. Numerical integration of this equation will give a profile which is varies as $(1/x)$ at small scales and goes over to $(1/x^2)$ and then to $(1/x^3)$, $(1/x^4)\ldots$ etc. with an asymptotic logarithmic dependence. In the regime $\bar{\xi}(a, x) > 1$, this will give results reasonably close to the earlier fitting function.

It should be noted that Eq. (40) reduces to an identity for any F, in the limit $\bar{\xi} \to 0$ since, in this limit $\mathcal{V}(z) \approx z$. This shows that the pseudo-linear profiles at large scales can be modified into any other form (essentially determined by the input linear power spectrum) without affecting any of the conclusions.

A different way of thinking about pseudolinear profiles which may be useful is as follows:

In studying the evolution of the density contrast $\delta(a, \mathbf{x})$, it is conventional to expand in in term of the plane wave modes as

$$\delta(a, \mathbf{x}) = \sum_{\mathbf{k}} \delta(a, \mathbf{k}) \exp(i\mathbf{k} \cdot \mathbf{x}) \tag{59}$$

In that case, the *exact* equation governing the evolution of $\delta(a, \mathbf{k})$ is given by [13]

$$\frac{d^2\delta_{\mathbf{k}}}{da^2} + \frac{3}{2a}\frac{d\delta_{\mathbf{k}}}{da} - \frac{3}{2a^2}\delta_{\mathbf{k}} = \mathscr{A} \tag{60}$$

where $\mathscr{A}$ denotes the terms responsible for the nonlinear coupling between different modes. The expansion in Eq. (59) is, of course, motivated by the fact that in the linear regime $\mathscr{A}$ can be ignored and each of the modes evolve independently. For the same reason, this expansion is not of much value in the highly nonlinear regime.

This prompts one to ask the question: Is it possible to choose some other set of basis functions $Q(\alpha, \mathbf{x})$, instead of exp $i\mathbf{k} \cdot \mathbf{x}$, and expand $\delta(a, \mathbf{x})$ in the form

$$\delta(a, \mathbf{x}) = \sum_{\alpha} \delta_\alpha(a)\, Q(\alpha, \mathbf{x}) \tag{61}$$

so that the nonlinear effects are minimised? Here α stands for a set of parameters describing the basis functions. This question is extremely difficult to answer, partly because it is ill-posed. To make any progress, we have to first give meaning to the concept of "minimising the effects of nonlinearity". One possible approach we would like to suggest is the following: It is known that when $\delta(a, \mathbf{x}) \ll 1$,then $\delta(a, \mathbf{x}) \propto a\,F(\mathbf{x})$ for *any* arbitrary $F(\mathbf{x})$; that is all power spectra grow as a^2 in the linear regime. In the intermediate and nonlinear regimes, no such general statement can be made. But it is conceivable that there exists certain *special* power spectra for which $P(\mathbf{k}, a)$ grows (at least approximately) as a^2 even in the nonlinear regime. For such a spectrum, the left hand side of (60) vanishes (approximately); hence the right hand side should also vanish. *Clearly, such power spectra are affected least by nonlinear effects.* Instead of looking for such a special $P(k, a)$ equivalently look for a particular form of $\bar{\xi}(a, x)$ which evolves as closely to the linear theory as possible. Such correlation functions and corresponding power spectra (which are the pseudo-linear profiles) must be capable of capturing most of the essence of nonlinear dynamics. *In this sense, the pseudo-linear profiles can be thought of as the basic building blocks of the nonlinear universe.* The fact that the correlation function is closely related to isothermal spheres, indicates a connection between local gravitational dynamics and large scale gravitational clustering.

Acknowledgements This article in its more detailed form was published in Astrophysical Journal (493) 1998.

References

1. A. Dekel, G. Ishai, A.A. Dutton, V.A. Maccio, arXiv:1610.0091
2. A.D. Ludlow, R.E. Angulo, arXiv:1610.04620
3. A.D. Ludlow, S. Bose, R.E. Angulo, L. Wang, W.A. Hellwing, J.F. Navarro, S. Cole, C.S. Frenk, MNRAS **460**, 1214 (2016)
4. A.D. Ludlow, J.F. Navarro, M. Boylan-Kolchin, P.E. Bett, R.E. Angule, M. Li, S.D.M. White, C.S. Frenk, V. Springel, MNRAS **432**, 1103 (2013)
5. J.F. Navarro, C.S. Frenk, S.D.M. White, ApJ **462**, 563 (1996)
6. A.J.S. Hamilton, P. Kumar, E. Lu, A. Mathews, ApJ **374**, L1 (1991)
7. R. Nityananda, T. Padmanabhan, MNRAS **271**, 976 (1994)
8. T. Padmanbhan, MNRAS **278**, L29 (1996)
9. H.J. Mo, B. Jain, S.M. White, MNRAS **276**, L25 (1995)
10. T. Padmanabhan, R. Cen, J.P. Ostriker, F.J. Summers, ApJ **466**, 604 (1996)
11. J.A. Peacock, S.J. Dodds, MNRAS **267**, 1020 (1994)
12. J.A. Peacock, S.J. Dodds, MNRAS **280**, L19 (1996)
13. P.J.E. Peebles, *Large Scale Structure of the Universe* (Princeton University Press, Princeton, 1980)
14. J.S. Bagla, T. Padmanabhan, MNRAS (1997)
15. J. McClelland, J. Silk, ApJ **217**, 331 (1977)
16. R. Sheth, B. Jain, MNRAS **285**, 231 (1997)
17. T. Padmanabhan, *Cosmology and Astrophysics - Through Problems* (Cambridge University Press, Cambridge, 1996), p. 410
18. Y. Lithwick, N. Dalal, arXiv:1010.3723
19. T. Padmanabhan, Phys. Rep. **188**, 285 (1990)

Self-similarity and Criticality in Gravitational Collapse

Pankaj S. Joshi

Abstract We examine here the question whether self-similarity always implies criticality in gravitational collapse. To this end, we consider one-parameter families of self-similar collapse models, namely the collapse of radiation shells as given by the Vaidya metric and inhomogeneous dust collapse. The regions of parameter space which give rise to the collapse final states as black holes or naked singularity are evaluated to conclude that a non-zero measure set of values give rise to each of these final states. This scenario suggests the possibility that while a zero measure of solutions may admit a visible singularity in collapse calculations such as the numerical simulation of one-parameter family of massless scalar fields, some other collapse models need not behave in the same manner even when the collapse is self-similar. In other words, self-similarity does not imply a zero measure of naked singularity solutions necessarily for all collapse models in general. We also make some remarks here on perfect fluid collapse, and implications on the genericity and stability aspects of collapse outcomes are discussed.

One of the most important issues in black hole physics, at the foundation of the basic theory as well as astrophysical applications, is that of genericity of gravitational collapse final states in terms of black holes and naked singularities. The only way to obtain a proper idea and perception on this issue is to examine gravitational collapse models in general relativity in detail, and examine the genericity of collapse outcomes with or without an event horizon. The point here is, general relativity implies that the space-time singularities must form as collapse final states, however, we have no idea as of today whether event horizons of gravity also must form so as to cover these singularities.

In this connection and to examine this question, a wide variety of gravitational collapse of various matter fields has been investigated in recent years, such as dust, perfect fluids, massless scalar fields and others, within the framework of Einstein gravity (see e.g. Joshi and Malafarina [13] for a review and references therein).

P.S. Joshi (✉)
Tata Institute of Fundamental Research, Homi Bhabha Road, Colaba,
Mumbai 400 005, India
e-mail: psj@tifr.res.in

J.S. Bagla and S. Engineer (eds.), *Gravity and the Quantum*,
Fundamental Theories of Physics 187, DOI 10.1007/978-3-319-51700-1_10

The general conclusion arrived at from these studies is, depending on the initial conditions for collapse in terms of the density and pressure profiles from which the collapse initiates, and in terms of the allowed evolutions of the Einstein equations, both black holes as well as naked singularities develop as collapse final states. From such a perspective, one of the main questions asked for gravitational collapse final states in terms of black holes and naked singularities has been, even if a naked singularity formed in collapse, if the 'measure' (to be suitably defined) of such solutions is vanishing within the space of all solutions, then effectively such naked singularities can be ignored.

Within such a context of genericity and stability of naked singularities, considerable discussion has taken place in recent years on the self-similar gravitational collapse of massless scalar fields. In particular, Christodoulou [6, 7] worked out collapse of self-similar massless scalar fields to show examples of naked singularity formation. Also, in a numerical simulation of the same matter model that assumed discrete self-similarity, Choptuik [5] examined gravitational collapse of a one-parameter family of massless scalar fields. It turned out in the numerical study of such a family that the naked singularity solution was a critical point, rest of these being either black holes or a dispersing away scalar field where no space-time singularity formed. Thus it was suggested that the naked singularity forming in these models is only a critical solution at the boundary of the black holes and dispersal, and is therefore a 'zero measure' set which is not generic or stable.

An important question then is: Does self-similarity imply criticality always in gravitational collapse, implying that the naked singularity solutions are only point-like, forming always a zero-measure set? while this is conjectured to be so, based on the example of massless scalar field collapse, we need to investigate if this is true in general, implying that the naked singularity whenever it formed will be non-generic in this sense.

Our purpose here is to show, by means of explicit examples of self-similar collapse models, namely the Vaidya radiation shells collapse and inhomogeneous dust collapse, that this is not the case in these one-parameter collapse models. We point out that the black hole and naked singularity regions in the corresponding initial data space have non-zero measures and each occur generically. Few remarks are made on the extension of these results to perfect fluid case.

As stated above, the collapse of a massless scalar field has been examined both analytically and numerically, within the context of self-similarity. This is a model problem of a single massless scalar field, minimally coupled to gravitational field, providing a useful scenario to investigate the nonlinearity effects of general relativity. On the analytic side, the results of Christodoulou show that when the scalar field is sufficiently weak, there is a regular solution or a global evolution for arbitrary long time, of the coupled Einstein and scalar field equations. There is a convergence towards the origin, and after a bounce the field would disperse to infinity. For strong enough field, the collapse would result in a space-time singularity, which for self-similar collapse could also be a naked singularity. The claim as discussed above, however, is that the initial conditions resulting into a naked singularity would be always a set of measure zero in the given parameter space. In that case, the naked singularity formation could be an unstable phenomenon. We can say that this is an approach that helps study the cosmic censorship problem [17] as an evolution problem in the sense of examining the global Cauchy development of a self-gravitating system outside an event horizon. As for numerical studies, Choptuik and others considered a family of scalar field solutions where a parameter p characterized the strength of the scalar field. The numerical calculations showed that for black hole formation, there is a critical limit $p \rightarrow p_*$ and the mass of the resulting black holes satisfy a power law $M_{bh} = (p - p_*)^\gamma$, where the critical exponent γ has value of about 0.37. It was then conjectured that such a critical behavior may be a general property of gravitational collapse, because similar behavior was found by Abrahams and Evans [1] for imploding axisymmetric gravitational waves, and also by Evans and Coleman [3] who considered collapse of radiation with an equation of state $p = /3$, assuming self-similarity for solutions. It is not clear if the critical parameter γ will have the same value for all forms of matter chosen and further investigation is needed to determine this. As the parameter p moves between the weak and strong range, very small mass black holes can form, in which case one can probe and receive messages from arbitrarily near to singularity which is a naked singularity like behavior.

In order to examine the self-similar collapse in some more detail, we note that a self-similar space-time is characterized by the existence of a homothetic Killing vector field in the space-time [4]. For example, a spherically symmetric space-time is self-similar if it admits a radial area coordinate r and an orthogonal time coordinate t such that for the metric components g_{tt} and g_{rr} we have

$$g_{tt}(ct, cr) = g_{tt}(t; r)\, ; \, g_{rr}(ct, cr) = g_{rr}(t, r) \tag{1}$$

for all $c > 0$. In such a case, along the integral curves of the Killing vector field all points are similar. A matching of a self-similar interior to a Schwarzschild exterior space-time can be done as smoothly as desired [16]. If the matching is sufficiently far from the center, the central region evolves in a self-similar manner without being affected by the matching.

We now consider first the collapse of inflowing radiation. In this case, a thick shell of radiation collapses at the center of symmetry in an otherwise empty universe asymptotically flat far away. This can be a relevant or interesting collapse model because a massive star, in the very final stages of collapse, would be largely radiation dominated. When one may regard a naked singularity forming in a gravitational collapse as an interesting situation which may have physical implications? Firstly, the singularity should be visible at least for a finite period of time to far away observers. If only a single null geodesic escaped the singularity, it would be only instantaneous exposure by means of a single wave front. To yield observable consequences, a family of future directed non-spacelike geodesics should terminate at the naked singularity in past. Also the singularity must be gravitationally strong so the space-time does not admit any continuous extension through the same, making it unavoidable. Finally, the form of matter should be reasonable in that it must satisfy a suitable energy condition ensuring the positivity of energy, collapsing from an initial spacelike hypersurface with a well-defined non-singular initial data.

Such an imploding radiation shell is described by the Vaidya space-time, given in (v, r, θ, ϕ) coordinates as,

$$ds^2 = -\left(1 - \frac{2m(v)}{r}\right) dv^2 + 2dvdr + r^2 d\Omega^2, \tag{2}$$

where $d\Omega^2 = d\theta^2 + \sin^2\theta\, d\phi^2$. The radiation collapses at the origin $v = 0, r = 0$. The null coordinate v denotes the advanced time and $m(v)$ is an arbitrary, non-negative increasing function. The stress-energy tensor for the radial radiation flux is,

$$T_{ij} = \rho k_i k_j = \frac{1}{4\pi r^2}\frac{dm}{dv} k_i k_j, \tag{3}$$

with

$$k_i = -\delta^v{}_i, \quad k_i k^i = 0, \tag{4}$$

which is radially inflowing radiation along the world lines $v =$ const.. Here $\frac{dm}{dv} \geq 0$ implies the weak energy condition is satisfied. Thus, a radially injected radiation by a distant source flows into an initially flat and empty region, focused into a central singularity of growing mass.

The source then turns off at a finite time T and the field then settles to the Schwarzschild space-time. In this case, the Minkowski space-time for $v < 0, m(v) = 0$ is joined to a Schwarzschild space-time for $v > T$ with mass $m_0 = m(T)$ through the Vaidya metric as given above.

Now assuming the mass function $m(v)$ to be a linear function results into the space-time being self-similar,

$$2m(v) = \lambda v, \tag{5}$$

with $\lambda > 0$ (see e.g. Joshi [11] and references therein). This is the Vaidya–Papapetrou space-time describing radiation collapse. Specifically we have,

$$m(v) = 0 \quad \text{for } v < 0, \;\; 2m(v) = \lambda v \quad \text{for } 0 < v < T, \;\; m(v) = m_0 \quad \text{for } v > T. \tag{6}$$

Then the mass for the final Schwarzschild black hole is M and the causal structure of the space-time would be determined by the values of the constants M, T, and λ. In this case, the Vaidya space-time region admits a homothetic Killing vector

$$\xi = v\left(\frac{\partial}{\partial v}\right) + r\left(\frac{\partial}{\partial r}\right),$$

which is given by the Lie derivative,

$$L_\xi g_{ij} = \xi_{i;j} + \xi_{j;i} = 2g_{ij}. \tag{7}$$

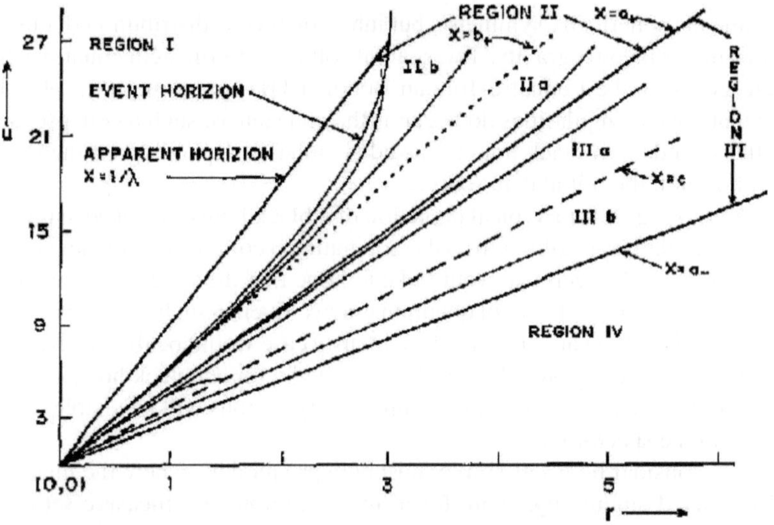

Fig. 1 The structure of escaping non-spacelike geodesics from the naked singularity in different regions in the Vaidya space-time. The event horizon and apparent horizon forming in the space-time are shown, which fail to cover the singularity unlike the black hole case when the mass parameter λ satisfies $\lambda < 1/8$. While some of the trajectories escape from the singularity, they fall back again into the horizon, however, other families escape to infinity, making the singularity visible to faraway observers (from Dwivedi and Joshi [9])

The central singularity $v = 0, r = 0$ can then be studied in detail and all families of future directed non-spacelike geodesics which terminate at the singularity $v = 0, r = 0$ in the past can be determined, thus producing a naked singularity of the space-time (see Fig. 1). Working out these families gives a good idea of the nature and structure of the naked singularity, and this also allows us to explicitly evaluate the curvature growth along these families in the limit of approach to the singularity. It turns out that this is a powerfully strong curvature naked singularity which is non-removable in that the space-time admits no extension through the same.

In summary, what we find is a naked singularity results when the collapse is sufficiently slow, and the radial as well as non-radial non-spacelike geodesics that emerge from the singularity can be studied in detail. Specifically, when $\lambda < 1/8$ then a naked singularity forms, and for the values of the mass function $\lambda \geq 1/8$ a black hole forms where the event horizon covers the singularity. It follows that, for a non-zero measure set of the parameter space values, the space-time singularity is naked in this self-similar collapse space-time.

We now consider another class of collapse models, namely the inhomogeneous dust collapse. Gravitational collapse of a sufficiently massive homogeneous dust ball leads to the formation of a black hole, as was indicated by the work of Oppenheimer and Snyder [15] and Datt [8]. The event horizon here covers the infinite density singularity forming at the center of the cloud. But realistic stars have inhomogeneous distribution of matter, with density somewhat peaked at the center. Thus it is useful to examine the spherically symmetric but inhomogeneous distribution of dust which collapses under its own gravity. The general solutions to Einstein equations for this case are given by the Lemaitre-Tolman-Bondi (LTB) spacetimes, [2, 14, 18] It is known that naked singularities do occur as the end state of such a collapse, and that the collapse ends in a black hole or naked singularity depending on the parameter values chosen in the initial data space.

We refer to Fig. 2, for a typical depiction of a black hole or a naked singularity as collapse final state, as resulting from the gravitational collapse of a matter cloud, such as a massive star. As such one would like to hope that the space-time singularity is resolved by some future theory of quantum gravity. Even in such a case, the difference between the black hole and the naked singularity case would be that, while the fuzzy resolved singularity region will be fully contained within a black hole, in the naked singularity case such ultra-strong quantum gravity regions will be visible to faraway observers in the space-time.

We now consider here self-similar dust collapse models, where the final collapse state is a naked singularity, again for a finite and non-zero measure set of values in the one-parameter family of initial data that determines the collapse endstates in terms of black hole or a naked singularity.

In comoving coordinates (t, r, θ, ϕ) the LTB metric for a spherically symmetric collapse of an inhomogeneous dust cloud is given as,

$$ds^2 = -dt^2 + \frac{R'^2}{1+f}dr^2 + R^2(t,r)(d\theta^2 + \sin^2\theta d\phi^2) \quad (8)$$

The energy momentum tensor is $T^{ij} = \varepsilon\delta^i_t\delta^j_t$, where ε is the energy density. Then the Einstein equations imply that

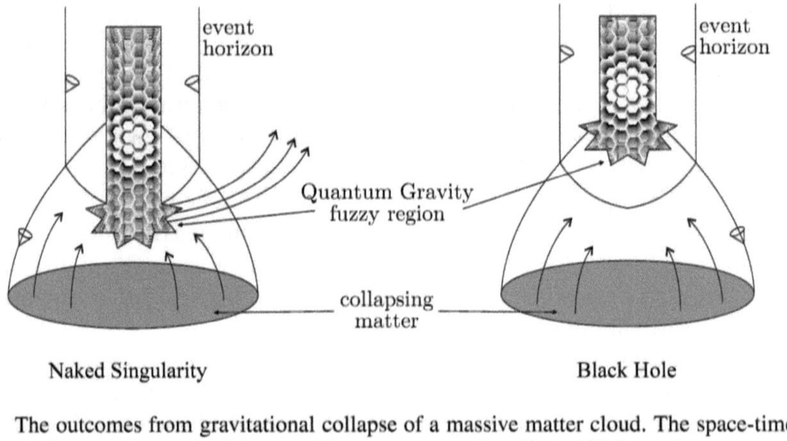

Fig. 2 The outcomes from gravitational collapse of a massive matter cloud. The space-time singularity is shown to be resolved by possible quantum gravity effects. While such a strong gravity region is fully contained within the horizon when a black hole forms, in the naked singularity case this could be visible to faraway observers in the space-time

$$\varepsilon = \varepsilon(r,t) = \frac{F'}{R^2 R'}, \qquad \dot{R}^2 = \frac{F}{R} + f \tag{9}$$

Here $8\pi G/c^4 = 1$, and the dot and prime denote partial derivatives with respect to t and r respectively. The function $F(r)$ is a free function from the integration of the Einstein equations and is interpreted physically as the total mass of the collapsing cloud within a comoving radius r. We take $F(r) \geq 0$. Here $R(t_1, r_1)$ denotes the physical radius of a shell of collapsing matter at a coordinate radius r_1 and on time slice $t = t_1$, and F and f are arbitrary functions of r. We consider here the class of solutions $f(r) \equiv 0$, which are called marginally bound LTB models. Here $R = 0$ denotes a physical singularity with the spherical shells of matter collapsing to zero radius and density blowing to infinity. The time $t = t_0(r)$ corresponds to $R = 0$, where the area of a shell of matter at a constant value of comoving coordinate r vanishes. This singularity curve $t = t_0(r)$ is the time when the matter shells meet the physical singularity. For a finite cloud of dust, there is a cut off at $r = r_b$, where the metric is matched smoothly with a Schwarzschild exterior.

Using the freedom of scaling to relabel the dust shells given by $r = const.$ on a given $t = const.$ epoch, we choose the scaling at $t = 0$ as given by $R(r, 0) = r$. Then the $\dot{R}$ equation (with $f = 0$) can be integrated to get,

$$R^{3/2}(r,t) = r^{3/2} - \frac{3}{2}\sqrt{F(r)}t, \tag{10}$$

and the energy equation becomes

$$\varepsilon(r,t) = \frac{4/3}{\left(t - \frac{2}{3}\frac{G(r)}{H(r)}\right)\left(t - \frac{2}{3}\frac{G'(r)}{H'(r)}\right)} \tag{11}$$

where $G(r) = r^{3/2}$, $G'(r) = (3/2)r^{1/2}$, and $H(r) = \sqrt{F(r)}$.

We now write $F(r) \equiv r\lambda(r)$ with $\lambda(r)$ being a finite positive function [10]. When $\lambda = const.$, then this gives us the class of all self-similar dust collapse models. The density at the center behaves with time as $\varepsilon(0,t) = 4/3t^2$. The central density becomes singular at $t = 0$, and the singularity is seen to arise from the evolution of dust collapse which had a finite density distribution in the past on an earlier non-singular initial epoch.

To check if the singularity is naked or within a black hole, we need to examine if future directed null geodesics could come out of the singularity at $t = 0, r = 0$. The equations of outgoing radial null geodesics, with k as affine parameter, are written as,

$$\frac{dK^t}{dk} + \dot{R}'K^r K^t = 0, \quad \frac{dt}{dr} = \frac{K^t}{K^r} = R', \tag{12}$$

where $K^t = dt/dk$ and $K^r = dr/dk$ are tangents to the outgoing radial null geodesics. The partial derivatives R' and $\dot{R}'$ which occur here can be worked out and are suitably written as,

$$R' = \eta P - \left[\frac{1-\eta}{\sqrt{\lambda}} + \eta\frac{t}{r}\right]\dot{R} \tag{13}$$

$$\dot{R}' = \frac{\lambda}{2rP^2}\left[\frac{1-\eta}{\sqrt{\lambda}} + \eta\frac{t}{r}\right] \tag{14}$$

where we introduce

$$R(r,t) = rP(r,t), \ \ \eta = \eta(r) = \frac{rF'}{F} \tag{15}$$

The functions $\eta(r)$ and $P(r)$ are introduced because they have a well-defined limit in the approach to the singularity.

If the outgoing null geodesics terminate in the past with a definite tangent at the singularity, which is the case of a naked singularity, then using above equations and l′ Hospital rule we get

$$X_0 = \lim_{t\to 0, r\to 0}\frac{t}{r} = \lim_{t\to 0, r\to 0}\frac{dt}{dr} = \lim_{t=0, r=0} R' \tag{16}$$

where $X = t/r$ is a new variable. The positive function $P(r,t) = P(X,r)$ is then given by

$$X - \frac{2}{3\sqrt{\lambda}} = -\frac{2P^{3/2}}{3\sqrt{\lambda}} \tag{17}$$

where we define $Q = Q(X) = P(X,0)$. If the future directed null geodesics come out of the singularity at $t = 0, r = 0$, which meet the singularity in the past with a definite tangent $X = X_0$ which is given above, then it follows from above that X_0 satisfies $X_0 < 2/3\sqrt{\lambda_0}$ We note that as $r \to 0$, $\eta \to 1$, and so $\lim \dot{R} = -\sqrt{\lambda_0/Q}$. Using these results in the expression for R' the condition above for outgoing null geodesics simplified to,

$$V(X_0) = 0; \ \ V(X) \equiv Q + X\sqrt{\frac{\lambda_0}{Q}} - X \tag{18}$$

To be the past end-point of outgoing null geodesics, at least one real positive value of X_0 must satisfy the above equation. In general, if the equation $V(X_0) = 0$ has a real positive root, the singularity would be naked, and in the case otherwise the collapse evolution leads to a black hole. For the self-similar collapse, typically the singularity when visible turns out to be globally naked.

Using the above equations, the condition $V(X_0) = 0$ can then be written as,

$$Y^3\left(Y - \frac{2}{3}\right) - \alpha(Y-2)^3 = 0 \tag{19}$$

where we put $Y = \sqrt{\lambda_0} X_0$ and $\alpha = \lambda_0^{3/2}/12$, with $F(r)$ and hence λ_0 being positive. Using standard results then it can be shown that this quartic equation has real positive roots if and only if $\alpha > \alpha_1$ or $\alpha < \alpha_2$, where

$$\alpha_1 = \frac{26}{3} + 5\sqrt{3} \approx 17.3269 \tag{20}$$

and,

$$\alpha_2 = \frac{26}{3} - 5\sqrt{3} \approx 6.4126 \times 10^{-3} \tag{21}$$

Here the larger range of α values are ruled out, because in that case the trajectories are no longer in the space-time (see e.g. Joshi and Singh [12]). It thus follows that a naked singularity arises if and only if $\alpha < \alpha_1$, or equivalently $\lambda_0 < 0.1809$. Whenever the limiting value λ_0 does not satisfy this constraint then the gravitational collapse must end in a black hole.

The physical interpretation for the quantity λ_0 can be easily seen by noting that it is in fact a combination of the central density ρ_0 and its derivative, at an initial epoch from where the collapse starts. Thus the interesting result that we have is, the occurrence of either a black hole or naked singularity is governed by the conditions on the initial central density and the initial density gradient at the center. Basically we have the situation that while homogeneous collapse leads to a black hole, a suitable amount of inhomogeneity, as represented by a non-zero density gradient, leads to a naked singularity.

What we have shown is, for the self-similar dust collapse again, the naked singularity occurs for a non-zero measure values of initial data in the given one-parameter space. In this sense, the self-similarity does not imply criticality or a single point naked singularity solution for this class of collapse models.

Finally, we note that similar results are expected to hold for a self-similar collapse of perfect fluids. The perfect fluid collapse with the barotropic equation of state $p = k\rho$, where k is a constant in the range 0 to 1 was analyzed in detail analytically [10], and the occurrence of a naked singularity was again reduced to the existence of real positive roots of a quartic equation, similar to what we discussed in the above case of dust collapse. To conclude, we may state that while self-similarity may imply criticality in some classes of collapse models such as the massless scalar fields, this need not be necessarily so for all physically reasonable classes of gravitational collapses.

Acknowledgements I like to thank the editors for their kind invite to contribute an article to this volume. I like to acknowledge here my many discussions with Paddy over past many years on various scientific as well as other themes. I was always very impressed with his scholarship, attention to details, and especially the methodology and approach to physics, which helped prepare many good students in the area of gravitation and cosmology. I take this opportunity to wish him many more years of fruitful research.

References

1. A.M. Abrahams, C.R. Evans, Critical behavior and scaling in vacuum axisymmetric gravitational collapse (1971); Phys. Rev. Lett. **70**, 2980 (1993)
2. H. Bondi, Spherically symmetric models in general relativity. Mon. Not. Astron. Soc. **107**, 400 (1948)
3. C.R. Evans, J.S. Coleman, Critical phenomena and self-similarity in the gravitational collapse of radiation fluid. Phys. Rev. Lett. **72**, 1782 (1994)
4. M.E. Cahill, A H. Taub, Spherically symmetric similarity solutions of the Einstein field equations for a perfect fluid. Commun. Math. Phys. **21**, 1 (1971)
5. M.W. Choptuik, Universality and scaling in gravitational collapse of a massless scalar field. Phys. Rev. Lett. **70**, 9 (1993)
6. D. Christodoulou, Examples of naked singularity formation in the gravitational collapse of a scalar field (1994); Ann. Maths. **140**, 607 (1994)
7. D. Christodoulou, The instability of naked singularities in the gravitational collapse of a scalar field (1999); Ann. Math. **149**, 183 (1999)
8. B. Datt, Uber eine Klasse von Lösungen der Gravitationsgleichungen der Relativität. Zeitschrift fur Physik **108**, 314 (1938)
9. I.H. Dwivedi, P.S. Joshi, Cosmic censorship violation in non-self-similar Tolman–Bondi models. Class. Quantum Grav. **9**, L69 (1992)
10. P.S. Joshi, I.H. Dwivedi, The structure of naked singularity in self-similar gravitational collapse. Commun. Math. Phys. **146**, 333 (1992)
11. P.S. Joshi, *Global Aspects in Gravitation and Cosmology* (Clarendon Press, OUP, Oxford, 1993)
12. P.S. Joshi, T.P. Singh, Phase transition in gravitational collapse of inhomogeneous dust. Phys. Rev. D **51**, 6778 (1995)
13. P.S. Joshi, D. Malafarina, Recent developments in gravitational collapse and spacetime singularities. Int. J. Mod. Phys. D **20**, 2641 (2011)
14. G. Lemaître, L'univers en expansion. Ann. Soc. Sci. Bruxelles I A **53**, 51 (1933)
15. J.R. Oppenheimer, H. Snyder, On continued gravitational contraction. Phys. Rev. **56**, 455 (1939)
16. A. Ori, T. Piran, Naked singularities and other features of self-similar general relativistic gravitational collapse. Phys. Rev. D **42**, 1068 (1990)
17. R. Penrose, Gravitational collapse: the role of general relativity. Nuovo Cimento Riv. Ser. **1**, 252 (1969)
18. R.C. Tolman, Effect of inhomogeneity on cosmological models. Proc. Natl. Acad. Sci. USA **20**, 169 (1934)

Notes on Semiclassical Weyl Gravity

Claus Kiefer and Branislav Nikolić

Abstract In any quantum theory of gravity, it is of the utmost importance to recover the limit of quantum theory in an external spacetime. In quantum geometrodynamics (quantization of general relativity in the Schrödinger picture), this leads in particular to the recovery of a semiclassical (WKB) time which governs the dynamics of non-gravitational fields in spacetime. Here, we first review this procedure with special emphasis on conceptual issues. We then turn to an alternative theory - Weyl (conformal) gravity, which is defined by a Lagrangian that is proportional to the square of the Weyl tensor. We present the canonical quantization of this theory and develop its semiclassical approximation. We discuss in particular the extent to which a semiclassical time can be recovered and contrast it with the situation in quantum geometrodynamics.

1 Notes on Semiclassical Einstein Gravity

Among Paddy's many interests in physics was always the deep desire to understand the relationship between classical and quantum gravity. In his paper "Notes on semiclassical gravity", written together with T.P. Singh in 1989, they write [29]:

> In the course of our investigation we came across a variety of methods for defining classical and semiclassical limits, apparently different, and all of which were possibly applicable to a quantum gravity. It then became necessary to compare these methods and to settle, once and for all, the relation of semiclassical gravity to quantum gravity.

The understanding of semiclassical gravity was also a long-term project by one of us, and we thus devote our festschrift contribution to this topic. More precisely, the topic is the recovery of quantum (field) theory in an external spacetime from

C. Kiefer (✉) · B. Nikolić
Institute for Theoretical Physics, University of Cologne,
Zülpicher Strasse 77, 50937 Köln, Germany
e-mail: kiefer@thp.uni-koeln.de

B. Nikolić
e-mail: nikolic@thp.uni-koeln.de

J.S. Bagla and S. Engineer (eds.), *Gravity and the Quantum*,
Fundamental Theories of Physics 187, DOI 10.1007/978-3-319-51700-1_11

canonical quantum gravity. We briefly review the standard procedure of obtaining this limit from the Wheeler–DeWitt equation of quantum general relativity (quantum geometrodynamics). In the next two sections, we then apply these methods to a different theory called Weyl gravity or conformal gravity. This is the main concern of our paper.

What is our motivation for doing so? Weyl gravity is a theory without intrinsic scale. It seems therefore not appropriate, by itself, to replace general relativity (GR) in the empirically tested macroscopic limit. It may, however, be appropriate to serve as a model for a fundamental conformally invariant theory, being of relevance in quantum gravity and its application to the very early universe. Many researchers entertain, in fact, the idea that Nature does not contain any scale at the most fundamental level; see, for example, [2, 31]. In these following sections, we shall outline the procedure for classical and quantum canonical Weyl gravity and perform the semiclassical limit. We shall point out in detail the similarities to and the differences from quantum GR. We shall see, in particular, that while a semiclassical time can be recovered, this time is of a different nature than the one recovered from quantum GR.

In canonical GR, the configuration variable is the three-metric $h_{ab}(\mathbf{x})$, while the canonical momentum $p^{cd}(\mathbf{x})$ is a linear function of the extrinsic curvature (second fundamental form) $K_{cd}(\mathbf{x})$. In the Dirac way of quantization, these variables are heuristically transformed into operators acting on wave functionals,

$$\hat{h}_{ab}(\mathbf{x})\Psi\left[h_{ab}(\mathbf{x})\right] = h_{ab}(\mathbf{x}) \cdot \Psi\left[h_{ab}(\mathbf{x})\right] , \tag{1}$$

$$\hat{p}^{cd}(\mathbf{x})\Psi\left[h_{ab}(\mathbf{x})\right] = \frac{\hbar}{\mathrm{i}}\frac{\delta}{\delta h_{cd}(\mathbf{x})}\Psi\left[h_{ab}(\mathbf{x})\right] . \tag{2}$$

The wave functionals are defined on the configuration space of all three-metrics (plus non-gravitational fields, which are not indicated here). In GR, one has four local constraints, the Hamiltonian constraint and the three diffeomorphism (momentum) constraints. They are implemented in the quantum theory as restrictions on physically allowed wave functionals [15],

$$\hat{\mathscr{H}}_{\perp}^{\mathrm{g}}\Psi := \left(-16\pi G\hbar^2 G_{abcd}\frac{\delta^2}{\delta h_{ab}\delta h_{cd}} - \frac{\sqrt{h}}{16\pi G}\left({}^{(3)}R - 2\Lambda\right)\right)\Psi = 0\,, \tag{3}$$

$$\hat{\mathscr{H}}_{a}^{\mathrm{g}}\Psi := -2D_b h_{ac}\frac{\hbar}{\mathrm{i}}\frac{\delta\Psi}{\delta h_{bc}} = 0\,. \tag{4}$$

The quantum Hamiltonian constraint (3) is called the Wheeler–DeWitt equation. The momentum constraints (4) guarantee that the wave functional remains unchanged (apart possibly from a phase) under a three-dimensional coordinate transformation. In the presence of non-gravitational fields, we need the corresponding contributions $\hat{\mathscr{H}}_{\perp}^{\mathrm{m}}$ for (3) and $\hat{\mathscr{H}}_{a}^{\mathrm{m}}$ for (4), see below.

The coefficients G_{abcd} in front of the kinetic term in (3) are the components of the DeWitt metric, which is the metric on configuration space. One of its important properties is its indefinite nature. Using instead of h_{ab} its scale part $\sqrt{h}$ (where h

denotes its determinant) and the conformal part $\bar{h}_{ab} = h^{-1/3}h_{ab}$, the Wheeler–DeWitt equation reads

$$\left(6\pi G\hbar^2\sqrt{h}\frac{\delta^2}{\delta(\sqrt{h})^2} - \frac{16\pi G\hbar^2}{\sqrt{h}}\bar{h}_{ac}\bar{h}_{bd}\frac{\delta^2}{\delta\bar{h}_{ab}\delta\bar{h}_{cd}} - \frac{\sqrt{h}}{16\pi G}({}^{(3)}R - 2\Lambda)\right)\Psi[\sqrt{h}, \bar{h}_{ab}] = 0\,. \quad (5)$$

One recognizes that the kinetic term connected with the local scale has a different sign. For this reason, the Wheeler–DeWitt equation is of a (local) hyperbolic nature and $\sqrt{h}$ can be interpreted as a local measure of *intrinsic time*. We shall introduce the scale and conformal parts of the metric also for the Weyl theory below, but as we shall see, the scale part (and thus the intrinsic time part) will be absent in the Weyl version of the Wheeler–DeWitt equation.

An important step in understanding the semiclassical limit for the above quantum equations is the WKB approximation [29]. One starts with the ansatz

$$\Psi[h_{ab}] = C[h_{ab}]\exp\left(\frac{\mathrm{i}}{\hbar}S[h_{ab}]\right) \quad (6)$$

and assumes that $C[h_{ab}]$ is a ‘slowly varying amplitude’ and $S[h_{ab}]$ is a ‘rapidly varying phase’. This corresponds to the substitution

$$p^{ab} \longrightarrow \frac{\delta S}{\delta h_{ab}}\,,$$

which is the classical relation for the canonical momentum. From (3) and (4) one finds then for $S[h_{ab}]$ the equations

$$16\pi G\, G_{abcd}\frac{\delta S}{\delta h_{ab}}\frac{\delta S}{\delta h_{cd}} - \frac{\sqrt{h}}{16\pi G}({}^{(3)}R - 2\Lambda) = 0\,, \quad (7)$$

$$D_a\frac{\delta S}{\delta h_{ab}} = 0\,. \quad (8)$$

In the presence of matter one has additional terms. The first Eq. (7) is the Hamilton–Jacobi equation for the gravitational field. One can prove that the four local Eqs. (7) and (8) are equivalent to all ten Einstein equations.

If non-gravitational fields are present, as we will now assume, a mixture of this WKB ansatz with the Born–Oppenheimer ansatz from molecular physics is appropriate [14, 15, 29]. One writes instead of (6) now

$$\Psi[h_{ab}, \phi] \equiv \exp\left(\frac{\mathrm{i}}{\hbar}S[h_{ab}, \phi]\right), \quad (9)$$

where $S[h_{ab}, \phi]$ here denotes a complex function that depends on both the three-metric h_{ab} and the non-gravitational fields denoted by ϕ (usually taken to be a scalar field). Plugging this ansatz into the quantum constraints (3) and (4) and performing an expansion scheme with respect to the square of the Planck mass $m_{\rm P} = \sqrt{\hbar/G}$,

$$S[h_{ab}, \phi] = m_{\rm P}^2 S_0 + S_1 + m_{\rm P}^{-2} S_2 + \dots, \tag{10}$$

one finds at highest order ($m_{\rm P}^2$) that S_0 depends only on the three-metric h_{ab} and that it obeys the Hamilton–Jacobi equations (7) and (8) for the pure gravitational field.

The next order ($m_{\rm P}^0$) gives a functional Schrödinger equation for a wave functional $\psi[h_{ab}, \phi]$ in the background spacetime defined from a solution S_0 to (7) and (8), where

$$\psi[h_{ab}, \phi] := D[h_{ab}] \exp\left(\frac{\mathrm{i}}{\hbar} S_1[h_{ab}, \phi]\right), \tag{11}$$

and D obeys the standard WKB prefactor equation (see e.g. Eq. (2.36) in [14]). This step yields a Tomonaga–Schwinger equation for $\psi[h_{ab}, \phi]$ with respect to a local time functional $\tau(\mathbf{x})$ that is defined from the solution S_0 by

$$\frac{\delta}{\delta \tau(\mathbf{x})} := G_{abcd} \frac{\delta S_0}{\delta h_{ab}} \frac{\delta}{\delta h_{cd}}. \tag{12}$$

In spite of its appearance, τ is not a scalar function [9]. The functional Schrödinger equation is obtained by evaluating $\psi[h_{ab}, \phi]$ along a solution of the classical Einstein equations, $h_{ab}(\mathbf{x}, t)$, that corresponds to a solution, $S_0[h_{ab}]$, of the Hamilton–Jacobi equation, $\psi[h_{ab}(\mathbf{x}, t), \phi]$. After a certain choice of lapse and shift functions, N and N^a, has been made, this solution is obtained from

$$\dot{h}_{ab} = N G_{abcd} \frac{\delta S_0}{\delta h_{cd}} + 2 D_{(a} N_{b)} \,. \tag{13}$$

Instead of $\psi[h_{ab}, \phi]$, we can write $|\psi[h_{ab}]\rangle$ to indicate (by the bra-ket notation) that one has a well-defined (standard) Hilbert space for the non-gravitational field ϕ. Defining

$$\frac{\partial}{\partial t} |\psi(t)\rangle := \int \mathrm{d}^3 x\, \dot{h}_{ab}(\mathbf{x}, t) \frac{\delta}{\delta h_{ab}(\mathbf{x})} |\psi[h_{ab}]\rangle \,, \tag{14}$$

one finds the functional Schrödinger equation for quantized non-gravitational fields in the chosen external classical gravitational field,

$$\begin{aligned} \mathrm{i}\hbar \frac{\partial}{\partial t} |\psi(t)\rangle &= \hat{H}^{\rm m} |\psi(t)\rangle \,, \\ \hat{H}^{\rm m} &:= \int \mathrm{d}^3 x \left\{ N(\mathbf{x}) \hat{\mathcal{H}}_\perp^{\rm m}(\mathbf{x}) + N^a(\mathbf{x}) \hat{\mathcal{H}}_a^{\rm m}(\mathbf{x}) \right\} \,. \end{aligned} \tag{15}$$

Here, $\hat{H}^{\mathrm{m}}$ is the non-gravitational Hamiltonian in the Schrödinger picture, parametrically depending on the metric coefficients of the curved spacetime background. This is the standard approach for obtaining the limit of quantum field theory in curved spacetime from canonical quantum gravity. Extending this scheme to higher orders in m_{P}^2, one arrives at quantum gravitational correction terms to this equation [18]. These terms can be used to calculate potentially observable effects such as corrections to the CMB anisotropy spectrum [4].

The Born–Oppenheimer approximation starting from (9) provides only part of the understanding why we observe a classical spacetime. The remaining part is provided by the process of decoherence. It was suggested in [32] and elaborated in [13] to using small inhomogeneities such as density perturbations or tiny gravitational waves as a "quantum environment" in configuration space, whose interaction with relevant degrees of freedom such as the global size of the universe gives rise to their classical appearance. Technically, this comes from tracing out these inhomogeneities in the globally entangled quantum states. In [24], Paddy has extended these investigations to more general situations and found that three-geometries with the same intrinsic metric but different size contribute decoherently to the reduced density matrix for the relevant degrees of freedom. He concludes his paper with the words

> ...the classical nature of the space-time will tend to disappear as we observe more and more matter modes. Probably, ignorance *is* bliss.

The recovery of time in semiclassical gravity raises the question whether time in quantum gravity can be recovered from a general solution of the Wheeler–DeWitt solution. The idea was followed independently by Paddy [25] and Greensite [10]. This generalized time is recovered from the phase of the wave function and used to define a Schrödinger-type inner product where all variables are integrated over *except for* this time. A necessary prerequisite for this to work is that the wave function is complex and that its phase is not a constant. One can then prove that the first Ehrenfest theorem is valid if this time variable and the corresponding inner product is used. Unfortunately, only a restricted class of solutions fulfills all consistency conditions (including the validity of the second Ehrenfest theorem), so one either has to abandon this proposal as a solution to the time problem or to use it as a new type of boundary condition to select physically allowed solutions [5].

2 Quantization of Conformal (Weyl) Gravity

The role of conformal transformations and of conformal symmetry is of central interest for gravitational systems at least since Hermann Weyl's pioneering work from 1918. Weyl suggested a theory in which not only the direction of a vector depends on the path along which the vector is transported through spacetime, but also its length. This means that space distances and time intervals depend on the path of rods and clocks through spacetime. In Weyl's theory there exists the freedom to re-scale ("gauge") rods and clocks; the metric can be multiplied by an arbitrary positive spacetime-dependent function,

$$g_{\mu\nu}(x) \quad \rightarrow \quad \tilde{g}_{\mu\nu}(x) = \Omega^2(x) g_{\mu\nu}(x) . \tag{16}$$

This transformation is called *conformal transformation* (later also *Weyl transformation*) and is an invariance in Weyl's theory. Connected with this freedom is a new quantity that Weyl identified with the electromagnetic four-potential, suggesting the idea of a unification between gravity and electromagnetism.[1]

Weyl's theory is impressive, but empirically wrong, as soon noticed by others, in particular Einstein. If it were true, spectral lines, for example, would depend on the history of the atomic worldlines, because an atom can be understood as constituting a clock.[2] Quite generally, a particle with rest mass m can be taken as a clock with frequency

$$\nu = m\frac{c^2}{h},$$

so a path-dependent frequency would correspond to a path-dependent rest mass, since c and h are universal units. This is definitely empirically wrong.

Weyl thus had to give up his theory, but later used essential elements of his idea to provide the foundation of modern gauge theory. Einstein, however, was speculating about the existence of a theory that, while preserving the conformal invariance of Weyl's theory, does not include a hypothesis about the transport of rods and clocks, thus avoiding the problems of Weyl's theory. In a paper entitled "Über eine naheliegende Ergänzung des Fundamentes der allgemeinen Relativitätstheorie" [7],[3] Einstein suggested to use the scalar

$$C_{\mu\nu\lambda\rho}C^{\mu\nu\lambda\rho} \tag{17}$$

formed from the Weyl tensor $C_{\mu\nu\lambda\rho}$ as the basis of this theory.[4] The Weyl tensor $C^{\mu}_{\nu\lambda\rho}$ (with one upper component) is invariant under the conformal transformations (16).

At the end of his article, Einstein intended to add the following short summary, which can be found in his hand-written manuscript, but which he deleted before submission. It reads (our translation from the German)[5]

> Short summary: it is shown that one can, following Weyl's basic ideas, develop a theory of invariants on the objective existence of lightcones (invariance of the equation $ds^2 = 0$) alone, which does not, in contrast to Weyl's theory, contain a hypothesis about transport of distances

[1]For a review and reference to original articles, see [11].

[2]Recall that the modern time standard is based on the hyperfine transitions in caesium-133.

[3]English translation: "On a Natural Addition to the Foundation of the General Theory of Relativity".

[4]In his paper, Einstein acknowledges the help of the Austrian mathematician Wilhelm Wirtinger in his attempt. In a letter to Einstein sent one day after Einstein's academy talk on which [7] is based, Wirtinger suggested as one possibility to use an action principle based on (17), see [8], p. 117.

[5]The φ_ν denote the components of the electromagnetic four-potential.

and in which the potentials φ_ν do not enter explicitly the equations. Later investigations must show whether the theory will be physically valid.[6]

In fact, even before Einstein, Rudolf Bach had considered an action based on (17) and derived and discussed the ensuing field equations [1]. When we talk here of conformal gravity or Weyl gravity, we do not mean Weyl's original gravitational gauge theory from 1918, but a theory that is based on the action suggested by Bach, Einstein, and Wirtinger. We write the action in the following form:

$$S^{\mathrm{W}} := -\frac{\alpha_{\mathrm{W}}\hbar}{4}\int \mathrm{d}^4x\sqrt{-g}\, C_{\mu\nu\lambda\rho}C^{\mu\nu\lambda\rho}\,, \tag{18}$$

where α_{W} is a dimensionless coupling constant. We have introduced Planck's constant $\hbar$ (which, of course, is irrelevant for the classical theory) for two reasons: first, it gives the correct dimensions for the action and renders the constant α_{W} dimensionless and, second, $S^{\mathrm{W}}/\hbar$ is the relevant quantity in the quantum theory on which we will focus in our paper; in fact, this will suggest the semiclassical expansion scheme with respect to α_{W} presented below.

The theory based on (18) was discussed at both the classical and quantum level [22]. At the classical level, it was used, for example, to explain galactic rotation curves without the need for dark matter, although this explanation has met severe criticism [26]. At the quantum level, it is a candidate for a renormalizable theory of quantum gravity, although it seems to violate unitarity.[7] We adopt here the point of view to take (18) as the starting point for a conformally invariant gravity theory, for which a semiclassical expansion scheme can be applied to its canonically quantized version and compared with the scheme for quantized GR. We do not assume that (18) is a candidate for an alternative to GR. In the following, we shall study the canonical structure of this theory. Our treatment is based on the more general treatment presented in [16].

In order to deal with higher-derivative theories such as (18),[8] it is convenient to reduce the order by introducing new independent variables. In our case, this is achieved by introducing the extrinsic curvature K_{ij}, which in general relativity is a function of the time derivative of the three-metric h_{ij}. This can be implemented in the canonical formalism by adding a constraint $\lambda^{ij}\left(2K_{ij} - \mathcal{L}_n h_{ij}\right)$, where λ^{ij} is a Lagrange multiplier. There are also other methods to "hide" the second derivative of the three-metric [3, 6].

[6]The original German reads ([8], p. 416): "Kurze Zusammenfassung: Es wird gezeigt, dass man entsprechend dem Weyl'schen Grundgedanken auf die objektive Existenz der Lichtkegel (Invarianz der Gleichung $ds^2 = 0$) alleine eine Invarianten-theorie gründen kann, die jedoch im Gegensatz zu Weyl's Theorie keine Hypothese über Streckenübertragung enthält und in welcher die Potentiale φ_ν nicht explizite in die Gleichungen eingehen. Ob die Theorie auf physikalische Gültigkeit Anspruch erheben kann, müssen spätere Untersuchungen ergeben.".

[7]We write "seems", because the ghosts connected with non-unitarity may be removable [21].

[8]For the history of such theories, see for example [28].

In order to manifestly reveal conformal invariance of the Weyl action, we will use here an irreducible decomposition of the 3-metric into its scale part[9]

$$a = (\sqrt{h})^{1/3} \tag{19}$$

and its conformally invariant (unimodular) part

$$\bar{h}_{ij} = a^{-2} h_{ij}. \tag{20}$$

In addition, we define the following variables [16]

$$\bar{N}^i = N^i\,, \quad \bar{N}_i = a^{-2} N_i\,, \quad \bar{N} = a^{-1} N\,, \tag{21}$$
$$\bar{K}^{\mathrm{T}}_{ij} = a^{-1} K^{\mathrm{T}}_{ij}\,, \qquad \bar{K} = \tfrac{aK}{3}\,, \tag{22}$$

where $\bar{K}^{\mathrm{T}}_{ij}$ and $\bar{K}$ are the rescaled traceless and trace parts of the extrinsic curvature, respectively; as a consequence of the decomposition of the three-metric (19) and (20) they are given by

$$\bar{K}^{\mathrm{T}}_{ij} = \frac{1}{\bar{N}}\left(\dot{\bar{h}}_{ij} - 2\left[D_{(i}\bar{N}_{j)}\right]^{\mathrm{T}}\right), \quad \bar{K} = \frac{1}{\bar{N}}\left(\frac{\dot{a}}{a} - \frac{1}{3} D_i N^i\right). \tag{23}$$

It can be shown by direct calculation that $\left[D_{(i}\bar{N}_{j)}\right]^{\mathrm{T}}$ is independent of a and that $\bar{K}^{\mathrm{T}}_{ij}$ is the conformally invariant part of the extrinsic curvature. We refer to the variables in (21) and (22) as *unimodular-conformal variables*, and we will formulate the canonical theory in terms of them. The advantage of using these variables is that only the scale a and the trace $\bar{K}$ transform under conformal transformation,

$$\bar{K} \quad\rightarrow\quad \bar{K} + \bar{n}^{\mu}\partial_{\mu}\log\Omega\,, \qquad a \quad\rightarrow\quad \Omega a\,, \tag{24}$$

where $\bar{n}^{\mu} = a n^{\mu}$ and $\bar{n}_{\mu} = a^{-1} n_{\mu}$. This significantly simplifies the canonical formulation and makes conformal invariance of the theory manifest, since the only two variables affected by conformal transformation completely vanish from the constraints, as will be shown below.

The canonical approach employs a $3+1$ decomposition of spacetime quantities. For GR, this is the standard ADM approach [15]. In the present case, one has to perform a $3+1$ decomposition of the Weyl tensor, which can be found, for example, in [12, 20]. The constrained $3+1$-decomposed Lagrangian density of the Weyl action in terms of the unimodular-conformal variables introduced above then becomes

[9]In quantum GR, there exist attempts to quantize solely the conformal factor [23]. Paddy has derived from this the interesting conclusion, that the Planck length provides a lower bound to measurable physical lengths. The situation will be different here, because Weyl gravity does not contain an intrinsic length scale.

$$\mathcal{L}_c^{\mathrm{W}} = \bar{N}\Big\{-\frac{\alpha_{\mathrm{W}}\hbar}{2}\bar{h}^{ia}\bar{h}^{jb}\bar{C}^{\mathrm{T}}_{ij}\bar{C}^{\mathrm{T}}_{ab} + \alpha_{\mathrm{W}}\hbar\bar{C}^2_{ijk} - a^5\lambda^{ij\mathrm{T}}\Big[2\bar{K}^{\mathrm{T}}_{ij} - \frac{1}{\bar{N}}\Big(\dot{\bar{h}}_{ij} - 2\big[D_{(i}\bar{N}_{j)}\big]^{\mathrm{T}}\Big)\Big]$$
$$-2a^3\lambda\Big[\bar{K} - \frac{1}{\bar{N}}\Big(\frac{\dot{a}}{a} - \frac{1}{3}D_a N^a\Big)\Big]\Big\}, \tag{25}$$

where $\lambda^{ij\mathrm{T}}$ and λ are traceless and trace parts of the Lagrange multiplier λ^{ij}, and

$$\bar{C}^{\mathrm{T}}_{ij} = \mathcal{L}_{\bar{n}}\bar{K}^{\mathrm{T}}_{ij} - \frac{2}{3}\bar{h}_{ij}\bar{K}^{\mathrm{T}}_{ab}\bar{h}^{an}\bar{h}^{bm}\bar{K}^{\mathrm{T}}_{nm} - {}^{(3)}R^{\mathrm{T}}_{ij} - \frac{1}{\bar{N}}\left[D_i D_j\right]^{\mathrm{T}}\bar{N} \tag{26}$$

is the "electric part" of the Weyl tensor, containing only velocities of the traceless part of the extrinsic curvature, and

$$\bar{C}_{ijk} = 2\delta_i^{[d}\left(\delta_j^{e]}\delta_k^f - \bar{h}_{jk}\bar{h}^{e]f}\right)D_d\bar{K}_{ef}\,,\quad \bar{C}^2_{ijk} \equiv \bar{C}_{ijk}\bar{h}^{ia}\bar{h}^{jb}\bar{h}^{kc}\bar{C}_{abc} \tag{27}$$

is related to the "magnetic part" of the Weyl tensor, as explained in [12]. The second expression in (27) should not contain any traces $\bar{K}$ and therefore be conformally invariant, but we assume this without proof. Each object with superscript "T" is traceless. It can be shown easily that the trace of the sum of the first two terms in (26) vanishes, that is, that $h^{ij}\mathcal{L}_n\bar{K}^{\mathrm{T}}_{ij} = 2a^{-2}\bar{K}^{\mathrm{T}}_{ab}\bar{h}^{an}\bar{h}^{bm}\bar{K}^{\mathrm{T}}_{nm}$.

We now take unimodular-conformal variables (19), (20), (21), and (22) and derive their conjugate momenta in the standard way,

$$p_{\bar{N}} = \frac{\partial\mathcal{L}_c^{\mathrm{W}}}{\partial\dot{\bar{N}}} \approx 0\,,\quad p^i = \frac{\partial\mathcal{L}_c^{\mathrm{W}}}{\partial\dot{N}^i} \approx 0\,,\quad \bar{P} = \frac{\partial\mathcal{L}_c^{\mathrm{W}}}{\partial\dot{\bar{K}}} \approx 0\,, \tag{28}$$

$$\bar{p}^{ij} = \frac{\partial\mathcal{L}_c^{\mathrm{W}}}{\partial\dot{\bar{h}}_{ij}} = a^5\lambda^{ij\mathrm{T}}\,,\quad p_a = \frac{\partial\mathcal{L}_c^{\mathrm{W}}}{\partial\dot{a}} = 2a^2\lambda\,, \tag{29}$$

$$\bar{P}^{ij} = \frac{\partial\mathcal{L}_c^{\mathrm{W}}}{\partial\dot{\bar{K}}^{\mathrm{T}}_{ij}} = -\alpha_{\mathrm{W}}\hbar\,\bar{h}^{ia}\bar{h}^{jb}\bar{C}^{\mathrm{T}}_{ab}\,. \tag{30}$$

Note that the momenta $\bar{p}^{ij}$ and $\bar{P}^{ij}$ are traceless. The novelty with respect to GR is the emergence of another primary constraint, $\bar{P} \approx 0$; this suggests that $\bar{K}$ is arbitrary, in the same manner as $p_{\bar{N}} \approx 0$ and $p_i \approx 0$ suggest that $\bar{N}$ and N^i are arbitrary.

It can easily be checked that the transformation from the original variables to the unimodular-conformal variables is a canonical one. The Poisson brackets (PBs) of the variables are

$$\left\{q^A_{ij}(\mathbf{x}), \Pi^{ab}_B(\mathbf{y})\right\} = \left(\delta^a_{(i}\delta^b_{j)} - \frac{1}{3}h_{ij}h^{ab}\right)\delta^A_B\delta(\mathbf{x},\mathbf{y})\,,\quad \left\{q^A(\mathbf{x}), \Pi_B(\mathbf{y})\right\} = \delta^A_B\delta(\mathbf{x},\mathbf{y})\,, \tag{31}$$

where $q^A_{ij} = (\bar{h}_{ij}, \bar{K}^{\mathrm{T}}_{ij})$, $\Pi^{ab}_B = (\bar{p}^{ab}, \bar{P}^{ab})$ are the variables in the conformally invariant subspace of phase space, and $q^A = (a, \bar{K})$, $\Pi_B = (p_a, \bar{P})$ is the scale-trace subspace of phase space (and similar for the lapse-shift sector). All other PBs vanish.

After performing the Legendre transformation (from which $\dot{\bar{K}}\bar{P}$ is absent, since $\dot{\bar{K}}$ does not appear in the Lagrangian) and investigating the emerging constraints, we can write the total Hamiltonian as

$$H^{\mathrm{W}} = \int \mathrm{d}^3x \left\{ \bar{N}\mathscr{H}_{\perp}^{\mathrm{W}} + N^i\mathscr{H}_i^{\mathrm{W}} + \left(\bar{N}\bar{K} - \frac{1}{3}D_iN^i\right)\mathscr{Q}^{\mathrm{W}} + \lambda_{\bar{N}}p_{\bar{N}} + \lambda_i p^i + \lambda_{\bar{P}}\bar{P} \right\} + H_{\mathrm{surf}}\,, \tag{32}$$

from which one finds the secondary constraints

$$\mathscr{H}_{\perp}^{\mathrm{W}} = -\frac{\bar{h}_{ik}\bar{h}_{jl}\bar{P}^{ij}\bar{P}^{kl}}{2\alpha_{\mathrm{W}}\hbar} + \left({}^{(3)}R_{ij}^{\mathrm{T}} + D_iD_j\right)\bar{P}^{ij} + 2\bar{K}_{ij}^{\mathrm{T}}\bar{p}^{ij} - \alpha_{\mathrm{W}}\hbar\bar{C}_{ijk}^2 \approx 0\,, \tag{33}$$

$$\mathscr{H}_i^{\mathrm{W}} = -2\partial_k\left(\bar{h}_{ij}\bar{p}^{jk}\right) + \partial_i\bar{h}_{jk}\bar{p}^{jk} - 2\partial_k\left(\bar{K}_{ij}^{\mathrm{T}}\bar{P}^{jk}\right) + \partial_i\bar{K}_{jk}^{\mathrm{T}}\bar{P}^{jk} \approx 0\,, \tag{34}$$

$$\mathscr{Q}^{\mathrm{W}} = ap_a \approx 0\,. \tag{35}$$

The first two are the Hamiltonian and momentum constraints, and they are analogous to (3) and (4), although the structure of the Hamiltonian constraint is significantly different. The new constraint (35) comes from the consistency condition for the primary constraint $\bar{P} \approx 0$,

$$\dot{\bar{P}} = \{\bar{P}, H^{\mathrm{W}}\} = -\frac{\partial H^{\mathrm{W}}}{\partial\bar{K}} = -\bar{N}a\,p_a \overset{!}{\approx} 0\,. \tag{36}$$

A brief inspection of constraints reveals that the Hamiltonian and momentum constraints are manifestly conformally invariant, due to the use of the unimodular-conformal variables[10] — the Hamiltonian and momentum constraints are independent of the scale a and trace $\bar{K}$. The constraints $\bar{P}$ and $\mathscr{Q}^{\mathrm{W}}$ commute, and they also commute with the rest of the constraints. The Hamiltonian and the momentum constraints close the same hypersurface foliation algebra as in GR, see [6]. This is expected for any reparametrization invariant metric theory, see [30], p. 57. Hence, all constraints are first class.

Therefore, the Hamiltonian and momentum constraints have the same meaning as in GR. The momentum constraint is extended to include the extrinsic curvature sector, since the components of K_{ij} are treated as independent variables in this higher-derivative theory. Thus the three-dimensional diffeomorphism invariance now includes changes of $\bar{K}_{ij}^{\mathrm{T}}$. But what is the meaning of the $\bar{P}$ and $\mathscr{Q}^{\mathrm{W}}$ constraints? It can be shown that these constraints comprise a generator of conformal *gauge* transformation, as shown in [12] in terms of the original variables (which also include the lapse, prone to conformal transformation). In unimodular-conformal variables, a procedure similar to [12] leads to the following generator of conformal transformation [16]:

$$G_{\omega}^{\mathrm{W}}[\omega,\dot{\omega}] = \int \mathrm{d}^3x\left(\mathscr{Q}^{\mathrm{W}}\omega + \bar{P}\mathscr{L}_{\bar{n}}\omega\right) = \int \mathrm{d}^3x\left(ap_a\omega + \bar{P}\mathscr{L}_{\bar{n}}\omega\right)\,, \tag{37}$$

[10]It can be shown that terms in $\left({}^{(3)}R_{ij}^{\mathrm{T}} + D_iD_j\right)\bar{P}^{ij}$ which depend on a cancel, making this expression conformally invariant.

which generates here a transformation only for the scale a and the trace $\bar{K}$. We emphasize that primary and secondary constraints have to appear together to ensure a correct transformation, as emphasized in particular by Pitts [27].

A closer look at the Hamiltonian constraint (33) reveals that the "intrinsic time" of GR contained in the scale part a is absent. This is not surprising, because we are dealing here with a conformally invariant theory. The "problem of time" in quantum gravity [15] is for the Weyl theory thus of a different nature than for GR. This difference will also be relevant for the recovery of semiclassical time discussed below.

Let us now turn to configuration space. In analogy to the Hamilton–Jacobi function of GR, Eq. (7), one can define a Hamilton–Jacobi functional in Weyl gravity as well, which is defined on full configuration space,

$$S^{\text{W}} = S^{\text{W}}[\bar{h}_{ij}, a, \bar{K}^{\text{T}}_{ij}, \bar{K}].$$

The conjugate momenta $\bar{p}^{ij}$ and $\bar{P}^{ij}$ follow form this functional in the usual way,

$$\bar{p}^{ij} = \frac{\delta S^{\text{W}}}{\delta \bar{h}_{ij}}\,, \quad \bar{P}^{ij} = \frac{\delta S^{\text{W}}}{\delta \bar{K}^{\text{T}}_{ij}}\,, \quad p_a = \frac{\delta S^{\text{W}}}{\delta a}\,, \quad \bar{P} = \frac{\delta S^{\text{W}}}{\delta \bar{K}}\,. \tag{38}$$

Due to the primary-secondary pair of constraints $\bar{P} \approx 0$ and $\mathcal{Q}^{\text{W}} \approx 0$, we can conclude, however, that the functional S^{W} does not depend on a and $\bar{K}$, since its infinitesimal conformal variations vanish,

$$\frac{\delta S^{\text{W}}}{\delta a} = 0\,, \quad \frac{\delta S^{\text{W}}}{\delta \bar{K}} = 0 \quad \Rightarrow \quad \delta_\omega S^{\text{W}} = \int \mathrm{d}^3x \left(\frac{\delta S^{\text{W}}}{\delta a}\,\delta a + \frac{\delta S^{\text{W}}}{\delta \bar{K}}\,\delta \bar{K} \right) = 0\,. \tag{39}$$

One can then interpret S^{W} as a conformally invariant functional solving the conformally invariant *Weyl–Hamilton–Jacobi equation* (WHJ equation) obtained from (33),

$$-\frac{1}{2\alpha_{\text{W}}\hbar}\bar{h}_{ik}\bar{h}_{jl}\frac{\delta S^{\text{W}}}{\delta \bar{K}^{\text{T}}_{ij}}\frac{\delta S^{\text{W}}}{\delta \bar{K}^{\text{T}}_{kl}} + \left({}^{(3)}\!R^{\text{T}}_{ij} + D_i D_j \right) \frac{\delta S^{\text{W}}}{\delta \bar{K}^{\text{T}}_{ij}} + 2\bar{K}^{\text{T}}_{ij}\frac{\delta S^{\text{W}}}{\delta \bar{h}_{ij}} - \alpha_{\text{W}}\hbar \bar{C}^2_{ijk} = 0\,. \tag{40}$$

We expect that S^{W}, as a solution to the above equation, gives a "classical trajectory" in the configuration subspace spanned by $\left\{\bar{h}_{ij}, \bar{K}^{\text{T}}_{ij}\right\}$. Due to (39), a tangent to this trajectory does not have components in the a and $\bar{K}$ directions of the configuration space. In other words, the classical state of this theory does not follow directions along changes of a and $\bar{K}$ in configuration space.

Quantization is now performed in the sense of Dirac by implementing the classical constraints as restrictions on physically allowed wave functionals on the full configuration space [15],

$$\Psi \equiv \Psi[\bar{h}_{ij}, a, \bar{K}^{\mathrm{T}}_{ij}, \bar{K}].$$

The canonical variables are promoted into operators in the standard way,

$$\hat{\bar{h}}_{ij}(x)\Psi = \bar{h}_{ij}(x)\Psi\,, \quad \hat{\bar{p}}^{ij}(x)\Psi = -i\hbar\frac{\delta}{\delta\bar{h}_{ij}(x)}\Psi\,, \tag{41}$$

$$\hat{\bar{K}}^{\mathrm{T}}_{ij}(x)\Psi = \bar{K}^{\mathrm{T}}_{ij}(x)\Psi\,, \quad \hat{\bar{P}}^{ij}(x)\Psi = -\mathrm{i}\hbar\frac{\delta}{\delta\bar{K}^{\mathrm{T}}_{ij}(x)}\Psi\,, \tag{42}$$

$$\hat{a}(x)\Psi = a(x)\Psi\,, \quad \hat{\bar{p}}_a(x)\Psi = -\mathrm{i}\hbar\frac{\delta}{\delta a(x)}\Psi\,, \tag{43}$$

$$\hat{\bar{K}}(x)\Psi = \bar{K}(x)\Psi\,, \quad \hat{\bar{P}}(x)\Psi = -\mathrm{i}\hbar\frac{\delta}{\delta\bar{K}(x)}\Psi. \tag{44}$$

The quantization of the constraints yields [17]

$$\hat{\mathscr{H}}^{\mathrm{w}}_{\perp}\Psi = 0\,, \quad \hat{\mathscr{H}}^{\mathrm{w}}_{i}\Psi = 0\,, \quad \hat{\bar{P}}\Psi = 0\,, \quad \hat{\mathscr{Q}}^{\mathrm{w}}\Psi = 0. \tag{45}$$

The first of these equations is the quantized Hamiltonian constraint, which replaces the WDW equation of quantum GR and which we will therefore call the "Weyl–Wheeler–DeWitt" (WWDW) equation. Neglecting here the ubiquitous factor ordering problem, it assumes the explicit form

$$\left[\frac{\hbar}{2\alpha_{\mathrm{w}}}\bar{h}_{ik}\bar{h}_{jl}\frac{\delta^2}{\delta\bar{K}^{\mathrm{T}}_{ij}\delta\bar{K}^{\mathrm{T}}_{kl}} - \mathrm{i}\hbar\left({}^{(3)}\!R^{\mathrm{T}}_{ij} + D^{\mathrm{T}}_{ij}\right)\frac{\delta}{\delta\bar{K}^{\mathrm{T}}_{ij}} - 2\mathrm{i}\hbar\bar{K}^{\mathrm{T}}_{ij}\frac{\delta}{\delta\bar{h}_{ij}}\right.$$
$$\left. -\alpha_{\mathrm{w}}\hbar\bar{C}^2_{ijk\perp} + \hat{\mathscr{H}}^{\mathrm{m}}_{\perp}\right]\Psi = 0\,. \tag{46}$$

One recognizes that the WWDW equation is structurally different from the WDW equation, since the wave functional does not depend only on the three-metric, but also on its evolution (the second fundamental form). There is also no scale a present and therefore no intrinsic time in the sense of the WDW equation; there is no indefinite "DeWitt metric". It is also interesting to see that $\hbar$ drops out after dividing the whole equation by $\hbar$. Formally this is due to our use of $\alpha_{\mathrm{W}}\hbar$ in the action instead of just α_{W}; re-scaling $\alpha_{\mathrm{W}} \to \alpha_{\mathrm{W}}/\hbar$ would bring back $\hbar$ at the places similar to the Wheeler–DeWitt equation (3), but the important point is that $\hbar$ can be made to disappear by a simple re-scaling. This is, of course, a property of the vacuum theory. If we add a matter Hamiltonian density to the WWDW equation, as we shall do below, $\hbar$ will not disappear when dividing the whole equation by $\hbar$.

The quantum momentum constraints read

$$\hat{\mathcal{H}}_i^{\mathrm{w}}\Psi = \mathrm{i}\hbar\left[2\partial_k\left(\bar{h}_{ij}\frac{\delta\Psi}{\delta\bar{h}_{jk}}\right) - \partial_i\bar{h}_{jk}\frac{\delta}{\delta\Psi\bar{h}_{jk}} + 2\partial_k\left(\bar{K}^{\mathrm{T}}_{ij}\frac{\delta\Psi}{\delta\bar{K}^{\mathrm{T}}_{jk}(x)}\right)\right.$$
$$\left. -\partial_i\bar{K}^{\mathrm{T}}_{jk} + \frac{\delta\Psi}{\delta\bar{K}^{\mathrm{T}}_{jk}} + \hat{\mathcal{H}}_i^{\mathrm{m}}\Psi\right] = 0, \tag{47}$$

or alternatively, in a manifestly covariant version,

$$\hat{\mathcal{H}}_i^{\mathrm{w}}\Psi = \mathrm{i}\hbar\left[2D_k\left(\bar{h}_{ij}\frac{\delta\Psi}{\delta\bar{h}_{jk}}\right) + 2D_k\left(\bar{K}^{\mathrm{T}}_{ij}\frac{\delta\Psi}{\delta\bar{K}^{\mathrm{T}}_{jk}}\right) - D_i\bar{K}^{\mathrm{T}}_{jk}\frac{\delta\Psi}{\delta\bar{K}^{\mathrm{T}}_{jk}} + \hat{\mathcal{H}}_i^{\mathrm{m}}\Psi\right] = 0\,. \tag{48}$$

Finally, the new quantum constraints read

$$\frac{\delta\Psi}{\delta\bar{K}} = 0\,, \quad a\frac{\delta\Psi}{\delta a} = 0\,. \tag{49}$$

The meaning of (49) is obvious: the wave functional does not depend on a and $\bar{K}$; hence, it is conformally invariant (apart from a possible phase factor). This is a direct consequence of the first class nature of the constraints $\bar{P} = 0$ and $\mathcal{Q}^{\mathrm{w}} = ap_a = 0$. Thus, we have a conformally invariant canonical quantum gravity theory derived from the Weyl action. Equivalently, one could have started from a reduced phase space without a and K and ended up with (46) and (47) only, with Ψ depending on 10 (instead of 12) configuration variables from the start.

Looking at the whole picture, we conclude that solutions to the WWDW equation are conformally invariant (scale and trace independent), and are indistinguishable for two three-metrics that are conformal to each other.

3 Semiclassical Weyl Gravity and the Recovery of Time

We consider quantum Weyl gravity with a conformally coupled matter field ϕ, for conformal matter does not spoil the first-class nature of constraints; it only modifies their explicit form. We can then quantize the theory while preserving its conformal invariance.

In the spirit of the semiclassical (Born–Oppenheimer type) expansion for the WDW equation, we make an ansatz for the wave functional in which the "heavy" part, being the pure gravitational part, is separated from the matter part [17]. We write for the full quantum state in analogy to (9)

$$\Psi\left[\bar{h}_{ij}, \bar{K}^{\mathrm{T}}_{ij}, \phi\right] \equiv \exp\left(\frac{\mathrm{i}}{\hbar}S\left[\bar{h}_{ij}, \bar{K}^{\mathrm{T}}_{ij}, \phi\right]\right). \tag{50}$$

Plugging (50) into the WWDW equation (46) gives

$$\frac{\mathrm{i}}{2\alpha_{\mathrm{W}}}\bar{h}_{ik}\bar{h}_{jl}\frac{\delta^2 S}{\delta \bar{K}^{\mathrm{T}}_{ij}\delta \bar{K}^{\mathrm{T}}_{kl}} - \frac{1}{2\alpha_{\mathrm{W}}\hbar}\bar{h}_{ik}\bar{h}_{jl}\frac{\delta S}{\delta \bar{K}^{\mathrm{T}}_{ij}}\frac{\delta S}{\delta \bar{K}^{\mathrm{T}}_{kl}} + \left({}^{(3)}R^{\mathrm{T}}_{ij} + D^{\mathrm{T}}_{ij}\right)\frac{\delta S}{\delta \bar{K}^{\mathrm{T}}_{ij}}$$
$$+2\bar{K}^{\mathrm{T}}_{ij}\frac{\delta S}{\delta \bar{h}_{ij}} - \alpha_{\mathrm{W}}\hbar \bar{C}^2_{ijk} + \frac{\left(\hat{\mathcal{H}}^{\mathrm{m}}_{\perp}\Psi\right)}{\Psi} = 0. \quad (51)$$

The expansion can be performed with respect to α_{W}^{-1}, for this coupling constant appears at the same place (in the kinetic term and in part of the potential) as m_{P}^2 appears in the WDW equation. The functional S can then be expanded in powers of $\alpha_{\mathrm{W}}^{-1} \ll 1$, assuming α_{W} to be large; this is similar to the Planck-mass expansion for quantum GR, see (10) above. Note that α_{W} is a dimensionless quantity, unlike the Planck mass in the case of the WDW equation. This is similar to the semiclassical expansion of quantum electrodynamics, with the (dimensionless) fine structure constant as the appropriate expansion parameter [19]. We thus write

$$S = \alpha_{\mathrm{W}} \sum_{n=0}^{\infty} \left(\frac{1}{\alpha_{\mathrm{W}}}\right)^n S^{\mathrm{W}}_n . \quad (52)$$

Note that $\left(\hat{\mathcal{H}}^{\mathrm{m}}_{\perp}\Psi\right)/\Psi$, when expanded in powers of α_{W}, is at most of the order α_{W}^2, since the highest derivative with respect to the matter field ϕ in $\hat{\mathcal{H}}^{\mathrm{m}}_{\perp}$ is the second order, which is the kinetic term (we assume it is the only one of that kind). We shall then denote with $\left(\left(\hat{\mathcal{H}}^{\mathrm{m}}_{\perp}\Psi\right)/\Psi\right)^{(n)}$, $n \le 2$, terms proportional to α_{W}^n.

Inserting the ansatz (52) into the WWDW equation and collecting the powers of α_{W}^2, we find

$$\alpha_{\mathrm{W}}^2 : \quad \left(\left(\hat{\mathcal{H}}^{m}_{\perp}\Psi\right)/\Psi\right)^{(2)} = 0 \quad \Rightarrow \quad \frac{\delta S^{\mathrm{W}}_0}{\delta\phi} = 0. \quad (53)$$

This is analogous to the situation in GR [18]. At the next order, α_{W}, we have

$$\alpha_{\mathrm{W}}^1 : \quad -\frac{1}{2\hbar}\bar{h}_{ik}\bar{h}_{jl}\frac{\delta S^{\mathrm{W}}_0}{\delta \bar{K}^{\mathrm{T}}_{ij}}\frac{\delta S^{\mathrm{W}}_0}{\delta \bar{K}^{\mathrm{T}}_{kl}} + \left({}^{(3)}R^{\mathrm{T}}_{ij} + D^{\mathrm{T}}_{ij}\right)\frac{\delta S^{\mathrm{W}}_0}{\delta \bar{K}^{\mathrm{T}}_{ij}} + 2\bar{K}^{\mathrm{T}}_{ij}\frac{\delta S^{\mathrm{W}}_0}{\delta \bar{h}_{ij}} - \hbar\bar{C}^2_{ijk} = 0\,, \quad (54)$$

which is nothing else than the Weyl-HJ equation (40), with $S^{\mathrm{W}} \equiv \alpha_{\mathrm{W}} S^{\mathrm{W}}_0$.

At the next order, (α_{W}^0), we obtain

$$\alpha_{\mathrm{W}}^0 : \quad \frac{\mathrm{i}}{2}\bar{h}_{ik}\bar{h}_{jl}\frac{\delta^2 S^{\mathrm{W}}_0}{\delta \bar{K}_{ij}\delta \bar{K}_{kl}} - \frac{1}{2\hbar}\bar{h}_{ik}\bar{h}_{jl}\frac{\delta S^{\mathrm{W}}_0}{\delta \bar{K}^{\mathrm{T}}_{ij}}\frac{\delta S^{\mathrm{W}}_1}{\delta \bar{K}^{\mathrm{T}}_{kl}} + \left({}^{(3)}R^{\mathrm{T}}_{ij} + D^{\mathrm{T}}_{ij}\right)\frac{\delta S^{\mathrm{W}}_1}{\delta \bar{K}^{\mathrm{T}}_{ij}}$$
$$+2\bar{K}_{ij}\frac{\delta S^{\mathrm{W}}_1}{\delta \bar{h}_{ij}} + \left(\left(\hat{\mathcal{H}}^{\mathrm{m}}_{\perp}\Psi\right)/\Psi\right)^{(0)} = 0\,. \quad (55)$$

A procedure analogous to the one used to arrive at the functional Schrödinger equation in quantum GR motivates us to propose the following functional:

$$f \equiv D[\bar{h}_{ij}, \bar{K}^{\mathrm{T}}_{ij}] \exp\left(\frac{\mathrm{i}}{\hbar} S^{\mathrm{W}}_1\right), \tag{56}$$

with a condition on the "WKB prefactor" D that will be derived below. We first calculate the following functional derivatives:

$$\begin{aligned}
&\mathrm{i}\bar{h}_{ik}\bar{h}_{jl}\frac{\delta S^{\mathrm{W}}_0}{\delta \bar{K}^{\mathrm{T}}_{ij}}\frac{\delta f}{\delta \bar{K}_{kl}} = \mathrm{i}\bar{h}_{ik}\bar{h}_{jl}\frac{\delta S^{\mathrm{W}}_0}{\delta \bar{K}^{\mathrm{T}}_{ij}}\frac{\delta D}{\delta \bar{K}^{\mathrm{T}}_{kl}}\frac{1}{D}f - \frac{1}{\hbar}\bar{h}_{ik}\bar{h}_{jl}\frac{\delta S^{\mathrm{W}}_0}{\delta \bar{K}^{\mathrm{T}}_{ij}}\frac{\delta S^{\mathrm{W}}_1}{\delta \bar{K}^{\mathrm{T}}_{kl}} f,\\
&-2\mathrm{i}\hbar\bar{K}^{\mathrm{T}}_{ij}\frac{\delta f}{\delta \bar{h}_{ij}} = -2\mathrm{i}\hbar\bar{K}^{\mathrm{T}}_{ij}\frac{\delta D}{\delta \bar{h}_{ij}}\frac{1}{D}f + 2\bar{K}^{\mathrm{T}}_{ij}\frac{\delta S^{\mathrm{W}}_1}{\delta \bar{h}_{ij}} f,\\
&-\mathrm{i}\hbar\left({}^{(3)}R^{\mathrm{T}}_{ij} + D^{\mathrm{T}}_{ij}\right)\frac{\delta f}{\delta \bar{K}^{\mathrm{T}}_{ij}} = -\mathrm{i}\hbar\left({}^{(3)}R^{\mathrm{T}}_{ij} + D^{\mathrm{T}}_{ij}\right)\frac{\delta D}{\delta \bar{K}^{\mathrm{T}}_{ij}}\frac{1}{D}f + \left({}^{(3)}R^{\mathrm{T}}_{ij} + D^{\mathrm{T}}_{ij}\right)\frac{\delta S^{\mathrm{W}}_1}{\delta \bar{K}^{\mathrm{T}}_{ij}} f.
\end{aligned}$$

These expressions are used in (55) to eliminate the second, third and fourth terms, after multiplying with f. As a result, one obtains

$$\begin{aligned}
&\frac{\mathrm{i}}{2}\bar{h}_{ik}\bar{h}_{jl}\frac{\delta S^{\mathrm{W}}_0}{\delta \bar{K}_{ij}}\frac{\delta f}{\delta \bar{K}^{\mathrm{T}}_{kl}} - \mathrm{i}\hbar\left({}^{(3)}R^{\mathrm{T}}_{ij} + D^{\mathrm{T}}_{ij}\right)\frac{\delta f}{\delta \bar{K}^{\mathrm{T}}_{ij}} - 2\mathrm{i}\hbar\bar{K}^{\mathrm{T}}_{ij}\frac{\delta f}{\delta \bar{h}_{ij}} + \hat{\mathscr{H}}^{\mathrm{m}}_{\perp} f\\
&+\left(\frac{\mathrm{i}}{2}\bar{h}_{ik}\bar{h}_{jl}\frac{\delta^2 S^{\mathrm{W}}_0}{\delta \bar{K}_{ij}\delta \bar{K}_{kl}} - \frac{\mathrm{i}}{2}\bar{h}_{ik}\bar{h}_{jl}\frac{\delta S^{\mathrm{W}}_0}{\delta \bar{K}^{\mathrm{T}}_{ij}}\frac{\delta D}{\delta \bar{K}^{\mathrm{T}}_{kl}}\frac{1}{D}\right.\\
&\left.+\mathrm{i}\hbar\left({}^{(3)}R^{\mathrm{T}}_{ij} + D^{\mathrm{T}}_{ij}\right)\frac{\delta D}{\delta \bar{K}^{\mathrm{T}}_{kl}}\frac{1}{D} + 2\mathrm{i}\bar{K}^{\mathrm{T}}_{ij}\frac{\delta D}{\delta \bar{h}_{kl}}\frac{1}{D}\right) f = 0,
\end{aligned} \tag{57}$$

where $\hat{\mathscr{H}}^{\mathrm{m}}_{\perp} f$ comes from $\left(\left(\hat{\mathscr{H}}^{\mathrm{m}}_{\perp}\Psi\right)/\Psi\right)^{(0)} f$. We now choose D such that the term in the parenthesis vanishes. This gives us the equation that defines D, in analogy to the situation in quantum GR [14]:

$$\frac{\mathrm{i}}{2}\bar{h}_{ik}\bar{h}_{jl}\frac{\delta^2 S^{\mathrm{W}}_0}{\delta \bar{K}_{ij}\delta \bar{K}_{kl}} - \frac{\mathrm{i}}{2}\bar{h}_{ik}\bar{h}_{jl}\frac{\delta S^{\mathrm{W}}_0}{\delta \bar{K}^{\mathrm{T}}_{ij}}\frac{\delta D}{\delta \bar{K}^{\mathrm{T}}_{kl}}\frac{1}{D} + \mathrm{i}\hbar\left({}^{(3)}R^{\mathrm{T}}_{ij} + D^{\mathrm{T}}_{ij}\right)\frac{\delta D}{\delta \bar{K}^{\mathrm{T}}_{kl}}\frac{1}{D} + 2\mathrm{i}\bar{K}^{\mathrm{T}}_{ij}\frac{\delta D}{\delta \bar{h}_{kl}}\frac{1}{D} = 0.$$

With this condition, (57) reduces to

$$\mathrm{i}\hbar\left[-\frac{1}{2\hbar}\bar{h}_{ik}\bar{h}_{jl}\frac{\delta S^{\mathrm{W}}_0}{\delta \bar{K}^{\mathrm{T}}_{ij}}\frac{\delta}{\delta \bar{K}^{\mathrm{T}}_{kl}} + \left({}^{(3)}R^{\mathrm{T}}_{ij} + D^{\mathrm{T}}_{ij}\right)\frac{\delta}{\delta \bar{K}^{\mathrm{T}}_{ij}} + 2\bar{K}^{\mathrm{T}}_{ij}\frac{\delta}{\delta \bar{h}_{ij}}\right] f = \hat{\mathscr{H}}^{\mathrm{m}}_{\perp} f. \tag{58}$$

Introducing a local "bubble" (Tomonaga–Schwinger) time functional by

$$\frac{\delta}{\delta\tau_{\mathrm{W}}(\mathbf{x})} := -\frac{1}{2\hbar}\bar{h}_{ik}\bar{h}_{jl}\frac{\delta S_0^{\mathrm{W}}}{\delta\bar{K}_{kl}^{\mathrm{T}}}\frac{\delta}{\delta\bar{K}_{ij}^{\mathrm{T}}} + \left({}^{(3)}R_{ij}^{\mathrm{T}} + D_i D_j\right)\frac{\delta}{\delta\bar{K}_{ij}^{\mathrm{T}}} + 2\bar{K}_{ij}^{\mathrm{T}}\frac{\delta}{\delta\bar{h}_{ij}}, \tag{59}$$

we arrive at the Tomonaga–Schwinger equation

$$\mathrm{i}\hbar\frac{\delta f}{\delta\tau_{\mathrm{W}}} = \hat{\mathscr{H}}_{\perp}^{\mathrm{m}} f\,. \tag{60}$$

Note that τ_{W} is, like its GR-counterpart (12), not a scalar function [9]. We emphasize that the wave function f is conformally invariant.

At a formal level, the Tomonaga–Schwinger equation (60) resembles the corresponding equation in quantum GR. We see, however, from the explicit expression for the WKB time (59) that it is defined only from the semiclassical shape degrees of freedom, since the traces (especially a) are absent. A functional Schrödinger equation of the form (15) can be derived from the Tomonaga–Schwinger equation by a procedure similar to the one in GR. This will involve a time parameter that should be identical with the time parameter of the classical solutions of Weyl gravity.

Proceeding with the Born–Oppenheimer scheme to higher orders in α_{W}, one arrives at quantum gravitational corrections terms proportional to α_{W}^{-1}, in analogy to the higher orders proportional to m_{P}^{-2} in quantum GR [18]. These may serve to study correction terms to the limit of quantum field theory in curved (Weyl) spacetime, but we will not discuss them here.

4 Outlook

Although there is not yet a consensus about the correct quantum theory of gravity, and about the need to quantize gravity, there exist several approaches within which concrete questions with potential observational relevance can be posed and answered. Among them is canonical quantum gravity in the metric formulation. If general relativity is quantized in this way, one arrives at the Wheeler–DeWitt equation and the momentum constraints. A semiclassical expansion leads to the recovery of quantum field theory in curved spacetime plus quantum gravitational corrections. The latter may be observationally tested, for example, in the CMB anisotropy spectrum.

Our concern here was to discuss canonical quantization and the semiclassical limit for an alternative theory based on the Weyl tensor. This "Weyl gravity" does not contain any scale, so it may be of empirical relevance only in the early Universe, where scales may be unimportant. Independent of this possibility, it is of structural interest to compare this theory in its quantum version with quantum general relativity. We have seen here that a semiclassical limit can be performed by a well defined approximation scheme, although the emerging semiclassical time has properties different from standard semiclassical time. In future investigations, we plan to apply a theory based on the sum of Weyl and Einstein-Hilbert action to the early Universe and to the understanding of spacetime structure at a fundamental level, topics that are also at the centre of Paddy's interest.

References

1. R. Bach, Math. Z. **9**, 110 (1921)
2. J. Barbour, in *Quantum Field Theory and Gravity*, ed. by F. Finster, et al. (Springer, Basel, 2012), p. 257
3. D. Boulware, in *Quantum Theory of Gravity*, ed. by S.M. Christensen (Adam Hilger Ltd., Bristol, 1984), p. 267
4. D. Brizuela, C. Kiefer, M. Krämer, Phys. Rev. D **93**, 104035 (2016)
5. T. Brotz, C. Kiefer, Nucl. Phys. B **475**, 339 (1996)
6. N. Deruelle, M. Sasaki, Y. Sendouda, D. Yamauchi, Prog. Theor. Phys. **123**, 169 (2010)
7. A. Einstein, Preuß. Akad. Wiss. (Berlin). Sitzungsberichte 261–264 (1921)
8. A. Einstein, *Collected Papers*, vol. 12 (Princeton University Press, Princeton, 2009)
9. D. Giulini, C. Kiefer, Class. Quantum Grav. **12**, 403 (1995)
10. J. Greensite, Nucl. Phys. B **351**, 749 (1991)
11. H.F.M. Goenner, Living Rev. Relativ. **7** (2016). http://www.livingreviews.org/lrr-2004-2. Accessed 13 July 2016
12. M. Irakleidou, I. Lovrekovic, F. Preis, Phys. Rev. D **91**, 104037 (2015)
13. C. Kiefer, Class. Quantum Grav. **4**, 1369 (1987)
14. C. Kiefer, in *Canonical Gravity: From Classical to Quantum*, ed. by J. Ehlers, H. Friedrich (Springer, Berlin, 1994), p. 170
15. C. Kiefer, *Quantum Gravity*, 3rd edn. (Oxford University Press, Oxford, 2012)
16. C. Kiefer, B. Nikolić, Conformal and Weyl-Einstein Gravity I. Classical Geometrodynamics (to be submitted) (2016)
17. C. Kiefer, B. Nikolić, Conformal and Weyl-Einstein Gravity II. Quantum Geometrodynamics (to be submitted) (2016)
18. C. Kiefer, T.P. Singh, Phys. Rev. D **44**, 1067 (1991)
19. C. Kiefer, T. Padmanabhan, T.P. Singh, Class. Quantum Grav. **8**, L185 (1991)
20. J. Klusoň, M. Oksanen, A. Tureanu, Phys. Rev. D **89**, 064043 (2014)
21. J. Maldacena, Einstein Gravity from Conformal Gravity. arXiv:1105.5632v2 [hep-th]
22. P.D. Mannheim, Found. Phys. **42**, 388 (2012)
23. T. Padmanabhan, Ann. Phys. (N.Y.) **165**, 38 (1985)
24. T. Padmanabhan, Phys. Rev. D **39**, 2924 (1989)
25. T. Padmanabhan, Pramana - J. Phys. **35**, L199 (1990)
26. V. Perlick, C. Xu, Astrophys. J. **449**, 47 (1995)
27. B. Pitts, Ann. Phys. (N.Y.) **351**, 382 (2014)
28. H.-J. Schmidt, Int. J. Geom. Methods Mod. Phys. **04**, 209 (2007)
29. T.P. Singh, T. Padmanabhan, Ann. Phys. (N.Y.) **196**, 296 (1989)
30. T. Thiemann, *Modern Canonical Quantum General Relativity* (Cambridge University Press, Cambridge, 2007)
31. G. 't Hooft, Int. J. Mod. Phys. D **24**, 1543001 (2015)
32. H.D. Zeh, Phys. Lett. A **116**, 9 (1986)

Accelerated Observers, Thermal Entropy, and Spacetime Curvature

Dawood Kothawala

Abstract Assuming that an accelerated observer with four-velocity $\boldsymbol{u}_{\mathrm{R}}$ in a curved spacetime attributes the standard Bekenstein–Hawking entropy and Unruh temperature to his "local Rindler horizon", I show that the *change* in horizon area under parametric displacements of the horizon has a very specific thermodynamic structure. Specifically, it entails information about the time–time component of the Einstein tensor: $\boldsymbol{G}(\boldsymbol{u}_{\mathrm{R}}, \boldsymbol{u}_{\mathrm{R}})$. Demanding that the result holds for all accelerated observers, this actually becomes a statement about the full Einstein tensor, $\boldsymbol{G}$. I also present some perspectives on the free fall with four-velocity $\boldsymbol{u}_{\mathrm{ff}}$ across the horizon that leads to such a loss of entropy for an accelerated observer. Motivated by results for some simple quantum systems at finite temperature T, we conjecture that at high temperatures, there exists a universal, system-independent curvature correction to partition function and thermal entropy of *any* freely falling system, characterised by the dimensionless quantity $\Delta = \boldsymbol{R}(\boldsymbol{u}_{\mathrm{ff}}, \boldsymbol{u}_{\mathrm{ff}})\,(\hbar c/kT)^2$.

1 Gravity and Thermodynamics

It has been well known for a long time that statistical mechanics in presence of gravitational interactions exhibits several peculiar features [1], deriving mostly from the fact that gravity couples to *everything*, and operates *unshielded* with an *infinite range*. Many of these peculiarities, such as negative specific heat, however, attracted attention only after they were encountered in the context of black holes. Indeed, existence of a horizon magnifies quantum effects in presence of a black hole, revealing a gamut of exotic features, the most famous being its thermal attributes [2]. This *gravity* $\leftrightarrow$ *quantum* $\leftrightarrow$ *thermodynamics* connection has been gaining increasing attention in recent years due to it's potential relevance for our understanding of gravity, and perhaps spacetime itself, at a fundamental level. In particular, the fact

D. Kothawala (✉)
Indian Institute of Technology Madras, Chennai, India
e-mail: dawood.ak@gmail.com

J.S. Bagla and S. Engineer (eds.), *Gravity and the Quantum*,
Fundamental Theories of Physics 187, DOI 10.1007/978-3-319-51700-1_12

that black holes have thermodynamic properties has time and again motivated the intriguing question: *is there a deeper clue hidden in the connection between gravitational dynamics and thermodynamics*? One is naturally lead to suspect whether the *dynamics* of gravity itself has a built in thermodynamic structure. Moreover, since these thermal attributes are essentially quantum mechanical in origin, they are particularly relevant in the context of quantum gravity.

This article zooms into two very specific questions in this context:

- What information about **spacetime curvature** can be obtained from **thermal properties** of local **acceleration** horizons?

  The first part of this article largely focuses on exploring the spacetime geometry in the vicinity of an accelerated observer in the hope that it will clarify the connection between gravity and thermodynamics. Specifically, I will try to clarify the relation between *area variation* and Einstein tensor

$$T \underset{?}{\delta \mathscr{A}} \underset{?}{\longleftrightarrow} \boldsymbol{G}(\boldsymbol{u}_{\mathrm{R}}, \boldsymbol{u}_{\mathrm{R}})$$

  by trying to answer each of the "?"s in the above expression. Such a relation has been known and discussed in various forms in several works; I hope the analysis given here would help clarify it's mathematical origin.
- What are the effects of **spacetime curvature** on **thermal properties** of a **freely falling** quantum system?

  I then present some speculations on geodesic free fall of a thermal system across the local horizon of such an observer. Motivated by results from a couple of examples, I conjecture that the partition function and thermal entropy of *any* quantum system at high temperatures acquire a system independent correction term characterised by the dimensional quantity

$$\Delta = \boldsymbol{R}(\boldsymbol{u}_{\mathrm{ff}}, \boldsymbol{u}_{\mathrm{ff}})\, (\hbar c/kT)^2$$

  which contributes an (entropy)/(degree of freedom) of

$$s_{\Delta} = (\text{const.})\, \boldsymbol{R}(\boldsymbol{u}_{\mathrm{ff}}, \boldsymbol{u}_{\mathrm{ff}})\, (\hbar c/kT)^2 \tag{1}$$

Each of the highlighted words in the above two questions carries physical significance of it's own; note that the only difference between the two is that of accelerated versus inertial motions.

I feel a deeper analysis of these questions is important for any research that uses the connection between gravity and thermodynamics as a starting point. One example would be the assertion that there is more to gravitational dynamics than the Einstein–Hilbert action – something that has been the focus of Paddy's research over the past decade. This article presents some of the lesser known results, both mathematical and physical, that could be important in a better understanding of such an assertion.[1]

2 Gravity $\Longleftrightarrow$ Accelerated Frames $\Longleftrightarrow$ Thermodynamics

Once we realise that certain surfaces - *horizons* - must be attributed entropy by certain class of observers, there is a tantalizing possibility of *imposing* the first law of thermodynamics and then see how it constrains the curvature of background spacetime. It is possible that such a study can also yield information about Einstein field equations themselves.

One physically relevant set-up for addressing this issue was given by Jacobson, based on usage of local acceleration (Rindler) horizons and the Raychaudhuri equation [3]. Subsequently, several generalisations as well as variations of this work have appeared in the literature. However, not all of them share the same physical and/or technical assumptions. Several such issues have been discussed in [4, 5]. At some level, some of these issues do become important when one actually probes the structure of gravitational field equations and look for some connection with some thermodynamic law, a route that was first taken by Padmanabhan and has subsequently been studied in much more generality [6]. The relevant structure of field equations (or technically, the Einstein tensor) is of obvious relevance to any argument(s) which attempt to reverse the logic and derive field equations from thermodynamic considerations. A clarification of some of the subtle mathematical issues involved in this program was presented in [4, 5].

In this note, we present a calculation directly in the local frame of the accelerated observer, described by Fermi coordinates, that illuminates several of the points concerning the structure of Einstein tensor and it's role in interpreting the field equations in thermodynamic terms. That is, we investigate the local spacetime structure by employing, as probes, accelerated trajectories and using the fact that such trajectories will have a local horizon attributed with the standard Unruh temperature [7].

This analysis is therefore complimentary, though not completely identical, to the one presented in [4]. It has the advantage that it very clearly demonstrates how various curvature components combine, in a rather specific manner, *through the transverse horizon area*, to reproduce the relevant (time–time) component of the Einstein tensor. *This mathematical fact lies at the heart of any attempt to relate gravitational dynamics with horizon thermodynamics.* From a broader perspective, the calculations presented here also re-emphasize the points made earlier in [4] concerning the difference(s), both technical and conceptual, between results arising from exploring thermodynamic structure of field equations and attempts to derive the field equations from a specific thermodynamic statement (see [5] for a more recent

---

[1]*Notation*: The metric signature is $(-, +, +, +)$; latin indices go from 0 to 3, greek indices from 1 to 3.

comprehensive discussion of these issues); in particular, the former is *not* implied by the latter

The key point clarified and highlighted by the analysis presented here is *how* and *why* the "time–time" component of the field equations (equivalently, of the Einstein tensor) acquires a direct thermodynamic relevance.[2] This forms the main focus of this paper. Appendix "A Simple Toy Model for a Point Mass Disappearing Across the Horizon" gives a simple toy model for quantifying the shift in location of the horizon as some form of energy crosses it. Appendix "Surface Term in Fermi Coordinates" presents the (leading order) form of the surface term in the Einstein–Hilbert action in the local coordinates of the accelerated observer in curved spacetime. This term is known to yield horizon entropy in flat spacetime, and hence one expects the curvature corrections to the same as also having direct thermodynamic significance.

2.1 Virtual Displacements of Local "Rindler" Horizons and Einstein Tensor

It is well known that accelerated observers in *flat spacetime* perceive the vacuumas a thermal bath; therefore, it is natural to start looking for the connection between gravity and thermodynamics by considering accelerated observers in *curved spacetime*. Consider, then, such an observer (or, more precisely, an accelerated timelike trajectory), and construct a locally inertial system of coordinates based on his/her trajectory. Such a coordinate system can be obtained by Fermi-Walker transporting the observer's orthonormal tetrad along the trajectory; the coordinate system so obtained is called the Fermi coordinate (FNC) system [9]. In FNC's (τ, x^μ), the metric acquires the following form:

$$
\begin{aligned}
g_{00} &= -\left[\left(1 + a_\mu x^\mu\right)^2 + R_{0\mu 0\nu} x^\mu x^\nu\right] + O(x^3) \\
g_{0\mu} &= -\frac{2}{3} R_{0\rho\mu\sigma} x^\rho x^\sigma + O(x^3) \\
g_{\mu\nu} &= \delta_{\mu\nu} - \frac{1}{3} R_{\mu\rho\nu\sigma} x^\rho x^\sigma + O(x^3)
\end{aligned}
\tag{2}
$$

where a^μ and R_{abcd} are all functions of τ. We shall assume that we are working in a sufficiently small region of space, and focussing on sufficiently small time scales, so as to ignore the time dependence of metric components. Technically, this is equivalent to assuming the existence of a static timelike Killing vector in the region of interest. For definiteness, we will take our observer to be moving along the x^3 direction; i.e., $a^\mu = a\delta^\mu_3$, where a is the norm of the acceleration. This can be always

---

[2]It is always this component that has appeared in works based on [6]; it's significance in related contexts is also becoming evident in some recent works (compare, for example, [3, 8]).

be done by an appropriate choice of the basis vectors: Given the four velocity $\boldsymbol{u}$ of the observer, one chooses the basis vectors such that, $\boldsymbol{e}_0 = \boldsymbol{u}$ and $\boldsymbol{e}_3 = (1/a)\ \boldsymbol{a}$, so that $\boldsymbol{e}_0 \cdot \boldsymbol{e}_3 = 0$. These basis vectors are then Fermi-Walker transported along the trajectory, $\nabla_{\boldsymbol{u}} \boldsymbol{e}_k = (\boldsymbol{u} \otimes \boldsymbol{a} - \boldsymbol{a} \otimes \boldsymbol{u}) \cdot \boldsymbol{e}_k$ for $k = 0 .. 3$. The remaining two basis vectors can be suitably chosen to be orthogonal to $\boldsymbol{e}_0$ and $\boldsymbol{e}_3$. In such a coordinate system, the only non-zero component of the acceleration 4-vector is $\boldsymbol{a}_3 = \boldsymbol{a} \cdot \boldsymbol{e}_3 = a$.

The FNC system (to quadratic order) describes the geometry in the neighbourhood of an accelerated observer to a very good accuracy for length scales $x \ll \min \{ |a|^{-1}, R^{-1/2}, |R/\partial R| \}$. We must now choose the acceleration such that the length scale $|a|^{-1}$ associated with it is much smaller than the curvature dependent terms above; then the presence of the horizon along with it's thermodynamic attributes will not be affected by curvature. Specifically, the horizon for such an observer is then located at $x^3 = -a^{-1} = z_0$, say.

The transverse area of the horizon 2-surface, $x^3 = z_0$, is given by

$$\sigma_{\mathrm{AB}} = \delta_{\mathrm{AB}} - \frac{1}{3} R_{\mathrm{A}\mu\mathrm{B}\nu} x^\mu x^\nu \tag{3}$$

so that

$$\sqrt{\sigma} = 1 - \frac{1}{6} R_{\mathrm{A}\mu\mathrm{A}\nu} x^\mu x^\nu \tag{4}$$

$$= 1 - \frac{1}{6} \left[R^{\mathrm{A3}}_{\ \ \mathrm{A3}} z_0^2 + 2 R^{\mathrm{A3}}_{\ \ \mathrm{AB}}\, z_0\, y^{\mathrm{B}} + R^{\mathrm{AC}}_{\ \ \mathrm{BD}} x^{\mathrm{C}} x^{\mathrm{D}} \right] \tag{5}$$

where we have used the Euclidean metric to raise the spatial indices on the curvature tensor, since the curvature tensor is evaluated on the observers' worldline.

We now wish to consider some physical process which produces a virtual displacement of the horizon normal to itself (that is, along the x^3 direction), and see how the horizon area changes in such a process (see Appendix "A Simple Toy Model for a Point Mass Disappearing Across the Horizon" for a toy model of such a process). The displacement is to be considered as a *virtual displacement*; we do *not* have in mind an observer with a different acceleration, which would require us to construct a different coordinate system based on the new worldline. Rather, the idea here is to see whether the differential equation governing the virtual displacement of the horizon has any physical meaning. We shall therefore consider the parametric dependence of the transverse area on $z_0 = -a^{-1}$, $\mathscr{A}(x, y; z_0)$, and then consider the variation,

$$\begin{aligned} z_0 &\rightarrow z_0 + \varepsilon \\ \varepsilon &= \Delta z_0 \end{aligned} \tag{6}$$

Since we generally associate an entropy $S \propto \mathscr{A}$ with a horizon surface, and a local Unruh temperature $T_U = a/(2\pi)$ with an accelerated observer, we expect the resultant equation for involving $\delta\mathscr{A}$ to have a thermodynamic interpretation.

We have,

$$\delta_{z_0}\sqrt{\sigma} = \sqrt{\sigma}|_{z_0+\varepsilon} - \sqrt{\sigma}|_{z_0}$$
$$= -\frac{1}{3}\left[R^{A3}_{\ A3} z_0 + R^{A3}_{\ AB}\, x^{B}\right]\varepsilon + O(\varepsilon^2) \tag{7}$$

where the subscript ε reminds us that we are dealing with a very specific variation. Now concentrate on a small patch of the horizon surface. It is then not unreasonable to assume that, upon integration over the transverse coordinates, the second term in the square brackets averages out to zero. (If it doesn't, we would have a preferred direction in the transverse horizon surface.) We shall nevertheless come back to this point later.

The change in area of this surface under the virtual displacement of the horizon is therefore given by

$$\delta_{z_0}\mathscr{A} = \int_{\mathscr{H}} \left(\delta_{z_0}\sqrt{\sigma}\right) \mathrm{d}^2 x_\perp \tag{8}$$

where $\mathrm{d}^2 x_\perp = \mathrm{d}x^1 \mathrm{d}x^2$, and $\mathscr{H}$ represents integration over the horizon surface $\tau =$ constant, $x^3 = z_0$. With $T_U = -1/(2\pi z_0)$ for $a > 0$, we have

$$T_U\ \delta_{z_0}\left(\frac{1}{4}\mathscr{A}\right) = \frac{\eta}{8\pi}\int_{x^3=z_0} R^{3A}_{\ 3A}\ \varepsilon\ \mathrm{d}^2 x_\perp \tag{9}$$

where $\eta = 1/3$. If one assumes the standard Bekenstein–Hawking expression for entropy of a horizon, $S = \mathscr{A}/4$, then the left hand side above is just $T_U\ \delta S$. To simplify the right hand side, we use the general decomposition of the Einstein tensor in terms of components of the curvature tensor. This is given by

$$G^0_{\ 0} = -\left(R^{12}_{\ 12} + R^{13}_{\ 13} + R^{23}_{\ 23}\right) \tag{10}$$

which implies

$$R^{3A}_{\ 3A} = -G^0_{\ 0} - R^{12}_{\ 12} \tag{11}$$

Then, we obtain

$$T_U\ \delta_{z_0} S = -\eta\left[\int_{\mathscr{H}} \frac{1}{8\pi}\, G^0_{\ 0}\, \varepsilon\, \mathrm{d}^2 x_\perp + \int_{\mathscr{H}} \frac{1}{16\pi}\, R_{\parallel}\, \varepsilon\, \mathrm{d}^2 x_\perp\right] \tag{12}$$

where $R_{\parallel} = R^{AB}_{\ AB}$ is the Ricci scalar of the in-horizon 2-surface. This is essentially the relation we wanted to establish. Note that, to the relevant order, one can consider

$d^2x_{\perp}$ on the RHS of Eq. (12) as the covariant measure of horizon area; since it is already multiplied by curvature components, any curvature corrections to volume would be higher order. The appearance of G^0_0 is already quite suggestive, and, in what follows, we argue that the second term also has a nice interpretation which qualifies it as a suitable thermodynamic variable.

2.1.1 Horizon Energy

We now provide a suitable interpretation for the second integral in Eq. (12). We begin by noting that the Euler character χ of a two dimensional manifold $\mathcal{M}_2$, in this case the horizon 2-surface, is given by

$$\chi\left(\mathcal{M}_2\right) = \frac{1}{4\pi}\int_{\mathcal{M}_2} R\ \mathrm{d}[\mathrm{vol}]\ +\ \text{boundary terms} \tag{13}$$

The second term in the square brackets in Eq. (12) therefore just $(\chi/4)\varepsilon$ plus the boundary term, which involves the integral of extrinsic curvature over the boundary of the region of integration. Keeping all these points in mind, we will simply call the second integral in Eq. (12) as $\hat{\chi}$. We want to interpret this term as change in "energy" associated with the horizon, say E_g. This expression is already well known in the context of quasilocal energy in spherically symmetric spacetimes. Let us consider the case of a general, spherically symmetric black hole. This term then essentially involves the curvature component, $R^{\theta\phi}{}_{\theta\phi} = (1/r^2)\ (1 - g^{rr})$ in standard coordinates. On the horizon $r = a$, g^{rr} vanishes and we obtain, after multiplying with the appropriate transverse area element, $E_g = a/2$, which is a very common expression for energy (called as the Misner–Sharp energy). In Appendix "A Simple Toy Model for a Point Mass Disappearing Across the Horizon", we give a toy model which further analyses this expression for (change in) energy in terms of horizon shift.

2.2 *Final Result and Discussion*

If one employs the Einstein field equations $G^0_0 = 8\pi T^0_0$ at this stage (where $T^0_0 = -\rho_m$ is the local energy density of matter in the instantaneous rest frame of the observer), Eq. (12) becomes

$$T_U\ \delta_\varepsilon S = -\eta\left[\left(\int_{\mathcal{H}} T^0_0 \mathrm{d}^2x_{\perp}\right)\ \Delta z_0 + \frac{\hat{\chi}}{2}\left(\frac{\Delta z_0}{2}\right)\right] \tag{14}$$

where the second equality is based on discussion of the previous section. This is our main result, which forms the basis for thermodynamic interpretation of field equations. For example, for static spacetimes, the above relation (apart from the factor $(-\eta)$), can be shown to have the form [10]

$$T\delta S = \bar{F}\Delta z_0 + \mathrm{d}E_g \tag{15}$$

where $\bar{F}$ is the average normal force on the horizon surface (see Eq. (21) of [10]).

It is gratifying to see that the final expression is in a form which can be readily expressed in any arbitrary coordinate system; it only depends upon the foliation provided by the accelerated observer. Given $\boldsymbol{u}$ and $\boldsymbol{a}$, one can identify the spacelike two-surface acting as local horizon, and the corresponding "heat" flow then depends only on $R_{\parallel}$ – the curvature scalar of the two surface, and $G^0_{\ 0} = -\boldsymbol{G}(\boldsymbol{u}, \boldsymbol{u})$ – which is the projection of Einstein tensor along the observer's time axis. This is as far as one can get trying to explore the connection between intrinsic properties of local Rindler horizons and gravitational dynamics.

To evaluate how rigorous the analysis is, let us take stock of the assumptions that have gone into the derivation:

(i) The acceleration length scale be small compared to any of the curvature length scales: This assumption is natural, given that the whole idea is to use solely the acceleration of probe observes and find constraints on background geometry - a natural physical setup to formulate the problem, sanctioned by the equivalence principle.
(ii) The assumption mentioned just below Eq. (7): This requires some consideration, since it is possible that the term contributes in a sensible manner to some form of energy associated with some geometric characteristic of the horizon. It might introduce additional stresses in the first law (similar to those resulting from, say, angular momentum of the horizon patch).

The only technical issue is that, this derivation, while as general as it can be, yields an extra factor of $-\eta = -1/3$ whose interpretation is unclear. Otherwise, the analysis very clearly indicates how the gravitational dynamics of a curved spacetime is intimately related to the property of spacelike two surfaces which can act as horizons for certain observers. In fact, as we have shown above, the gravitational dynamics (or more specifically, the Einstein tensor) is completely encoded in the transverse geometry of the spacelike two surface and it's normal variation, to which a physical interpretation can be attached via thermodynamic quantities such as entropy and temperature associated with the horizon [11].

It is indeed curious that we can make any statement at all concerning the dynamics of gravity by (i) assuming a curved Lorentzian spacetime with it's associated light-cones, and (ii) a temperature associated with local Rindler horizons. Curiously, the curvature tensor makes it's appearance through the thermodynamic variables T_U and δS, rather than the conventional *tidal* terms. The reason is that, apart from the laboratory length scale, we now also have the length scale c^2/a associated with acceleration. The presence of a causal horizon at $z_0 = -1/a$ and it's associated thermodynamics is what brings in the curvature information through *quantum* effects. The non-trivial part being that it brings in just the right combination of curvature components to facilitate making a general statement about the dynamics of gravity.

More succinctly, just as ***accelerated frames*** *played a key role in arriving at a* ***kinematic*** *description of gravity in terms of spacetime geometry,* ***virtual displacements of acceleration horizons****, through associated quantum effects, might play a key role in understanding better the* ***dynamics*** *of gravity.*

We hope we have provided one of the most straightforward demonstrations of the above ideas. In the next section, we focus attention to the fate of actual thermal systems in free fall in a curved spacetime, and conjecture on a universal tiny contribution to thermal entropy of *any* system at sufficiently high temperatures.

3 Thermal Entropy and Spacetime Curvature

Most of the work investigating the connection between gravity and thermodynamics so far has focussed on accelerated observers and the thermal effects associated with their horizons. However, all such analyses invariably are based on a scenario in which a system disappears across the acceleration horizon, carrying entropy with it that is *lost* to the accelerated observer. Indeed, it is this question (posed in the context of black holes) of Wheeler's that had prompted Bekenstein to investigate the physical basis behind the laws of black hole mechanics [12]. Unfortunately, however, not much attention has been given to the statistical mechanics of quantum systems that are (in some suitably defined sense) in a free fall in a curved spacetime.

It is precisely this problem that motivated the analysis in [13], two of the key points of interest being: (i) to study the interplay between quantum and thermal fluctuations in presence of spacetime curvature, (ii) to analyse whether this can have any implications for understanding the physical nature of black hole entropy. Here, I will focus only on point (i). Since a general analysis of statistical mechanics of a quantum system at finite temperature, in a curved spacetime, is understandably difficult, the above questions were addressed in [13] in the context of a very simple system: non-relativistic ideal gas enclosed in a box whose (geometric center) follows a geodesic $\boldsymbol{u}_{\rm ff}$. The choice of central geodesic only serves to define the reference measure for energy of the system; in particular, the Hamiltonian of the particles is defined as

$$H = -c\, p_{\widehat{0}} \tag{16}$$

where $\widehat{0}$ is the time coordinate defined by $\boldsymbol{u}_{\rm ff}$. More discussion on the physical significance of this Hamiltonian, as well as it's detailed form, can be found in [13].

In the non-relativistic limit, one obtains

$$H - mc^2 \approx \frac{p^2}{2m} + \frac{1}{2} mc^2 R_{\hat{0}\mu\hat{0}\nu} y^\mu y^\nu \tag{17}$$

I will also ignore the time dependence carried by curvature components, a reasonable assumption if the time scale $\mathscr{R}/\dot{\mathscr{R}}$ is much larger compared to typical time scale associated with the system; one expects this to be the case at high temperature (see below).

One can now use the above Hamiltonian to evaluate the energy eigenvalues $E\left[\{n\}\right]$ of constituent particles of a system (where $\{n\}$ represents the set of all relevant quantum numbers) at temperature $kT = 1/\beta$, and therefore the partition function

$$Z(\beta) = \sum_{\{n_i\}} e^{-\beta E[\{n_i\}]} \tag{18}$$

The one-particle energy eigenvalues have the form

$$E_1\left[\{n_i\}\right] = E_{1,0}\left[\{n_i\}\right] + \Delta E_1\left[\{n_i\}\right] \tag{19}$$

where the first term on RHS represents unperturbed energy eigenvalues, while the second term is the perturbation cause by spacetime curvature. The corresponding partition has the form The one-particle energy eigenvalues have the form

$$Z(\beta) = Z_0(\beta) + \Delta Z\left(\beta, R_{abcd}, \{\varkappa\}\right) \tag{20}$$

where again, the first term on RHS is the flat spacetime expression and the second term the first order modification due to spacetime curvature; $\{\varkappa\}$ symbolically denotes all the system specific parameters, such as mass, physical dimensions etc.

Our key conjecture would be that:

$$\lim_{\text{large } T} \Delta Z\left(\beta, R_{abcd}, \{\varkappa\}\right) \tag{21}$$

generically contains a term independent of $\{\varkappa\}$, and has the form

$$\lim_{\text{large } T} \Delta Z\left(\beta, R_{abcd}, \{\varkappa\}\right) = (\text{const.}) R_{00} \Lambda^2 + \{\varkappa\} \text{ dependent terms} \tag{22}$$

Once again, such a result would imply that there is a universal, inherent, Ricci contribution to thermodynamic quantities of a system in curved spacetime, which, for obvious reasons, can have great implications when a system at finite temperature disappears across a causal horizon of some observer.

Box of Ideal Gas

The calculation mentioned above was carried out in [13] for a box of ideal gas freely falling in a curved spacetime, and it was shown that, in the regime where the analysis was valid ($\lambda^3/V \ll 1$ where $\lambda = h/\sqrt{2\pi mkT}$ is the thermal de Broglie wavelength), the partition function has the form:

$$\ln\left(\frac{Z}{Z_{\rm F}}\right) = -\frac{1}{4} N R_{\widehat{0}\widehat{0}} \Lambda^2 + \text{terms depending on curvature and box details} \qquad (23)$$

where $\ln Z_{\rm F} = \ln(V^N \lambda^{-3N}/N!)$ is the flat space expression, and $\Lambda = \beta\hbar c$ – a length scale independent of box dimensions L_i and mass m.

We can now obtain corrections to various thermodynamic quantities: $U_{\rm corr} = U - U_{\rm F}$ and $S_{\rm corr} = S - S_{\rm F}$, where $U_{\rm F} = 3N/2\beta$ and $S_{\rm F} = 3N/2 + N\ln\left(eV/N\lambda^3\right)$ are standard flat space expressions. Using standard definitions $U = -\partial_\beta \ln Z$ and $S = \ln Z + \beta U$ to evaluate $U_{\rm corr}$, $S_{\rm corr}$ and heat capacity at constant volume, $C_V = -\beta^2\partial_\beta U = 3N/2 + C_{V{\rm corr}}$, we obtain

$$\left.\begin{aligned} 2S_{\rm corr}/N &= \underbrace{+(1/2) R_{\widehat{0}\widehat{0}} \Lambda^2}_{2s_\Delta/N} \\ \beta U_{\rm corr}/N &= \underbrace{+(1/2) R_{\widehat{0}\widehat{0}} \Lambda^2}_{\beta u_\Delta/N} \\ C_{V{\rm corr}}/N &= \underbrace{-(1/2) R_{\widehat{0}\widehat{0}} \Lambda^2}_{c_\Delta/N} \end{aligned}\right\} + \text{system dependent terms}$$

Before proceeding, it is worth pausing to check whether the approximations made to arrive at the above result(s) are not mutually inconsistent. This is an important issue, and is discussed at length in Appendix "Validity of the Approximations for the Ideal Gas Calculation".

Simple Harmonic Oscillator

One can also do the above analysis for a bunch of simple harmonic oscillators with frequency ω [14] (which might be physically more relevant system for obvious reasons); in this case, one obtains, in the high temperature limit $\beta\hbar\omega \ll 1$

$$\left.\begin{aligned} 2S_{\rm corr}/N &= \underbrace{+(1/12) R_{\widehat{0}\widehat{0}} \Lambda^2}_{2s_\Delta/N} \\ \beta U_{\rm corr}/N &= \underbrace{+(1/12) R_{\widehat{0}\widehat{0}} \Lambda^2}_{\beta u_\Delta/N} \\ C_{V{\rm corr}}/N &= \underbrace{-(1/12) R_{\widehat{0}\widehat{0}} \Lambda^2}_{c_\Delta/N} \end{aligned}\right\} + \text{system dependent terms}$$

The system dependent terms not indicated above can be found in [13, 14], and they essentially involve terms that depend on m, L_i in the case of ideal gas, and the frequency ω in the case of harmonic oscillator.

I now wish to highlight some of the key features of this result:

- Perhaps the most important point to be noted is the following: For the example of ideal gas, the curvature dependent correction term $\Delta E_1\,[\{n_i\}]$ in Eq. (19) turns out to be independent of $\hbar$! (The full expression is given in [13].) The quantum*ness* of this term is solely due to the $\{n_i\}$ dependence of $\Delta E_1\,[\{n_i\}]$, and manifests itself in the final expressions since $\Lambda \propto \hbar$. This is a nice demonstration of how the interplay between quantum and thermal fluctuations induced by a background spacetime curvature can be non-trivial.
- Further, s_Δ and u_Δ satisfy the relation: $s_\Delta = (1/2)\beta u_\Delta$ with s_Δ as mentioned in the Introduction (Eq. (1)). This is a Euler relation of homogeneity two, well known from black hole thermodynamics; in particular, black hole horizons have temperature β_H^{-1}, entropy $S_{\rm bh}$ and (Komar) energy $U_{\rm bh}$ which also satisfy $S_{\rm bh} = (1/2)\beta_H U_{\rm bh}$. Relevance of such Euler relation and area scaling of entropy for self-gravitating systems has already been emphasized in [15]. This relation also plays an important role in the *emergent gravity* paradigm, leading to an *equipartition law* for microscopic degrees of freedom associated with spacetime horizons [16].
- The Δ contribution to specific heat is *negative* if the condition ($R_{\widehat{0}\widehat{0}} \geq 0$) holds. (This condition is, of course, tied to the strong-energy condition if Einstein equations are assumed.) Also, $c_\Delta = -\beta u_\Delta = -2 s_\Delta$, which are again the same as the relations satisfied by a Schwarzschild black hole.
- The appearance of the length scale $\Lambda = \hbar c/kT$ in the non-relativistic limit is curious, and it would be interesting to understand the physical meaning of this length scale at the basic level [17].

3.1 Speculation

All the above points are extremely suggestive as far as the role of Ricci correction to thermodynamic properties of arbitrary systems is concerned. In fact, a similar analysis can be done for a harmonic oscillator, and it can again be shown that at sufficiently high temperatures, thermodynamic quantities acquire specific correction terms which are independent of the frequency ω of the oscillator [14].

In particular, based on these, one can speculate the following:

Entropy of a system at temperature T generically acquires a system independent contribution in a curved spacetime characterized by the dimensionless quantity

$$\Delta = \boldsymbol{R}(\boldsymbol{u}_{\rm ff}, \boldsymbol{u}_{\rm ff})\,(\hbar c/kT)^2 \tag{24}$$

at sufficiently large temperatures T.

This should motivate further study of thermal systems in curved spacetime along the suggested route. Note that the important features associated with the Ricci corrections would not appear if: (i) one studies the problem using *classical* statistical mechanics – since then the modification of energy eigenvalues, $1/n^2$ in present case, is missed), or (ii) assume a priori that *finite size corrections* would necessarily depend only on area, perimeter etc. – the Ricci term here in fact does not involve dimensions of the box at all! For the result to have any deeper significance, it's main *qualitative* aspects (in the high temperature limit) must, of course, survive further generalizations (relativistic gas, different statistics etc.), insofar as the *form* of the Ricci term is concerned. Some preliminary calculations do seem to point to this [18].

3.2 *Possible Implications*

So, what could be possible implications of this?

Well, for one, if there does exist a universal term in entropy of systems at high temperature that depends on the Ricci tensor, it might lead to some insight into understanding the interplay between thermal and quantum fluctuations in a curved spacetime.

Second, such analyses can add interesting (and non-trivial) physics to arguments given by Bekenstein in his original paper [12] in support of the so called generalised second law (GSL). Most of Bekenstein's supporting analysis used expressions for thermal energy and entropy of various systems in flat spacetime, added to the (minimum) change in black hole entropy when such a system falls across the black hole horizon. It is, of course, important and interesting to know how curvature corrections to thermodynamic attributes of a system affect this analysis.

More specifically, one would like to know *quantitatively* how and where the Δ term appears in the proposed GSL:

$$\Delta S_{\rm BH} + \Delta S_{\rm ext} > 0$$

where $\Delta S_{\rm BH}$ represents the change in entropy of the black hole, and $\Delta S_{\rm ext} = \Delta S_{\rm ext}\,(\beta,\, R_{abcd},\, \{\varkappa\})$ is the change in common entropy in the region exterior to the black hole. This is work under progress.

And lastly, appearance of such a term can be of direct relevance for understanding of thermodynamical aspects of gravitational dynamics.

4 Concluding Remarks

> Now, here, you see, it takes all the running you can do, to keep in the same place. If you want to get somewhere else, you must run at least twice as fast as that!
>
> — Lewis Carroll, *Through the Looking Glass*

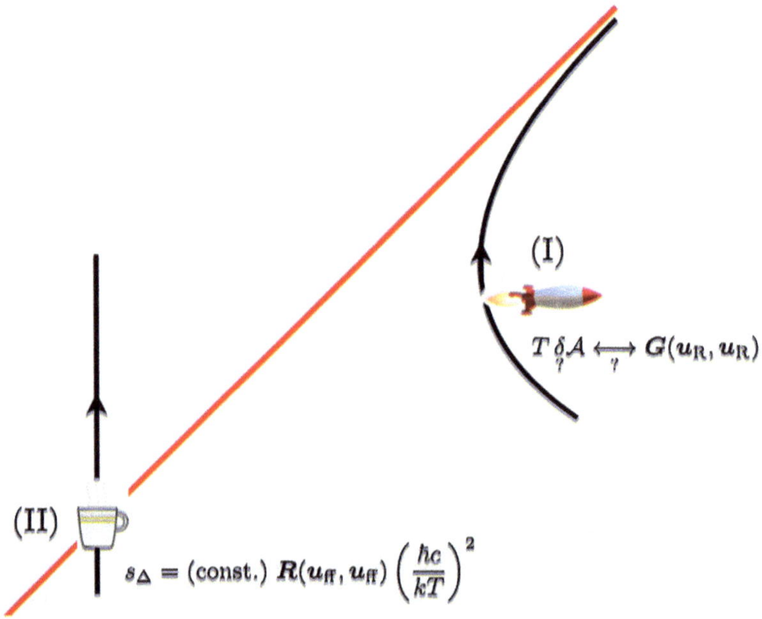

Fig. 1 A combined description of a freely falling thermal system *and* the thermodynamics associated with an accelerated observer can yield physically useful insights into interplay between quantum mechanics, thermodynamics, and spacetime curvature

Since all the relevant comments and remarks have been given in the respective sections, I conclude with a pictorial summary of the theme of this article, depicted in Fig. 1.

Study of thermodynamic aspects of gravity, which derives it's motivation from semi-classical results such as the Hawking and Unruh effects, does provide an elegant route to make some deep observations concerning the nature of gravity. However, such considerations, by themselves, might turn out to be *too elegant* to be of any use, unless coupled with a deeper study of the structure of statistical mechanics in a curved spacetime.

Acknowledgements I thank Paddy for many discussions and comments on these and related topics over a course of almost ten years. The support of Department of Science and Technology (DST), India, through it's INSPIRE Faculty Award, is gratefully acknowledged.

Appendix

A Simple Toy Model for a Point Mass Disappearing Across the Horizon

In this section, let us try to model the loss of energy across a local Rindler horizon via a particle of mass m that disappears across the horizon. We will present the analysis for the case when the background spacetime is flat in the limit $m \to 0$; that is, m is the only source of curvature. As we shall comment in the end, this suffices as long as one is working to first order in background curvature.

A complete dynamical description of this process is expected to be complicated, but since we are only interested in shift of the horizon as the particle disappears across it, we propose the following simplified scenario: we consider the particle when it is on the verge of crossing the horizon, that is, at $x^\mu \approx -(1/a)\delta_3^\mu - 0, t = t_0$, and compare this with a situation when it is no longer in the causal domain of the accelerated observer. The shift in horizon position can then be evaluated by using an orthonormal tetrad for this particle that maps smoothly to that of the accelerated observer (in an asymptotic sense, see below). This is most easily done by using the Schwarzschild metric in isotropic coordinates y^μ for the particle; the metric near m in these coordinates is given by

$$ds^2 = -\left(\frac{1-\frac{m}{2r}}{1+\frac{m}{2r}}\right)^2 dt^2 + \left(1+\frac{m}{2r}\right)^4 dl_{\text{flat}}^2 \tag{25}$$

where

$$dl_{\text{flat}}^2 = \delta_{\mu\nu} dy^\mu dy^\nu \tag{26}$$

$$r = \sqrt{\delta_{\mu\nu} y^\mu y^\nu} \tag{27}$$

and we shall be interested in $r \gg m$. In the limit $r \to \infty$, the curvature components in these coordinates are given by (with $r_0 = 2m$, and *no summation* over repeated indices)

$$R_{0\mu0\mu} \to -\frac{r_0}{r^3} + O\left(\frac{1}{r^4}\right) \qquad \text{(no summation)}$$

$$R_{0\mu0\nu} \to -\frac{3}{2}\left(\frac{r_0}{r^5}\right) y_\mu y_\nu + O\left(\frac{1}{r^6}\right) \qquad (\mu \neq \nu)$$

$$R_{\mu\nu\mu\nu} \to -\frac{r_0}{2r^3} + O\left(\frac{1}{r^4}\right) \qquad (\mu \neq \nu)$$

$$R_{\mu\nu\mu\lambda} \to -\frac{3}{2}\left(\frac{r_0}{r^5}\right) y_\nu y_\lambda + O\left(\frac{1}{r^6}\right) \qquad (\mu \neq \nu \neq \lambda) \tag{28}$$

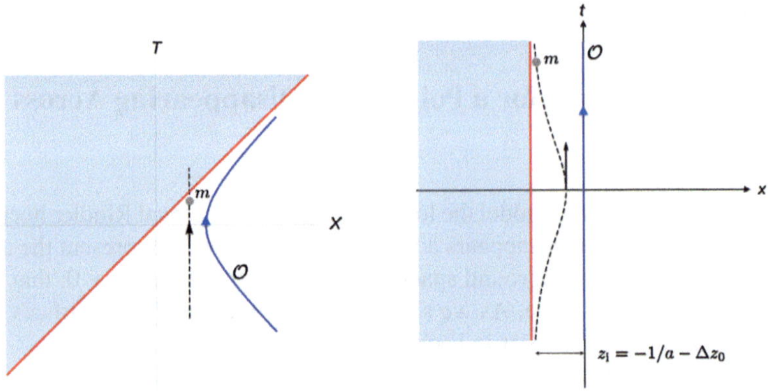

(a) Inertial coordinates (b) Rindler coordinates

Fig. 2 A particle of mass m crossing a local Rindler horizon of $\mathcal{O}$; Inertial and Rindler perspectives

We now wish to compare two situations: an accelerated observer in flat spacetime, and an accelerated observer *with this same acceleration* but in presence of the mass m. We expect that such a comparison would provide a natural setting to study what happens when m disappears from the causal domain of the accelerated observer, when it crosses the horizon; in effect, we are accounting for the effect of the particle by considering how the curvature produced by it changes the horizon location (see Fig. 2). With this in mind, we can go ahead with the calculation. We are mainly interested in knowing how the horizon location changes when the perturbing mass m moves "across" the horizon (see Fig. 3). To lowest order, this can be done by setting $g_{00} = 0$, and it is easy to see that one obtains,

$$z_{0\pm} = -\frac{1}{a}\left(P \mp \sqrt{P^2 - Q}\right) \tag{29}$$

where

$$P = 1 + a^{-1} R_{030\mathrm{A}} y^{\mathrm{A}} - a^{-2} R_{0303} \tag{30}$$

$$Q = 1 + R_{0\mathrm{A}0\mathrm{B}} y^{\mathrm{A}} y^{\mathrm{B}} - a^{-2} R_{0303} \tag{31}$$

From Eqs. (28), we see that in the limit $r \to \infty$, it is only the R_{0303} term that matters. Also, we make a further *assumption* of using the average of the two roots above, $z_{0\,\mathrm{avg}} = -P/a$, to quantify horizon displacement (without this assumption, it is unclear which root to pick, and further, it is not clear what terms containing square root of curvature tensor would mean). We then obtain (see Fig. 3)

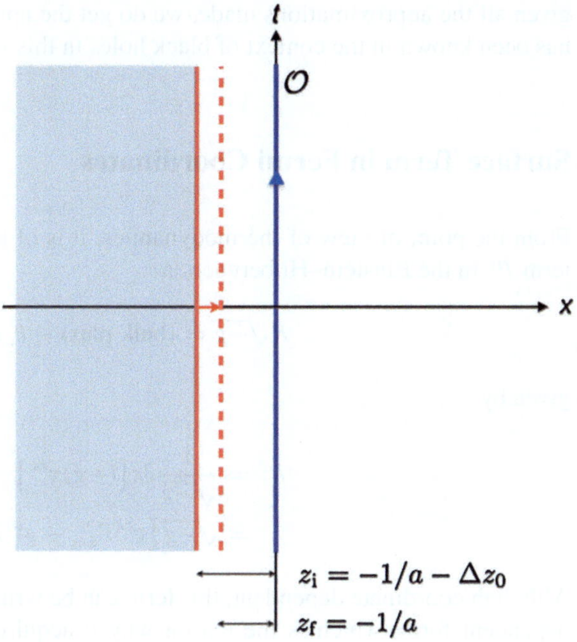

Fig. 3 The shift of horizon due to "loss" of mass m (see text for details)

$$\begin{aligned}\Delta z_0 &= z_f - z_i \\ &= -1/a - z_{0\,\text{avg}} \\ &= \left.\frac{r_0}{(ra)^3}\right|_{r=1/a} \\ &= r_0 \end{aligned} \tag{32}$$

Since $r_0 = 2m$, we therefore get

$$m = \frac{\Delta z_0}{2} \tag{33}$$

The above result is in exactly similar to the case of particle capture by a Schwarzschild black hole. In fact, for spherically symmetric black holes, the above result is equivalent to attributing an energy $E = r_h/2$ to the horizon (with radius r_h), so that $\Delta E = \Delta r_h/2$ is the change in energy when a particle falls into the black hole. Indeed, this is the energy that appears in [6]. (In General Relativity, this definition of energy is equivalent to the so called Misner–Sharp energy associated with the black hole.)

Aside: Note that one can not argue for the above result on dimensional grounds alone, since more than one length scales are involved. In fact, the scaling of Δz_0 is an outcome of our (admittedly adhoc) choice of $z_{0\,\text{avg}}$. What is remarkable is that, given all the approximations made, we do get the appropriate numerical factor that has been known in the context of black holes in this very simple model.

Surface Term in Fermi Coordinates

From the point of view of thermodynamics, it is of interest to evaluate the surface term P^c in the Einstein–Hilbert action

$$R\sqrt{-g} = (\text{bulk part}) - \partial_c P^c \tag{34}$$

given by

$$\begin{aligned} P^c &= \frac{1}{\sqrt{-g}} \partial_b \left[(-g) g^{bc} \right] \\ &= \sqrt{-g} \left[g^{ck} \Gamma^m_{km} - g^{ik} \Gamma^c_{ik} \right] \end{aligned} \tag{35}$$

Although coordinate dependent, this term can be written in a covariant but observer dependent form, which is the reason why it acquires relevance in the context of the relationship between gravity and thermodynamics. We will calculate this term in the local coordinates based on the worldline of our accelerated observer; such a coordinate system is unique upto general Lorentz transformations on the observer worldline. Since the surface term, to relevant order (which we shall make precise soon) is Lorentz invariant, we are effectively probing the space time with the worldlines of such observers. Any local information about the space time geometry should be then encoded in the surface term of the action (in fact, being made up of second derivatives of the metric, it is only this term that is expected to be relevant in locally inertial coordinates) [7].

For convenience, we first define three spatial tensors formed from the curvature tensor

$$\mathscr{S}_{\alpha\beta} = R_{0\alpha0\beta} = \mathscr{S}_{\beta\alpha} \tag{36}$$

$$\mathscr{E}_{\alpha\beta} = (1/4)\varepsilon_{\alpha\gamma\sigma}\varepsilon_{\beta\lambda\mu}R_{\gamma\sigma\lambda\mu} = \mathscr{E}_{\beta\alpha} \tag{37}$$

$$\mathscr{B}_{\alpha\beta} = (1/2)\varepsilon_{\alpha\gamma\sigma}R_{0\beta\gamma\sigma} \tag{38}$$

in terms of which the FNC metric becomes

$$\begin{aligned} \mathrm{d}s^2 = &- \left[\left(1 + a_\mu y^\mu \right)^2 + \mathscr{S}_{\mu\nu} y^\mu y^\nu \right] \mathrm{d}\tau^2 + 2 \left[-\frac{2}{3} \varepsilon_{\rho\alpha\nu} \mathscr{B}_{\rho\mu} y^\mu y^\nu \right] \mathrm{d}\tau \mathrm{d}y^\alpha \\ &+ \left[\delta_{\alpha\beta} - \frac{1}{3} \varepsilon_{\rho\alpha\mu} \varepsilon_{\sigma\beta\nu} \mathscr{E}_{\rho\sigma} y^\mu y^\nu \right] \mathrm{d}y^\alpha \mathrm{d}y^\beta \end{aligned} \tag{39}$$

Note that,

$$\mathscr{S}^{\alpha}_{\ \alpha} = -R^{0}_{\ 0}\,, \quad \mathscr{E}^{\alpha}_{\ \alpha} = -G^{0}_{\ 0} \tag{40}$$

The coordinate system itself is good for length scales

$$y \ll \min\{a^{-1}, \mathscr{R}^{-1/2}, \mathscr{R}/\partial\mathscr{R}, 1/(a\mathscr{R})^{1/3}\}$$

(symbolically). One can always choose observers for whom the length scale set by acceleration, a^{-1} is much smaller than the curvature dependent terms above. We then effectively have a local Rindler observer, with time dependent acceleration, and a horizon determined by $a_\alpha(y^0)$.

The expression for P^μ in terms of above tensors can be shown to be

$$\begin{aligned} P^0 &= (2/3)\, N^{-1} R_{0\mu\nu\mu} y^\nu + O_1 \\ P_\mu &= 2a_\mu + 2N^{-1}\mathscr{S}_{\mu\nu} y^\nu + N\left(\mathscr{E}_{\mu\nu} - \mathscr{E}_{\alpha\alpha}\delta_{\mu\nu}\right) y^\nu + O_2 \end{aligned} \tag{41}$$

where $N = 1 + a_\mu y^\mu$, and O_1, O_2 represent terms which are higher order in curvature, and/or involve time derivatives of metric components, and/or quadratic in y^μ [19]. Note that the terms given above are not necessarily linear in y^μ (due to the presence of N); rather, these are the only terms which can lead to terms linear in y^μ, and hence we have stated them as it is. Using identities given above, it is straightforward to see that

$$\partial_\mu P^\mu \Big|_{y^\mu = 0} = -R \tag{42}$$

In flat spacetime, $R_{abcd} = 0$, and the contribution on any $\tau =$constant surface becomes

$$\int \mathrm{d}\tau \int \mathrm{d}^2 y_\perp \ (2 n_\sigma a^\sigma) \tag{43}$$

in obvious notation. For the well known case of a Rindler observer in flat spacetime, this gives the standard contribution of one-quarter of transverse area, when evaluated on the horizon.

As an example, let us consider a closed tubular neighbourhood of the trajectory in curved spacetime (Fig. 4). That is, at each τ, one sends out geodesics of constant length, say s_0, to form a tube, and close this surface at the $\tau = \tau_0$ and $\tau = \tau_0 + \Delta\tau$ to obtain a closed surface. The contribution of P^μ on the curved timelike surface $\mathscr{C}$, which has normal $n^\mu = y^\mu/s_0$, is given (to relevant order) by

$$\begin{aligned} P_\mu n^\mu \Big|_{\mathscr{C}} &= 2a_\mu n^\mu + 2N^{-1} s_0 \mathscr{S}_{\mu\nu} n^\mu n^\nu \\ &\quad + N s_0 \left(\mathscr{E}_{\mu\nu} - \mathscr{E}_{\alpha\alpha}\delta_{\mu\nu}\right) n^\mu n^\nu \end{aligned} \tag{44}$$

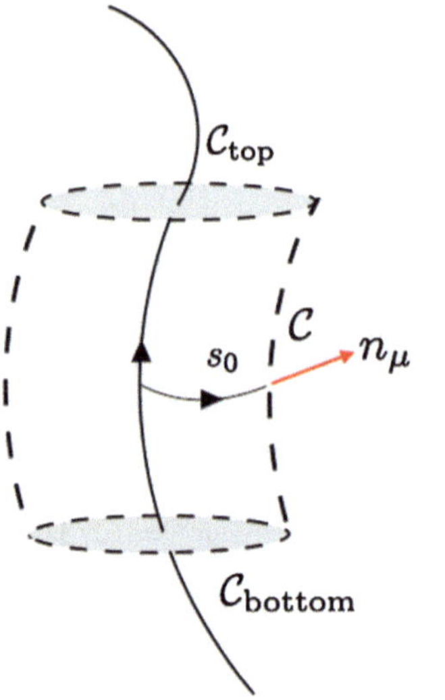

Fig. 4 A tubular neighborhood of the trajectory, with boundary $\mathscr{C} \cup \mathscr{C}_{\rm top} \cup \mathscr{C}_{\rm bottom}$ (see text)

It would be very interesting to explore the detailed mathematical structure of the above expression(s), since it might yield insights into the horizon entropy in curved spacetime.

Validity of the Approximations for the Ideal Gas Calculation

In this appendix, I give some numerical estimates to illustrate how well the various approximations made in the manuscript hold.

I consider a box of Nitrogen gas, with $m = 4.6 \times 10^{-26}$ kg, with approximately $N = 6.022 \times 10^{23}$ molecules, at room temperature $kT = 4.11 \times 10^{-21}$ J. I will also take, for the background curvature, the typical magnitude of curvature, say $\mathscr{R}$, produced by the Sun at the location of Earth's orbit [20]. Since the Sun-Earth distance is 1.496×10^{11} m and Sun's Schwarzschild radius is 3 km, this gives the curvature length scale as

$$\begin{aligned} L_{\mathscr{R}} &\approx \sqrt{(1.496 \times 10^{11})^3/3000\,\text{m}.} \\ &= 1.056 \times 10^{15}\,\text{m}. \end{aligned} \tag{45}$$

For definiteness, we consider a box of size $L = 100$ m. In this case, we get the following hierarchy of energy scales

$$\underbrace{E_1 = (mc^2) \times \mathscr{R}\lambda_c^2}_{2.17\times 10^{-73}\,\text{J}} \ll \underbrace{E_2 = \frac{\hbar^2}{mL^2}}_{2.42\times 10^{-47}\,\text{J}} \ll \underbrace{E_3 = \frac{1}{\beta} = kT}_{4.11\times 10^{-21}\,\text{J}} \ll \underbrace{E_4 = mc^2}_{4.14\times 10^{-9}\,\text{J}} \tag{46}$$

where $\lambda_c = \hbar/mc = 7.64 \times 10^{-18}$ m is the Compton wavelength. Just a glimpse at these numbers illustrate quite clearly how excellently do the various approximations made in the analysis hold. In fact, one gets a more intuitive understanding of the various numbers and their inter-relationships above by forming their dimensionless ratios.

$$\begin{aligned} E_1/E_2 &= \mathscr{R}L^2 \\ E_2/E_3 &= (\lambda/L)^2 \\ E_3/E_4 &= \left(\beta mc^2\right)^{-1} \\ E_1/E_3 &= \mathscr{R}\lambda^2 \\ E_2/E_4 &= (\lambda_c/L)^2 \end{aligned} \tag{47}$$

which, along with the fact that $(\lambda_c/\lambda)^2 = (\beta mc^2)^{-1}$, nicely illustrates the self-consistency of the approximations used, viz:

1. non-relativistic: $\beta mc^2 \gg 1$
2. validity of Fermi coordinates: $\mathscr{R}L^2 \ll 1$
3. use of Boltzmann distribution: $\lambda \ll L$

which imply that the first three of the energy ratios (47) above are *small* compared to unity, and the smallness of the last two follow from them. We have therefore shown that our 3 approximations are sufficient to ensure the above mentioned hierarchy of energy scales, which, as illustrated, holds very well for the typical case of N_2 gas in a box of size $L = 100$ m under the assumption that the background curvature is that produced by Sun at location of Earth's orbit. In fact, for our example

1. $\beta mc^2 = 1.01 \times 10^{12}$
2. $\mathscr{R}L^2 = 8.96 \times 10^{-27}$
3. $\lambda/L = 1.92 \times 10^{-13}$

Backreaction Due to Box Contents:
However, we need to take into account some additional constraints, which turn out to be conceptually trickier and more restrictive. If the box size L is reduced too much, the density of gas inside the box increases, which has following implications:

1. The gas can no longer be treated as Maxwell–Boltzmann (as was done in [13]).
2. The curvature produced by the box contents itself might become stronger than the background curvature, in which case the Fermi metric based on background curvature can not be used. And finally, a related fact that,
3. The energy content of the box might result in a black hole and engulf the box if the Schwarzschild radius L_{schw} corresponding to box's energy content exceeds L.

I discuss the constraints corresponding to (2) and (3) above, and leave (1) for future work since treatment of Bose-Einstein or Fermi-Dirac statistics for this problem requires much further work. Since mc^2 is the largest of all energies (per particle), one can use it to make the required estimates. Taking

$$L_{\text{schw}} \approx \frac{2G(Nm)}{c^2}$$

condition (3) above requires $L_{\text{schw}} < L$. On the other hand, Einstein equations imply, for the curvature produced by the box contents, an estimate $\mathscr{R}_{\text{box}} \approx L_{\text{schw}}/L^3$. To satisfy condition (2), we need $\mathscr{R}_{\text{box}} \ll \mathscr{R}$, which becomes equivalent to $L_{\text{schw}}/L \ll \mathscr{R}L^2$. Since we require $\mathscr{R}L^2 \ll 1$ for Fermi coordinates based on background curvature to be applicable, the above condition therefore provides a quite stringent upper bound on precisely *how small* must L_{schw}/L. The physics is quite clear: the smaller the box dimensions, the better the quadratic expansion in Fermi coordinates becomes, but this also increases the density of gas in the box, which might no longer allow the box to be treated as a perturbation over a given, fixed background spacetime.

However, the above conditions are possible to satisfy, and indeed are satisfied excellently in our example. Following is the hierarchy of length scales which illustrate the numbers involved (all lengths are in meters):

$$\begin{aligned} L &= 10, 100, 1000 \\ L_{\mathscr{R}_{\text{box}}} &= 6.98 \times 10^{15}, 2.21 \times 10^{17}, 6.98 \times 10^{18} \\ L_{\mathscr{R}} &= 1.06 \times 10^{15} \\ \Lambda &= 7.70 \times 10^{-6} \\ \lambda &= 1.92 \times 10^{-11} \\ \lambda_c &= 7.64 \times 10^{-18} \\ L_{\text{schw}} &= 4.11 \times 10^{-29} \end{aligned}$$

The magnitudes of $L_{\mathscr{R}_{\text{box}}}$ and $L_{\mathscr{R}}$ indicate the sensitivity of the issue of backreaction, discussed above, to the size L of the box. For $L = 10\,\text{m}$, the curvature length scales of the background and the box are of the same order ($L_{\mathscr{R}_{\text{box}}} \sim L_{\mathscr{R}}$), hence a larger box is essential for self-consistency of the approximations used. For $L > 100\,\text{m}$, $L_{\mathscr{R}_{\text{box}}} \gg L_{\mathscr{R}}$, and hence one can safely use the Fermi coordinate expansion based on background curvature.

References

1. D. Lynden-Bell, R. Wood, Mon. Not. R. Astron. Soc. **138**, 495 (1968); T. Padmanabhan, Phys. Rep. **188**, 285 (1990)
2. R. Wald, Living Rev. Rel. **4**, 6 (2001); T. Padmanabhan, Rep. Prog. Phys. **73**, 046901 (2010)
3. T. Jacobson, Phys. Rev. Lett. **75**, 1260 (1995). arXiv:gr-qc/9504004
4. D. Kothawala, Phys. Rev. D **83**, 024026 (2011). arXiv:1010.2207
5. S. Chakraborty, K. Parattu, T. Padmanabhan, J. High Energy Phys. **10**, 097 (2015). arXiv:1505.05297; See also T. Padmanabhan, arXiv:1602.01474 for a recent comparison and critique of various approaches linking gravitational dynamics with thermodynamics
6. T. Padmanabhan, Class. Quantum Gravity **19**, 5387 (2002). arXiv:gr-qc/0204019

7. Most of the calculations presented here were done as a part of (and partially included in) the author's Ph.D. thesis: D. Kothawala, Semiclassical gravity and thermodynamics of horizons, Ph.D. thesis (2010)
8. T. Jacobson, arXiv:1505.04753
9. F. Manasse, C. Misner, J. Math. Phys. **4**, 735 (1963); E. Poisson, A. Pound, I. Vega, Living Rev. Relativ.**14**, 7 (2011)
10. D. Kothawala, T. Padmanabhan, Phys. Rev. D **79**, 104020 (2009). arXiv:0904.0215
11. Analysis along somewhat similar lines, involving dynamics of spacelike two-surfaces, can be found in: J. Makela, A. Peltola, Int. J. Mod. Phys. D **18**, 669 (2009). arXiv:gr-qc/0612078. Discussion of first law in presence of asymptotic acceleration horizons in black hole spacetimes can be found in: K. Dutta, S. Ray, J. Traschen, Class. Quantum Gravity **23**, 335 (2006). arXiv:hep-th/0508041
12. J. Bekenstein, Phys. Rev. D **7**, 2333 (1973)
13. D. Kothawala, Phys. Letts. B **720**, 410 (2013). arXiv:1108.6301
14. D. Kothawala, work in progress
15. F. Pretorius, D. Vollick, W. Israel, Phys. Rev. D **57**, 6311 (1998). arXiv:9712085; J. Oppenheim, Phys. Rev. D **65**, 024020 (2001). arXiv:0105101; S. Kolekar, D. Kothawala, T. Padmanabhan, Phys. Rev. D **85**, 064031 (2012). arXiv:1111.0973
16. T. Padmanabhan, Class. Quantum Gravity **21**, 4485 (2004). Mod. Phys. Lett. A **25**, 1129 (2010)
17. It is worth mentioning here that the time scale $t_o = \Lambda/c = \hbar/kT$ has recently been discussed by Haggard and Rovelli as the average time a system in thermal equilibrium takes to move from a state to the next distinguishable state, thereby making t_o universal and independent of any details of the system other than its temperature; see: H. Haggard, C. Rovelli, arXiv:1302.0724
18. D. Kothawala, unpublished work
19. Incidentally, the weak field version of this result, $P_\mu = 2a_\mu$, was discussed in: T. Padmanabhan, Braz. J. Phys.**35**, 362 (2005). arXiv:gr-qc/0412068. See discussion around Eq. (10) there
20. This is simply a choice to provide an estimate of curvature of a typical background spacetime. One could, of course, imagine a more physical situation in which a box is actually present near Earth, in which case the curvature of Earth would be more dominant. In fact, the curvature due to Earth in such a case would be $\sim 10^8$ times that of Sun. However, it is clear from the numerical values in Eq. (46) that even this big a factor makes no difference to the arguments!

100 Years of the Cosmological Constant: Past, Present and Future

Ofer Lahav

Abstract The Cosmological Constant Λ, in different incarnations, has been with us for 100 years. Many surveys of dark energy are underway, indicating so far that the data are consistent with a dark energy equation of state of $w = -1$, i.e. a Λ term in Einstein's equation, although time variation of w is not yet ruled out. The ball is now back in the theoreticians' court, to explain the physical meaning of Λ. We discuss sociological aspects of this field, in particular to what extent the agreement on the cold dark matter + Λ concordance model is a result of the globalization of research over-communication.

1 Introduction

The year 2017 marks not only that Paddy is 60 years old, but also 100 years of the Cosmological Constant Λ. One of the greatest mysteries in the whole of science is the prospect that 70% of the universe is made from a mysterious substance known as 'dark energy', which causes an acceleration of the cosmic expansion. A further 25% of the universe is made from invisible 'cold dark matter' that can only be detected through its gravitational effects, with the ordinary atomic matter making up the remaining 5% (see the Planck Collaboration [13] study and references therein). This "Λ + cold dark matter" (ΛCDM) paradigm and its extensions pose fundamental questions about the origins of the universe. If dark matter and dark energy truly exist, we must understand their nature. Alternatively, General Relativity and related assumptions may need radical modifications. These topics have been flagged as key problems by researchers and by advisory panels around the world, and significant funding has been allocated towards large surveys of dark energy. Commonly, dark energy is quantified by an equation of state parameter, w, which is the ratio of pressure to density. The case $w = -1$ corresponds to Einstein's Cosmological Constant in General Relativity, but in principle w may vary with cosmic epoch, e.g. in the case of scalar fields. Essentially, w affects both the geometry of the universe and the growth

O. Lahav (✉)
University College London, London, UK
e-mail: o.lahav@ucl.ac.uk

J.S. Bagla and S. Engineer (eds.), *Gravity and the Quantum*,
Fundamental Theories of Physics 187, DOI 10.1007/978-3-319-51700-1_13

rate of structures. These effects can be observed via a range of cosmological probes, including the Cosmic Microwave Background (CMB), galaxy clustering, clusters of galaxies, and weak gravitational lensing, in addition to Supernovae Ia. The Hubble diagram of Type Ia Supernova [12, 14], for which the 2011 Nobel Prize in Physics was awarded, revealed that our universe is not only expanding but is also accelerating in its expansion. The main problem is that we still have no clue as to what is causing the acceleration, and what dark matter and dark energy are. Many cosmologists have puzzled over the meaning of Λ during the past 100 years, and it is not surprising that Paddy, with his deep insight into the foundations of physics, has written many inspiring books and papers on this and related topics (e.g. Padmanabhan [10]).

2 Background

It is well known that Einstein added in 1917 the Cosmological Constant Λ to his equations in order to have a static universe. His full equation is then:

$$R_{\mu\nu} - \frac{1}{2} R g_{\mu\nu} + \Lambda g_{\mu\nu} = \frac{8\pi G}{c^4} T_{\mu\nu} \; . \tag{1}$$

The big question is if Λ should be on the left hand side, as part of the curvature, or on the right hand side, as part of the stress-energy tensor $T_{\mu\nu}$, for example associated with the vacuum energy $\Lambda = 8\pi G \rho_{vac}/c^2$. In fact a prediction for the amount of vacuum energy is expected to be 10^{120} times the observed value; that is a challenging problem by itself (e.g. Weinberg [15]).

In the weak-field limit the equation of motion is:

$$\frac{d^2 r}{dt^2} = -\frac{GM}{r^2} + \frac{c^2}{3} \Lambda r \; . \tag{2}$$

A linear force was actually already discussed by Newton in Principia in addition to the more famous inverse square law.[1] A somewhat intuitive way to think about dark energy is as a repulsive linear force, opposing the inverse squared gravitational force. It is interesting that such a force can be noticeable on the Mpc scale. For example the mass of the Local Group would be estimated to be 13% higher in the presence of a Cosmological Constant.[2]

Should a discrepancy between data and the existing cosmological theory be resolved by adding new entities such as dark matter and dark energy, or by modifying the underlying theory? This reminds us of two cases in our own Solar System: the

[1] See e.g. Calder and Lahav [2, 3] for review.

[2] See e.g. Binney and Tremaine [1], Partridge, Lahav and Hoffman [11], McLeod et al. [9].

perturbed orbit of Uranus was explained by adding a new planet, Neptune, within the existing Newtonian model. On the other hand, understanding the perihelion of Mercury required an entirely new theory, General Relativity.[3]

There is still the possibility of another paradigm shift in our understanding of the cosmos, including the following options:

- Violation of the Copernican Principle: for example, if we happen to be living in the middle of a large void;
- Dark Energy being something different to vacuum energy: although vacuum energy is mathematically equivalent to Λ, the value predicted by fundamental theory is as much as 10^{120} times larger than observations permit;
- Modifications to gravity: it may be that General Relativity requires revision to a more complete theory of gravity;
- Multiverse: if Λ is large and positive, it would have prevented gravity from forming large galaxies, and life would never have emerged. Using this anthropic reasoning to explain the Cosmological Constant problems suggests a large number of universes ('multiverse') in which Λ and other cosmological parameters take on all possible values. We happen to live in one of the universes, that is fortunately 'habitable'.

3 The Dark Energy Survey

Many ongoing and planned imaging and spectroscopic surveys aim at measuring dark energy and other cosmological parameters. As an example we focus here on the Dark Energy Survey (DES).[4] I have chosen DES as it has already accumulated data, and I happen to have been involved in the project since its early days back in 2004, in particular as co-chair of its Science Committee (until 2016).

DES is an imaging survey of 5000 square degrees of the Southern sky, utilising a 570 mega-pixel camera on the 4 m Blanco telescope in Chile. Photometric redshifts are obtained from the multi-band photometry to produce a three dimensional map of 300 million galaxies. The main goal of DES is to determine the dark energy equation of state w and other key cosmological parameters to high precision. DES will measure w using four complementary techniques in a single survey: counts of galaxy clusters, weak gravitational lensing, galaxy distributions and thousands of type Ia supernovae in a 'time domain' survey over 27 sq. deg. DES is an international collaboration, with more than 500 scientists from the US, the UK, Spain, Brazil, Germany, Switzerland and Australia involved. The DES science is coordinated by a Science Committee composed of eleven Science Working Groups (SWGs). Core dark energy SWGs include large scale structure, clusters, weak lensing and supernovae Ia. Additional SWGs focusing on the primary science are photometric and spectroscopic redshifts,

---

[3] See e.g. a discussion in Lahav and Massimi [7] and references therein.

[4] http://www.darkenergysurvey.org/.

simulations, and theory and combined probes. The Non-cosmology SWGs focus on Milky Way science, galaxy evolution and quasars, strong lensing, and transients and moving objects.

The survey had its first light in September 2012 and started observations in September 2013. Observations are running for 525 nights spread over five years. The performances of several photo-z methods applied to Science Verification data were evaluated and the best methods yielded scatter $\sigma_{68} = 0.08$ (defined as the 68% width about the median of $\Delta z = z_{\text{spec}} - z_{\text{phot}}$). Regarding the image quality, the achieved median seeing FWHM is about 0.9" in filters *riz*, as expected when designing the survey for weak lensing analyses. DES has already 'seen' dark matter via weak gravitational lensing [4], and the analysis towards measuring and characterising dark energy is underway. The camera (DECam) can also capture many other celestial objects. This has resulted in both expected and unexpected discoveries [5] including solar system objects, 17 new Milky Way companions, galaxy evolution, galaxy clusters, high-redshift objects and gravitational wave follow ups.

We highlight here two contributions of DES to new research frontiers. Firstly, it has recently been suggested that there might be a ninth planet in the solar system. One of the six minor planets to predict 'Planet 9' was discovered by DES. The DECam camera is well placed to monitor other minor planets that would help in constraining Planet 9, and of course to search for Planet 9 itself. Secondly, the LIGO collaboration [8] reported the first detection of gravitational waves, resulting from the merger of two black holes. This remarkable measurement confirms another of Einstein's prediction of 100 years ago. DES provided optical follow up to this event. There were no optical detections, which is not surprising, as in the conventional model a binary black hole merger is not expected to have any optical counterparts, and the DES observations covered only part of the sky where the event was likely to happen. However, DES will be vital for future LIGO follow ups.

DES is also providing valuable experience and training of early career scientists for on-going and future large surveys, including the Hyper Suprime Cam (HSC),[5] the Kilo-Degree Survey (KiDS),[6] the Large Synoptic Survey Telescope (LSST),[7] *Euclid*,[8] the Wide-Field Infrared Survey Telescope (WFIRST), the Subaru Prime Focus Spectrograph (PFS),[9] the Dark Energy Spectroscopic Instrument (DESI)[10] and 4MOST.[11]

[5]http://www.naoj.org/Projects/HSC/.

[6]http://kids.strw.leidenuniv.nl/.

[7]http://www.lsst.org/.

[8]http://www.euclid-ec.org/.

[9]http://pfs.ipmu.jp/factsheet/.

[10]http://desi.lbl.gov/.

[11]http://www.4most.eu/.

4 The Globalization of Research: Pros and Cons

It may well be that the ΛCDM model is indeed the best description of our universe, with dark matter and dark energy ingredients that eventually will be detected independently. But there is also a chance that this is the 'modern Ether' and future generations will adopt an entirely different description of the universe. It is also possible that the community has converged on a single preferred model due to 'over communication'.[12] The society at large is going through a globalization process. There is a diversity of definitions for globalization, some in positive context, others with negative connotations. The sociologist Anthony Giddens defines globalization as "decoupling of space and time - emphasizing that with instantaneous communications, knowledge and culture can be shared around the world simultaneously." Another definition given in the same website sees globalization as being "an undeniably capitalist process. It has taken off as a concept in the wake of the collapse of the Soviet Union and of socialism as a viable alternate form of economic organization." A further discussion on globalization can be found in Thomas Friedman's book (2005) *The World is Flat* (an interesting title in the context of cosmology!). He questions whether "the world has got too small and too flat for us to adjust".

Research in academia is of course a human activity that is affected, like any other sector, by social and technological changes and trends. The advantages of globalization to academic research are numerous: open access to data sources for all (e.g. via the World Wide Web), rapid exchange of ideas, and international research teams. These aspects make science more democratic and they enable faster achievements. The numerous conferences, electronic archives and teleconferences generate a global village of thinkers. While this could lead to a faster convergence in answering fundamental questions, there is also the risk of preventing independent and original ideas from developing, as most researchers might be too influenced by the consensus view.

Let us consider the above mentioned 'concordance' model of cosmology. The two main ingredients, dark matter and dark energy, are still poorly understood. We do not know if they are 'real' or they are the modern 'epicycles' which just help to fit the data better, until a new theory greatly simplifies our understanding of the observations. A disturbing question is whether the popular cosmological 'concordance model' is a result of globalization? It is interesting to contrast the present day research in cosmology with the research in the 1970s and 1980s. This was the period of the 'cold war' between the former Soviet Union and the West. During the 1970s the Russian school of cosmology, led by Yakov Zeldovich, advocated massive neutrinos, 'hot dark matter', as the prime candidate for dark matter. As neutrinos were relativistic when they decoupled, they moved very fast and wiped out structure on small scales. This led to the 'top–down' scenario of structure formation. In this picture 'Zeldovich pancakes' of the size of superclusters formed first, and then they fragmented into clusters and galaxies. This was in conflict with observations, and cosmologists concluded that neutrinos are not massive enough to make up all of the dark matter. The downfall of the top-down 'hot dark matter' scenario of structure formation, and the

[12]The discussion below is based on Lahav [6].

lack of evidence for neutrino masses from terrestrial experiments made this model unpopular. The Western school of cosmology, led by Jim Peebles and others, advocated a 'bottom up' scenario, a framework that later became known as the popular 'cold dark matter'. However the detection of neutrino oscillations showed that neutrinos indeed have a mass, i.e. hot dark matter does exist, even if in small quantities. Current upper limits from a combination of cosmic probes are about 0.2 eV, while the lower limit from neutrino oscillations is 0.06 eV. Therefore both forms, cold dark matter and hot dark matter, probably exist in nature. This example illustrates that having two independent schools of thoughts was actually beneficial for progress in cosmology. Paddy has taught us numerous times how to think 'outside the box'. We wish him many more years of original research in cosmology.

Acknowledgements I thank my DES and UCL collaborators for many inspiring discussions on this topic over the years, and for support from a European Research Council Advanced Grant FP7/291329.

References

1. J. Binney, S. Tremaine, *Galactic Dynamics* (Princeton University Press, Princeton, 2008)
2. L. Calder, O. Lahav, Astron. Geophys. **49**, 1.13 (2008)
3. L. Calder, O. Lahav, Phys. World **32** (2010)
4. DES Collaboration (2015). arXiv:1507.05552
5. DES Collaboration, MNRAS **460**, 1270 (2016)
6. O. Lahav, in *Organizations and Strategies in Astronomy II*, ed. by A. Heck (Kluwer Acad. Publ., 2002). arXiv:astroEph/0105351
7. O. Lahav, M. Massimi, Astron. Geophys. **55**, 3.12 (2014)
8. LIGO Scientific Collaboration, PRL **116**, 061102 (2016)
9. M. McLeod, N. Libeskind, O. Lahav, Y. Hoffman (2016). arXiv:1606.02694
10. T. Padmanabhan (2016). arXiv:1611.03505
11. C. Partridge, O. Lahav, Y. Hoffman, MNRAS **436**, L45 (2013)
12. S. Perlmutter, ApJ **116**, 1009 (1999)
13. Planck collaboration (2015). arXiv:1502.01589
14. A. Riess, ApJ **517**, 565 (1998)
15. S. Weinberg, Rev. Mod. Phys. **61**, 1 (1989)

Pedagogical and Real Physics

Malcolm S. Longair

Abstract Paddy is a brilliant scientist, as well as being a wonderful teacher and expositor of physics and theoretical physics. Over the years, we have regularly discussed the problems of relating how we teach physics and theoretical physics to how we actually carry out physics in a research context. In my experience, these are rather different activities. To celebrate Paddy's 60th birthday, I expand upon this theme, giving some examples to illustrate the dichotomy between pedagogical and real physics. We need to ensure that the message is communicated to all the wonderful students we are privileged to teach.

1 Introduction

Every time I meet Paddy, our conversation quickly converges on the problems of teaching and understanding physics. Paddy's deep understandings of physics and theoretical physics are a joy and stimulus to probing deeper and deeper into the essence of these disciplines. One of our favourite topics is the issue of 'pedagogical' as opposed to 'real' physics and this is the theme I explore in this essay. Physics students want to understand the profound depths of the discipline and, as we all know, this is a highly non-trivial business. There is just so much technical material which has to be assimilated and mastered that the technical issues can overwhelm the physics concepts and the imaginative use of the tools which are at the core of the discipline.

2 A Personal Prelude

Let me recount my own experience. As an undergraduate, I took a combined Electrical Engineering and Physics course, called Electronic Physics, at Queen's College,

M.S. Longair (✉)
Cavendish Laboratory, J.J. Thomson Avenue, Cambridge CB3 0HE, UK
e-mail: msl1000@cam.ac.uk

J.S. Bagla and S. Engineer (eds.), *Gravity and the Quantum*,
Fundamental Theories of Physics 187, DOI 10.1007/978-3-319-51700-1_14

Dundee, which was then part of the University of St. Andrews. I already knew I wanted to study physics, but the teaching was variable and so this combined course, which was based in the rather good Electrical Engineering Department, was a good solution – I could read both courses simultaneously. But when I arrived in Cambridge to begin a PhD in the Cavendish Laboratory in 1963, my physics preparation was weak and way behind my fellow graduate students, many of whom had been through the rigours of the Cambridge Physics Tripos.

In due course, I was appointed a demonstrator, as assistant lecturers were known in Cambridge in those days, and it quickly sank in just how little I really understood. I had no alternative but to go back to the beginning and relearn my physics from scratch. In many ways, this was the best thing that could have happened. I had to go back to the beginning and ask much deeper questions about what I was teaching and what it was really all about. The result was that my lectures had a somewhat different flavour from the received Cambridge tradition in which the lecture notes and problem sheets tended to be handed down from one generation of lecturers to the next.

In 1975, I gave for the first time a non-examinable new course entitled *Theoretical Concepts in Physics* which was designed to give final-year undergraduates some appreciation of how the physics came about and the relation between the discoveries in physics and the mathematical tools needed to describe them. This later became my book of the same name [16]. In due course, I followed this up with books which elaborated the same themes – *The Cosmic Century* [17] on the development of astrophysical and cosmological understanding and the sequel to the Theoretical Concepts book on *Quantum Concepts in Physics* [19]. All these books involved rereading many of the original papers and working out how the great experimental and theoretical physicists achieved what they did. The one thing which was immediately obvious was that this was not pedagogical physics. The real thing is much more complex, tentative, subjective, intuitive and exciting. This is the challenge Paddy and I have discussed over the years.

I am under no illusion about the difficulty of getting the right balance between the pedagogical and the inspirational. The problem is that there is just so much hard work involved in getting on top of the technical aspects of the discipline that the leaps of imagination needed to make it come about can be squeezed out. In the last course I gave before my retirement, I was asked to deliver an introductory 22-lecture course on Astrophysics and Cosmology to third-year physics students. I aimed to make it the most exciting and inspirational course I could. I enjoyed preparing the lectures based upon 45 years of working at the frontiers of these disciplines. The course received the worst ratings I had ever had. The complaints were familiar 'Too much material', 'What do we need to know?', 'Was that examinable?' These were uniformly fabulous students, but even after three years of really demanding physics instruction, they had not grasped the essential message of what physics was about. I am unrepentant - it was a great course!

And, of course, the students had a point - you cannot teach experience. Every serious physicist has to go through the same process of re-assimilation which I was forced to go through – and it keeps recurring throughout a lifetime. We need to instil the messages that there are no shortcuts, the subject is really quite hard and you just have to find your own way of understanding the material in your own terms. But I also feel strongly that some appreciation of how the great discoveries came about in a non-trivial way, rather than the sanitised, regurgitated material of so many modern books, provide glorious examples of the creativity and beauty of physics.

In what follows, I discuss a few examples of the types of material which tends to get squeezed out of our teaching, but which illustrate how physics actually works. It would be wonderful to expose our students to this slightly different approach to physical thinking.

3 Galileo and the Nature of the Physical Sciences

In 1595, Galileo began to take the Copernican model of the Solar System seriously in order to explain the origin of tides in the Adriatic. To quantify the problem, he needed to develop a better quantitative understanding of the nature of velocity, acceleration and inertia. In the early 1600s, he began a magnificent set of experiments to elucidate the nature of motion [3]. The three great achievements were:

- The law of acceleration, $x = \frac{1}{2}at^2$, where a is the constant acceleration of the body,
- Galileo's law - the time for free fall down the diameter of a circle equals the time to roll down a chord,
- The period of the swing of a long pendulum is independent of its amplitude.

In 1608, the invention of the telescope was announced by Hans Lipperhey. Galileo heard about this invention in July 1609 and began constructing a series of telescopes of increasing magnifying power. By August he had achieved 9 times magnification, already three times better than Lipperhay and by the end of the year a magnification of 30 times was attained. In late August 1609 he presented his telescope to Doge Doná and the Venetian Senate who recognised its military importance. The observations of distant ships were made from the top of the Campanile in Venice. Note the immediate practical application of a major technical discovery.

Galileo's brilliant astronomical observations only occupied a relatively short period of his scientific career, roughly 1609 to 1612, but they were to resonate down the centuries. Among these, the two most important for physics were the phases of the planet Venus which were consistent with the Copernican rather than the Ptolemaic picture of the Universe and that Jupiter has four satellites orbiting the planet, exactly like a miniature Copernican Solar System. Galileo's remarkable skills in experiment as well as in theory should be emphasised. As a result of these observations, he was a strong proponent of the Copernican model of the Solar System.

Following an exchange of letters with the Grand Duchess Christina about reconciling his astronomical findings and the tenets of the Catholic faith, Galileo came under suspicion of propagating heretical dogma, resulting in his trial for heresy. Physicists remember the ruling of the Congregation of the Index of 1616 that Copernicanism was philosophically and scientifically untenable and theologically heretical. Galileo was exonerated but it was in effect a defeat for Copernicanism. In 1632, he was tried for a second time, resulting in his condemnation and house arrest for violating the instruction that he should not advocate the Copernican picture.

What is less well publicised was that there were serious physical deficiencies with the Copernican Theory on the basis of received understanding of natural philosophy at that time. In Finocchiero's summary [12],

1. If the Earth moves, falling bodies should not fall vertically.
2. The speed of projectiles fired in the direction of motion of the Earth and in the opposite direction should be different if the Earth were rotating.
3. Objects placed on a rotating potter's wheel are flung off if they are not held down. Why are we not flung off the Earth if it is rotating?
4. The Copernican picture was inconsistent with Aristotelian physics.
5. If Aristotelian physics was to be rejected, what was going to replace it?

Recall that the laws of motion were not yet formulated and there was no understanding of the dependence of the gravitation force upon distance. But what was more serious from our perspective is that, by asserting that Copernicus was right and Ptolomy wrong, Galileo had made a basic logical error. This was gently pointed out to him by Cardinal Roberto Bellarmine in 1616 who wrote:

> …it appears to me that Your Reverence (Foscarini) and Signor Galileo did prudently to content yourself with speaking hypothetically and not positively, as I have always believed Copernicus did. For to say that, assuming the Earth moves and the Sun stands still serves all appearances better than eccentrics and epicycles, is to speak well. This has no danger in it, and it suffices for mathematicians.
>
> But to wish to affirm that the Sun is really fixed in the centre of the heavens …is a dangerous thing, not only by irritating all the theologians and scholastic philosophers, but also by injuring our holy faith …

What Bellarmine was pointing out was the hypothetical nature of Galileo's argument. At the heart of it is the difference between *deductive* and *inductive reasoning*. Here is an example:

Deduction	Induction
If it is raining, the streets are wet	If it is raining, the streets are wet
It is raining	The streets are wet
Therefore, the streets are wet	Therefore, it is raining

The inductive argument does not prove that it is raining – the streets could have just been washed! In the same way, Galileo could not prove that the Copernican model was correct on the basis of his observations. For example, his observations of Venus would have been consistent with Tycho's synthesis of the Ptolemaic and Copernican pictures of the world. Even a sufficiently complicated Ptolemaic model could have been devised to explain the observations.

The key point is that physics is a *hypothetico-deductive process*. We make hypotheses and see how economically we can explain observed physical phenomena. The best theories are those which can explain large amounts of independent data quantitatively and make predictions to new circumstances. Intriguingly, in Newton's great *Principia Mathematica*, on pages 12 and 13, the title of the chapter is 'Axioms or the Laws of Motion' – to my way of thinking, 'Axioms' is a better description of the underlying assumptions than 'Laws of Motion' – the former designation recognises the hypothetical nature of the assumptions, whereas the term 'Laws of Motion' lend the axioms a much greater authoritative status.

I prefer to use the word *model* to describe this process rather than asserting that it is in some sense the *truth*. Galileo's enormous achievement was to realise that the models to describe nature could be put on a rigorous mathematical basis. In perhaps his most famous remark, he stated in his treatise *Il Saggiatore (The Assayer)* of 1624:

> Book of Nature ...is written in mathematical characters.

This was the great achievement of the Galilean revolution. Notice that even the apparently elementary facts established by Galileo required an extraordinary degree of imaginative abstraction.

My reason for relating this story is that it encapsulates the process of establishing the laws of physics. Many of the great advances in physics come from changing the axioms which underlie our description of the physical world - Newtonian space and time to Einsteinian space-time, Newtonian gravity to General Relativity, classical to quantum physics, and so on. But, do the students appreciate these profound lessons? There is no greater change of perspective than the discovery of fields to replace the mechanistic Newtonian world picture.

4 Maxwell, Reduced Momenta, Differential Gears and Electromagnetism

The great revolution in physics of the late 19th Century was the shift in perspective from the Newtonian mechanistic world view to a one in which fields became the fundamental entities for the description of nature. James Clerk Maxwell was at the heart of this revolution through his discovery of the field equations of electromagnetism. This great transition was to culminate in the work of Einstein and all succeeding generations of theorists. As Freeman Dyson has written:

> Maxwell's theory had to wait for the next generation of physicists, Hertz and Lorentz and Einstein, to reveal its power and clarify its concepts. The next generation grew up with Maxwell's equations and was at home with a Universe built out of fields. The primacy of fields was as natural to Einstein as the primacy of mechanical structures had been for Maxwell.

I recently reviewed the route Maxwell took to the discovery of his equations in a paper celebrating the 350th Anniversary of the founding of the *Philosophical*

Transactions of the Royal Society [20]. Maxwell's paper of 1865 is a wonderful exposition of the theory and is not so hard to follow once the notation is translated into modern usage [23]. His reliance upon mechanical analogues was only too apparent in his earlier papers of 1856 [21] and 1861–1862 [22]. In his assessment of the significance of Maxwell's 1865 paper, Whittaker famously remarked that

> In this, the architecture of his system was displayed, stripped of the scaffolding by aid of which it had been first erected.

and this is the general position adopted by subsequent commentators [37]. And yet, however well disguised, the mechanical roots are present, particularly in the slightly opaque Sect. 2 of the paper, which is generally passed over. In fact, this section illuminates Maxwell's deep understanding of mechanics and electromagnetism and his remarkable use of analogy in formulating his equations [14, 36].

Here is the entire section (24) in Sect. 2 of the 1865 paper. It is entitled *Dynamical Illustration of Reduced Momentum.*[1]

> (24) As a dynamical illustration, let us suppose a body C [of mass $M_{\rm C}$] so connected with two independent driving-points A and B that its velocity is p times that of A together with q times that of B. Let $v_{\rm A}$ be the velocity of A, $v_{\rm B}$ that of B, and $v_{\rm C}$ that of C, and let δx, δy, δz be their simultaneous displacements, then by the general equation of dynamics,[2]
>
> $$M_{\rm C}\frac{{\rm d}v_{\rm C}}{{\rm d}t}\,\delta z = F_{\rm A}\,\delta x + F_{\rm B}\,\delta y,$$
>
> where $F_{\rm A}$ and $F_{\rm B}$ are the forces acting at A and B.
>
> But
>
> $$\frac{{\rm d}v_{\rm C}}{{\rm d}t} = p\frac{{\rm d}v_{\rm A}}{{\rm d}t} + q\frac{{\rm d}v_{\rm B}}{{\rm d}t},$$
>
> and
>
> $$\delta z = p\,\delta x + q\,\delta y\,. \tag{1}$$
>
> Substituting, and remembering that δx and δy are independent,
>
> $$\left.\begin{aligned} F_{\rm A} &= \frac{\rm d}{{\rm d}t}(M_{\rm C}p^2v_{\rm A} + M_{\rm C}pqv_{\rm B})\,, \\ F_{\rm B} &= \frac{\rm d}{{\rm d}t}(M_{\rm C}pqv_{\rm A} + M_{\rm C}q^2v_{\rm B}). \end{aligned}\right\} \tag{2}$$
>
> We may call $M_{\rm C}p^2v_{\rm A} + M_{\rm C}pqv_{\rm B}$ the momentum of C referred to A, and $M_{\rm C}pqv_{\rm A} + M_{\rm C}q^2v_{\rm B}$ its momentum referred to B ; then we may say that the effect of the force $F_{\rm A}$ is to increase the momentum of C referred to A, and that of $F_{\rm B}$ to increase its momentum referred to B.
>
> If there are many bodies connected with A and B in a similar way but with different values of p and q, we may treat the question in the same way by assuming
>
> $$L_{\rm A} = \sum(M_{\rm C}p^2), \quad M_{\rm AB} = \sum(M_{\rm C}pq), \quad \text{and} \quad L_{\rm B} = \sum(M_{\rm C}q^2)\,,$$
>
> where the summation is extended to all the bodies with their proper values of [$M_{\rm C}$], p, and q. Then the momentum of the system referred to A is

---

[1] I have slightly altered the notation to make the argument more transparent.

[2] LAGRANGE, Méc. Anal. ii. 2. Sect. 5.

$$L_A v_A + M_{AB} v_B \ ,$$

and referred to B,

$$M_{AB} v_A + L_B v_B \ ,$$

and we shall have

$$\left.\begin{aligned} F_A &= \frac{d}{dt}(L_A v_A + M_{AB} v_B) \ , \\ F_B &= \frac{d}{dt}(M_{AB} v_A + L_B v_B) \ , \end{aligned}\right\} \quad (3)$$

where F_A and F_B are the external forces acting on A and B.

There are many remarkable features of this argument. The first is Maxwell's explicit use of Lagrangian techniques, with which he had been familiar since his days as an undergraduate at Edinburgh University and then reinforced by his studies at Cambridge. His starting point is the application of two forces to an extended rigid body at two separate 'driving points'. In accordance with the Lagrangian approach, the action of the forces is described in generalised coordinates. The key point is that these equations describe the generalised work done on the rigid body and hence the increase in its total kinetic energy. By working in terms of energy and changes in velocity, Maxwell arrives by straightforward algebra at the key result (2) and then, by extension to many forces acting on the body, to (3).

It is only at this point that Maxwell introduces Newton's second law $\mathbf{F} = d\mathbf{p}/dt$, where $\mathbf{p}$ is the momentum. The quantities $(L_A v_A + M_{AB} v_B)$ and $(M_{AB} v_A + L_B v_B)$ are then *defined* as the *reduced momenta* which will accurately describe the combined action of the two forces applied at different points A and B to the extended rigid body. It is important to appreciate that these momenta are not 'real momenta', any more than a 'reduced mass' is a real mass in the description of two-body systems. The reduced mass $\mu = m_1 m_2/(m_1 + m_2)$ converts a two-body problem into a one-body problem. For example, in the orbital motion of an electron about at atomic nucleus, the motion about the common centre of momentum of the two bodies is replaced by the circular motion of a single body with reduced mass μ. This then gives the correct frequency of orbital motion of the combined system.

The key point about Maxwell's analysis is that the two forces are acting on a single extended rigid body and so, although the forces are independent, they are coupled and the reduced momentum at A effects that at B and vice versa. If there were only a single force acting on the body, $M_{AB} = 0$, we would recover Newton's law with the usual meaning of momentum. But the forces are coupled because they are acting simultaneously on a single rigid body.

To complete the analogy, in Section (25), Maxwell next includes a frictional force acting on the motion of the body. Here is that section in full.

> (25) To make the illustration more complete we have only to suppose that the motion of A is resisted by a force proportional to its velocity, which we may call $R_A v_A$, and that of B by a similar force, which we may call $R_B v_B$, R_A and R_B being coefficients of resistance. Then if F'_A and F'_B are the forces on A and B

$$\left.\begin{aligned} F'_A &= F_A + R_A v_A = R_A v_A + \frac{d}{dt}(L_A v_A + M_{AB} v_B)\,, \\ F'_B &= F_B + R_B v_B = R_B v_B + \frac{d}{dt}(M_{AB} v_A + L_B v_B)\,. \end{aligned}\right\} \tag{4}$$

If the velocity of A be increased at the rate dv_A/dt, then in order to prevent B from moving a force, $F'_B = d/dt(M_{AB} v_A)$ must be applied to it.

This effect on B, due to an increase of the velocity of A, corresponds to the electromotive force on one circuit arising from an increase in the strength of a neighbouring circuit.

This dynamical illustration is to be considered merely as assisting the reader to understand what is meant in mechanics by *Reduced Momentum*. The facts of the induction of currents as depending on the variations of the quantity called Electromagnetic Momentum, or Electrotonic State, rest on the experiments of FARADAY,[3] FELICI,[4] etc.

Maxwell now makes the following identifications: the force F becomes the electromotive force $\mathscr{E}$, the velocities v become the currents I, the Ls and M become the self and mutual inductances respectively and R becomes the resistance. Then, the equations of electromagnetic induction between two current-carrying conductors are

$$\mathscr{E}_A = R_A I_A + \frac{d}{dt}(L_A I_A + M_{AB} I_B)\,, \tag{5}$$

and

$$\mathscr{E}_B = R_B I_B + \frac{d}{dt}(M_{AB} I_A + L_B I_B)\,. \tag{6}$$

From these, he goes on to derive the expressions for work and energy, the heat produced by the currents, the intrinsic energy of the currents and the mechanical action between conductors, and much more.

In typical Maxwellian fashion, he designed a mechanical model which precisely illustrates the rules of induction in mechanical terms (Fig. 1) [25]. Figure 1a shows the illustration which appears on p. 228, Volume 2 of the third (1891) edition of Maxwell's *Treatise*, edited by J.J. Thomson. The model, built by Messrs Elliot Brothers of London in 1876, is shown in Fig. 1b. The extended rigid body C is a flywheel which consists of two long steel rods at right angles to each other to which heavy weights are attached towards the four ends of the rods, giving the flywheel a large moment of inertia. Forces are applied to the flywheel to cause it to rotate through the differential gear arrangement shown in Fig. 1a. To make the arrangement clearer, I have redrawn the differential gearing schematically in Fig. 2a, b.

A and B are attached to separate axles which have bevel gears at their ends. They mesh with the horizontal bevel gear X, as shown in Fig. 2a, which is attached to the flywheel but which is free to rotate about its own axis. The discs A and B are the origins of the forces F_A and F_B and their 'independent driving points' are the opposite sides of the bevel gear X.

[3] Experimental Researches, Series I., IX.

[4] Annales de Chimie, sor. 3. xxxiv. (1852) p. 64.

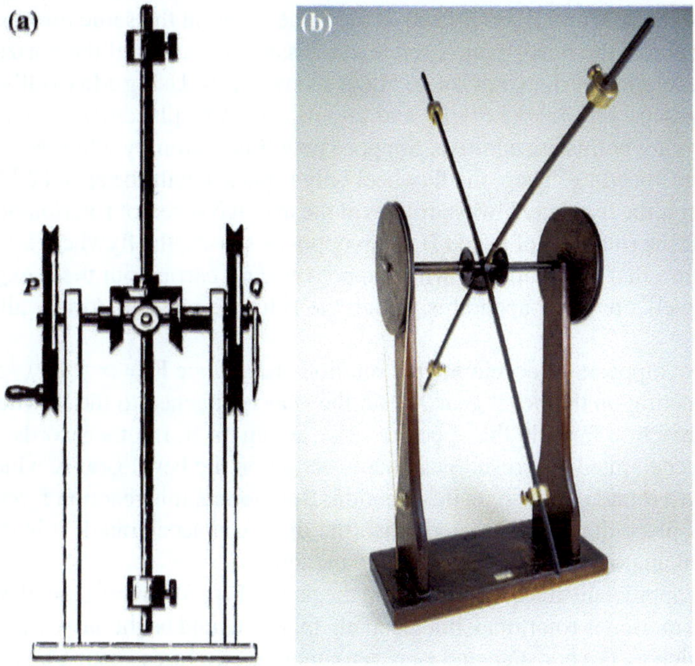

Fig. 1 **a** The diagram from Maxwell's *Treatise on Electricity and Magnetism, 3rd edition* showing the mechanical model he had built to illustrate the principles of electromagnetic induction. **b** The original model belonging to the Cavendish Laboratory and now on loan to the Whipple Museum, Cambridge University

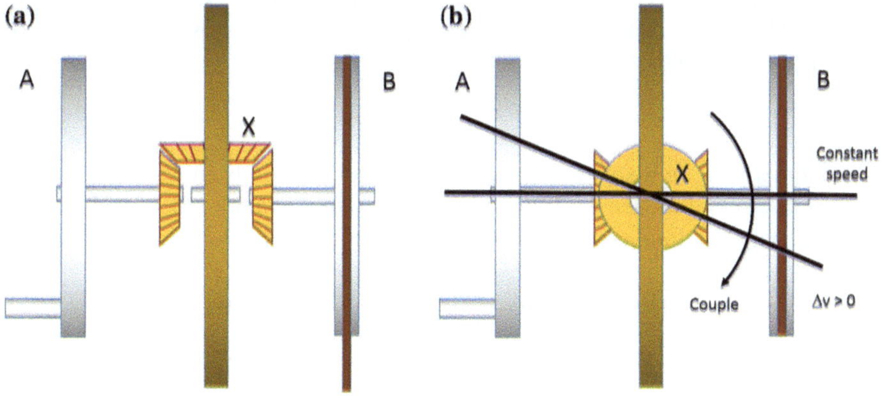

Fig. 2 **a** A schematic diagram showing more clearly than Fig. 1 the arrangement of the bevel gear which transmits the forces to the flywheel. **b** Illustrating the origin of the couple acting on the system when the disc A is accelerated

Suppose both A and B are rotated at the same speed in the same direction. Then, the flywheel and the bevel gear X rotate at the same speed about the horizontal axis and the bevel gear X does not rotate about its own axis. Using Maxwell's equation (1) with $p = q = 1$, $\delta z = \delta x + \delta y$ and the impressed displacements contributed at the two driving points are additive. Suppose now B is stationary while A is rotated at the constant speed v_{A}. Then, the flywheel only rotates at half the speed of A. In fact, by geometry, the flywheel always rotates at the average speed of rotation of the discs A and B. If the rotations of A and B are in opposite senses, the flywheel is stationary. I can confirm that this is indeed what happens when I carried out these experiments with Maxwell's original apparatus, which the Whipple museum kindly allowed me to operate.

But now suppose we accelerate the rotation of A. Since $\mathbf{F} = m_{\mathrm{A}}(\Delta v_{\mathrm{A}}/\Delta t)$, there is a force acting on the bevel gear X. But the gear is attached to the flywheel and so there is a reaction force in the opposite sense acting on B. In other words, when the disc A is accelerated, the result is a couple acting on the bevel gear X which causes the disc B to rotate in the opposite direction. But, notice, this reaction force at B can only takes place during the period when the disc A is accelerated. It is the perfect mechanical analogue for electromagnetic induction.

We can appreciate this behaviour from the pair of Eq. (3). In this case of Maxwell's model, the motion is rotational, but the mathematics would be the same. In the simple case in which v_{A} is a constant and v_{B} is zero, there are no forces acting on the system. But, if we now accelerate the rotation of the disc A, there is a force at A of $L_{\mathrm{A}}\ \Delta v/\Delta t$ and a force $\mathrm{d/dt}(M_{\mathrm{AB}}v_{\mathrm{A}})$ acting at B in the opposite sense, as illustrated in Fig. 2b. This is the origin of Maxwell's remark in (24)

> in order to prevent B from moving a force, $F'_{\mathrm{B}} = \mathrm{d}/\mathrm{d}t(M_{\mathrm{AB}}v_{\mathrm{A}})$ must be applied to it.

What is so remarkable about this analysis is that Maxwell's electromagnetic momentum is precisely what we would now call the *vector potential*, $\mathbf{A} \equiv [A_x, A_y, A_z]$. The origin of this identification is apparent from above Eqs. (5) and (6) which is no more than

$$\mathscr{E} = \frac{\partial \mathbf{A}}{\partial t}\,, \tag{7}$$

in one dimension. In his great paper, Maxwell works entirely in terms of $\mathbf{A}$ rather than the magnetic flux density $\mathbf{B}$ which is found from the relation $\mathbf{B} = \mathrm{curl}\,\mathbf{A}$. Thus, retaining only the four equations which reduce to his equations in modern form, Maxwell writes:

$$\mathbf{E} = \mu(\mathbf{v} \times \mathbf{H}) - \frac{\mathrm{d}\mathbf{A}}{\mathrm{d}t} - \nabla\phi\,, \qquad \text{(D)}$$

$$\mathrm{curl}\,\mathbf{H} = 4\pi\left(\mathbf{J} + \frac{\mathrm{d}\mathbf{D}}{\mathrm{d}t}\right), \qquad \text{(C)}$$

$$\rho_{\mathrm{e}} + \nabla \cdot \mathbf{D} = 0\,, \qquad \text{(G)}$$

$$\mu\mathbf{H} = \mathrm{curl}\,\mathbf{A}. \qquad \text{(B)}$$

To obtain precisely the modern version we need only take the curl of (D) and the divergence of (B). Note that I have retained Maxwell's alphabetical numbering of the equations and that the MKS units are unrationalised.

I find this all very thought-provoking.

- First of all, what is behind the identification of the mechanical analogue of the flywheel and the process of electromagnetic induction? The key is that the flywheel is a rigid body. The forces acting at the different 'independent driving points' are not unconstrained since they are coupled by the rigid body itself. Thus, the forces, although independent, are coupled and there is a reaction to the application of driving forces at A on B and vice versa. This is where the cross term comes from mechanically. In the electromagnetic case, the coupling is through the fields themselves which Maxwell assumed were embedded in the aether.
- Notice that using **A** rather than **B**, we are dealing with vectors rather than tensors in the description of the electromagnetic field.
- Maxwell's development presages the four–vector notation of special relativity in which the four–vector for the electromagnetic four potential is $[\phi/c, A_x, A_y, A_z]$, with the dimensions of momentum divided by the electric charge. Notice also the appearance of the same four-vector in the four-dimensional derivative in (D), which is no more than the derivative of the four-vector potential with respect to four-vector displacement $c\mathrm{d}t$, $\mathrm{d}x$, $\mathrm{d}y$, $\mathrm{d}z$.
- Maxwell makes liberal use of the vector potential in his development of the equations, in contrast to contemporary practice of working with the fields **E**, **D**, **B**, **H** and **J**.

For me, this analysis deepens the understanding of the role of vector potential in classical physics. The general pedagogical opinion is that **A** has no real significance in classical physics beyond the fact that, when curled, it gives **B**. And yet, as soon as you tackle serious problems in electromagnetism, it is always best to start with **A**, find the distribution of **A** and in the end curl it. And, of course, in quantum mechanics, it is **A** which has to be quantised from the very beginning and which has real physical significance in phenomena such as the Aharonov–Bohm effect. Maxwell, as usual, was far ahead of his time in understanding these remarkable relations between electromagnetic phenomena at a very deep level.

An important footnote to this story is that the Eq. (4) reappear in Maxwell's important, pioneering paper On Governors of 1868 [24]. This paper on the conditions for the stability of control systems in regarded by engineers as the origin of the science of cybernetics.

Should students know about this? Most text books would regard this as 'antiquarian' physics and yet what magic Maxwell distills from his technique of working and thinking by analogy. And note how naturally he uses Lagrangian methods. But most of all, the origin of the concept of fields was based upon his Newtonian mechanical model for the key process of electromagnetic induction. If nothing else, students should know that this is how the mind of a genius works.

5 Einstein and Statistics

I remember vividly an occasion in Edinburgh when Wilson Poon and I were talking about Einstein's achievements and he made the perceptive remark, 'the two things which Einstein really understood were invariance and statistics'. The role of invariance in Einstein's thinking needs little elaboration, but statistical arguments played a key role in his thinking throughout his career and perhaps receives less attention. Yet this statistical approach led to some of his most spectacular insights.[5]

5.1 Planck and Quantisation

In the first of Planck's two key papers of 1900, he established the form of the spectrum of black body radiation by empirical arguments. His arguments had a strong thermodynamic flavour and, by requiring the formula for the entropy of an oscillator to have the correct temperature dependence in the high and low frequency limit, derived the correct form for what we now refer to as the *Planck spectrum* [33]. In his scientific biography, Planck wrote:

> On the very day when I formulated this law, I began to devote myself to the task of investing [the Planck spectrum] with a true physical meaning. This quest automatically led me to study the interrelation of entropy and probability – in other words, to pursue the line of thought inaugurated by Boltzmann. [35]

In his second paper, Planck's analysis began by following Boltzmann's procedure [34]. There is a fixed total energy E to be divided among the N oscillators and energy elements ε are introduced. Therefore, there are $r = E/\varepsilon$ energy elements to be shared among the oscillators. Rather than following Boltzmann's procedure in statistical physics, however, Planck simply worked out the *total number of ways* in which r energy elements can be distributed over the N oscillators. The answer is

$$\frac{(N+r-1)!}{r!(r-1)!} \,. \tag{8}$$

Now Planck *defined* (8) to be the probability p to be used in Boltzmann's expression for the entropy $S = C \ln p$. This led to the expression for the entropy of an oscillator which was identical to that he had derived in his earlier paper.

$$S = a\left[\left(1+\frac{\overline{E}}{b}\right)\ln\left(1+\frac{\overline{E}}{b}\right) - \frac{\overline{E}}{b}\ln\frac{\overline{E}}{b}\right] , \tag{9}$$

[5] In this section, I have used material which is discussed in much more detail in my book *Quantum Concepts in Physics* [19].

with the requirement $b \propto \nu$ where $\overline{E}$ is the average energy of the oscillator. Thus, the energy elements ε must be proportional to frequency and Planck wrote this requirement in the familiar form $\varepsilon = h\nu$, the first appearance of *Planck's constant h* in the history of physics. This is the origin of the concept of *quantisation*. According to the procedures of classical statistical mechanics, we ought now to allow $\varepsilon \to 0$, but evidently we cannot obtain the expression for the entropy of an oscillator unless the energy elements do *not* disappear, but have finite magnitude $\varepsilon = h\nu$. Therefore, the expression for the energy density of black-body radiation is

$$u(\nu) = \frac{8\pi h\nu^3}{c^3} \frac{1}{\mathrm{e}^{h\nu/kT} - 1} . \tag{10}$$

What is one to make of Planck's argument? He certainly had not followed Boltzmann's procedure for finding the equilibrium energy distribution of the oscillators. What he defines as a probability is not really a probability of anything drawn from any parent population. Planck had no illusions about this. In his own words:

> In my opinion, this stipulation basically amounts to a definition of the probability W; for we have absolutely no point of departure, in the assumptions which underlie the electromagnetic theory of radiation, for talking about such a probability with a definite meaning.

Einstein repeatedly pointed out this weak point in Planck's argument

> The manner in which Mr Planck uses Boltzmann's equation is rather strange to me in that a probability of a state W is introduced without a physical definition of this quantity. If one accepts this, then Boltzmann's equation simply has no physical meaning. [4]

Why did Planck find the correct expression for the radiation spectrum, despite the fact that the statistical procedures he used were more than a little suspect? It seems quite likely that Planck worked backwards. It was suggested by Rosenfeld, and endorsed by Klein on the basis of an article by Planck of 1943, that he started with the expression for the entropy of an oscillator and worked backwards to find W from $\exp(S/k)$. This results in the permutation formula (8) for large values of N and r. The expression (8) was a well-known formula in permutation theory and appears early in Boltzmann's exposition of the fundamentals of statistical physics. Planck then regarded (9) as the definition of entropy according to statistical physics. If this is indeed what happened, it in no sense diminishes Planck's achievement in establishing the necessity of quantisation through the finite value of $h\nu$.

5.2 Einstein on Brownian Motion

Einstein's virtuosity in statistical physics is splendidly illustrated by his paper of 1905, more familiarly known by the title of a subsequent paper published in 1906 entitled *On the theory of Brownian motion* [6]. Brownian motion is the irregular motion of microscopic particles in fluids and had been studied in detail in 1828 by the botanist Robert Brown. The motion results from the statistical effect of very large numbers of collisions between molecules of the fluid and the microscopic particles. Although each impact is very small, the net result of a very large number of them randomly colliding with the particle is a 'drunken man's walk'. Einstein was not certain about the applicability of his analysis to Brownian motion, writing in the introduction to his paper:

> It is possible that the movements to be discussed here are identical with the so-called 'Brownian molecular motion'; however, the information available to me regarding the latter is so lacking in precision that I can form no judgment in the matter.

Einstein begins with Stokes' formula for the force acting on a sphere of radius a moving at speed $\mathcal{V}$ through a medium of kinematic viscosity ν, $F = 6\pi \nu a \mathcal{V}$, where a is the radius of the sphere and ν the coefficient of kinematic viscosity of the fluid. Considering the one-dimensional diffusion of the particles in the steady state, he found the diffusion coefficient for the particles in the medium, $D = kT/6\pi \nu a$ and from this the one-dimensional distance the particle diffuses $< \lambda_x^2 >= 2Dt$ in time t. The result is his famous formula for the mean squared distance travelled by the particle in time t in one dimension,

$$\langle \lambda_x^2 \rangle = \frac{kTt}{3\pi \nu a}, \tag{11}$$

where T is the temperature and k Boltzmann's constant. Crucially, Einstein had discovered the relation between the molecular properties of fluids and the observed diffusion of macroscopic particles.

In 1908, Jean Perrin [31] carried out a meticulous series of brilliant experiments which confirmed in detail all Einstein's predictions. This work convinced everyone, even the sceptics, of the reality of molecules. In Perrin's delightful words,

> I think it is impossible that a mind free from all preconception can reflect upon the extreme diversity of the phenomena which thus converge to the same result without experiencing a strong impression, and I think it will henceforth be difficult to defend by rational arguments a hostile attitude to molecular hypotheses. [32]

Einstein was well aware of the importance of this calculation for the theory of heat – the agitational motion of the particles observed in Brownian motion *is* heat, the macroscopic particles reflecting the motion of the molecules on the microscopic scale.

5.3 The Statistical Origin of Einstein's Discovery of Light Quanta

Einstein's great paper of 1905 is commonly referred to as his paper on the photoelectric effect but that scarcely does justice to its profundity – his argument is based upon the thermodynamics of radiation and the statistics of particles and waves in a box with perfectly reflecting walls. Einstein's approach differed radically from Planck's discovery of quantisation in which the 'energy elements' $\varepsilon = h\nu$ are associated with *oscillators* in thermal equilibrium at temperature T. These same oscillators are the source of the electromagnetic radiation in the black-body spectrum, but Planck had absolutely nothing to say about the radiation emitted by them. In contrast, Einstein proposed that the radiation field itself should be quantised [5].

Einstein derived a suitable form for the entropy of black-body radiation using only thermodynamics and the observed form of the radiation spectrum with the result

$$S = V\phi\,\Delta\nu = -\frac{\varepsilon}{\beta\nu}\left(\ln\frac{\varepsilon c^3}{8\pi\alpha\nu^3 V\,\Delta\nu} - 1\right). \tag{12}$$

Suppose the volume changes from V_0 to V, while the total energy remains constant. Then, the entropy change is

$$S - S_0 = \frac{\varepsilon}{\beta\nu}\ln(V/V_0)\,. \tag{13}$$

But this entropy change is exactly the same as that found in the Joule expansion of a perfect gas according to elementary statistical mechanics, $S - S_0 = kN\,\ln(V/V_0)$, which is simply derived from the statistics of particles in a box. Einstein immediately concluded that the radiation behaves thermodynamically as if it consisted of discrete particles, their number N being equal to $\varepsilon/k\beta\nu$. In Einstein's own words,

> Monochromatic radiation of low density (within the limits of validity of Wien's radiation formula) behaves thermodynamically as though it consisted of a number of independent energy quanta of magnitude $k\beta\nu$.

Rewriting this result in Planck's notation, since $\beta = h/k$, the energy of each quantum is $h\nu$.

Einstein finally considers three phenomena which cannot be explained by classical electromagnetic theory.

1. *Stokes' rule* is the observation that the frequency of photoluminescent emission is less than the frequency of the incident light. If the incoming quanta each have energy $h\nu$, the re-emitted quanta can at most have this energy.
2. *The photoelectric effect*. One of the remarkable features of the effect was Lenard's discovery that the energies of the electrons emitted from the metal surface are *independent of the intensity of the incident radiation*. Einstein's proposal provided an immediate solution to this problem. Radiation of a given frequency consists of quanta of the same energy $h\nu$. If one of these is absorbed by the material, the electron may receive sufficient energy to remove it from the surface against the forces which bind it to the material. If the intensity of the light is increased, more electrons are ejected, but their energies remain unchanged. The maximum kinetic energy which the ejected electron can have, E_k, is

$$E_k = h\nu - W, \tag{14}$$

where W is the amount of work necessary to remove the electron from the surface of the material, its *work function*. Nothing was known about the dependence of the photoelectric effect upon the frequency of the incident radiation at that time. It was only in 1916 that Millikan's meticulous experiments confirmed precisely Einstein's prediction [26].

3. *Photoionisation of gases*. The third piece of experimental evidence was the fact that the energy of each photon has to be greater than the ionisation potential of the gas if photoionisation is to take place. He showed that the smallest energy quanta for the ionisation of air were approximately equal to the ionisation potential determined independently by Stark.

This is the work described in Einstein's Nobel Prize citation of 1921.

5.4 *Fluctuations of Particles and Waves - Einstein (1909)*

Most of the major figures in physics rejected the idea that light could be considered to be made up of discrete quanta. Einstein, however, never deviated from his conviction about the reality of quanta and continued to find other ways in which the experimental features of black body radiation lead inevitably to the conclusion that light consists of quanta. In one of his most impressive papers written in 1909, he showed how fluctuations in the intensity of the black-body radiation spectrum provide further evidence on the quantum nature of light [7].[6]

In the case of *particles in a box*, we first divide it into N equal cells and a large number of particles n is distributed randomly among them. If n is very large, the mean number of particles in each cell is roughly the same, but there is a real scatter about the mean value because of statistical fluctuations. If N large, the variance of the fluctuations is $\sigma^2 = n/N$ which is also the average number of particles in each cell. This is the well-known result that, for large values of n and N, the mean is equal to the variance.

In the case of *fluctuations of randomly superposed waves*, the result is different. Since the phases of the waves are random, $\langle E_x^* E_x \rangle = N\xi^2 \propto u_x$, where ξ is the amplitude of the waves. For incoherent radiation, the waves have random phases and the total energy density is the sum of the energies in all the waves.

To find the fluctuations in the average energy density of the waves, we work out the quantity $\langle (E_x^* E_x)^2 \rangle$ with respect to the mean energy. The answer is $\Delta u_x^2 = u_x^2$, that is, *the fluctuations in the energy density are of the same magnitude as the energy density of the radiation field itself.* The physical meaning of this calculation is clear. Every pair of waves of frequency ν interferes to produce fluctuations in intensity of the radiation $\Delta u \approx u$. This analysis refers to waves of random phase ϕ and of a particular angular frequency $\omega = 2\pi\nu$, what we would refer to as waves corresponding to a single mode.

---

[6]I have given a detailed treatment of the theory of fluctuations in the number densities of particles and waves in my book *Theoretical Concepts in Physics*.

Einstein begins by reversing Boltzmann's relation between entropy and probability, $W = \exp(S/k)$. In terms of fractional fluctuations, Einstein finds the result

$$\frac{\sigma^2}{\varepsilon^2} = \left(\frac{h\nu}{\varepsilon} + \frac{c^3}{8\pi\nu^2 V\,\mathrm{d}\nu}\right). \tag{15}$$

The first term on the right-hand side originates from the Wien part of the spectrum and, if we suppose the radiation consists of photons, each of energy $h\nu$, it corresponds to the statement that the fractional fluctuation in the intensity is just $1/N^{1/2}$ where N is the number of photons, that is, $\Delta N/N = 1/N^{1/2}$. This is exactly the result expected if light consists of discrete particles.

The second term originates from the Rayleigh–Jeans part of the spectrum. According to an entirely classical calculation, notably used by Rayleigh in his derivation of the black body spectrum, there are $8\pi\nu^2 V\,\mathrm{d}\nu/c^3$ modes in the frequency interval ν to $\nu + \mathrm{d}\nu$ and the fluctuations associated with each wave mode have magnitude $\Delta\varepsilon^2 = \varepsilon^2$. When we add together randomly all the independent modes in the frequency interval ν to $\nu + \mathrm{d}\nu$, we add their variances and hence

$$\frac{\langle\delta E^2\rangle}{E^2} = \frac{1}{N_{\text{mode}}} = \frac{c^3}{8\pi\nu^2 V\,\mathrm{d}\nu},$$

which is exactly the second term on the right-hand side of (15).

Thus, the two parts of the fluctuation spectrum in (15) correspond to particle and wave statistics, the former to the Wien region of the spectrum and the latter to the Rayleigh–Jeans part. We recall that we add together the variances due to independent causes and so Eq. (15) states that *we should add independently the variances of the 'wave' and 'particle' fluctuations of the radiation field to find the total magnitude of the fluctuations*. This is insight of the very highest order.

5.5 Einstein on Stimulated Emission

The paper of 1916 was a further contribution to Einstein's crusade to convince his colleagues of the reality of light quanta [8]. The paper is best remembered today for its introduction of what are now known as *Einstein's A and B coefficients*, but the bulk of the paper is about the statistical nature of momentum transfer between photons and electrons.

He begins by noting the formal similarity between the Maxwell–Boltzmann distribution for the velocity distribution of the molecules in a gas and Planck's formula for the black-body spectrum. Einstein shows how these distributions can be reconciled through his new derivation of the Planck spectrum, which gives insight into what he refers to as the 'still unclear processes of emission and absorption of radiation by matter.'

There is no need to go through the standard derivation. Einstein finds that the equilibrium radiation spectrum u can be written

$$u = \frac{A_m^n / B_m^n}{\exp\left(\frac{\varepsilon_m - \varepsilon_n}{kT}\right) - 1} \ . \tag{16}$$

This is Planck's radiation law. He then finds the relations between the As and Bs,

$$\frac{A_m^n}{B_m^n} \propto \nu^3, \quad \varepsilon_m - \varepsilon_n = h\nu \ . \tag{17}$$

The value of the constants in (16) can be found from the Rayleigh–Jeans limit of the black-body spectrum, $\varepsilon_m - \varepsilon_n / kT \ll 1$. It follows that

$$u(\nu) = \frac{8\pi \nu^2}{c^3} kT = \frac{A_m^n}{B_m^n} \frac{kT}{h\nu} \quad \text{and so} \quad \frac{A_m^n}{B_m^n} = \frac{8\pi h \nu^3}{c^3} \ . \tag{18}$$

The A_m^n and B_m^n coefficients are associated with atomic processes at the microscopic level. Einstein wrote exuberantly to his friend Michele Besso on 11 August 1916,

> A splendid flash came to me concerning the absorption and emission of radiation …A surprisingly simple derivation of Planck's formula, I would say *the* derivation. Everything completely quantum.

This analysis occupies only the first three sections of Einstein's paper. The remainder concerns the transfer of momentum as well as energy between matter and radiation. According to standard kinetic theory, when molecules collide in a gas in thermal equilibrium, there are fluctuations in the momentum transfer between molecules which amounts to

$$\frac{\overline{\Delta^2}}{\tau} = 2RkT \ , \tag{19}$$

where Δ is the momentum transfer to a molecule during a short time interval τ and R is a constant related to the 'frictional' force acting upon the moving molecules. Note the striking similarity with Einstein's formula (11) for the diffusion of microscopic particles undergoing Brownian motion.

Now suppose the density of atoms is reduced to an extremely low value so that the momentum transfer is dominated by collisions between photons and the few remaining atoms, which can be considered collisionless. According to Einstein's quantum hypothesis, the photons have energy $h\nu$ and momentum $h\nu/c$. Therefore, the particles should be brought into thermal equilibrium at temperature T entirely through collisions between photons and particles. This was the reason that Einstein needed his equations for spontaneous emission and induced absorption and emission of radiation since these determine the rate of transfer of energy and momentum between the particles and the radiation. Einstein showed that, assuming the momentum transfers occur randomly in directional collisions between photons and electrons, the variance of the fluctuations in the momentum transfer is exactly the same expression as (19). This could not happen if the energy re-radiated by the particles was isotropic because then there would be no random component in the momentum transfer process. The key result was that, when a molecule emits or absorbs a quantum $h\nu$, there must be a positive or negative change in the momentum of the molecule of magnitude $|h\nu/c|$, even in the case of spontaneous emission. In Einstein's words,

> Outgoing radiation in the form of spherical waves does not exist. During the elementary process of radiative loss, the molecule suffers a recoil of magnitude $h\nu/c$ in a direction which is determined only by 'chance', according to the present state of the theory.

His view of the importance of the calculation is summarised in the version of the paper published in 1917.

> The most important thing seems to me to be the momenta transferred to the molecule [atom] in the processes of absorption and emission. If any of our assumptions concerning the transferred momenta were changed (19) would be violated. It hardly seems possible to reach agreement with this relation, which is demanded by the [kinetic] theory of heat, in any other way than on the basis of our assumption. [9]

5.6 Bose–Einstein Statistics

One of the intriguing questions about Planck's derivation of the black-body energy distribution is why he obtained the correct answer using 'wrong' statistical procedures. In fact, Planck had stumbled by accident upon the correct method of evaluating the statistics for *indistinguishable particles*. These procedures were first demonstrated by the Indian mathematical physicist Satyendra Nath Bose in a manuscript entitled *Planck's Law and the Hypothesis of Light Quanta*, which he sent to Einstein in 1924. Einstein immediately appreciated its deep significance, translated it into German himself and arranged for it to be published in the *Zeitschrift für Physik* [1].

To paraphrase Pais's account of the paper [30], Bose introduced three new features into statistical physics:

(i) Photon number is not conserved.
(ii) Bose divides phase space into coarse-grained cells. The counting of the numbers of particles per cell explicitly requires that, because the photons are taken to be identical, each possible distribution of states should be counted only once. Thus, Boltzmann's axiom of the distinguishability of particles is gone.
(iii) Because of this method of counting, the statistical independence of particles has also gone.

These are profound differences as compared with classical Boltzmann statistics. The differences were to find an explanation in quantum mechanics and are associated with the symmetries of the wavefunctions for particles of different spins. As Pais remarks,

> The astonishing fact is that Bose was correct on all three counts. (In his paper, he commented on none of them.) I believe there had been no such successful shot in the dark since Planck introduced the quantum in 1900 [30].

Consider one of the cells, which we label k and which has energy ε_k and degeneracy g_k, meaning the number of available states with the same energy ε_k within that cell. Now suppose there are n_k particles to be distributed over these g_k states and that the particles are identical. Then, the number of different ways the n_k particles can be distributed over these states is

$$\frac{(n_k + g_k - 1)!}{n_k!(g_k - 1)!} \,. \tag{20}$$

using standard procedures in statistical physics. This is the key step in the argument and differs markedly from the corresponding Boltzmann result. In (20) duplications of the same distribution are eliminated because of the factorials in the denominator. In quantum mechanical terms, the explanation for this distinction is neatly summarised by Huang who remarks:

> The classical way of counting in effect accepts all wave functions regardless of their symmetry properties under the interchange of coordinates. The set of acceptable wave functions is far greater than the union of the two quantum cases [the Fermi-Dirac and Bose-Einstein cases]. [13]

Notice that this is the point at which the statistical independence of the particles is abandonned. The particles cannot be placed randomly in all the cells since duplication of configurations are not allowed.

The result (20) refers only to a single cell in phase space and we need to extend it to all the cells which make up the phase space. Carrying out that standard calculation, we find the result

$$n_k = \frac{g_k}{e^{\alpha + \beta\varepsilon_k} - 1} \,. \tag{21}$$

This is the *Bose–Einstein distribution* and is the correct statistics for counting indistinguishable particles.

In the case of black-body radiation, we do not need to specify the number of photons present. The distribution is determined solely by one parameter – the total energy, or the temperature of the system. Therefore, in the method of undetermined multipliers, we can drop the restriction on the total number of particles. The distribution automatically readjusts to the total amount of energy present, and so $\alpha = 0$. Therefore,

$$n_k = \frac{g_k}{e^{\beta\varepsilon_k} - 1} \,. \tag{22}$$

By inspecting the low-frequency behaviour of the Planck spectrum, we find that $\beta = 1/kT$, as in the classical case.

Finally, the degeneracy of the cells in phase space g_k for radiation in the frequency interval ν to $\nu + d\nu$ is needed. One of the reasons for Einstein's enthusiasm for Bose's paper was that Bose derived this factor entirely by considering the phase-space available to the photons, rather than appealing to Planck's or Rayleigh's approaches, which relied upon results from classical electromagnetism. Bose considered the photons to have momenta $p = h\nu/c$ and so the volume of momentum, or phase, space for photons in the energy range $h\nu$ to $h(\nu + d\nu)$ is, using the standard procedure,

$$dV_p = V\, dp_x\, dp_y\, dp_z = V 4\pi p^2\, dp = \frac{4\pi h^3 \nu^2\, d\nu}{c^3} V\,, \tag{23}$$

where V is the volume of real space. Recalling that there are two polarisation states of the photon, Rayleigh's result was recovered,

$$dN = \frac{8\pi \nu^2}{c^3}\, d\nu \quad \text{with} \quad \varepsilon_k = h\nu\,, \tag{24}$$

and the spectral energy density of the radiation is

$$u(\nu)\, d\nu = \frac{8\pi h\nu^3}{c^3} \frac{1}{e^{h\nu/kT} - 1}\, d\nu\,. \tag{25}$$

Planck's expression for the black-body radiation spectrum has been derived using Bose–Einstein statistics for indistinguishable particles.

Einstein went on to apply these new procedures to the statistical mechanics of an ideal gas [10, 11]. As he stated, the application of these statistics to a monatomic gas leads to a 'far-reaching formal relationship between radiation and gas'. In particular, he realised that the expression he had derived for the fluctuations in black-body radiation in his paper of 1909 must apply equally for the statistics of monatomic gases as well:

$$\frac{\sigma^2}{\varepsilon^2} = \left(\frac{h\nu}{\varepsilon} + \frac{c^3}{8\pi \nu^2 V\, d\nu}\right)\,. \tag{26}$$

This is a dramatic result. Simply from the statistics of indistinguishable particles, the expression for the fluctuations consists of one term associated with the statistics of non-interacting particles according to the Maxwell–Boltzmann prescription and a second term associated with interference phenomena due to the wave properties of the particles. For these reasons, Einstein was particularly intrigued by the work of Louis de Broglie, which was reported by Langevin at the fourth Solvay conference in April 1924. De Broglie had made another 'shot in the dark', by ascribing wave properties to the electron. This conjecture was to have profound implications for the development of quantum mechanics in the hands of Schrödinger.

5.7 *Reflections*

My reason for telling this story in some detail is that these examples represent statistics at the very heart of fundamental physics. Statistics is normally introduced in the context of the analysis of experimental data, which is of course essential. But look how rich statistical ideas were in the hands of a genius like Einstein. In my view, we could introduce students to many of these statistical concepts through their role in basic physics rather than as drill exercises in 'errors'.

6 Cosmological Conundrums—Space–time Diagrams for the Standard World Models

The application of the established laws of physics to the Universe as a whole has been extraordinarily successful. Not only have the laws resulted in deep understandings of how our Universe has evolved, but also new physics has been discovered. For example, limits to the number of neutrino species were derived from studies of the primordial synthesis of the light elements and then these limits were confirmed by the LEP experiments at CERN.

We are able to pursue these endeavours because of the extraordinary isotropy of the Universe on the large scale. Measurements of the isotropy of the cosmic background radiation over the sky as observed by the WMAP and Planck space missions show that the Universe looks the same in all directions on the large scale to better than one part in a hundred thousand. This results in an tremendous simplification in the construction of cosmological models. As a result, one bit of Universe is just as good as any other for the construction of cosmological models. The galaxies are also moving apart according to Hubble's Law, the observation that the velocities of recessions of galaxies are proportional to their distances. Together, these observations mean that the Universe as a whole is expanding uniformly. At a particular cosmic epoch, every observer in the Universe can move in such a way that they observe an isotropic, expanding Universe with the same Hubble's law - such observers are known as *fundamental observers*. This is why the simple Newtonian construction by Milne and McCrea [27, 28] for the large scale dynamics of the Universe works so well.

But we have to be careful about the meanings of the coordinates we use.[7] The significance of the various ways of characterising time and space in these models is quite subtle and pedagogically often not as clearly enunciated as I believe they should be. It is helpful to represent the various scales on space-time diagrams for the standard world models and to discuss some of their somewhat surprising features.[8]

---

[7]I have dealt with many of the issues in this Section in my book *Galaxy Formation* [18].

[8]This section was inspired by the illuminating papers by Davis and Lineweaver [2, 15].

First, let us summarise the various times and distances introduced in the construction of cosmological models.

Cosmic time Cosmic time t is defined to be time measured by a fundamental observer who reads time on a standard clock.

$$t = \int_0^t \mathrm{d}t = \int_0^a \frac{\mathrm{d}a}{\dot{a}}, \tag{27}$$

where the scale factor a describes how the distance between fundamental observers changes with cosmic time, normalised to unity at the present cosmic time t_0. Notice that $a(t)$ describes the kinematics of the expanding universe.

Comoving radial distance coordinate When we make observations of the Universe, we look into the past along our *past light cone* (Fig. 3). The conditions of isotropy and homogeneity can only be applied at a particular cosmic epoch and so, to define a self-consistent distance measure at cosmic time t, we project the proper distances along our past light cone to a reference epoch which can be taken to be the present epoch t_0. Then, the comoving radial distance coordinate r is defined as

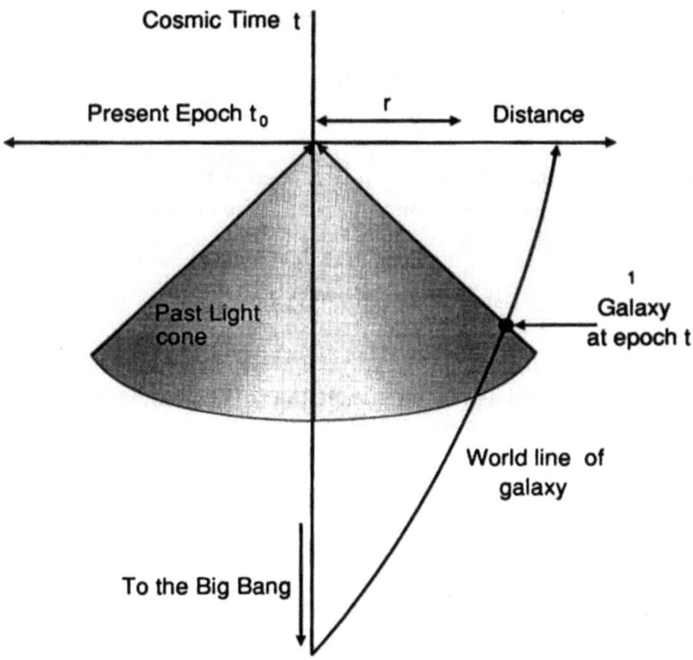

Fig. 3 A simple space-time diagram illustrating the definition of comoving radial distance coordinate

$$r = \int_t^{t_0} \frac{c\,\mathrm{d}t}{a} = \int_a^1 \frac{c\,\mathrm{d}a}{a\dot{a}}\,. \tag{28}$$

Note that this is a fictitious distance and depends upon the choice of cosmological model through the scale factor $a(t)$.

Conformal time The *conformal time* τ has similarities to the definition of comoving radial distance coordinate in that time intervals are projected forward to present epoch using the definition

$$\mathrm{d}t_{\mathrm{conf}} = \mathrm{d}\tau = \frac{\mathrm{d}t}{a}\,. \tag{29}$$

According the cosmological time dilation formula $\mathrm{d}t_0 = \mathrm{d}t/a$ and so the interval of conformal time $\mathrm{d}\tau$ is the time measured by a fundamental observer at the present epoch t_0. At any epoch, the conformal time has value

$$\tau = \int_0^t \frac{\mathrm{d}t}{a} = \int_0^a \frac{\mathrm{d}a}{a\dot{a}}\,. \tag{30}$$

Proper radial distance coordinate To define a proper distance at some earlier epoch, we *define* the proper radial distance r_{prop} to be the comoving radial distance coordinate projected back to the epoch t. Then,

$$r_{\mathrm{prop}} = a \int_t^{t_0} \frac{c\,\mathrm{d}t}{a} = a \int_a^1 \frac{c\,\mathrm{d}a}{a\dot{a}}\,. \tag{31}$$

Particle horizon The particle horizon r_{H} is defined as the maximum proper distance over which there can be causal communication at the epoch t

$$r_{\mathrm{H}} = a \int_0^t \frac{c\,\mathrm{d}t}{a} = a \int_0^a \frac{c\,\mathrm{d}a}{a\dot{a}}\,. \tag{32}$$

It immediately follows that, in a space-time diagram in which comoving radial distance coordinate is plotted against conformal time, the particle horizon is a straight line with slope equal to the speed of light.

Event horizon The event horizon r_{E} is the greatest proper radial distance an object can have if it is ever to be observable by an observer at cosmic time t_1,

$$r_{\mathrm{E}} = a \int_{t_1}^{t_{\max}} \frac{c\,\mathrm{d}t}{a(t)} = a \int_{a_1}^{a_{\max}} \frac{c\,\mathrm{d}a}{a\dot{a}}\,. \tag{33}$$

6.1 The Past Light Cone

Because of the assumptions of isotropy and homogeneity, Hubble's linear relation $v = H_0 r$ applies at the present epoch *to recessions speeds which exceed the speed of light* where r is the 'artificial' comoving radial distance coordinate. We can imagine measuring this distance by lining up a very large number of fundamental observers who measure increments of distance Δr at the present epoch t_0 and who are moving apart at speed $H_0\,\Delta r$. Thus, if the fundamental observers are far enough apart, this speed can exceed the speed of light. There is nothing in this argument which contradicts the special theory of relativity – it is simply a geometric result of the requirements of isotropy and homogeneity.

Consider the familiar analogy of the surface of an expanding spherical balloon. As the balloon inflates, a linear velocity-distance relation is found on the surface of the sphere, not only about any point on the sphere, but also at arbitrarily large distances on its surface. Hence at very large distances, the speed of separation can be greater than the speed of light, but there is no causal connection between these points – they are simply partaking in the uniform expansion of what Bondi calls the substratum, the underlying space-time geometry of the Universe.

The proper distance between two fundamental observers at cosmic time t is

$$r_{\rm prop} = a(t) r \,, \tag{34}$$

where r is comoving radial distance. Differentiating with respect to cosmic time,

$$\frac{{\rm d}r_{\rm prop}}{{\rm d}t} = \dot{a} r + a \frac{{\rm d}r}{{\rm d}t} \,. \tag{35}$$

The first term on the right-hand side represents the motion of the substratum and, at the present epoch, becomes $H_0 r$. Consider, for example, the case of a very distant object in the critical world model, $\Omega_0 = 1$, $\Omega_\Lambda = 0$. As a tends to zero, the comoving radial distance coordinates tends to $r = 2c/H_0$. Therefore, the local rest frame of objects at these large distances moves at three the speed of light relative to our local frame of reference *at the present epoch*, since the age of the model is $(2/3)/H_0$. At the epoch at which the light signal was emitted along our past light cone, the recessional velocity of the local rest frame $v_{\rm rec} = \dot{a} r$ was greater than this value, because $\dot{a} \propto a^{-1/2}$.

The second term on the right-hand side of (35) corresponds to the velocity of peculiar motions in the local rest frame at r, since it corresponds to changes of the comoving radial distance coordinate. The element of proper radial distance is $a{\rm d}r$ and so, if we consider a light wave travelling along our past light cone towards the observer at the origin, we find from (35)

$$v_{\rm tot} = \dot{a} r - c \,. \tag{36}$$

This result defines the propagation of light from the source to the observer on space-time diagrams for the expanding Universe.

We can now plot the trajectories of light rays from their source to the observer at t_0. The proper distance from the observer at $r = 0$ along the past light cone $r_{\rm PLC}$ is

$$r_{\rm PLC} = \int_0^t v_{\rm tot}\,{\rm d}t = \int_0^a \frac{v_{\rm tot}\,{\rm d}a}{\dot{a}}\,. \tag{37}$$

Initially the light rays from distant objects are propagating away from the observer – this is because the local isotropic cosmological rest frame is moving away from the observer at $r = 0$ at a speed greater than that of light. The light waves are propagated to the observer at the present epoch through local inertial frames which expand with progressively smaller velocities until they cross the *Hubble sphere* at which the recession velocity of the local frame of reference is the speed of light. The definition of the radius of the Hubble sphere $r_{\rm HS}$ at epoch t is

$$c = H(t)\,r_{\rm HS} = \frac{\dot{a}}{a} r_{\rm HS} \qquad \text{or} \qquad r_{\rm HS} = \frac{ac}{\dot{a}}\,. \tag{38}$$

$r_{\rm HS}$ is a proper radial distance. From this epoch onwards, propagation is towards the observer until, as $t \to t_0$, the speed of propagation towards the observer is the speed of light.

6.2 *Application to Cosmological Models*

We consider first the critical world model and then a reference Λ model.

The Critical World Model $\Omega_0 = 1$, $\Omega_\Lambda = 0$

The space-time diagrams shown below are presented with time measured in units of H_0^{-1} and distance in units of c/H_0. The advantage of studying this simple case first is that there are simple analytic relations for the various quantities which appear in Fig. 4. These are listed in Table 1.

Different versions of the space-time diagrams for the critical world model are shown in Fig. 4a–c. In all three presentations, the world lines of galaxies having redshifts 0.5, 1, 2 and 3 are shown. When plotted against comoving radial distance coordinate in Fig. 4b, c, these are vertical lines. The Hubble sphere and particle horizon, as well as the past light cone, are shown in all three diagrams. There is no event horizon in this model.

These diagrams illustrate a number of interesting features.

- Figure 4a is the most intuitive diagram. It illustrates clearly many of the points discussed above. For example, the Hubble sphere intersects the past light cone at the point where the $v_{\rm tot} = 0$ and the tangent to the past light cone at that point is vertical.

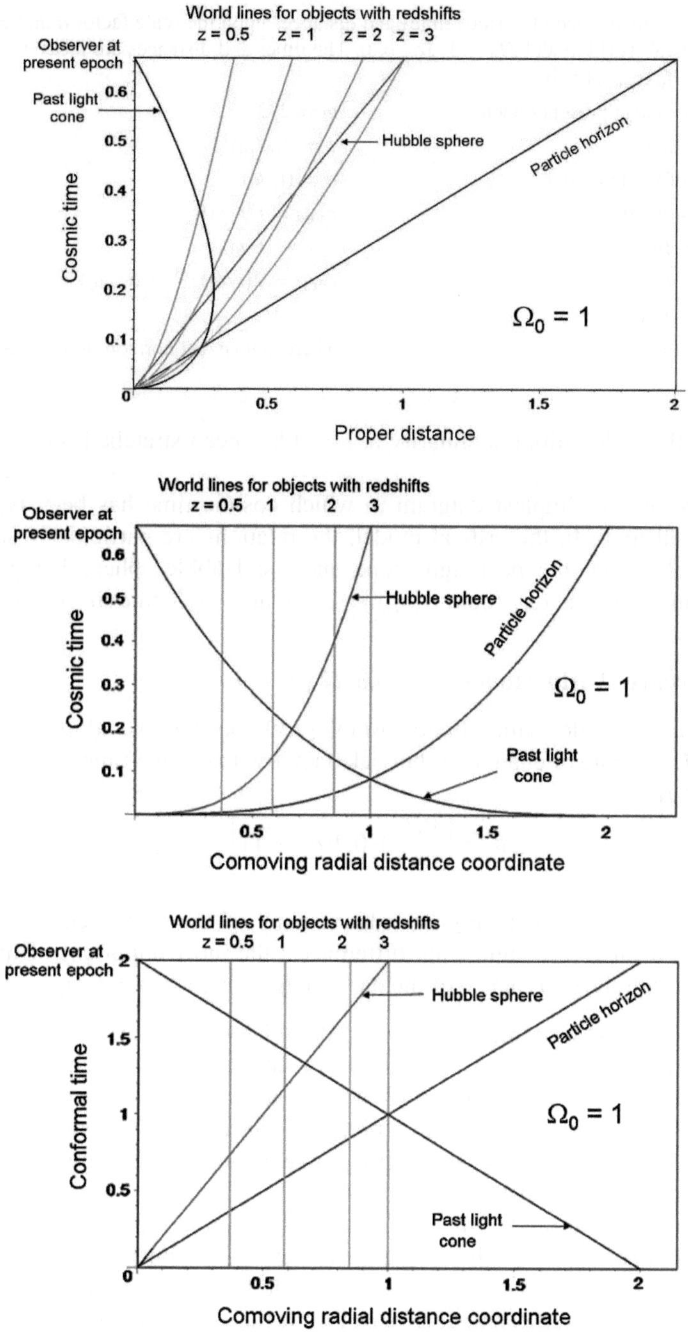

Fig. 4 Space-time diagrams for the critical cosmological model, $\Omega_0 = 1$, $\Omega_\Lambda = 0$. The times and distances are measured in units of H_0^{-1} and c/H_0 respectively

Table 1 The dependence of various times and distances upon the scale factor a and cosmic time t for the critical world model $\Omega_0 = 1$, $\Omega_\Lambda = 0$. The times and distances are measured in units of H_0^{-1} and c/H_0 respectively

Age of Universe at present epoch	$t_0 = 2/3$
Conformal time	$\tau = 2(t/t_0)^{1/3}$
Dynamics of world model	$a = (t/t_0)^{2/3}$
World lines of galaxies	$r_{\rm prop} = r(t/t_0)^{2/3}$
Hubble sphere	$r_{\rm HS} = (t/t_0)$
Past light cone	$r_{\rm PLC} = 2(t/t_0)^{2/3} - 2(t/t_0)$
Particle horizon	$r_{\rm H} = 3t$
Event horizon	There is no event horizon in this model

- In Fig. 4b, c, the initial singularity at $t = 0$ has been stretched out to become a singular line.
- Figure 4c is the simplest diagram in which cosmic time has been replaced by conformal time. In the critical model, the relations are particularly simple, the particle horizon, the past light cone and the Hubble sphere being given by $r_{\rm H}(\text{comoving}) = \tau$, $r_{\rm PLC}(\text{comoving}) = 2 - \tau$ and $r_{\rm HS}(\text{comoving}) = \tau/2$ respectively.

The Reference World Model $\Omega_0 = 0.3$, $\Omega_\Lambda = 0.7$

For a reference model with a finite cosmological constant, we adopt $\Omega_0 = 0.3$ and $\Omega_\Lambda = 0.7$. The rate of change of the scale factor with cosmic time in units in which $c = 1$ and $H_0 = 1$ is

$$\dot{a} = \left[\frac{0.3}{a} + 0.7(a^2 - 1)\right]^{1/2} . \tag{39}$$

The diagrams shown in Fig. 5a–c have many of the same general features as Fig. 4a–c, but there are significant differences, the most important of these being associated with the dominance of the dark energy term Ω_Λ at late epochs.

- First, the cosmic time-scale is stretched out relative to the critical model.
- The world lines of galaxies begin to diverge at the present epoch as the repulsive effect of the dark energy dominates over the attractive force of gravity.
- The Hubble sphere converges to a proper distance of 1.12 in units of c/H_0. The reason for this is that the expansion becomes exponential in the future and Hubble's constant tends to a constant value of $\Omega_\Lambda^{1/2}$.
- Unlike the critical model, there is an event horizon in the reference model. The reason is that, although the geometry is flat, the exponential expansion drives galaxies beyond distances at which there could be causal communication with an observer at epoch t. It can be seen from Fig. 5a that the event horizon tends towards the same asymptotic value of 1.12 in proper distance units as the Hubble sphere. In Fig. 5b, c, the comoving distance coordinates for the Hubble sphere and the event horizon tend to zero as $t \to \infty$.

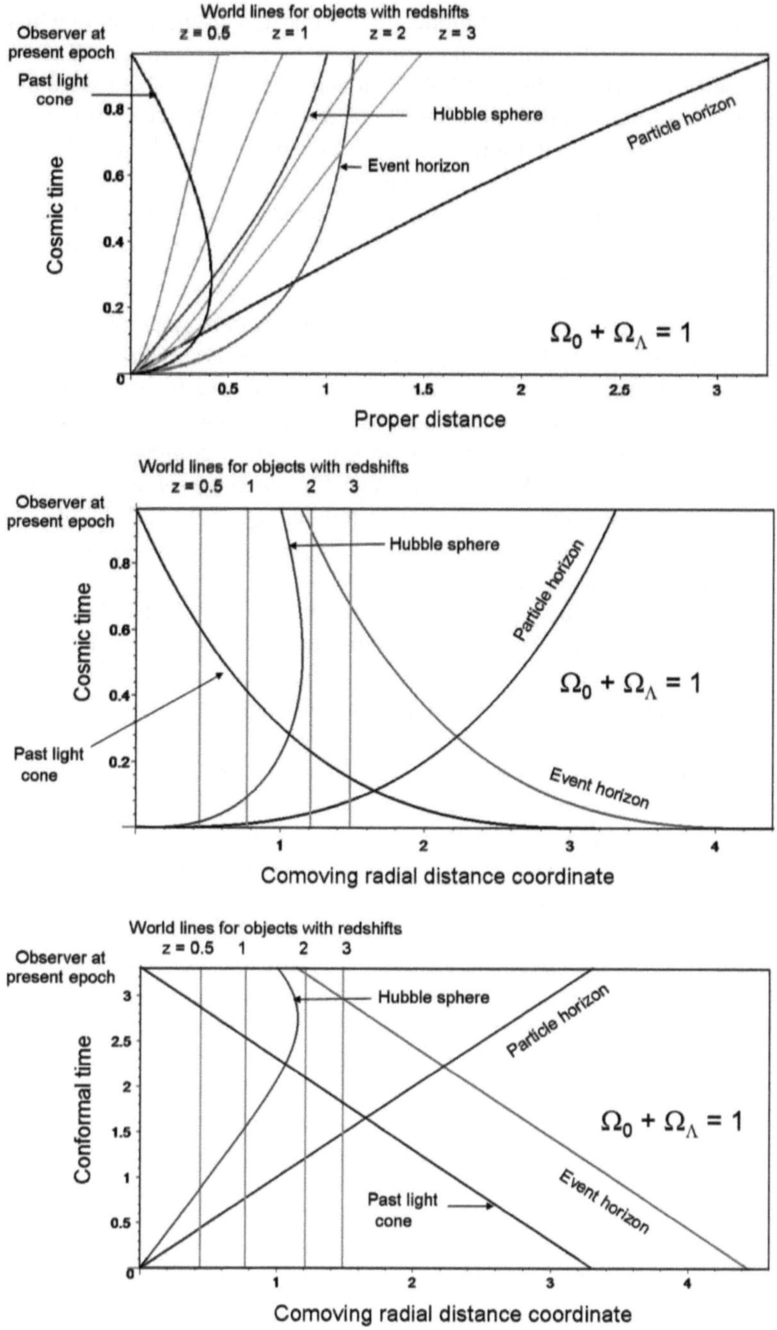

Fig. 5 Space-time diagrams for the reference cosmological model, $\Omega_0 = 0.3$, $\Omega_\Lambda = 0.7$. The times and distances are measured in units of H_0^{-1} and c/H_0 respectively [2]

- Just as in the case of the critical model, the simplest diagram is that in which conformal time is plotted against comoving radial distance coordinate. The relations for the particle horizon, the past light cone and the event horizon were all given in proper coordinates and so they have to be divided by a to convert to comoving coordinates. Using these definitions, it is a simple exercise to show that the various lines are $r_{\rm H}(\text{comoving}) = \tau$, $r_{\rm PLC}(\text{comoving}) = \tau_0 - \tau$, $r_{\rm E}(\text{comoving}) = r_0 - \tau$, where $\tau_0 = 3.305$ and $r_0 = 4.446$ for our reference cosmological model. These forms of the relations in terms of comoving distance coordinate and conformal time are true for all models.

There are two extensions of Fig. 5c which help elucidate some of the features of the standard world models. In Fig. 6a, the redshift of 1000 is shown corresponding to the last scattering surface from which the Cosmic Microwave Background Radiation originates. The intersection with our past light cone is shown and then a past light cone from the last scattering surface to the singularity at conformal time $\tau = 0$ is shown as a shaded triangle. This demonstrates the *horizon problem* – the region of causal contact is very small compared with moving an angle of 180° over the sky which would correspond to twice the distance between the origin and the comoving radial distance coordinate at 3.09.

In Fig. 6b, the end of the inflation era is taken as the zero of time for the standard Big Bang and the diagram has been extended back to negative conformal times. Thus, we shift the zero of conformal time very slightly to, say, 10^{-32} s and then we can extend the light cones to incorporate the inflationary expansion of the very early universe.

This construction provides a way of understanding how the inflationary picture resolves the causality problem. Light cones have unit slope in the conformal diagram and so we draw them from the ends of the element of comoving radial distance at $\tau = 0$ from the last scattering surface. These are shown in the diagram and it can be seen that projecting far enough back in time, the light cones from opposite directions on the sky overlap, meaning causal contact in the early Universe.

We have to distinguish between the Hubble sphere and the particle horizon. The latter is defined as the maximum distance over which causal contact could have been made from the time of the singularity to a given epoch. In other words, it is not just what happened at a particular epoch which is important, but the history along the past light cone. In contrast, the Hubble radius is the distance of causal contact *at a particular epoch*. It is the distance at which the velocity in the velocity-distance relation at that epoch is equal to the speed of light. Writing the exponential inflationary expansion of the scale factor as $a = a_0 \exp[H(t - t_{\rm i})]$, where a_0 is the scale factor when the inflationary expansion began at $\tau_{\rm i}$, $r_{\rm HS} = c/H$ and the comoving Hubble sphere has radius $r_{\rm HS}(\text{com}) = c/(Ha)$. Since H is a constant throughout most of the inflationary era, it follows that the comoving Hubble sphere *decreases* as the inflationary expansion proceeds.

We now join this evolution of the comoving Hubble sphere onto its behaviour after the end of inflation. The expression for conformal time during the inflationary era is

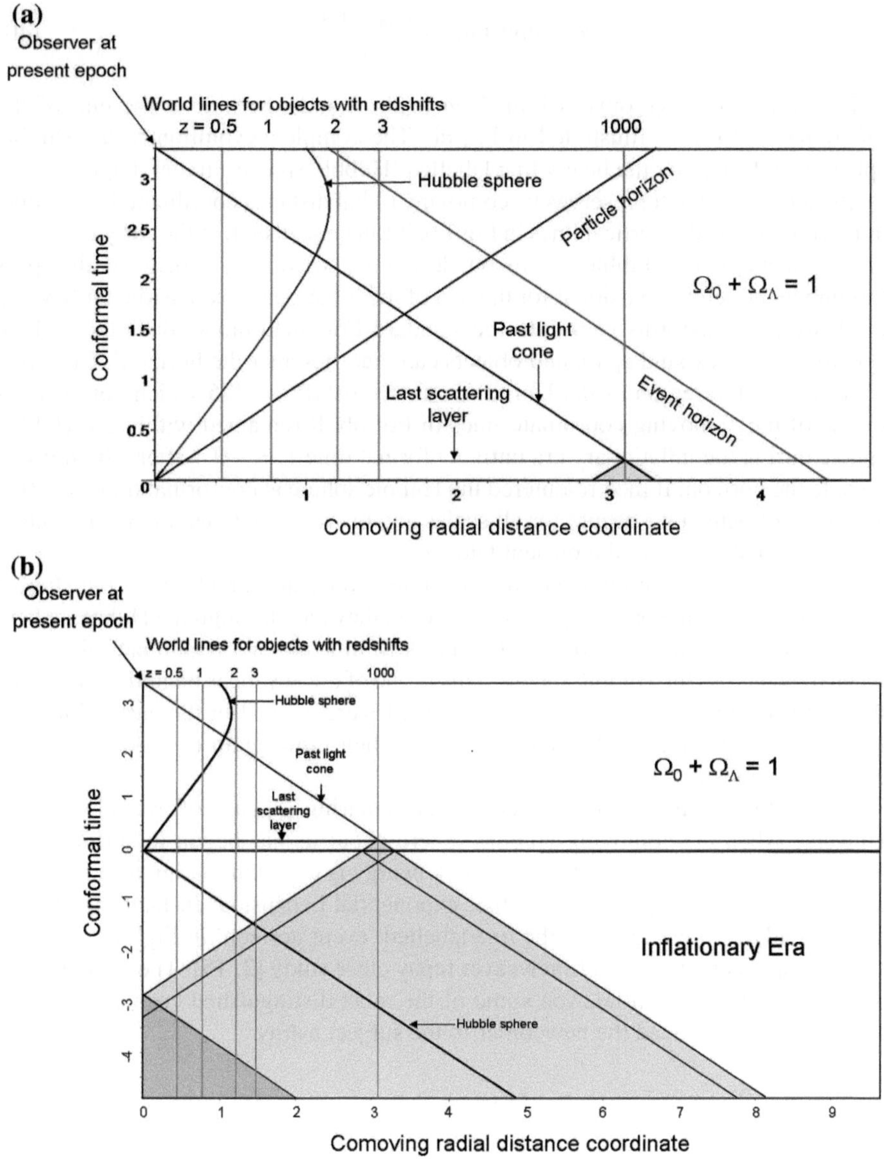

Fig. 6 **a** A repeat of conformal diagram Fig. 5c in which conformal time is plotted against comoving radial distance coordinate. Now, the last scattering surface at the epoch of recombination is shown as well as the past light cone from the point at which our past light cone reaches the last scattering layer. **b** An extended conformal diagram now showing the inflationary era. The time coordinate is set to zero at the end of the inflationary era and evolution of the Hubble sphere and the past light cone at recombination extrapolated back to the inflationary era

$$\tau = \text{constant} - \frac{r_{\rm HS}(\text{com})}{c} \,. \tag{40}$$

This solution for $r_{\rm HS}(\text{com})$ is joined on to the standard result at the end of the inflationary epoch, as illustrated in Fig. 6b. The complete evolution of the Hubble sphere is indicated by the heavy line labelled 'Hubble sphere' in that diagram.

Because any object preserves its comoving radial distance coordinate for all time, as represented by the vertical lines in Fig. 6b, it can be seen that, in the early Universe, objects lie within the Hubble sphere, but during the inflationary expansion, they pass through it and remain outside it for the rest of the inflationary expansion. Only when the Universe transforms back into the standard Friedman model does the Hubble sphere begin to expand again and objects can then 're-enter the horizon'. Consider, for example, the region of the Universe out to redshift $z = 0.5$ which corresponds to one of the comoving coordinate lines in Fig. 6b. It remained within the Hubble sphere during the inflationary era until conformal time $\tau = -0.4$ after which it was outside the horizon. It then re-entered the Hubble sphere at conformal time $\tau = 0.8$. This type of behaviour occurs for all scales and masses of interest in understanding the origin of structure in the present Universe.

Since causal connection is no longer possible on scales greater than the Hubble sphere, it follows that objects 'freeze out' when they pass through the Hubble sphere during the inflationary era, but they come back in again and regain causal contact when they recross the Hubble sphere. This is one of the key ideas behind the idea that the perturbations from which galaxies formed were created in the early Universe, froze out on crossing the Hubble sphere and then grew again on re-entering it at conformal times $\tau > 0$.

Notice that, at the present epoch, we are entering a phase of evolution of the Universe when the comoving Hubble sphere about us has begun to shrink again. This is entirely due to the fact that the dark energy is now dominating the expansion and its dynamics are precisely another exponential expansion. In fact, the Hubble sphere tends asymptotically to the line labelled 'event horizon' in Fig. 5a.

The papers by Davis and Lineweaver repay close study [2, 15]. Their remarkable Appendix B indicates how even some of the most distinguished cosmologists and astrophysicists can lead the newcomer to the subject astray.

7 Epilogue

I have selected only a few of the many topics which Paddy and I love to explore and which are rich in physical content, illustrating how real physics can be understood. The intention and spirit are similar to Paddy's delightful *Sleeping Beauties in Theoretical Physics* [29]. Long may Paddy continue to inspire us with these remarkable insights into how physics really works.

References

1. S.N. Bose, Planck's Gesetz und Lichtquantenhypothese (Planck's Law and the Hypothesis of Light Quanta). Zeitschrift für Physik **26**, 178–181 (1924)
2. T.M. Davis, C.H. Lineweaver, Expanding confusion: common misconceptions of cosmological horizons and the superluminal expansion of the Universe. Publ. Astron. Soc. Aust. **21**, 97–109 (2004)
3. S. Drake, *Galileo: Pioneer Scientist* (University of Toronto Press, Toronto, 1990)
4. A. Einstein, in *The Theory of Radiation and Quanta, Proceedings of the First Solvay Conference*, ed. by P. Langevin, M. de Broglie (Paris, Gautier-Villars, 1912). p. 115
5. A. Einstein, Über einen die Erzeugung und Verwandlung des Lichtes betreffenden heuristischen Gesichtspunkt (On a heuristic point of view concerning the production and transformation of light). Annalen der Physik **4**(17), 132–148 (1906)
6. A. Einstein, Zur Theorie der Brownschen Bewegung. Annalen der Physik **4**(19), 371–381 (1906)
7. A. Einstein, Zum gegenwärtigen Stand des Strahlungsproblems. Physikalische Zeitschrift **10**, 185–193 (1909)
8. A. Einstein, Quantentheorie der Strahlung. Mitteilungen der Physikalischen Gesellschaft Zürich **16**, 47–62 (1916)
9. A. Einstein, Zur Quantentheorie der Strahlung. Physikalische Zeitschrift **18**, 121–128 (1917)
10. A. Einstein, *Quantentheorie des einatomigen idealen Gases*, Sitzungberichte der (Kgl.) Preussischen Akademie der Wissenschaften (Berlin, 1924), pp. 261–267
11. A. Einstein, *Quantentheorie des einatomigen idealen Gases*, Sitzungberichte der (Kgl.) Preussischen Akademie der Wissenschaften (Berlin, 1925), pp. 3–14
12. M.A. Finocchiaro, *The Galileo Affair: A Documnetary History* (University of California Press, Berkeley, 1989)
13. K. Huang, *Introduction to Statistical Physics* (Macmillan and Company, London, 2001)
14. C. Lazaroff-Puck, Gearing up for Lagrangian dynamics:the flywheel analogy in Maxwell's 1865 paper on electrodynamics. Arch. Hist. Exact Sci. **69**, 455–490 (2015)
15. C.H. Lineweaver, Inflation and the cosmic microwave background, in *The New Cosmology*, ed. by M. Colless, (2005), pp. 31–65
16. M.S. Longair, *Theoretical Concepts in Physics*, 2nd edn. (Cambridge University Press, Cambridge, 2003)
17. M.S. Longair, *The Cosmic Century: A History of Astrophysics and Cosmology* (Cambridge University Press, Cambridge, 2006)
18. M.S. Longair, *Galaxy Formation*, 2nd edn. (Springer, Berlin, 2008)
19. M.S. Longair, *Quantum Concepts in Physics*, 2nd edn. (Cambridge University Press, Cambridge, 2013)
20. M.S. Longair, '..a paper..I hold to be great guns': a commentary on Maxwell (1865) 'A dynamical theory of the electromagnetic field'. Philos. Trans. R. Soc. **A373**, 20140473 (2015)
21. J.C Maxwell, *On Faradays lines of force*. Philosophical Magazine, **XI**, part 4, 404-405 (1856). This is the abstract of Maxwells' paper. It was printed in full in 1864
22. J.C. Maxwell, *On physical lines of force*. Philosophical Magazine: *I. The theory of molecular vortices applied to magnetic phenomena*, **21**, 161-175 (1861); *II. The theory of molecular vortices applied to electric currents*, **21**, 281–291 (1861); *III. The theory of molecular vortices applied to statical electricity*, **23**, 12–24, (1862); *IV. The theory of molecular vortices applied to the action of magnetism on polarised light*, **23**, 85–95, (1862)
23. J.C. Maxwell, A dynamical theory of the electromagnetic field. Philos. Trans. R. Soc. **155**, 459–512 (1865)
24. J.C. Maxwell, On Governors. Proc. R. Soc. Lond. **16**, 270–283 (1868)
25. J.C. Maxwell, *Treatise on Electricity and Magnetism (in two volumes)* (Clarendon Press, Oxford, 1873)
26. R.Á. Millikan, A direct photoelectric determination of Planck's h. Phys. Rev. **7**, 355–388 (1916)

27. E.A. Milne, W.H. McCrea, Newtonian expanding Universe. Q. J. Math. **5**, 64–72 (1934)
28. E.A. Milne, W.H. McCrea, Newtonian universes and the curvature of space. Q. J. Math. **5**, 73–80 (1934)
29. T. Padmanabhan, *Sleeping Beauties in Theoretical Physics* (Springer, Berlin, 2015)
30. A. Pais, *Inward Bound* (Clarendon Press, Oxford, 1985)
31. J.B. Perrin, Mouvement brownien et réalité moléculaire. Annales de Chemie et de Physique (Viii) **18**, 5–114 (1909)
32. J.B. Perrin, *Brownian Movement and Physical Reality* (Taylor and Francis, London, 1910)
33. M. Planck, Über eine Verbesserung der Wien'schen Spektralgleichung (On an improvement of Wien's spectral distribution. Verhandlungen der Deutschen Physikalischen Gesellschaft **2**, 202–204 (1900)
34. M. Planck, Zur Theorie des Gestzes der Energieverteilung im Normalspektrum. Verhandlungen der Deutschen Physikalischen Gesellschaft **2**, 237–245 (1900)
35. M. Planck, *Scientific autobiogrphy and Other Papers* (Williams and Norgate, London, 1950)
36. T.K. Simpson, *Maxwell on the Electromagnetic Field: A Guided Study* (Rutgers University Press, New Brunswick, New Jersey, 1997)
37. E. Whittaker, *A History of the Theories of the Aether and Electricity* (Thomas Nelson and Sons, London, 1951)

A Local Stress Tensor for Gravity Fields

D. Lynden-Bell

Abstract After a discussion of what ought to be meant by a *local* conservation law in General Relativity, we show physically that the energy density of the gravity field of static spherical systems can be deduced from natural axioms that we state. The result demonstrates that in these cases the energy density in both field and matter can be deduced from the spatial metric but not from the gravitational potential. We demonstrate that although the total energy can be determined from the gravitational potential at large r, nevertheless the distribution of that energy is not contained in the potential. We then derive all components of the stress-energy tensor of the gravity field of spherical systems. Finally we find a three dimensional flat space associated with all asymptotically flat static spaces.

1 Introduction

In physics global conservation laws follow from local conservation laws. The density of the conserved quantity integrated over all space is the same for all time because the decrease in density in each small volume is compensated by the outflow of the corresponding flux through the surface of that volume. When the strict requirements of causality are considered it is hard to imagine that anything could be conserved globally without there being an underlying local conservation law. However in General Relativity we meet global conservation laws of energy momentum and angular momentum without the existence of any "local" conservation law. The pseudo-tensors are sometimes invoked to fill this gap but they are made from unphysical quantities which change when we consider different coordinates. When I say there are no corresponding "local" conservation laws I am using the standard idea that by "local" we mean a quantity that can be constructed from the metric and its derivatives at any chosen point. It could be that it is this definition of what is to be called local, that is the seat of the problem. Consider an asymptotically flat space-time. Before we can decide what is to be called the total energy or the total momentum we must

D. Lynden-Bell (✉)
Institute of Astronomy, Madingley Rd, Cambridge, UK
e-mail: dlb@ast.cam.ac.uk

J.S. Bagla and S. Engineer (eds.), *Gravity and the Quantum*,
Fundamental Theories of Physics 187, DOI 10.1007/978-3-319-51700-1_15

specify the Lorentz frame at infinity. Even when this can be done consistently we shall face a much more challenging problem in deciding how such a frame should be constructed within the changing curved space. If we fail to specify which parts of even the stress-energy tensor are contributors to the energy and which parts are contributors to momentum then it is perhaps no real surprise that the mathematics is unable to produce a local density for either. We can specify the Lorentz frame at infinity via four orthogonal unit displacement vectors w_a^μ. Here μ is the vector index while a is a Lorentz index that tells us which of the vectors we consider. Changes in the Lorentz frame at infinity can be made without changing the (possibly curvilinear) coordinates there. Under such a Lorentz transformation the w_a^μ will transform linearly not merely at infinity but everywhere. Thus $\tilde{w}_b^\mu = L_b^a w_a^\mu$. Where the L_b^a define the Lorentz transformation and are independent of position. Implicit in using this transformation even in the curved regions of space is the idea that the w_a^μ must obey a linear equation. These vectors should be asymptotic Killing vectors of the Minkowski space at spatial infinity. That fact might suggest that to extend these vectors into the curved part of space-time, we need the w_a^μ to be 'Willing' vectors defined to obey some softened but linear form of Killing's equation, which allows solutions in spaces without any symmetry. While the late Joseph Katz and I spent some effort exploring the willing vector concept, we did not find success in that direction. In those very special cases where there is a global time-like Killing vector the gravitational part of the energy will be conserved separately from the material energy, but we do not want that. We merely want conservation of the sum. While I still believe there is a future in discovering a way to bring the Lorentz frame at infinity into the curved part of space, our efforts in this direction have not yet led to the exact tensorial local conservation laws we seek. Here I use local in the generalised sense that involves the w_a^μ at the point considered as well as the metric there. There is a real challenge here. What linear equation should the willing vectors w_a^μ obey in order that there be energy and momentum conservation laws $D_\mu(\Theta_\nu^\mu w_a^\nu) = 0$? Here Θ_ν^μ is a stress tensor that describes both the material and the gravitational stress. The latter may possibly depend on the willing vectors.

When there is no current success in solving a great problem one may look for crumbs of comfort in special cases. Consider the simplest case, static spaces. Here there is general agreement that the local time-like Killing vector ξ^μ should be normalised to unity at infinity. This is actually a non-local operation, but few object to expressions for energy density that may involve ξ^μ and its existence everywhere gives the static Lorentz frame. We shall hereafter concentrate on defining the gravitational field's energy density for this very special situation. There are some who deny the existence of an energy density for gravitational fields and others who give formulae for it, which are not generally in agreement with one another. We therefore consider the special case of static spherical symmetry, and state axioms from which we derive the energy density of the gravity field. This answer is negative definite and is contained in the spatial part of the metric. We then tabulate some books and papers that agree and others that disagree with our result in this special case and show which axioms are denied by those who disagree. In Sect. 4 we consider two spherical distributions with the same density everywhere one of which has no radial component to its pressure tensor (so it is held up by its tangential pressures) and the other has the usual isotropic pressure at each point. Both have the same spatial metric but they have different potentials although those agree outside the matter. We deduce that a knowledge of the gravitational potential is insufficient for a determination of the gravitational field's energy density. We then show that our axioms also give the gravitational field's energy between any two equipotentials for any static space, even one with no spatial symmetry.

In electrodynamics the energy density of the fields is $(E^2 + B^2)/(8\pi)$. Any surface distribution of charge or current has no electrical energy stored in the surface itself since it has no volume. A shell of surface charge density, σ with no field incident from below has an external normal component of field $E_n = 4\pi\sigma$.

Gravitational waves carry energy, momentum and angular momentum so there is certainly energy in gravitational fields. We shall assume that no gravitational field energy resides in surface densities of matter since, as in the electrical case, they have no volume. A static particle has rest energy mc^2 and a moving particle has energy $mc^2(1 - v^2/c^2)^{-1/2}$. Many such particles may contribute to the stress energy tensor $T_{\mu\nu}$. This describes the matter whose gravity field we are about to study. It is not itself a part of that gravity. To a static observer in a static space-time it contributes a material energy density $\rho c^2 = T_0^0 = w_\mu T^{\mu\nu} w_\nu$ where w_μ is the static observer's 4-velocity, ξ_μ/ξ and ξ_μ is the Killing vector of length $\xi = \sqrt{\xi^\mu \xi_\mu}$. The matter-energy density in an elementary 3-volume dV is also the material energy flux through the corresponding element of space-time surface $d\Sigma_\nu$ which has magnitude $w_\mu T^{\mu\nu}\sqrt{-g}d\Sigma_\nu$. Summing such contributions and taking no account of gravitation we find a material energy within some boundary S of $E_m = \int \rho c^2 dV$ where the integration is over the volume within S. Notice that this definition of the material energy allows us to split the total energy into the material energy which includes any electrodynamic energy, and the rest which we attribute to gravitation. We shall also assume that a static flat region of space-time has no energy and that the total energy of an asymptotically flat space-time can be determined from the asymptotic form of the metric. We summarise the discussion above in the following four axioms:

AXIOMS

1/There is no gravitational field energy in a surface itself, not even in a surface distribution of matter, because such surfaces have no 3-volume.
2/The material energy in a static matter distribution is $\int \rho c^2 dV$ or for a surface distribution $\int \sigma c^2 dS$.
3/There is no energy in a static flat region of space-time.
4/The total energy of stationary asymptotically flat space-times can be determined externally.

2 Gravity Field's Energy Density for Spheres

We now show that these axioms are sufficient to determine uniquely the gravitational field's energy density in static spherical space-times and furthermore the total field energy between any two equipotentials in any static space-time. The method was invented by Lynden-Bell and Katz [10] [see also Grøn [4] and Katz et al. [6]] but unfortunately we used an incorrect formula for the material energy which disobeys axiom 2, so those results are WRONG. Indeed the referee of our 1985 paper Dr. Schutz questioned us on precisely this point and he was very right to do so! We start with the spherical case with some matter density $\rho(r) = T_0^0, \quad c = 1$ in the spherical metric

$$ds^2 = e^{-2\psi}dt^2 - (e^{2\lambda}dr^2 + r^2 d\hat{\mathbf{r}}^2), \tag{1}$$

where $\hat{\mathbf{r}}$ is the unit Cartesian radial vector. Once some pole is chosen we may introduce spherical polar coordinates and then $d\hat{\mathbf{r}}^2 = d\theta^2 + \sin^2\theta d\phi^2$. Both ψ and λ are functions of r. We now compare this metric with one which is truncated beneath some equipotential $\psi = \Psi$ which corresponds to some sphere $r = a$, thus $\Psi = \psi(a)$. See Fig. 1. In $r < a$ we have flat space. In $r > a$ we leave the space unchanged. From the gradient discontinuities at $r = a$, we can discover what surface distribution of matter and stress are needed on the shell that now replaces whatever was formerly within its radius. Since everything outside $r = a$ has been left unchanged and there is now

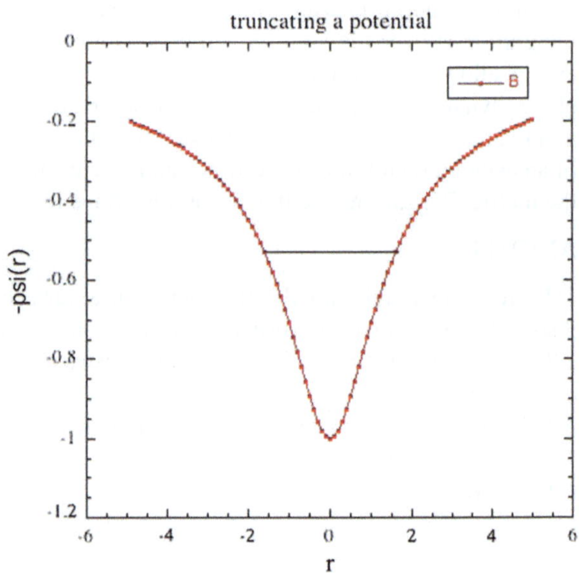

Fig. 1 Any static gravitational potential ψ is truncated at some equipotential leaving a flat 3-space inside. The shell created at the truncation has a material energy equal to the total material and field energy inside that equipotential in the original space. The flat 3-space of Sect. 6 is obtained by superposing the internal flat spaces for all equipotential truncations

flat space in $r < a$ the energy in the spherical shell itself which is all material energy by axiom 1 must equal the energy that was formerly in $r \leq a$ which will be partly material and partly gravitational field energy. However we already have formulae for the material energy so we can find the gravitational field energy that is in $r < a$ in the original metric. By differentiating that expression with respect to a and dividing by the volume $4\pi a^2 e^{\lambda(a)} da$ we can find the gravitational field's energy density at $r = a$. As the value of a can be chosen at will this gives us this field energy density everywhere. From the metric (1) we calculate the non-zero Christoffel symbols, $\Gamma^{\alpha}_{\beta\gamma} = \Gamma^{\alpha}_{\gamma\beta}$

$$\Gamma^0_{01} = -\psi', \quad \Gamma^1_{00} = -\psi' e^{-2(\psi+\lambda)}, \quad \Gamma^1_{11} = \lambda', \quad \Gamma^1_{22} = -re^{-2\lambda},$$
$$\Gamma^1_{33} = -r\sin^2\theta e^{-2\lambda}, \quad \Gamma^2_{33} = -\sin\theta\cos\theta, \; \Gamma^2_{12} = \Gamma^3_{13} = 1/r, \; \Gamma^3_{23} = \cot\theta. \tag{2}$$

The flat space metric inside the shell has $\psi = \psi(a)$, $\lambda = 0$ so with those substitutions we get the Christoffels inside. To calculate the stress tensor of the shell by Israel's method we follow Goldwirth and Katz [3] in using normals that point into the spaces in which the external curvatures are calculated. The unit normal outside the shell is $n_\mu = (0, -e^{\lambda}, 0, 0)$. On the sphere we use coordinates $\theta_0 = t, \; \theta_2 = \theta, \; \theta_3 = \phi$, that is θ_b with $b = 0, 2, 3$. The external curvature is given by $K_{a,b} = -\frac{\partial x^\mu \partial x^\nu}{\partial\theta^a \partial\theta^b} D_\nu n_\mu = \Gamma^1_{a,b} n_1$, where D_ν denotes the covariant derivative. Thus

$$K^0_0 = \psi' \, e^{-\lambda}, \quad K^2_2 = K^3_3 = -e^{-\lambda}/a, \quad K = K^b_b = (a\psi' - 2)e^{-\lambda}/a. \tag{3}$$

From these we construct the Lancos tensor $L^b_a = K\delta^b_a - K^b_a$.

$$L^0_0 = -2e^{-\lambda}/a, \quad L^2_2 = L^3_3 = (a\psi' - 1)e^{-\lambda}/a. \tag{4}$$

Similarly for the flat space inside the spherical shell with unit inward normal $\overline{n}_\mu = (0, 1, 0, 0)$

$$\overline{L^0_0} = 2/a, \quad \overline{L^2_2} = \overline{L^3_3} = 1/a. \tag{5}$$

The surface energy tensor of the shell is given by $\kappa\tau^b_a = L^b_a + \overline{L^b_a}$ so,

$$\kappa\tau^0_0 = (2/a)(1 - e^{-\lambda}), \quad \kappa\tau^2_2 = \kappa\tau^3_3 = (1/a)(1 - e^{-\lambda}) + \psi' \, e^{-\lambda}. \tag{6}$$

The total energy that is in $r \leq a$ in the original space is the same as the matter energy in the shell so, calling the gravitational field's energy density Π^0_0

$$4\pi a^2\tau_0^0 = \int_0^a 4\pi r^2 (T_0^0 + \Pi_0^0) e^\lambda dr.$$

$$\kappa\Pi_0^0 = -\kappa T_0^0 + a^{-2}e^{-\lambda}\frac{d}{da}(a^2\kappa\tau_0^0) = -\kappa T_0^0 + 2a^{-2}e^{-\lambda}[(1-e^{-\lambda}) + a\lambda' e^{-\lambda}]. \quad (7)$$

But from Einstein's equations in spherical symmetry

$$\kappa T_0^0 = [e^{-2\lambda}(2a\lambda' - 1) + 1]/a^2; \quad (8)$$

Hence

$$\kappa\Pi_0^0 = \; -(1-e^{-\lambda})^2/a^2. \quad (9)$$

It is nice that the terms in λ' cancel and it is interesting that it is the spatial metric not the potential that determines the field energy. It is also minus a perfect square and thus negative definite as befits the binding energy of the matter. However it is not a sum of squares of the gravomagnetic fields $\mathcal{E}, \mathcal{B}$ that we introduced elsewhere. We pointed out in LKB [11] that it is closely related to the 3-conformal factor that transforms the spherical 3-space to a flat 3-space. Integrating this expression over the volume outside the matter we find

$$E_G(a) = \kappa^{-1}\int_a^\infty \Pi_0^0 4\pi r^2 e^\lambda dr = -\frac{2m^2}{Ga(1+\sqrt{1-2m/a})^2}. \quad (10)$$

We remark that a dust shell with rest mass M and radius $a(\tau)$ falling under its own gravity obeys the energy equation

$$M\sqrt{1+\dot{a}^2} - \tfrac{1}{2}GM^2/a = m/G, \quad (11)$$

where m is the gravitational mass seen at infinity. The first term is the rest mass plus kinetic energy, so it is natural to regard the second term as the gravitational potential energy for the motion of the shell. However $-\frac{1}{2}GM^2/a$ is not the same as the expression for the total gravitational field energy derived in Eq. (10) unless $\dot{a} = 0$. This may be considered a PARADOX worthy of further elucidation. See the discussion of energy in Wald's book [16].

The basic argument is not limited to spherical symmetry but can be applied to any static space. We create the cut on any equipotential and replace the interior by flat space. Thus we can determine the field energy between any two equipotentials however, except for the symmetrical spherical case, we do not get the distribution of field energy over an equipotential from these physical arguments. We discuss this generalisation in Sect. 6.

3 Gravity's Energy Density Discussed

Date	Authors	Agrees	Axiom	Ref
1916	Einstein pseudo-tensor in isotropic coords	Yes		[2]
1959	Landau and Lifshitz pseudo-t. ditto	Yes		[9]
1959	Komar	No	1	[8]
1964	Misner, Sharpe	No	1	[12]
1973	Misner Thorne, Wheeler	No	1	[13]
1972	Møller			[14]
1990	Brown and York	Yes		[1]
1985–87	L*K*; Grøn; K*L*Israel	No	2	[4, 6, 10]
2005	Katz	Yes		[5]
2006–7	K*L*B*,L*K*B*	Yes		[7, 11]

L* = Lynden-Bell, K* = Katz, B* = Bicak

4 Potentials Do Not Yield Energy Densities

Consider two static spherical systems with the same density distribution $\rho(r)$. The first is held static by an isotropic pressure $p(r)$, but the second has no radial component to its stress tensor, each sphere being self-supported by its tangential stresses. Their metrics will be of the form

$$\begin{aligned} ds_1^2 &= e^{-2\psi_1(r)}dt^2 - (1-2m(r)/r)^{-1}dr^2 - r^2 d\hat{\mathbf{r}}^2, \\ ds_2^2 &= e^{-2\psi_2(r)}dt^2 - (1-2m(r)/r)^{-1}dr^2 - r^2 d\hat{\mathbf{r}}^2, \\ 2m(r) &= \int_0^r r^2 \kappa\rho(r)dr. \end{aligned} \tag{12}$$

The spatial metrics are the same by Einstein's equations but the potentials $\psi(r)$ differ to take into account the different stresses. Nevertheless outside the matter each will give the same form $\psi = -\frac{1}{2}\ln(1-2GM/r)$. Thus although the total mass-energy can be deduced from the potential the detailed distribution of mass is not so contained and it is this distribution that gives the energy density of the gravity field.

5 Gravitational Field Stresses

The stress tensor $T_{\mu\nu}$ describes the stresses in the material and its energy density. For static systems the three dimensional divergence of the material stress gives the non-material i.e. gravitational force needed to maintain the equilibrium. Instead of thinking of this force as like the classical $\rho\nabla\psi$, we would like to follow Maxwell's electrostatics and rewrite the gravitational force as the three dimensional divergence of a gravitational stress tensor Π_ν^μ. While it is always possible to express the vector force-density as the divergence of a tensor that tensor may not always be expressible in terms of the field as it is in the both electrostatics and Newtonian gravity. In the static case the contracted Bianchi identity $D_\mu T^\mu_\nu = 0$ yields

$$(\xi\sqrt{\gamma})^{-1}\partial_\mu(\xi\sqrt{\gamma}T^\mu_\nu) - \Gamma^\sigma_{\mu\nu}T^\mu_\sigma = 0; \quad \xi = e^{-\psi}. \tag{13}$$

This equation can be rewritten in terms of the three dimensional divergence $\overline{D}_m T^m_n$ which gives the gravitational force density

$$\overline{D}_m T^m_n = \partial_m\psi(T^m_n - \delta^m_n T^0_0). \tag{14}$$

We notice that the first term on the right arrises from the bending of the 3-space in the 4-space. If we try to follow Maxwell's use of Poisson's equation to rewrite the sources (T^μ_ν) on the right in terms of field gradients, we fail. While we can use Einstein's equations to re-express the T^m_n in terms of second derivatives of the metric, they are not normally expressible in terms of derivatives of $\partial_m\psi$ or its square. For statics the relativistic equivalent of Poisson's equation is

$$(-g)^{-1/2}\partial_\mu((-g)^{1/2}g^{\mu\nu}\partial_\nu\psi) = \nabla^2\psi - \nabla\psi.\nabla\psi = -\kappa(T^0_0 - \tfrac{1}{2}T). \tag{15}$$

where ∇^2 is the operator in the three dimensional gamma space. Even were the relevant component of $T^m_n - \delta^m_n T^0_0$ a multiple of $T^0_0 - \frac{1}{2}T$ the second term in the relativistic Poisson equation would stop us converting the right hand side of (15) into the 3-divergence of a stress tensor made from $\mathcal{E}_k = \partial_k\psi$. The fact that the energy density was unrelated to $\mathcal{E}$ should have warned us of this difficulty. For the spherical case we can still find expressions for the radial and tangential field stresses. Replacing the gravitational forces, those on the right hand side of (15), by the divergence $-\overline{D}_m\Pi^m_n$ and considering the stresses on the shell introduced in Sect. 2 the two external forces are the radial material stress T^1_1 and the radial gravitational stress Π^1_1. These are balanced by the effect of the tangential stresses in the shell so

$$T^1_1 + \Pi^1_1 = (2/a)\tau^2_2, \tag{16}$$

but we already found the latter in Eq. (6) and Einstein's equations give

$$\kappa T^1_1 = [1 - e^{-2\lambda}(1 - 2a\psi')]/a^2, \tag{17}$$

so writing r for a we deduce

$$\kappa\Pi^1_1 = (1 - e^{-\lambda})^2/r^2 + 2\psi'\,e^{-\lambda}(1 - e^{-\lambda})/r. \tag{18}$$

Whereas the first term of this expression might be expected from our expression for Π^0_0 the second is linear in the field $\mathcal{E} = \nabla\psi = d\psi/dr$ and is not of a form familiar from the electrostatic analogy. This last term may be written $(\psi'/\lambda')r^{-1}\partial_r[(1 - e^{-\lambda})^2]$, and outside matter just $r^{-1}\partial_r[(1 - e^{-\lambda})^2]$. Re-writing (14) with gravitational forces $-\overline{D}_m \Pi^m_n$

$$\overline{D}_m(T^m_n + \Pi^m_n) = 0 = r^{-2}\frac{d}{dr}[r^2(T^1_1 + \Pi^1_1)] - (2/r)(T^2_2 + \Pi^2_2). \qquad (19)$$

The other components are automatically satisfied due to the symmetry. Einstein's equations give us

$$\kappa T^2_2 = e^{-2\lambda}(\psi'' - \psi'^{\,2} - \psi'\lambda' + \psi'/r + \lambda'/r). \qquad (20)$$

We already know $\Pi^1_1 + T^1_1$ from (16) so we solve Eq. (19) for Π^2_2 obtaining

$$\begin{aligned}\kappa\Pi^2_2 &= e^{-\lambda}(1 - e^{-\lambda})(\psi'' - \psi'^{\,2} - \psi'\lambda' + \psi'/a + \lambda'/a) + e^{-\lambda}\psi'^{\,2} \\ &= (e^{\lambda} - 1)\kappa T^2_2 + e^{-\lambda}\psi'^2.\end{aligned} \qquad (21)$$

Equation (21) contains second derivatives unlike the stress tensors of electromagnetism which contain only squares of first derivatives of the potentials. This completes our derivation of the stress tensor which is purely diagonal in spherical coordinates with Π^0_0 given by (9), Π^1_1 by (18), $\Pi^2_2 = \Pi^3_3$ by (21).

6 A Flat 3-Space for Every Static Space

One seldom encounters a flat space intimately associated with the spaces under study. In the belief that such a rarity will prove useful possibly as a map of the true space into which tensors and even Einstein's equations themselves may be translated I here give the construction of such a space. It follows naturally from our method of determining the energy between any two equipotentials and no doubt holds the clue as to how the energy density is distributed. As pointed out in Sect. 2 the energy within any equipotential Ψ can be determined by cutting the metric on that equipotential and replacing the space inside with a flat space. From the gradient discontinuity so generated we read out the surface density and surface stresses that result. The material energy in this surface distribution will equal the sum of the material energy and the gravitational field's energy that was within this equipotential in the original space. By doing this for each equipotential we generate a function $E(\psi)$. By adding together the material energy-densities within each equipotential we can also calculate $E_m(\psi)$ and by subtraction we find the gravitational field's energy $E_G(\psi) = E - E_m$ within each equipotential. At each step we generate a flat internal 3-space that exists only out to the equipotential concerned, see Fig. 1. In some sense each of these internal flat spaces fits inside the next one. Up to now we have pictured them as being at different heights along an axis that measures gravitational potential but this height displacement is merely a useful thought picture. We shall now think of them as different parts of a single three-dimensional flat space in which three functions are defined ψ, $E(\psi)$, $E_G(\psi)$. To each point of this flat space there corresponds a point in the original curved space and vice-versa so the metric functions of the original space can also be evaluated at each point in the flat space. It is my hope that the distribution of gravitational energy over each equipotential can be defined using the properties of this flat space. If the metric of the curved space is written using ψ as a coordinate

$$ds^2 = e^{-2\psi}dt^2 - [\gamma_{11}d\psi^2 + 2\gamma_{1,a}d\psi dx^a + \gamma_{ab}dx^a dx^b], \quad a, b = 2, 3, \tag{22}$$

then on the equipotentials the spatial metric reduces to $\gamma_{ab}dx^a dx^b$. By construction this will be the metric on each equipotential in the flat space. While its metric and internal curvature is the same as that of the equipotential of the curved space, its external curvatures will be different. The spatial metric in the flat space will be of the form

$$[\overline{\gamma}_{11}d\psi^2 + 2\overline{\gamma}_{1,a}d\psi dx^a + \gamma_{ab}dx^a dx^b], \tag{23}$$

and the map of the points of the flat space to those of the curved space and vice versa is given by taking equal values of these coordinates (ψ, x^2, x^3). This sketch of an argument is incomplete; the exact way in which the different flat space pieces are to be superposed into one space is left unanswered but if there is axial symmetry then the axis must be made to coincide and if there is also a symmetry about an equatorial plane then its intersection with the axis gives a centre which is seen in each space and when both axis and centre are made to coincide the map is fixed up to a trivial symmetry group. While I expect the map of less symmetrical spaces can be determined I leave this to be decided by those who like Paddy see things clearly and can explain them concisely.

7 Thanks Paddy

I have much enjoyed, almost yearly, deep discussions of many subtle issues in both Astronomy and relativity physics with Paddy. As compared to my muddled gropings for insight he has an eye that sees the issues with remarkable clarity and a mind that can explain them to the fascination of both us and his students (see for example his fine book on relativity [15]). He is also a master of the mathematical formalisms of many subjects. His books on the subjects that I know, are authoritative, accurate and insightful. His inspiration is Landau and I believe that like Landau his greatest legacy to Science will lie therein. As we now know there are many productive years past the age of sixty and I encourage Paddy to enjoy using them to the full. There is still much to be discovered and elucidated, even the simple paradox given here!

References

1. J.D. Brown, J.W. York, Phys. Rev. D **47**, 1407 (1993)
2. A. Einstein, The Principle of Relativity, ed. by Einstein, Minkowski, Weyl (Dover reprint from Methuen Publ, 1923) (1916)
3. D. Goldwirth, J. Katz, Class. Quant. Grav. **13**, 769 (1995)
4. Ø. Grøn, Gen. Rel. Gravit. **18**, 889 (1986)
5. J. Katz, Class. Quant. Grav. **22**, 5169 (2005)
6. J. Katz, D. Lynden-Bell, W. Israel, Class. Quant. Grav. **5**, 971 (1988)
7. J. Katz, D. Lynden-Bell, J. Bicak, Class. Quant. Grav. **23**, 7111 (2006)
8. A. Komar, Phys. Rev. **113**, 934 (1959)
9. L.D. Landau, E.M. Lifshitz, *The Classical Theory of Fields* (MIR, Moscow, 1959)
10. D. Lynden-Bell, J. Katz, MNRAS **213**, 21p (1985)
11. D. Lynden-Bell, J. Katz, J. Bicak, Phys. Rev. D **75**, 024040 (2007)
12. C.W. Misner, D.H. Sharpe, Phys. Rev. B **136**, 571 (1964)
13. C.W. Misner, K.S. Thorne, J.A. Wheeler, *Gravitation* (Freeman, San Francisco, 1973)
14. C. Møller, *The Theory of Relativity*, 2nd edn. (Oxford, 1972)
15. T. Padmanabhan, *Gravitation* (Cambridge U.P, Cambridge, 2010)
16. R. Wald, *General Relativity* (Chicago, 1984)

Nonlocal Infrared Modifications of Gravity. A Review

Michele Maggiore

Abstract We review an approach developed in the last few years by our group in which GR is modified in the infrared, at an effective level, by nonlocal terms associated to a mass scale. We begin by recalling the notion of quantum effective action and its associated nonlocalities, illustrating some of their features with the anomaly-induced effective actions in $D = 2$ and $D = 4$. We examine conceptual issues of nonlocal theories such as causality, degrees of freedoms and ghosts, stressing the importance of the fact that these nonlocalities only emerge at the effective level. We discuss a particular class of nonlocal theories where the nonlocal operator is associated to a mass scale, and we show that they perform very well in the comparison with cosmological observations, to the extent that they fit CMB, supernovae, BAO and structure formation data at a level fully competitive with ΛCDM, with the same number of free parameters. We explore some extensions of these 'minimal' models, and we finally discuss some directions of investigation for deriving the required effective nonlocality from a fundamental local QFT.

1 Introduction

I am very glad to contribute to this Volume in honor of prof. Padmanabhan (Paddy, to his friends), on the occasion of his 60th birthday. I will take this opportunity to give a self-contained account of the work done in the last few years by our group in Geneva, on nonlocal modifications of gravity.

Our motivation comes from cosmology. In particular, the observation of the accelerated expansion of the Universe [75, 80] has revealed the existence of dark energy (DE). The simplest explanation for dark energy is provided by a cosmological constant. Indeed, ΛCDM has gradually established itself as the cosmological paradigm, since it accurately fits all cosmological data, with a limited set of parameters. From a theoretical point of view, however, the model is not fully satisfying, because a cos-

M. Maggiore (✉)
Département de Physique Théorique and Center for Astroparticle Physics,
Université de Genève, 24 quai Ansermet, 1211 Genève 4, Switzerland
e-mail: Michele.Maggiore@unige.ch

J.S. Bagla and S. Engineer (eds.), *Gravity and the Quantum*,
Fundamental Theories of Physics 187, DOI 10.1007/978-3-319-51700-1_16

mological constant is not technically natural from the point of view of the stability under radiative corrections. Independently of such theoretical 'prejudices', the really crucial fact is that, with the present and forthcoming cosmological data, alternatives to ΛCDM are testable, and it is therefore worthwhile to explore them.

At the fundamental level QFT is local, and in our approach we will not depart from this basic principle. However, both in classical and in quantum field theory, at an effective level nonlocal terms are unavoidably generated. Classically, this happens when one integrates out some degree of freedom to obtain an effective dynamics for the remaining degrees of freedom. Consider for instance a system with two degrees of freedom ϕ and ψ, described classically by two coupled equations of the generic form $\Box\phi = j(\psi)$ and $\Box\psi = f(\phi)$. The first equation is solved by $\phi = \Box^{-1} j(\psi)$. This solutions can then be re-injected in the equation for the remaining degree of freedom ψ, leading to a nonlocal equations involving only ψ. In QFT, nonlocalities appear in the quantum effective action, as we will review below. The appearance of nonlocal terms involving inverse powers of the d'Alembertian is potentially interesting from a cosmological point of view, since we expect that the $\Box^{-1}$ operator becomes relevant in the infrared (IR).

This review is organized as follows. In Sect. 2 we recall the notion of quantum effective action, in particular in gravity, and we discuss the associated nonlocalities. In Sect. 3 we examine two particularly important nonlocal quantum effective actions, the anomaly-induced effective actions in $D = 2$ (i.e. the Polyakov quantum effective action) and in $D = 4$. In Sect. 4 we introduce a class of nonlocal theories in which the nonlocality is associated to a mass scale. In Sect. 5, building also on the experience gained in Sect. 3 with the anomaly-induced effective actions, we discuss conceptual issues of nonlocal theories, such as causality and degrees of freedom, emphasizing the importance of dealing with them as quantum effective actions derived from a fundamental local QFT. In Sect. 6 we discuss how nonlocal theories can be formally put in a local form, and we examine the conceptual subtleties associated to the localization procedure concerning the actual propagating degrees of freedom of the theory.

The cosmological consequences of these nonlocal models are studied in Sect. 7.1 at the level of background evolution, while in Sect. 7.2 we study the cosmological perturbations and in Sect. 7.3 we present the results of a full Bayesian parameter estimation and the comparison with observational data and with ΛCDM. In Sect. 7.4 we discuss further possible extensions of the 'minimal models', and their phenomenology.

As we will see, these nonlocal models turn out to be phenomenologically very successful. The next step will then be understanding how these nonlocalities emerge. Possible directions of investigations for deriving the required nonlocality from a fundamental theory are briefly explored in Sect. 8, although this part is still largely work in progress.

We use units $\hbar = c = 1$, and MTW conventions [69] for the curvature and signature, so in particular $\eta_{\mu\nu} = (-, +, +, +)$.

2 Nonlocality and Quantum Effective Actions

At the quantum level nonlocalities are generated when massless or light particles run into quantum loops. The effect of loop corrections can be summarized into a quantum effective action which, used at tree level, takes into account the effect of quantum loops. The quantum effective action is a nonlocal object. For instance in QED, if we are interested in amplitudes where only photons appear in the external legs, we can integrate out the electron. The corresponding quantum effective action $\Gamma_{\rm QED}$ is given by

$$\begin{aligned} e^{i\Gamma_{\rm QED}[A_\mu]} &= \int \mathscr{D}\psi\mathscr{D}\overline{\psi}\, \exp\left\{i\int d^4x \left[-\frac{1}{4e^2}F_{\mu\nu}F^{\mu\nu} + \overline{\psi}(i\not{D} - m_e + i\varepsilon)\psi\right]\right\} \\ &= e^{-\frac{i}{4e^2}\int d^4x\, F_{\mu\nu}F^{\mu\nu}}\det(i\not{D} - m_e + i\varepsilon)\,. \end{aligned} \tag{1}$$

To quadratic order in the electromagnetic field this gives

$$\Gamma_{\rm QED}[A_\mu] = -\frac{1}{4}\int d^4x \left[F_{\mu\nu}\frac{1}{e^2(\Box)}F^{\mu\nu} + \mathcal{O}(F^4)\right], \tag{2}$$

where, to one-loop order and in the $\overline{\rm MS}$ scheme [28],

$$\frac{1}{e^2(\Box)} = \frac{1}{e^2(\mu)} - \frac{1}{8\pi^2}\int_0^1 dt\,(1-t^2)\log\left[\frac{m_e^2 - \frac{1}{4}(1-t^2)\Box}{\mu^2}\right]. \tag{3}$$

Here μ is the renormalization scale and $e(\mu)$ is the renormalized charge at the scale μ. In the limit $|\Box/m_e^2| \gg 1$, i.e. when the electron is light with respect to the relevant energy scale, the form factor $1/e^2(\Box)$ becomes

$$\frac{1}{e^2(\Box)} \simeq \frac{1}{e^2(\mu)} - \beta_0 \log\left(\frac{-\Box}{\mu^2}\right), \tag{4}$$

where $\beta_0 = 1/(12\pi^2)$. The logarithm of the d'Alembertian is a nonlocal operator defined by

$$\log\left(\frac{-\Box}{\mu^2}\right) = \int_0^\infty dm^2 \left[\frac{1}{m^2+\mu^2} - \frac{1}{m^2-\Box}\right]. \tag{5}$$

Thus, in this case the nonlocality of the effective action is just the running of the coupling constant, expressed in coordinate space. In the opposite limit $|\Box/m_e^2| \ll 1$ the form factor (3) becomes local,

$$\frac{1}{e^2(\Box)} \simeq \frac{1}{e^2(\mu)} - \beta_0 \log\left(\frac{m_e^2}{\mu^2}\right). \tag{6}$$

Observe that the corresponding beta function, which is obtained by taking the derivative with respect to $\log \mu$, is independent of the fermion mass, so in particular in a theory with several fermions even the heavy fermions would contributes to the beta function, and would not decouple. Actually, this is a pathology of the $\overline{\rm MS}$ subtraction scheme, and is related to the fact that, when m_e^2 is large, Eq. (6) develops large logarithms $\log m_e^2/\mu^2$, so in this scheme perturbation theory breaks down for particles heavy with respect to the relevant energy scales. To study the limit $|\Box/m_e^2| \ll 1$ it can be more convenient to use a mass-dependent subtraction scheme, such as subtracting from a divergent graph its value at an Euclidean momentum $p^2 = -\mu^2$. Then, in the limit $|\Box/m_e^2| \ll 1$,

$$\frac{1}{e^2(\Box)} \simeq \frac{1}{e^2(\mu)} + \frac{4}{15\,(4\pi)^2}\,\frac{\Box}{m_e^2}\,, \tag{7}$$

so the contribution of a fermion with mass m_e to the beta function is suppressed by a factor $|\Box/m_e^2|$, so the decoupling of heavy particles is explicit [68].[1] Thus, using a mass-dependent subtraction scheme, the effect of a heavy fermion with mass m_e, at quadratic order in the fields, is to produce the local higher-derivative operator $F_{\mu\nu}\Box F_{\mu\nu}$, suppressed by a factor $1/m_e^2$. Adding to this also the terms of order $F_{\mu\nu}^4$ gives the well-known local Euler-Heisenberg effective action (see e.g. [41] for the explicit computation), valid in the limit $|\Box/m_e^2| \ll 1$,

$$\begin{aligned}\Gamma_{\rm QED}[A_\mu] \simeq \int d^4x \Big[&-\frac{1}{4e^2(\mu)}F_{\mu\nu}F^{\mu\nu} - \frac{1}{15\,(4\pi)^2}\,\frac{1}{m_e^2}\,F_{\mu\nu}\Box F^{\mu\nu} \\ &+\frac{e^2(\mu)}{90(4\pi)^2}\,\frac{1}{m_e^4}\left((F^{\mu\nu}F_{\mu\nu})^2 + \frac{7}{4}(F^{\mu\nu}\tilde{F}_{\mu\nu})^2\right)\Big]\,. \end{aligned}\tag{8}$$

To sum up, nonlocalities emerge in the quantum effective action when we integrate out a particle which is light compared to the relevant energy scale. In contrast, heavy particles give local contributions which, if computed in a mass-dependent subtraction scheme, are encoded in higher-dimension local operators suppressed by inverse powers of the particle mass.

The quantum effective action is a particularly useful tool in gravity, where the integration over matter fields gives the quantum effective action for the metric (see e.g. [15, 19, 71, 82] for pedagogical introductions). Let us denote collectively all matter fields as ϕ, and the fundamental matter action by $S_m[g_{\mu\nu}, \phi]$. Then the quantum effective action Γ is given by

[1] Alternatively, in a theory with N fermion fields, one can still use the $\overline{\rm MS}$ scheme. However, if m_f is the mass of the heaviest among the N fermions, at energies $E < m_f$, one must use the theory without the heavy fermion of mass m_f, and impose appropriate matching conditions at $E = m_f$ between the theory with N fermions at $E > m_f$ and the theory with $N-1$ fermions at $E < m_f$. One proceeds similarly whenever, lowering the energy, we reach the mass of any of the other fermions. This is the standard way of treating weak interactions at low energies, 'integrating out' the heavy quarks, see Sects. 6 and 7 of [68].

$$e^{i\Gamma[g_{\mu\nu}]} = e^{iS_{\rm EH}[g_{\mu\nu}]} \int \mathscr{D}\phi \, e^{iS_m[g_{\mu\nu},\phi]} , \tag{9}$$

where $S_{\rm EH}$ is the Einstein–Hilbert action.[2] The effective quantum action Γ determines the dynamics of the metric, including the backreaction from quantum loops of matter fields. Even if the fundamental action $S_m[g_{\mu\nu}, \phi]$ is local, again the quantum effective action for gravity is unavoidably nonlocal. Its nonlocal part describes the running of coupling constants, as in Eq. (2), and other effects such as particle production in the external gravitational field.

The matter energy-momentum tensor $T^{\mu\nu}$ is given by the variation of the fundamental action, according to the standard GR expression $T^{\mu\nu} = (2/\sqrt{-g})\delta S_m/\delta g_{\mu\nu}$. In contrast, the variation of the effective quantum action gives the *vacuum expectation value* of the energy-momentum tensor,

$$\langle 0|T^{\mu\nu}|0\rangle = \frac{2}{\sqrt{-g}}\frac{\delta\Gamma}{\delta g_{\mu\nu}} . \tag{10}$$

More precisely, the in-out expectation value $\langle 0_{\rm out}|T^{\mu\nu}|0_{\rm in}\rangle$ is obtained when the path-integral in Eq. (9) is the standard Feynman path-integral, while using the Schwinger-Keldish path integral gives the in-in expectation value $\langle 0_{\rm in}|T^{\mu\nu}|0_{\rm in}\rangle$. This point will be important for the discussion of the causality of the effective nonlocal theory, and we will get back to it in Sect. 5.1.

In principle, in Eq. (9) one could expand $g_{\mu\nu} = \eta_{\mu\nu} + h_{\mu\nu}$ and compute perturbatively in $h_{\mu\nu}$. A much more powerful and explicitly covariant computational method is based on the heat-kernel technique (see e.g. [71] for review), combined with an expansion in powers of the curvature. In this way Barvinsky and Vilkovisky [12, 13] have developed a formalisms that allows one to compute, in a covariant manner, the gravitational effective action as an expansion in powers of the curvature, including the nonlocal terms. The resulting quantum effective action, up to terms quadratic in the curvature, has the form

$$\Gamma = \frac{m_{\rm Pl}^2}{2}\int d^4x\sqrt{-g}\,R + \frac{1}{2(4\pi)^2}\int d^4x\sqrt{-g}\left[R\,k_R(\Box)R + \frac{1}{2}C_{\mu\nu\rho\sigma}k_W(\Box)C^{\mu\nu\rho\sigma}\right] , \tag{11}$$

where $m_{\rm Pl}$ is the reduced Planck mass, ${m_{\rm Pl}}^2 = 1/(8\pi G)$, $C_{\mu\nu\rho\sigma}$ is the Weyl tensor, and we used as a basis for the quadratic term R^2, $C_{\mu\nu\rho\sigma}C^{\mu\nu\rho\sigma}$ and the Gauss-Bonnet term, that we have not written explicitly. Just as in Eq. (4), in the case of loops of massless particles the form factors $k_R(\Box)$ and $k_W(\Box)$ only contain logarithmic terms plus finite parts, i.e. $k_{R,W}(\Box) = c_{R,W}\log(\Box/\mu^2)$, where now $\Box$ is the generally-covariant d'Alembertian, μ is the renormalization point, and c_R, c_W are known coefficients that depend on the number of matter species and on their spin. The form factors generated

---

[2] Depending on the conventions, Γ can be defined so that it includes $S_{\rm EH}$, or just as the term to be added to $S_{\rm EH}$.

by loops of a massive particles are more complicated. For instance, for a massive scalar field with mass m_s and action

$$S_s = -\frac{1}{2}\int d^4x\,\sqrt{-g}\,\left(g^{\mu\nu}\partial_\mu\phi\partial_\nu\phi + m_s^2\phi^2 + \xi R\phi^2\right), \tag{12}$$

the form factors $k_R(-\Box/m_s^2)$ and $k_W(-\Box/m_s^2)$ in Eq. (11) were computed in [50, 51] in closed form, for $(\Box/m_s^2)$ generic, in a mass-dependent subtraction scheme where the decoupling of heavy particles is explicit. After subtracting the divergent part, the result is

$$k_W(-\Box/m_s^2) = \frac{8A}{15\,a^4} + \frac{2}{45\,a^2} + \frac{1}{150} + \frac{1}{60}\log\frac{\mu^2}{m_s^2}, \tag{13}$$

$$\begin{aligned} k_R(-\Box/m_s^2) = {} & \bar{\xi}^2 A + \left(\frac{2A}{3a^2} - \frac{A}{6} + \frac{1}{18}\right)\bar{\xi} + A\left(\frac{1}{9a^4} - \frac{1}{18a^2} + \frac{1}{144}\right) \\ & + \frac{1}{108\,a^2} - \frac{7}{2160} + \frac{1}{2}\bar{\xi}^2\log\frac{\mu^2}{m_s^2}, \end{aligned} \tag{14}$$

where $\bar{\xi} = \xi - (1/6)$, and

$$A = 1 - \frac{1}{a}\log\left(\frac{2+a}{2-a}\right), \qquad a^2 = \frac{4\Box}{\Box - 4m_s^2}. \tag{15}$$

In the limit $|\Box/m_s^2| \gg 1$ (i.e. in the limit in which the particle is very light compared to the typical energy or curvature scales), Eq. (14) has the expansion

$$k_R\left(\frac{-\Box}{m_s^2}\right) = \alpha\log\left(\frac{-\Box}{m_s^2}\right) + \beta\frac{m_s^2}{\Box} + \gamma\,\frac{m_s^2}{\Box}\log\left(\frac{-\Box}{m_s^2}\right) + \delta\,\frac{m_s^4}{\Box^2} + \ldots, \tag{16}$$

and similarly for k_W. This result has also been re-obtained with effective field theory techniques [23, 42, 43]. Similar results can also be obtained for different spins, so in the end the coefficients $\alpha, \beta, \gamma, \delta$ depend on the number and type of massive particles.

The result further simplifies for a massless conformally-invariant scalar field. Taking the limit $m_s \to 0$, $\xi \to 1/6$ in Eq. (11) one finds that the terms involving $\log m_s^2$ cancel and the form factor $k_R(\Box)$ becomes local, $k_R = -1/1080$, while $k_W(\Box) \to -(1/60)\log(-\Box/\mu^2)$. Similar results, with different coefficients, are obtained from massless vectors and spinor fields. So, for conformal matter, the one-loop effective action has the form

$$\Gamma_{\text{conf. matter}} = \int d^4x\sqrt{-g}\left[\frac{m_{\rm Pl}^2}{2}R + c_1 R^2 + c_2 C_{\mu\nu\rho\sigma}\log(-\Box/\mu^2)C^{\mu\nu\rho\sigma} + \mathcal{O}(R^3_{\mu\nu\rho\sigma})\right], \tag{17}$$

where c_1, c_2 are known coefficients that depends on the number and type of conformal matter fields, and we have stressed that the computation leading to Eq. (17) has been performed only up to terms quadratic in the curvature.

In contrast, when the particle is heavy compared to the relevant energy or curvature scales, i.e. in the limit $-\Box/m_s^2 \ll 1$, the form factors in Eqs. (13) and (14) become local,

$$k_W(-\Box/m_s^2),\ k_R(-\Box/m_s^2) = \mathcal{O}(\Box/m_s^2)\,. \tag{18}$$

Again, this expresses the fact that particles which are massive compared to the relevant energy scale decouple, leaving a local contribution to the effective action proportional to higher derivatives, and suppressed by inverse powers of the mass. This decoupling is explicit in the mass-dependent subtraction scheme used in Refs. [50, 51].

3 The Anomaly-Induced Effective Action

In a theory with massless, conformally-coupled matter fields, in $D = 2$ space-time dimensions, the quantum effective action can be computed *exactly*, at all perturbative orders, by integrating the conformal anomaly. In $D = 4$ one can obtain in this way, again exactly, the part of the quantum effective action that depends on the conformal mode of the metric.

These examples of quantum effective actions for the gravitational field will be relevant for us when we discuss how the nonlocal models that we will propose can emerge from a fundamental local theory. They also provide an explicit example of the fact that effective quantum actions must be treated differently from fundamental QFT, otherwise one might be fooled into believing that they contain, e.g., ghost-like degrees of freedom, when in fact the fundamental theories from which they are derived are perfectly healthy. We will then devote this section to recalling basic facts on the anomaly-induced effective action, both in $D = 2$ and in $D = 4$ (see e.g. [6, 8, 15, 19, 71, 82] for reviews).

3.1 The Anomaly-Induced Effective Action in D = 2

Consider 2D gravity coupled to N_s conformally-coupled massless scalars [i.e. $m_s = 0$ and $\xi = 1/6$ in Eq. (12)] and N_f massless Dirac fermions. We take these fields to be free, apart from their interaction with gravity. For conformal matter fields, classically the trace T_a^a of the energy-momentum tensor vanishes [in $D = 2$ we use $a = 0, 1$ as Lorentz indices, and signature $\eta_{ab} = (-, +)$]. However, at the quantum level the vacuum expectation value of T_a^a is non-zero, and is given by

$$\langle 0|T_a^a|0\rangle = \frac{N}{24\pi}R\,, \tag{19}$$

where $N = N_s + N_f$. Equation (19) is the trace anomaly. The crucial point about this result is that, even if it can be obtained with a one-loop computation, it is actually *exact*.[3] No contribution to the trace anomaly comes from higher loops. We can now find the effective action that reproduces the trace anomaly, by integrating Eq. (10). We write

$$g_{ab} = e^{2\sigma}\,\bar{g}_{ab}\,, \tag{20}$$

where $\bar{g}_{ab}$ is a fixed reference metric. The corresponding Ricci scalar is

$$R = e^{-2\sigma}(\overline{R} - 2\Box\sigma)\,, \tag{21}$$

where the overbars denotes the quantities computed with the metric $\bar{g}_{ab}$. In $D = 2$, Eq. (10) gives

$$\delta\Gamma = \frac{1}{2}\int d^2x\,\sqrt{-g}\,\langle 0|T^{ab}|0\rangle\delta g_{ab} = \int d^2x\,\sqrt{-g}\,\langle 0|T^{ab}|0\rangle g_{ab}\delta\sigma\,. \tag{22}$$

Therefore

$$\frac{\delta\Gamma}{\delta\sigma} = 2g_{ab}\frac{\delta\Gamma}{\delta g_{ab}} = \sqrt{-g}\,\langle 0|T^a_a|0\rangle\,, \tag{23}$$

where $T^a_a = g_{ab}T^{ab}$. In $D = 2$, without loss of generality, locally we can always write the metric as $g_{ab} = e^{2\sigma}\eta_{ab}$, i.e. we can chose $\bar{g}_{ab} = \eta_{ab}$. In this case, from Eq. (21),

$$R = -2e^{-2\sigma}\Box_\eta\sigma\,, \tag{24}$$

where $\Box_\eta$ is the flat-space d'Alembertian, $\Box_\eta = \eta^{ab}\partial_a\partial_b$. Then, inserting Eq. (19) into Eq. (23) and using $\sqrt{-g} = e^{2\sigma}$, we get

$$\frac{\delta\Gamma}{\delta\sigma} = -\frac{N}{12\pi}\Box\sigma\,. \tag{25}$$

This can be integrated to obtain

$$\Gamma[\sigma] - \Gamma[0] = -\frac{N}{24\pi}\int d^2x\,\sigma\Box_\eta\sigma\,. \tag{26}$$

We see that, in general, the trace anomaly determines the effective action only modulo a term $\Gamma[0]$ independent of the conformal mode. However, in the special case $D = 2$, when $\sigma = 0$ we can choose the coordinates so that, locally, $g_{ab} = \eta_{ab}$. Thus, all curvature invariants vanish when $\sigma = 0$, and therefore $\Gamma[0] = 0$. Therefore, in $D = 2$ the trace anomaly determines *exactly* the quantum effective action, at all perturbative orders! Finally, we can rewrite this effective action in a generally-covariant

---

[3] For the trace anomaly (19), this can be shown using the Seeley-DeWitt expansion of the heat kernel, see Sect. 14.3 of [71].

but non-local form observing that $\Box_g = e^{-2\sigma}\Box_\eta$, where $\Box_g$ is the d'Alembertian computed with the full metric $g_{ab} = e^{2\sigma}\eta_{ab}$. Then, from Eq. (24), $R = -2\Box_g\sigma$, which can be inverted to give $\sigma = -(1/2)\Box_g^{-1}R$, so that

$$\begin{aligned}\Gamma[g_{\mu\nu}] &= -\frac{N}{24\pi}\int d^2x\, e^{2\sigma}\sigma\Box_g\sigma \\ &= -\frac{N}{96\pi}\int d^2x\,\sqrt{-g}\,R\Box_g^{-1}R\,. \end{aligned} \tag{27}$$

This is the Polyakov quantum effective action. The remarkable fact about this effective quantum action is that, even if it has been obtained from the one-loop computation of the trace anomaly, it is the *exact* quantum effective action, to all perturbative orders.

In the above derivation we have studied matter fields in a fixed gravitational background. We now add the dynamics for the metric itself, i.e. we consider 2D gravity, including also a cosmological constant λ, coupled to N massless matter fields,

$$S = \int d^2x\,\sqrt{-g}(\kappa R - \lambda) + S_m\,, \tag{28}$$

where S_m is the the action describing $N = N_S + N_F$ conformally-coupled massless scalar and massless Dirac fermion fields. In 2D the Einstein–Hilbert term is a topological invariant and, once we integrate out the massless matter field, all the gravitational dynamics comes from the anomaly-induced effective action. The contribution of the N matter fields is given by the Polyakov effective action (27). Diff invariance fixes locally $g_{ab} = e^{2\sigma}\bar{g}_{ab}$, where $\bar{g}_{ab}$ is a reference metric. In a theory with dynamical gravity, where in the path integral we also integrate over g_{ab}, this is now a gauge fixing condition, and the corresponding reparametrization ghosts give a contribution -26 to be added to N, while the conformal factor σ gives a contribution $+1$ [29, 40, 59]. Then, after dropping the topologically-invariant Einstein–Hilbert term, the exact quantum effective action of $2D$ gravity reads

$$\Gamma = -\frac{N-25}{96\pi}\int d^2x\sqrt{-g}\,R\frac{1}{\Box}R - \lambda\int d^2x\,\sqrt{-g}\,, \tag{29}$$

with an overall factor in the nonlocal term proportional to $(N-25)$.[4] Using Eq. (21) and dropping a σ-independent term $\sqrt{-\bar{g}}\,\bar{R}\bar{\Box}^{-1}\bar{R}$ we see that, in terms of the conformal mode, Eq. (29) becomes local,

---

[4]In bosonic string theory $\lambda = 0$ and, beside diff invariance, one also has Weyl invariance on the world-sheet. This allows one to eliminate also σ, so one only has the contribution -26 from the reparametrization ghosts, together with the contribution from the $N = D$ matter fields $X^\mu(\sigma_1, \sigma_2)$ living in the world-sheet, where $\mu = 0, \dots, D-1$ and D is the number of spacetime dimensions of the target space. Then the coefficient in the anomaly-induced effective action is proportional to $D - 26$, leading to the condition $D = 26$ for the anomaly cancellation, necessary for the elimination of the ghost-like X^0 field.

$$\Gamma = \int d^2x\,\sqrt{-\bar{g}}\left[\frac{N-25}{24\pi}\bar{g}^{ab}\partial_a\sigma\,\partial_b\sigma + \frac{N-25}{24\pi}\bar{R}\sigma - \lambda e^{2\sigma}\right], \tag{30}$$

which is the action of Liouville field theory.

Equation (30) also allows us to illustrate an issue that will emerge later, in the context of the nonlocal model that we will propose. If we try to read the spectrum of the quantum theory from Eq. (30), treating it as if it were the fundamental action of a QFT, we would conclude that, for $N \neq 25$, there is one dynamical degree of freedom, σ. Recalling that our signature is $\eta_{ab} = (-, +)$, we would also conclude that for $N > 25$ this degree of freedom is a ghost and for $N < 25$ it has a normal kinetic term.

However, this conclusion is wrong. Equation (30) is the quantum effective action of a fundamental theory which is just 2D gravity coupled to N healthy fields, in which there is no ghost in the spectrum of the fundamental theory. If we perform the quantization of the fundamental theory in the conformal gauge (20), the fields involved are the matter fields, the reparametrization ghosts, and the only surviving component of the metric once we have fixed the conformal gauge, i.e. the conformal factor σ. Each of them has its own creation and annihilation operators, which generate the full Hilbert space of the theory. However, as always in theories with a local invariance (in this case diff invariance) the physical Hilbert space is a subset of the full Hilbert space. The condition on physical states can be obtained requiring that the amplitude $\langle f|i\rangle$ between an initial state $|i\rangle$ and a final state $|f\rangle$ is invariant under a change of gauge fixing (see e.g. Chap. 4 of [77] for a discussion in the context of bosonic string theory). From this it follows that two states $|s\rangle$ and $|s'\rangle$ are physical if and only if

$$\langle s'|T^{ab}_{\rm tot}|s\rangle = 0\,, \tag{31}$$

where $T^{ab}_{\rm tot}$ is the sum of the energy-momentum tensors of matter, ghosts and σ. This condition (or, more, precisely, the condition that physical states must by BRST invariant) eliminates from the physical spectrum both the states associated with the reparametrization ghosts, and the states generated by the creation operators of the conformal mode, as explicitly proven in [76]. Of course, the physical-state condition (31) is the analogous of the physical-state condition

$$\langle s'|\partial_\mu A^\mu|s\rangle = 0 \tag{32}$$

in the Gupta–Bleuler quantization of electrodynamics, which again eliminates from the physical spectrum the would-be ghost states associated to A_0.

What we learn from this example is that, if we start from a theory such as (30), e.g. to explore its cosmological consequences, there is a huge difference between the situation in which we take it to be a fundamental QFT, and the situation in which we consider it as the quantum effective action of some underlying fundamental theory. In the former case, in the theory (30) we would treat σ as a scalar field living in 2D, and the theory would have one degree of freedom, which is a ghost for $N > 25$ and a healthy scalar for $N < 25$, while for $N = 25$ there would be no dynamics at all. In contrast, when Eq. (30) is treated as the effective quantum action derived from the fundamental QFT theory (28), the interpretation is completely different. The field σ is not just a scalar field living in 2D, but the component of the 2D metric that remains after gauge fixing. The physical spectrum of the fundamental theory is given by the quanta of the N healthy matter fields, which are no longer visible in (30) because they have been integrated out. There is no ghost, independently of the value of N, and there are no physical quanta associated to σ, because they are removed by the physical-state condition associated to the diff invariance of the underlying fundamental theory.

As a final remark, observe that the fact that no physical quanta are associated to σ does not mean that the field σ itself has no physical effects. The situation is again the same as in electrodynamics, where there are no physical quanta associated to A_0, but still the interaction mediated by A_0 generates the Coulomb potential between static charges. In other words, the quanta associated to σ (or to A_0 in QED) cannot appear in the external lines of Feynman diagram, since there are no physical states associated to them, but do appear in the internal lines.

3.2 The Anomaly-Induced Effective Action in $D = 4$

Let us now follow the same strategy in $D = 4$ space-time dimensions, again for massless conformally-coupled matter fields. As we will see, in this case we will not be able to compute the quantum effective action exactly, but still we will be able to obtain valuable non-perturbative information from the trace anomaly. In $D = 4$ the trace anomaly is

$$\langle 0|T^{\mu}_{\mu}|0\rangle = b_1 C^2 + b_2\left(E - \frac{2}{3}\Box R\right) + b_3\Box R\,, \tag{33}$$

where C^2 is the square of the Weyl tensor, E the Gauss-Bonnet term, and it is convenient to use as independent combinations $[E - (2/3)\Box R]$ and $\Box R$, rather than E and $\Box R$. The coefficients b_1, b_2, b_3 are known constants that depend on the number of massless conformally-coupled scalars, massless fermions and massless vector fields. Once again, the anomaly receives contribution only at one loop order, so Eq. (33) is *exact*. Let us now write again

$$g_{\mu\nu} = e^{2\sigma}\bar{g}_{\mu\nu}\,. \tag{34}$$

A crucial difference compared to the $2D$ case is that in $D = 4$ diff invariance no longer allows us to set $\bar{g}_{\mu\nu} = \eta_{\mu\nu}$. Equation (23) still holds, so the anomaly-induced effective action satisfies

$$\frac{\delta\Gamma_{\rm anom}}{\delta\sigma} = \sqrt{-g}\left[b_1 C^2 + b_2\left(E - \frac{2}{3}\Box R\right) + b_3\Box R\right]. \tag{35}$$

We have added the subscript 'anom' to stress that this is the part of the effective action which is obtained from the anomaly. The total quantum effective action is obtained adding $\Gamma_{\rm anom}$ to the classical Einstein–Hilbert term.

To integrate Eq. (35) we first of all observe that the $\Box R$ term can be obtained from the variation of a local R^2 term,

$$g_{\mu\nu}\frac{\delta}{\delta g_{\mu\nu}}\int d^4x\,\sqrt{-g}\,R^2=-6\sqrt{-g}\,\Box R\,. \tag{36}$$

To integrate the other terms we observe that

$$\sqrt{-g}\,C^2=\sqrt{-\bar g}\,\bar C^2\,, \tag{37}$$

$$\sqrt{-g}\left(E-\frac{2}{3}\Box R\right)=\sqrt{-\bar g}\left(\bar E-\frac{2}{3}\bar\Box\,\bar R+4\bar\Delta_4\sigma\right), \tag{38}$$

where the overbars denotes the quantities computed with the metric $\bar g_{\mu\nu}$, and Δ_4 is the Paneitz operator

$$\Delta_4\equiv\Box^2+2R^{\mu\nu}\nabla_\mu\nabla_\nu-\frac{2}{3}R\Box+\frac{1}{3}g^{\mu\nu}\nabla_\mu R\nabla_\nu\,. \tag{39}$$

Thus, we get

$$\begin{aligned}\Gamma_{\rm anom}[g_{\mu\nu}]&=\Gamma_{\rm anom}[\bar g_{\mu\nu}]-\frac{b_3}{12}\int d^4x\,\sqrt{-g}\,R^2\\&\quad+\int d^4x\,\sqrt{-\bar g}\left[b_1\sigma\bar C^2+b_2\sigma\left(\bar E-\frac{2}{3}\bar\Box\,\bar R\right)+2b_2\sigma\bar\Delta_4\sigma\right],\end{aligned} \tag{40}$$

where $\Gamma_{\rm anom}[\bar g_{\mu\nu}]$ is an undetermined integration 'constant', i.e. a term independent of σ, equal to $\Gamma_{\rm anom}[g_{\mu\nu}]$ evaluated at $\sigma=0$. We will discuss below the possible covariantizations of the term in the second line. First, we can rewrite everything in terms of σ and $\bar g_{\mu\nu}$ using

$$R=e^{-2\sigma}\left[\bar R-6\bar\Box\sigma-6\bar\nabla_\mu\sigma\bar\nabla^\mu\sigma\right]. \tag{41}$$

Then

$$\begin{aligned}\Gamma_{\rm anom}[g_{\mu\nu}]&=\Gamma_{\rm anom}[\bar g_{\mu\nu}]-\frac{b_3}{12}\int d^4x\,\sqrt{-\bar g}\left[\bar R-6\bar\Box\sigma-6\bar\nabla_\mu\sigma\bar\nabla^\mu\sigma\right]^2\\&\quad+\int d^4x\,\sqrt{-\bar g}\left[b_1\sigma\bar C^2+b_2\sigma\left(\bar E-\frac{2}{3}\bar\Box\,\bar R\right)+2b_2\sigma\bar\Delta_4\sigma\right].\end{aligned} \tag{42}$$

Once again, the trace anomaly allowed us to determine *exactly* the dependence of the action on the conformal mode σ. However, we cannot determine in this way the

σ-independent part of the effective action, $\Gamma_{\rm anom}[\bar{g}_{\mu\nu}]$. This is an important difference compared to the $D=2$ case, where we could show that $\Gamma_{\rm anom}[\bar{g}_{ab}]=0$ using the fact that locally we can always choose $g_{ab}=\eta_{ab}$. In the end, the effective action must be a function of $\bar{g}_{\mu\nu}$ and σ only in the combination $g_{\mu\nu}=e^{2\sigma}\bar{g}_{\mu\nu}$, so the σ-independent term $\Gamma_{\rm anom}[\bar{g}_{\mu\nu}]$ is just the conformally-invariant part of the effective action, $\Gamma_{\rm c}[g_{\mu\nu}]$, which by definition satisfies

$$\Gamma_c[e^{2\sigma}\bar{g}_{\mu\nu}]=\Gamma_c[\bar{g}_{\mu\nu}]\,. \tag{43}$$

It is interesting to compare the anomaly-induced effective action (42) with the conformal limit of the explicit one-loop computation given in Eq. (17). First of all, the anomaly-induced effective action has a local R^2 term, coming both from the explicit b_3R^2 term and from the term $(-2/3)b_2\sigma\Box\bar{R}$, corresponding to the two terms proportional to $\Box R$ in Eq. (35). The value of its overall coefficient $-[b_3-(2/3)b_2]/12$, obtained from the trace anomaly as a function of the number of conformal massless scalar, massless spinor and massless vector fields, agrees with the coefficient c_1 obtained from the one-loop computation, as it should. Consider now the Weyl-square term in Eq. (17). Recall that Eq. (17) is valid only up to second order in the curvature. Thus, strictly speaking, in the term $C_{\mu\nu\rho\sigma}\log(-\Box/\mu^2)C^{\mu\nu\rho\sigma}$, the $\Box$ operator is the flat-space d'Alembertian. If one would compute to higher orders in the curvature, this term should naturally become a covariant d'Alembertian acting on a tensor such as $C^{\mu\nu\rho\sigma}$. The covariantization of the $\log(\Box)$ operator acting on such a tensor is a non-trivial problem, see the discussion in [33, 44]. In any case we expect that, at least in the simple case of $g_{\mu\nu}=e^{2\sigma}\bar{g}_{\mu\nu}$ with σ constant, we will have $\Box_g=e^{-2\sigma}\Box_{\bar{g}}$, just as for the scalar d'Alembertian. Then,

$$C_{\mu\nu\rho\sigma}\log(-\Box/\mu^2)C^{\mu\nu\rho\sigma}=-2\sigma C^2+C_{\mu\nu\rho\sigma}\log(-\bar{\Box}/\mu^2)C^{\mu\nu\rho\sigma}\,. \tag{44}$$

The second term on the right-hand side, once multiplied by $\sqrt{-g}$, is independent of σ and therefore belongs to $\Gamma_c[\bar{g}_{\mu\nu}]$. On the other hand, the term proportional to $\sqrt{-g}\,\sigma C^2=\sqrt{-\bar{g}}\,\sigma\bar{C}^2$ is just the term proportional to b_1 in Eq. (42). Once again, one can check that the numerical value of the coefficient from the explicit one-loop computation and from the trace anomaly agree. We see that the anomaly-induced effective action and the explicit one-loop computation give complementary information. The anomaly-induced effective action misses all terms independent of σ, such as the term proportional to $C_{\mu\nu\rho\sigma}\log(-\bar{\Box}/\mu^2)C^{\mu\nu\rho\sigma}$ that gives the logarithmic running of the coupling constant associated to C^2. However, the terms that depend on the conformal mode are obtained *exactly*, without any restriction to quadratic order in the curvature.

One can now look for a covariantization of Eq. (40), in which everything is written in terms of $g_{\mu\nu}=e^{2\sigma}\bar{g}_{\mu\nu}$. In general, the covariantization of an expression is not unique. A possible covariantization is given by the Riegert action [79]

$$\Gamma_{\rm anom}[g_{\mu\nu}] = \Gamma_c[g_{\mu\nu}] - \frac{b_3}{12}\int d^4x\,\sqrt{-g}\,R^2 \tag{45}$$
$$+\frac{1}{8}\int d^4x\sqrt{-g}\left(E-\frac{2}{3}\Box R\right)\Delta_4^{-1}\left[b_2\left(E-\frac{2}{3}\Box R\right)+2b_1C^2\right].$$

Just as for the Polyakov action, even if the anomaly-induced action is local when written in terms of the conformal factor, it becomes nonlocal when written in terms of curvature tensors. In this covariantization, as we have seen, the $\log\Box$ form factor in Eq. (17) is not really visible since the term $C_{\mu\nu\rho\sigma}\log(-\bar{\Box}/\mu^2)C^{\mu\nu\rho\sigma}$ is hidden in $\Gamma_{\rm conf}[g_{\mu\nu}]$. Alternative ways of covariantizing the $\log\Box$ operator are discussed in [33, 44]. In any case, in the approximation in which one is interested only in the dynamics of the conformal mode one can use the effective action in the form (42), simply dropping the σ-independent term $\Gamma[\bar{g}_{\mu\nu}]$, independently of the covariantization chosen.

Once again, if one uses Eq. (42) as if it were a fundamental QFT, one would reach the conclusion that this theory contains a ghost. This would be an unavoidable consequence of the presence of the four-derivative term $\sigma\bar{\Delta}_4\sigma$ in Eq. (42) which, expanding over flat space and after integrations by parts, is simply $(\Box\sigma)^2$. As a fundamental QFT, the theory defined by Eq. (42) would then be hopelessly sick. In contrast, we have seen that Eq. (42) is the quantum effective action derived from a fundamental and healthy quantum theory, with no ghost. One could still wander whether the appearance a four-derivative term $\sigma\bar{\Delta}_4\sigma$ signals the fact that a new ghost-like state emerges in the theory because of quantum fluctuations. To answer this question one can quantize the theory (42), and see which states survive the physical-state condition, analogous to Eq. (31) in $D=2$, which reflects the diff-invariance of the underlying fundamental theory. This analysis has been carried out in [5] and it was found that, once one imposes the physical state condition, there is no local propagating degree of freedom associated to σ. Rather, we remain with an infinite tower of *discrete* states, one for each level, all with positive norm. In the limit $Q^2/(4\pi)^2 \equiv -2b_2 \to \infty$, these states have the form $\int d^4x\sqrt{-g}\,R^n|0\rangle$.

4 Nonlocality and Mass Terms

In this section we introduce a class of nonlocal theories where the nonlocality is associated to a mass term. In Sect. 5, using also the experience gained with the study of the anomaly-induced effective action, we will discuss some conceptual issues (such as causality and ghosts) in these theories. A different class of nonlocal models, which do not feature an explicit mass scale, has been introduced in [35, 36], and reviewed in [85]. In this review we will rather focus on the nonlocal models where the nonlocal terms are associated to a mass scale.

4.1 Nonlocal Terms and Massive Gauge Theories

A simple and instructive example of how a nonlocal term can appear in the description of a massive gauge theory is given by massive electrodynamics. Consider the the Proca action with an external conserved current j^μ

$$S = \int d^4x \left[-\frac{1}{4} F_{\mu\nu} F^{\mu\nu} - \frac{1}{2} m_\gamma^2 \, A_\mu A^\mu - j_\mu A^\mu \right] . \tag{46}$$

The equations of motion obtained from (46) are

$$\partial_\mu F^{\mu\nu} - m_\gamma^2 A^\nu = j^\nu . \tag{47}$$

Acting with ∂_ν on both sides and using $\partial_\nu j^\nu = 0$, Eq. (47) gives

$$m_\gamma^2 \, \partial_\nu A^\nu = 0 . \tag{48}$$

Thus, if $m_\gamma \neq 0$, we get the condition $\partial_\nu A^\nu = 0$ dynamically, as a consequence of the equation of motion, and we have eliminated one degree of freedom. Making use of Eq. (48), Eq. (47) becomes

$$(\Box - m_\gamma^2) A^\mu = j^\mu . \tag{49}$$

Equations (48) and (49) together describe the three degrees of freedom of a massive photon. In this formulation locality is manifest, while the $U(1)$ gauge invariance of the massless theory is lost, because of the non gauge-invariant term $m_\gamma^2 \, A_\mu A^\mu$ in the Lagrangian. However, as shown in [46], this theory can be rewritten in a gauge-invariant but nonlocal form. Consider in fact the equation of motion

$$\left(1 - \frac{m_\gamma^2}{\Box}\right) \partial_\mu F^{\mu\nu} = j^\nu , \tag{50}$$

or, rewriting it in terms of A_μ,

$$(\Box - m_\gamma^2) A^\nu = \left(1 - \frac{m_\gamma^2}{\Box}\right) \partial^\nu \partial_\mu A^\mu + j^\nu . \tag{51}$$

Equation (50) is clearly gauge invariant. We can therefore chose the gauge $\partial_\mu A^\mu = 0$. As we see more easily from Eq. (51), in this gauge the nonlocal term vanishes, and Eq. (51) reduces to the local equation $(\Box - m_\gamma^2) A^\nu = j^\nu$. Thus, we end up with the same equations as in Proca theory, $(\Box - m_\gamma^2) A^\mu = j^\mu$ and $\partial_\mu A^\mu = 0$. Note however that they were obtained in a different way: in the Proca theory there is no gauge invariance to be fixed, but Eq. (48) comes out dynamically, as a consequence of the equations of motion, while in the theory (50) there is a gauge invariance and $\partial_\mu A^\mu = 0$ can be imposed as a gauge condition. In any case, since the equations of motions are finally the same, we see that the theory defined by (50) is classical equivalent to the theory defined by Eq. (46). Observe also that Eq. (50) can be formally obtained by taking the variation of the nonlocal action

$$S = -\frac{1}{4}\int d^4x \left[F_{\mu\nu}\left(1 - \frac{m_\gamma^2}{\Box}\right) F^{\mu\nu} - j_\mu A^\mu \right], \tag{52}$$

(apart from a subtlety in the variation of $\Box^{-1}$, that we will discuss in Sect. 5.1).[5] Thus, Eq. (52) provides an alternative description of a massive photon which is explicitly gauge invariant, at the price of nonlocality. In this case, however, the nonlocality is only apparent, since we see from Eq. (51) that the nonlocal term can be removed with a suitable gauge choice. In the following we will study similar theories, in which however the nonlocality cannot be simply gauged away.

An interesting aspect of the nonlocal reformulation of massive electrodynamics is that it also allows us to generate the mass term dynamically, through a non-vanishing gauge-invariant condensate $\langle F_{\mu\nu}\Box^{-1}F^{\mu\nu}\rangle \neq 0$. In the $U(1)$ theory we do not expect non-perturbative effects described by vacuum condensates. However, these considerations can be generalized to non-abelian gauge theories. Indeed, in pure Yang-Mills theory the introduction in the action of a nonlocal term

$$\frac{m^2}{2}\text{Tr}\int d^4x\, F_{\mu\nu}\frac{1}{D^2}F^{\mu\nu}, \tag{53}$$

(where $D_\mu^{ab} = \delta^{ab}\partial_\mu - gf^{abc}A_\mu^c$ is the covariant derivative and m is a mass scale) correctly reproduces the results on the non-perturbative gluon propagator in the IR, obtained from operator product expansions and lattice QCD [17, 21, 45]. In this case this term is generated in the IR dynamically by the strong interactions. In other words, because of non-perturbative effects in the IR, at large distances we have

$$\langle \text{Tr}\,[F_{\mu\nu}D^{-2}F^{\mu\nu}]\rangle \neq 0\,, \tag{54}$$

which amounts to dynamically generating a mass term for the gluons.

4.2 *Effective Nonlocal Modifications of GR*

We next apply a similar strategy to GR. We will begin with a purely phenomenological approach, trying to construct potentially interesting IR modifications of GR by playing with nonlocal operators such as $m^2/\Box$, and exploring different possibilities.

[5] The equivalence of the two theories can also be directly proved using the "Stückelberg trick": one introduces a scalar field φ and replaces $A_\mu \to A_\mu + (1/m_\gamma)\partial_\mu\varphi$ in the action. The equation of motion of this new action $S[A_\mu, \varphi]$, obtained performing the variation with respect to φ, is $\Box\varphi + m_\gamma\partial_\mu A^\mu = 0$, which can be formally solved by $\varphi(x) = -m_\gamma\Box^{-1}(\partial_\mu A^\mu)$. Inserting this expression for φ into $S[A_\mu, \varphi]$ one gets Eq. (52), see [46].

When one tries to construct an infrared modification of GR, usually the aims that one has in mind is the construction of a fundamental QFT (possibly valid up to a UV cutoff, beyond which it needs a suitable UV completion). In that case a crucial requirement is the absence of ghosts, at least up to the cutoff of the UV completion, as in the dRGT theory of massive gravity [30, 31, 53], or in ghost-free bigravity [52]. In the following we will instead take a different path, and present these models as *effective* nonlocal modification of GR, such as a quantum effective action. This change of perspective, from a fundamental action to an effective quantum action, is important since (as we already saw for the anomaly-induced effective action, and as we will see in Sect. 6 for the nonlocal theories that we will propose) the presence of an apparent ghost in the effective quantum action does not imply that a ghost is truly present in the physical spectrum of the theory. Similarly, we will see in Sect. 5.1 that the issue of causality is different for a nonlocal fundamental QFT and a nonlocal quantum effective action.

A nonlinear completion of the degravitation model. As a first example we consider the theory defined by the effective nonlocal equation of motion

$$\left(1-\frac{m^2}{\Box}\right) G_{\mu\nu} = 8\pi G\, T_{\mu\nu}\,, \tag{55}$$

where $\Box$ is the fully covariant d'Alembertian. Equation (55) is the most straightforward generalization of Eq. (50) to GR. This model was proposed in [9] to introduce the degravitation idea. Indeed, at least performing naively the inversion of the nonlocal operator, Eq. (55) can be rewritten as $G_{\mu\nu} = 8\pi G\,[\Box/(\Box - m^2)]T_{\mu\nu}$. Therefore the low-momentum modes of $T_{\mu\nu}$, with $|k^2| \ll m^2$, are filtered out and in particular a constant term in $T_{\mu\nu}$, such as that due to a cosmological constant, does not contribute.[6]

The degravitation idea is very interesting, but Eq. (55) has the problem that the energy-momentum tensor is no longer automatically conserved, since in curved space the covariant derivatives ∇_μ do not commute, so $[\nabla_\mu, \Box] \neq 0$ and therefore also $[\nabla_\mu, \Box^{-1}] \neq 0$. Therefore the Bianchi identity $\nabla^\mu G_{\mu\nu} = 0$ no longer ensures $\nabla^\mu T_{\mu\nu} = 0$. In [54] it was however observed that it is possible to cure this problem, by making use of the fact that any symmetric tensor $S_{\mu\nu}$ can be decomposed as

$$S_{\mu\nu} = S^{\rm T}_{\mu\nu} + \frac{1}{2}(\nabla_\mu S_\nu + \nabla_\nu S_\mu)\,, \tag{56}$$

[6] Observe however that the inversion of the nonlocal operator is more subtle. Indeed, by definition, $\Box^{-1}$ is such that, on any differentiable function $f(x)$, $\Box\Box^{-1} f = f$, i.e. $\Box\Box^{-1} = \mathbb{1}$. In contrast, from $\Box^{-1}\Box f = g$ it does not follows $f = g$. Rather, applying $\Box$ to both sides and using $\Box\Box^{-1} = \mathbb{1}$ we get $\Box(f - g) = 0$, so $f = g + h$ where h is any function such that $\Box h = 0$. Therefore, $\Box^{-1}\Box \neq \mathbb{1}$. The same holds for the inversion of $(\Box - m^2)$. Thus, more precisely, the inversion of Eq. (55) is $G_{\mu\nu} = 8\pi G\,(\Box - m^2)^{-1}\Box T_{\mu\nu} + S_{\mu\nu}$, where $S_{\mu\nu}$ is any tensor that satisfies $(\Box - m^2)S_{\mu\nu} = 0$. In any case, a constant vacuum energy term $T_{\mu\nu} = -\rho_{\rm vac}\eta_{\mu\nu}$ does not contribute, because of the $\Box$ operator acting on $T_{\mu\nu}$, while $S_{\mu\nu}$ only has modes with $k^2 = -m^2$, so it cannot contribute to a constant vacuum energy.

where $S^{\rm T}_{\mu\nu}$ is the transverse part of $S_{\mu\nu}$, i.e. it satisfies $\nabla^\mu S^{\rm T}_{\mu\nu}=0$. Such a decomposition can be performed in a generic curved space-time [32, 87]. The extraction of the transverse part of a tensor is itself a nonlocal operation, which is the reason why it never appears in the equations of motions of a local field theory.[7] Here however we are already admitting nonlocalities, so we can make use of this operation. Then, in [54] (following a similar treatment in the context of nonlocal massive gravity in [78]) it was proposed to modify Eq. (55) into

$$G_{\mu\nu}-m^2\left(\Box^{-1}G_{\mu\nu}\right)^{\rm T}=8\pi G\,T_{\mu\nu}\,, \tag{58}$$

so that energy-momentum conservation $\nabla^\mu T_{\mu\nu}=0$ is automatically ensured. This model can be considered as a nonlinear completion of the original degravitation idea. Furthermore, Eq. (58) still admits a degravitating solution [54]. Indeed, consider a modification of Eq. (58) of the form

$$G_{\mu\nu}-m^2\left[(\Box-\mu^2)^{-1}\,G_{\mu\nu}\right]^{\rm T}=8\pi G\,T_{\mu\nu}\,, \tag{59}$$

with μ is a regularization parameter to be eventually sent to zero. If we set $T_{\mu\nu}=-\rho_{\rm vac}g_{\mu\nu}$, Eq. (59) admits a de Sitter solution $G_{\mu\nu}=-\Lambda g_{\mu\nu}$ with $\Lambda=8\pi G\,[\mu^2/(m^2+\mu^2)]\,\rho_{\rm vac}$. In the limit $\mu\to 0$ we get $\Lambda\to 0$, so the vacuum energy has been completely degravitated. However, the cosmological evolution of this model induced by the remaining cosmological fluid, such as radiation or non-relativistic matter, turns out to be unstable, already at the background level [49, 64]. We will see in Sect. 7 how such an instability emerges. In any case, this means that the model (58) is not phenomenologically viable.

The RT and RR models. The first phenomenologically successful nonlocal model of this type was then proposed in [64], where it was noticed that the instability is specific to the form of the $\Box^{-1}$ operator on a tensor such as $R_{\mu\nu}$ or $G_{\mu\nu}$, and does not appear when $\Box^{-1}$ is applied to a scalar, such as the Ricci scalar R. Thus, in [64] it was proposed a model based on the nonlocal equation

$$G_{\mu\nu}-\frac{m^2}{3}\left(g_{\mu\nu}\Box^{-1}R\right)^{\rm T}=8\pi G\,T_{\mu\nu}\,, \tag{60}$$

where the factor 1/3 is a useful normalization for the mass parameter m. We will discuss its phenomenological consequences in Sect. 7. We will denote it as the "RT" model, where R stands for the Ricci scalar and T for the extraction of the transverse part. A closed form for the action corresponding to Eq. (60) is currently not known.

---

[7] In flat space $\nabla_\mu\to\partial_\mu$ and, applying to both sides of Eq. (56) ∂^μ and $\partial^\mu\partial^\nu$ we find that

$$S^{\rm T}_{\mu\nu}=S_{\mu\nu}-\Box^{-1}(\partial_\mu\partial^\rho S_{\rho\nu}+\partial_\nu\partial^\rho S_{\rho\mu})+\Box^{-2}\partial_\mu\partial_\nu\partial^\rho\partial^\sigma S_{\rho\sigma}\,. \tag{57}$$

In a generic curved spacetime there is no such a simple formula, because $[\nabla_\mu,\nabla_\nu]\neq 0$, but we will see in Sect. 6 how to deal, in practice, with the extraction of the transverse part.

This model is however closely related to another nonlocal model, proposed in [67], and defined by the effective action

$$\Gamma_{\rm RR} = \frac{m_{\rm Pl}^2}{2} \int d^4x \sqrt{-g} \left[R - \frac{m^2}{6} R \frac{1}{\Box^2} R \right] . \tag{61}$$

Again, we will see that this model is phenomenologically viable, and we will refer to it as the RR model. The RT and RR models are related by the fact that, if we compute the equations of motion from Eq. (61) and we linearize them over Minkowski space, we find the same equations of motion obtained by linearizing Eq. (60). However, at the full nonlinear level, or linearizing over a background different from Minkowski, the two models are different.

We have seen above that nonlocal terms of this sort may be related to a mass for some degree of freedom. One might then ask whether this is the case also for the RR and RT models. In fact, the answer is quite interesting: the nonlocal terms in Eq. (60) or (61) correspond to a mass term for the conformal mode of the metric [65, 66]. Indeed, consider the conformal mode $\sigma(x)$, defined choosing a fixed fiducial metric $\bar{g}_{\mu\nu}$ and writing $g_{\mu\nu}(x) = e^{2\sigma(x)}\bar{g}_{\mu\nu}(x)$. Let us restrict the dynamics to the conformal mode, and choose for simplicity a flat fiducial metric $\bar{g}_{\mu\nu} = \eta_{\mu\nu}$. The Ricci scalar computed from the metric $g_{\mu\nu} = e^{2\sigma(x)}\eta_{\mu\nu}$ is then

$$R = -6e^{-2\sigma} \left(\Box\sigma + \partial_\mu\sigma\partial^\mu\sigma \right) . \tag{62}$$

Therefore, to linear order in σ, $R = -6\Box\sigma + \mathcal{O}(\sigma^2)$ and (upon integration by parts)

$$R \frac{1}{\Box^2} R = 36\sigma^2 + \mathcal{O}(\sigma^3) . \tag{63}$$

Thus, the $R\Box^{-2}R$ terms gives a nonlocal but diff-invariant mass term for the conformal mode, plus higher-order interaction terms (which are nonlocal even in σ) which are required to reconstruct a diff-invariant quantity. The same is true for the nonlocal term in the RT model, since the RR and RT models coincide when linearized over Minkowski space.

5 How Not to Deal with Effective Nonlocal Theories

In this section we discuss some conceptual aspects of general nonlocal theories, that involve some subtleties. The bottomline is that quantum field theory must be played according to its rules and, as we have already seen in Sect. 3 with the explicit example of the anomaly-induced effective action, the rules for quantum effective actions are different from the rules for the fundamental action of a QFT.

5.1 Causality

We begin by examining causality in nonlocal theories (we follow the discussion in Appendix A of [26]; see also [14, 35, 36, 47, 48, 84, 85] for related discussions). In a fundamental QFT with a nonlocal action, the standard variational principle produces acausal equations of motion. Consider for instance a nonlocal term $\int dx\, \phi\Box^{-1}\phi$ in the action of a scalar field ϕ, where $\Box^{-1}$ is defined with respect to some Green's function $G(x; x')$. Then

$$\frac{\delta}{\delta\phi(x)}\int dx'\phi(x')(\Box^{-1}\phi)(x') = \frac{\delta}{\delta\phi(x)}\int dx' dx''\phi(x')G(x'; x'')\phi(x'')$$
$$= \int dx'[G(x; x') + G(x'; x)]\phi(x') . \qquad (64)$$

Thus, the variation symmetrizes the Green's function. However, the retarded Green's function is not symmetric; rather, $G_{\rm ret}(x'; x) = G_{\rm adv}(x; x')$, and therefore it cannot be obtained from such a variation. In a fundamental action, nonlocality implies the loss of causality, already at the classical level (unless, as in Eq. (51), we have a gauge symmetry that allows us to gauge away the nonlocal term in the equations of motion).

However, quantum effective actions are in general nonlocal, as in Eq. (2), (27) or (45). Of course, this does not mean that they describe acausal physics. These nonlocal effective actions are just a way to express, with an action that can be used at tree level, the result of a quantum computation in fundamental theories which are local and causal. Therefore, it is clear that their nonlocality has nothing to do with acausality. Simply, to reach the correct conclusions one must play QFT according to its rules. The variation of the quantum effective action does not give the classical equations of motion of the field. Rather, it provides the time evolution, or equivalently the equations of motion, obeyed by the *vacuum expectation values* of the corresponding operators, as in Eq. (10). These equations of motion are obtained in a different way depending on whether we consider the in-in or the in-out matrix elements. The in-out expectation values are obtained using the Feynman path integral in Eq. (9), and are indeed acausal. Of course, there is nothing wrong with it. The in-out matrix element are not observable quantities, but just auxiliary objects which enter in intermediate steps in the computation of scattering amplitudes, and the Feynman propagator, which is acausal, enters everywhere in QFT computations.

The physical quantities, which can be interpreted as physical observables, are instead the in-in expectation values. For instance, $\langle 0_{\rm in}|\hat{g}_{\mu\nu}|0_{\rm in}\rangle$ can be interpreted as a semiclassical metric, while $\langle 0_{\rm out}|\hat{g}_{\mu\nu}|0_{\rm in}\rangle$ is not even a real quantity. The equations of motion of the in-in expectation values are obtained from the Schwinger–Keldysh path integral, which automatically provides nonlocal *but causal* equations [20, 57]. In practice, the equations of motion obtained from the Schwinger–Keldysh path integral turn out to be the same that one would obtain by treating formally the $\Box^{-1}$

operator in the variation, without specifying the Green's function, and replacing in the end $\Box^{-1} \to \Box_{\rm ret}^{-1}$ in the equations of motion (see e.g. [71]).[8]

Thus nonlocal actions, interpreted as quantum effective actions, provide causal evolution equations for the in-in matrix elements.

5.2 Degrees of Freedom and Ghosts

Another subtle issue concerns the number of degrees of freedom described by a nonlocal theory such as (61). Let us at first treat it as we would do for a fundamental action. We write $g_{\mu\nu} = \eta_{\mu\nu} + h_{\mu\nu}$ and expand the quantum effective action to quadratic order over flat space.[9] The corresponding flat-space action is [67]

$$\Gamma_{\rm RR}^{(2)} = \int d^4x \left[\frac{1}{2} h_{\mu\nu}\mathcal{E}^{\mu\nu,\rho\sigma} h_{\rho\sigma} - \frac{1}{3} m^2 h_{\mu\nu} P^{\mu\nu} P^{\rho\sigma} h_{\rho\sigma}\right], \tag{65}$$

where

$$P^{\mu\nu} = \eta^{\mu\nu} - \frac{\partial^\mu \partial^\nu}{\Box}, \tag{66}$$

where now $\Box$ is the flat-space d'Alembertian. We then add the usual gauge fixing term of linearized massless gravity, $\mathcal{L}_{\rm gf} = -(\partial^\nu \bar{h}_{\mu\nu})(\partial_\rho \bar{h}^{\rho\mu})$, where $\bar{h}_{\mu\nu} = h_{\mu\nu} - (1/2)h\eta_{\mu\nu}$. Inverting the quadratic form we get the propagator $\tilde{D}^{\mu\nu\rho\sigma}(k) = -i\Delta^{\mu\nu\rho\sigma}(k)$, where

$$\Delta^{\mu\nu\rho\sigma}(k) = \frac{1}{2k^2}\left(\eta^{\mu\rho}\eta^{\nu\sigma} + \eta^{\mu\sigma}\eta^{\nu\rho} - \eta^{\mu\nu}\eta^{\rho\sigma}\right) + \frac{1}{6}\left(\frac{1}{k^2} - \frac{1}{k^2 - m^2}\right)\eta^{\mu\nu}\eta^{\rho\sigma}, \tag{67}$$

plus terms proportional to $k^\mu k^\nu$, $k^\rho k^\sigma$ and $k^\mu k^\nu k^\rho k^\sigma$, that give zero when contracted with a conserved energy-momentum tensor. The term in the second line in Eq. (67) gives an extra contribution to $\tilde{T}_{\mu\nu}(-k)\tilde{D}^{\mu\nu\rho\sigma}(k)\tilde{T}_{\rho\sigma}(k)$, equal to

$$\frac{1}{6}\tilde{T}(-k)\left[-\frac{i}{k^2} + \frac{i}{k^2 - m^2}\right]\tilde{T}(k). \tag{68}$$

[8] In the in-in formalism the equations of motions are more easily obtained using the tadpole method, i.e. writing a generic field ϕ as $\phi = \phi_{\rm cl} + \varphi$, where φ are the quantum fluctuations over a classical configuration $\phi_{\rm cl}$, and requiring that $\langle 0_{\rm in}|\varphi|0_{\rm in}\rangle = 0$. See [18, 25] for an instructive computation, showing explicit how nonlocal but causal terms emerge in the in-in equations of motion.

[9] The same treatment holds for the RT model, since at the level of the equations of motion linearized over flat space the RR and RT model are identical.

This term apparently describes the exchange of a healthy massless scalar plus a ghostlike massive scalar. The presence of a ghost in the spectrum of the quantum theory would be fatal to the consistency of the model. However, once again, this conclusion comes from a confusion between the concepts of fundamental action and quantum effective action.

To begin, let us observe that it is important to distinguish between the effect of a ghost in the classical theory and its effect in the quantum theory. Let us consider first the classical theory. At linear order, the interaction between the metric perturbation and an external conserved energy-momentum tensor $T_{\mu\nu}$ is given by

$$S_{\rm int} = \int d^4x\, h_{\mu\nu} T^{\mu\nu}\,, \tag{69}$$

where $h_{\mu\nu}$ is the solution of the equations of motion derived from Eq. (65). Solving them explicitly and inserting the solution for $h_{\mu\nu}$ in Eq. (69) one finds [64]

$$S_{\rm int} = 16\pi G \int \frac{d^4k}{(2\pi)^4}\, \tilde{T}_{\mu\nu}(-k) \Delta^{\mu\nu\rho\sigma}(k) \tilde{T}_{\rho\sigma}(k)\,, \tag{70}$$

with $\Delta^{\mu\nu\rho\sigma}(k)$ given by Eq. (67). The quantity $\Delta^{\mu\nu\rho\sigma}(k)$ therefore plays the role of the propagator in the classical theory [and differs by a factor of $-i$ from the quantity usually called the propagator in the quantum theory, $\tilde{D}^{\mu\nu\rho\sigma}(k) = -i\Delta^{\mu\nu\rho\sigma}(k)$]. A 'wrong' sign in the term proportional to $1/(k^2 - m^2)$ in Eq. (67) might then result in a classical instability. Whether this is acceptable or not must be studied on a case-by-case basis. For instance, taking $m = \mathcal{O}(H_0)$, as we will do below, the instability will only develop on cosmological timescales. Therefore, it must be studied in the context of a FRW cosmology, where it will also compete with damping due to the Hubble friction. Whether this will result or not in a viable cosmological evolution, both at the level of background evolution and of cosmological perturbations, can only be deduced an explicit quantitative study of the solutions of these cosmological equations. We will indeed see in Sect. 7 that the cosmological evolution obtained from this model is perfectly satisfying.

A different issues is the presence of a ghost in the spectrum of the quantum theory. After quantization a ghost carries negative energy, and induces vacuum decay through the associated production of ghosts and normal particles, which would be fatal to the consistency of the theory. However, here we must be aware of the fact that the spectrum of the quantum theory can be read from the free part of the *fundamental* action of the quantum theory. To apply blindly the same procedure to the quantum effective action is simply wrong. We have already seen this in Sect. 3 for the anomaly-induced effective action, where the action (30) with $N > 25$, or the action (42), naively seem to have a ghost, but in fact are perfectly healthy effective quantum actions, derived from fundamental QFTs that have no ghost. Another example that illustrates the sort of nonsense that one obtains if one tries to read the spectrum of the quantum theory from the quantum effective action Γ, consider for instance the one-loop effective action of QED, Eq. (2). If we proceed blindly and quantize it as if it were a fundamental action, we would add to Eq. (2) a gauge fixing term $\mathscr{L}_{\rm gf} = -(1/2)(\partial_\mu A^\mu)^2$ and invert the resulting quadratic form. We would then obtain, for the propagator in the $m_e \to 0$ limit,

$$\tilde{D}^{\mu\nu}(k) = -i\frac{\eta^{\mu\nu}}{k^2}\left[1 - e^2(\mu)\beta_0 \log\frac{k^2}{\mu^2}\right], \tag{71}$$

plus terms proportional to $k^\mu k^\nu$ that cancel when contracted with a conserved current j^μ.[10] Using the identities

$$\log\frac{k^2}{\mu^2} = \int_0^\infty dm^2 \left(\frac{1}{m^2-\mu^2} - \frac{1}{k^2+m^2}\right) \tag{72}$$

and

$$\frac{m^2}{k^2(k^2+m^2)} = \frac{1}{k^2} - \frac{1}{k^2+m^2} \tag{73}$$

we see that the "propagator" (71) has the standard pole of the electromagnetic field, proportional to $-i\eta^{\mu\nu}/k^2$ with a positive coefficient, plus a continuous set of ghost-like poles proportional to $+i\eta^{\mu\nu}/(k^2+m^2)$, with m an integration variable. We would then conclude that QED as a continuous spectrum of ghosts! Of course this is nonsense, and it is just an artifact of having applied to the quantum effective action a procedure that only makes sense for the fundamental action of a QFT. In fact, the proper interpretation of Eq. (71) is that $\log(k^2/\mu^2)$ develops an imaginary part for $k^2 < 0$ (e.g. for $k_0 \neq 0$, $\mathbf{k} = 0$, i.e. for a spatially uniform but time-varying electromagnetic field). This is due to the fact that, in the limit $m_e \to 0$ in which we are working (or, more generally, for $-k^2 > 4m_e^2$), in such an external electromagnetic field there is a rate of creation of electron-positron pairs, and the imaginary part of the effective action describes the rate of pair creation [41].

These general considerations show that the spectrum of the theory cannot be read naively from the quantum effective action. Thus, in particular, from the presence of a 'ghost-like' pole obtained from the effective quantum action (65), one cannot jump to the conclusion that the underlying fundamental theory has a ghost. In the next section we will be more specific, and try to understand the origin of this 'wrong-sign' pole in the RR and RT theories.

6 Localization of Nonlocal Theories

Nonlocal models can be formally written in a local form introducing auxiliary fields, as discussed in similar contexts in [14, 36, 56, 60–62, 74]. This reformulation is quite useful both for the numerical study of the equations of motion, and for understanding

[10] Actually, the terms $k^\mu k^\nu$ can be made to vanish if we take also the gauge fixing as nonlocal, and given by $(-1/2)(\partial_\mu A^\mu)[1/e^2(\Box)](\partial_\nu A^\nu)$. The same could be done for the propagator in Eq. (67).

exactly why the ghosts-like poles in Eq. (67) do not correspond to states in the spectrum of the quantum theory. It is useful to first illustrate the argument for the Polyakov effective action, for which we know that it is the effective quantum action of a perfectly healthy fundamental theory.

Localization of the Polyakov action. In $D = 2$ the Polyakov action becomes local when written in terms of the conformal factor. Let us however introduce a different localization procedure, that can be generalized to 4D. We start from Eq. (27),

$$\Gamma = c \int d^2x \sqrt{-g}\, R \Box^{-1} R\,, \tag{74}$$

where we used the notation $c = -N/(96\pi)$. We now introduce an auxiliary field U defined by $U = -\Box^{-1} R$. At the level of the action, this can be implemented by introducing a Lagrange multiplier ξ, and writing

$$\Gamma = \int d^2x \sqrt{-g}\, [-cRU + \xi(\Box U + R)]\,. \tag{75}$$

The variation with respect to ξ gives

$$\Box U = -R\,, \tag{76}$$

so it enforces $U = -\Box^{-1} R$, while the variation with respect to U gives $\Box\xi = cR$ and therefore $\xi = c\Box^{-1} R = -cU$. This is an algebraic equation that can be put back in the action so that, after an integration by parts, Γ can be rewritten as [6]

$$\Gamma = c \int d^2x \sqrt{-g}\, \left[\partial_a U \partial^a U - 2UR\right]. \tag{77}$$

The theories defined by Eqs. (74) and (77) are classically equivalent. As a check, one can compute the energy-momentum tensor from Eq. (77), and verify that its classical trace is given by $T = 4c\Box U = -4cR$. So Eq. (77), used as a classical action, correctly reproduces the quantum trace anomaly (19) [6]. We can further manipulate the action (77) writing $g_{ab} = e^{2\sigma}\eta_{ab}$. Using Eq. (24) and introducing a new field φ from $U = 2(\varphi + \sigma)$ to diagonalize the action, we get

$$\Gamma = 4c \int d^2x \left(\eta^{ab}\partial_a\varphi\partial_b\varphi - \eta^{ab}\partial_a\sigma\partial_b\sigma\right). \tag{78}$$

Taken literally, this action seems to suggest that in the theory there are two dynamical fields, φ and σ. For $c > 0$, φ would be a ghost and σ a healthy field, and viceversa if $c < 0$ (in the Polyakov action (74) $c = -N/(96\pi) < 0$, but exactly the same computation could be performed with the action (29), where $c = -(N - 25)/(96\pi)$ can take both signs). Of course, we know that this conclusion is wrong, since we know exactly the spectrum of the quantum theory at the fundamental level, which is made uniquely by the quanta of the conformal matter fields. As we mentioned, even taking into account the anomaly-induced effective action, still σ has no quanta in the physical spectrum, since they are eliminated by the physical-state condition [76]. As for the auxiliary field φ, or equivalently U, there is no trace of its quanta in the physical spectrum. U is an artificial field which has been introduced by the localization procedure, and there are no quanta associated with it.

This can also be understood purely classically, using the fact that, in $D = 2$, the Polyakov action becomes local when written in terms of the conformal factor. Therefore, the classical evolution of the model is fully determined once we give the initial conditions on σ, i.e. $\sigma(t_i, x)$ and $\dot{\sigma}(t_i, x)$ at an initial time. Thus, once we localize the theory introducing U, the initial conditions on U are not arbitrary. Rather, they are uniquely fixed by the condition that the classical evolution, in the formulation obtained from Eq. (77), must be equivalent to that in the original theory (27). In other words, U is not the most general solution of Eq. (76), which would be given by a particular solution of the inhomogeneous equation plus the most general solution of the associated homogeneous equation $\Box U = 0$. Rather, it is just one specific solution, with given boundary conditions, such as $U = 0$ when $R = 0$ in Eq. (76). Thus, if we are for instance in flat space, there are no arbitrary plane waves associated to U, whose coefficients $a_{\mathbf{k}}$ and $a^*_{\mathbf{k}}$ would be promoted to creation and annihilation operators in the quantum theory. In this sense, the situation is different with respect to the conformal mode σ: the conformal mode, at the quantum level, is a quantum field with its own creation and annihilation operators, but the corresponding quantum states do no survive the imposition of the physical-state condition, and therefore do not belong to the physical Hilbert space. The U field, instead, is a classical auxiliary field and has not even creation and annihilation operators associated to it.

Localization of the RR theory. We next consider the RR model. To put the theory in a local form we introducing two auxiliary fields U and S, defined by

$$U = -\Box^{-1} R \,, \qquad S = -\Box^{-1} U \,. \tag{79}$$

This can be implemented at the Lagrangian level by introducing two Lagrange multipliers ξ_1, ξ_2, and rewriting Eq. (61) as

$$\Gamma_{\rm RR} = \frac{m_{\rm Pl}^2}{2} \int d^4x \sqrt{-g} \left[R \left(1 - \frac{m^2}{6} S \right) - \xi_1 (\Box U + R) - \xi_2 (\Box S + U) \right] .$$

The equations of motion derived performing the variation of this action with respect to $h_{\mu\nu}$ is

$$G_{\mu\nu} = \frac{m^2}{6} K_{\mu\nu} + 8\pi G T_{\mu\nu} \,, \tag{80}$$

where

$$K_{\mu\nu} = 2SG_{\mu\nu} - 2\nabla_\mu\partial_\nu S + g_{\mu\nu}[-2U + \partial_\rho S\partial^\rho U - (1/2)U^2] - (\partial_\mu S\partial_\nu U + \partial_\nu S\partial_\mu U)\,. \tag{81}$$

At the same time, the definitions (79) imply that U and S satisfy

$$\Box U = -R\,, \tag{82}$$

$$\Box S = -U\,. \tag{83}$$

Using the equations of motion we can check explicitly that $\nabla^\mu K_{\mu\nu} = 0$, as it should, since the equations of motion has been derived from a diff-invariant action. Linearizing Eq. (81) over flat space we get

$$\mathcal{E}^{\mu\nu,\rho\sigma}h_{\rho\sigma} - \frac{2}{3}m^2 P^{\mu\nu}P^{\rho\sigma}h_{\rho\sigma} = -16\pi G T^{\mu\nu}\,, \tag{84}$$

Let us we restrict to the scalar sector, which is the most interesting for our purposes. We proceed as in GR, and use the diff-invariance of the nonlocal theory to fix the Newtonian gauge

$$h_{00} = -2\Psi\,, \qquad h_{0i} = 0\,, \qquad h_{ij} = 2\Phi\delta_{ij}\,. \tag{85}$$

We also write the energy-momentum tensor in the scalar sector as

$$T_{00} = \rho\,, \qquad T_{0i} = \partial_i\Sigma\,, \tag{86}$$

$$T_{ij} = P\delta_{ij} + [\partial_i\partial_j - (1/3)\delta_{ij}\nabla^2]\Pi\,. \tag{87}$$

A straightforward generalization of the standard computation performed in GR (see e.g. [55]) gives four independent equations for the four scalar variables Φ, Ψ, U and S. For the Bardeen variables Φ and Ψ we get [67][11]

$$\nabla^2\left[\Phi - (m^2/6)S\right] = -4\pi G\rho\,, \tag{88}$$

$$\Phi + \Psi - (m^2/3)S = -8\pi G\Pi\,. \tag{89}$$

Thus, just as in GR, Φ and Ψ remain non-radiative degrees of freedom, with a dynamics governed by a Poisson equation rather than by a Klein–Gordon equation. This should be contrasted with what happens when one linearizes massive gravity with a Fierz–Pauli mass term. In that case Φ becomes a radiative field that satisfies $(\Box - m^2)\Phi = 0$ [3, 34, 55], and the corresponding jump in the number of radiative degrees of freedom of the linearized theory is just the vDVZ discontinuity. Further-

[11] Compared to [67], in Eq. (85) we have changed the sign in the definition of Ψ, in order to be consistent with the convention that we used in [37] when studying the cosmological perturbations of this model, compare with Eq. (134) below.

more, in local massive gravity with a mass term that does not satisfies the Fierz–Pauli tuning, in the Lagrangian also appears a term $(\Box\Phi)^2$ [55], signaling the presence of a dynamical ghost.

To linearize Eq. (82) we first observe that, taking the trace of Eq. (84), we get

$$R^{(1)} - m^2 P^{\mu\nu} h_{\mu\nu} = 8\pi G(\rho - 3P)\,, \tag{90}$$

where

$$R^{(1)} = \partial_\mu \partial_\nu (h^{\mu\nu} - \eta^{\mu\nu} h) \tag{91}$$

is the linearized Ricci scalar. From Eq. (66),

$$P^{\mu\nu} h_{\mu\nu} = \frac{1}{\Box}(\Box h - \partial^\mu \partial^\nu h_{\mu\nu}) = -\frac{1}{\Box} R^{(1)}\,. \tag{92}$$

Therefore, Eq. (90) can also be rewritten in the suggestive form

$$\left(1 + \frac{m^2}{\Box}\right) R^{(1)} = 8\pi G(\rho - 3P)\,. \tag{93}$$

Equation (92) also implies that, to linear order,

$$P^{\mu\nu} h_{\mu\nu} = U\,, \tag{94}$$

and therefore Eq. (90) can be rewritten as

$$R^{(1)} = 8\pi G(\rho - 3P) + m^2 U\,. \tag{95}$$

Inserting this into Eq. (82) we finally get

$$(\Box + m^2)U = -8\pi G(\rho - 3P)\,, \tag{96}$$

where, in all the linearized equations, $\Box = -\partial_0^2 + \nabla^2$ is the flat-space d'Alembertian. Similarly the linearized equation for S is just given by Eq. (83), again with the flat-space d'Alembertian.

Thus, in the end, in the scalar sector we have two fields Φ and Ψ which obey Eqs. (88) and (89) and are therefore non-radiative, just as, in GR. Furthermore, we have two fields U and S that satisfy Klein–Gordon equations with sources. In particular U satisfies the massive KG equation (96), so is clearly the field responsible for the ghost-like $1/(k^2 - m^2)$ pole in Eq. (68), while S satisfies a massless KG with source, and is the field responsible for the healthy $1/k^2$ pole in Eq. (68). This analysis shows that the potential source of problems is not one of the physical fields Φ and Ψ, but rather the auxiliary field U. However, at this point the solution of the potential problem becomes clear (see in particular the discussions in [14, 36, 61, 62] in different nonlocal models, and in [48, 64, 67] for the RR and RT models), and is in fact completely analogous to the situation that we have found for the Polyakov effective action. In general, an equation such as $\Box U = -R$ is solved by $U = -\Box^{-1} R$, where

$$\Box^{-1} R = U_{\rm hom}(x) - \int d^4x' \sqrt{-g(x')}\, G(x;x') R(x') \,, \tag{97}$$

with $U_{\rm hom}(x)$ any solution of $\Box U_{\rm hom} = 0$, and $G(x;x')$ a Green's function of the $\Box$ operator. The choice of the homogeneous solution is part of the definition of the $\Box^{-1}$ operator and therefore of the original nonlocal effective theory. In principle, the appropriate prescription would emerge once one knows the fundamental theory behind. In any case, there will be one prescription for what $\Box^{-1}$ means in the effective theory. This means that the auxiliary field U is not *the most general* solution of $\Box U = -R$, which is given by a solution of the inhomogeneous equation plus the most general solution of the associated homogeneous equation $\Box U = 0$. Rather, it is just a single, specific, solution. In other words, the boundary conditions of the equation $\Box U = -R$ are fixed. Whatever the choice made in the definition of $\Box^{-1}$, the corresponding homogeneous solution is fixed. For instance, in flat space this homogeneous solution is a superposition of plane waves, and the coefficients $a_{\mathbf{k}}, a^*_{\mathbf{k}}$ are fixed by the definition of $\Box^{-1}$ (e.g. at the value $a_{\mathbf{k}} = a^*_{\mathbf{k}} = 0$ if the definition of $\Box^{-1}$ is such that $U_{\rm hom} = 0$). They are not free parameters of the theory, and at the quantum level it makes no sense to promote them to annihilation and creation operators. There is no quantum degree of freedom associated to them.

To conclude this section, it is interesting to observe that the need of imposing boundary conditions on some classical fields, in order to recover the correct Hilbert state at the quantum level, is not specific to nonlocal effective actions. Indeed, GR itself can be formulated in such a way that requires the imposition of similar conditions [48, 55]. Indeed, let us consider GR linearized over flat space. To quadratic order, adding to the Einstein–Hilbert action the interaction term with a conserved energy-momentum tensor, we have

$$S^{(2)}_{\rm EH} + S_{\rm int} = \int d^4x \left[\frac{1}{2} h_{\mu\nu} \mathcal{E}^{\mu\nu,\rho\sigma} h_{\rho\sigma} + \frac{\kappa}{2} h_{\mu\nu} T^{\mu\nu}\right] . \tag{98}$$

We decompose the metric as

$$h_{\mu\nu} = h^{\rm TT}_{\mu\nu} + (\partial_\mu \varepsilon_\nu + \partial_\nu \varepsilon_\mu) + \frac{1}{3}\eta_{\mu\nu} s \,, \tag{99}$$

where $h^{\rm TT}_{\mu\nu}$ is transverse and traceless,

$$\partial^\mu h^{\rm TT}_{\mu\nu} = 0 \,, \qquad \eta^{\mu\nu} h^{\rm TT}_{\mu\nu} = 0 \,. \tag{100}$$

Thus, the 10 components of $h_{\mu\nu}$ are split into the 5 components of the TT tensor $h^{\rm TT}_{\mu\nu}$, the four components of ε_μ, and the scalar s. Under a linearized diffeomorphism $h_{\mu\nu} \to h_{\mu\nu} - (\partial_\mu \xi_\nu + \partial_\nu \xi_\mu)$, the four-vector ε_μ transforms as $\varepsilon_\mu \to \varepsilon_\mu - \xi_\mu$, while

$h_{\mu\nu}^{\rm TT}$ and s are gauge invariant. We similarly decompose $T_{\mu\nu}$. Plugging Eq. (99) into Eq. (98) ε_μ cancels (as it is obvious from the fact that Eq. (98) is invariant under linearized diffeomorphisms and ε_μ is a pure gauge mode), and we get

$$S_{\rm EH}^{(2)} + S_{\rm int} = \int d^4x\, \frac{1}{2}\left[h_{\mu\nu}^{\rm TT}\Box(h^{\mu\nu})^{\rm TT} - \frac{2}{3}s\Box s\right] + \frac{\kappa}{2}\left[h_{\mu\nu}^{\rm TT}(T^{\mu\nu})^{\rm TT} + \frac{1}{3}sT\right]. \tag{101}$$

The equations of motion derived from $S_{\rm EH}^{(2)} + S_{\rm int}$ are

$$\Box h_{\mu\nu}^{\rm TT} = -\frac{\kappa}{2}T_{\mu\nu}^{\rm TT}, \qquad \Box s = +\frac{\kappa}{4}T. \tag{102}$$

This result seems to suggest that in ordinary massless GR we have six propagating degrees of freedom: the five components of the transverse-traceless tensor $h_{\mu\nu}^{\rm TT}$, plus the scalar s. Note that $h_{\mu\nu}^{\rm TT}$ and s are gauge invariant, so they cannot be gauged away. Furthermore, from Eq. (101) the scalar s seems a ghost!

Of course, we know that in GR only the two components with helicities ± 2 are true propagating degrees of freedom. In fact, the resolution of this apparent puzzle is that the variables $h_{\mu\nu}^{\rm TT}$ and s are nonlocal functions of the original metric. Indeed, inverting Eq. (99), one finds

$$s = P^{\mu\nu}h_{\mu\nu}, \tag{103}$$

$$h_{\mu\nu}^{\rm TT} = h_{\mu\nu} - \frac{1}{3}P_{\mu\nu}h - \frac{1}{\Box}(\partial_\mu\partial^\rho h_{\nu\rho} + \partial_\nu\partial^\rho h_{\mu\rho}) + \frac{1}{3}\eta_{\mu\nu}\frac{1}{\Box}\partial^\rho\partial^\sigma h_{\rho\sigma} + \frac{2}{3}\frac{1}{\Box^2}\partial_\mu\partial_\nu\partial^\rho\partial^\sigma h_{\rho\sigma}, \tag{104}$$

where $P^{\mu\nu}$ is the nonlocal operator (66). Observe that the nonlocality is not just in space but also in time. Therefore, giving initial conditions on a given time slice for the metric is not the same as providing the initial conditions on $h_{\mu\nu}^{\rm TT}$ and s, and the proper counting of dynamical degrees of freedom gets mixed up. If we want to study GR in terms of the variables $h_{\mu\nu}^{\rm TT}$ and s, which are nonlocal functions of the original variables $h_{\mu\nu}$, we can do it, but we have to be careful that the number of independent initial conditions that we impose to evolve the system must remains the same as in the standard Hamiltonian formulation of GR. This means in particular that the initial conditions on s and on the components of $h_{\mu\nu}^{\rm TT}$ with helicities $0, \pm 1$ cannot be freely chosen, and in particular the solution of the homogeneous equations $\Box s = 0$ associated to the equation $\Box s = (\kappa/4)T$ is not arbitrary. It is fixed, e.g. by the condition that $s = 0$ when $T = 0$. Just as for the auxiliary field U discussed above, there are no quanta associated to s (nor to the components of $h_{\mu\nu}^{\rm TT}$ with helicities $0, \pm 1$), just as in the standard $3+1$ decomposition of the metric there are no quanta associated to the Bardeen potentials Φ and Ψ.

The similarity between the absence of quanta for the field U in the localization procedure of the RR model, and the absence of quanta for s in GR, is in fact more than an analogy. Comparing Eqs. (94) and (103) we see that, at the level of the linearized theory, U reduces just to s in the $m = 0$ limit. The boundary condition that eliminates the quanta of U in the RR theory therefore just reduces to the boundary condition that eliminates the quanta of s in GR.

The bottomline of this discussion is that the 'wrong-sign' pole in Eq. (68) is not due to a ghost in the quantum spectrum of the underlying fundamental theory. It is simply due to an auxiliary field that enters the dynamics at the classical level, but has no associated quanta in the physical spectrum of the theory. A different question is whether this auxiliary field might induce instabilities in the classical evolution. Since we will take m of order of the Hubble parameter today, H_0, any such instability would only develop on cosmological timescale, so it must be studied on a FRW background, which we will do in the next section.

The above analysis was performed for the RR model. For the RT model the details of the localization procedure are technically different [58, 64]. In that case we define again $U = -\Box^{-1} R$, and we also introduce $S_{\mu\nu} = -U g_{\mu\nu} = g_{\mu\nu}\Box^{-1} R$. We then compute $S^T_{\mu\nu}$ using Eq. (56). Thus, Eq. (60) is localized in terms of an auxiliary scalar field U and the auxiliary four-vector field S_μ that enters through Eq. (56), obeying the coupled system

$$G_{\mu\nu} + \frac{m^2}{6}\left(2U g_{\mu\nu} + \nabla_\mu S_\nu + \nabla_\nu S_\mu\right) = 8\pi G\, T_{\mu\nu}\,, \tag{105}$$

$$\Box U = -R\,, \tag{106}$$

$$(\delta^\mu_\nu \Box + \nabla^\mu \nabla_\nu) S_\mu = -2\partial_\nu U\,, \tag{107}$$

where the latter equation is obtained by taking the divergence of Eq. (56). We see that, at the full nonlinear level, the RT model is different from the RR model. However, linearizing over flat space they become the same. In fact in this case, using Eq. (92), to linear order we have

$$S_{\mu\nu} \equiv g_{\mu\nu}\Box^{-1} R \simeq -\eta_{\mu\nu} P^{\rho\sigma} h_{\rho\sigma}\,. \tag{108}$$

In flat space the extraction of the transverse part can be easily performed using Eq. (57), without the need of introducing auxiliary fields. This gives, again to linear order, $S^T_{\mu\nu} = -P_{\mu\nu} P^{\rho\sigma} h_{\rho\sigma}$. Using the fact that, to linear order, $G^{(1)}_{\mu\nu} = -(1/2)\mathcal{E}_{\mu\nu,\rho\sigma} h^{\rho\sigma}$, we see that the linearization of Eq. (60) over flat space gives the same equation as Eq. (84). Thus, the RR and RT model coincide at linear order over flat space, but not on a general background (nor at linear order over a non-trivial background, such as FRW).

It should also be stressed that the RR and RT models are not theories of massive gravity. The graviton remains massless in these theories. Observe also, from Eq. (67), that when we linearize over flat space the limit $m \to 0$ of the propagator is smooth, and there is no vDVZ discontinuity, contrary to what happens in massive gravity. The continuity with GR has also been explicitly verified for the Schwarzschild solution [58].[12]

[12] See app. B of [39] for the discussion of a related issue on the comparison with Lunar Laser Ranging, raised in [11].

7 Cosmological Consequences

We can now explore the cosmological consequences of the RT and RR models, as well as of some of their extensions that we will present below, beginning with the background evolution, and then moving to cosmological perturbation theory and to the comparison with cosmological data.

7.1 *Background Evolution and Self-Acceleration*

We begin with the background evolution (we closely follow the original discussions in [49, 64] for the RT model and [67] for the RR model). It is convenient to use the localization procedure discussed in Sect. 6, so we deal with a set of coupled differential equations, rather than with the original integro-differential equations.

7.1.1 The RT Model

Let us begin with the RT model. In FRW, at the level of background evolution, for symmetry reasons the spatial component S_i of the auxiliary field S_μ vanish, and the only variables are $U(t)$ and $S_0(t)$, together with the scale factor $a(t)$. Equations (105)–(107) then become

$$H^2 - \frac{m^2}{9}(U - \dot{S}_0) = \frac{8\pi G}{3}\rho \tag{109}$$

$$\ddot{U} + 3H\dot{U} = 6\dot{H} + 12H^2\,, \tag{110}$$

$$\ddot{S}_0 + 3H\dot{S}_0 - 3H^2 S_0 = \dot{U}\,. \tag{111}$$

We supplement these equations with the initial conditions

$$U(t_*) = \dot{U}(t_*) = S_0(t_*) = \dot{S}_0(t_*) = 0\,, \tag{112}$$

at some time t_* deep in the radiation dominated (RD) phase. We will come back below to how the results depend on this choice. Observe that we do not include a cosmological constant term. Indeed, our aim is first of all to see if the nonlocal term produces a self-accelerated solution, without the need of a cosmological constant.

It is convenient to pass to dimensionless variables, using $x \equiv \ln a(t)$ instead of t to parametrize the temporal evolution. We denote $df/dx = f'$, and we define $Y = U - \dot{S}_0$, $h = H/H_0$, $\Omega_i(t) = \rho_i(t)/\rho_c(t)$ (where i labels radiation, matter and dark energy), and $\Omega_i \equiv \Omega_i(t_0)$, where t_0 is the present value of cosmic time. Then the Friedmann equation reads

$$h^2(x) = \Omega_M e^{-3x} + \Omega_R e^{-4x} + \gamma Y(x)\,, \tag{113}$$

where $\gamma \equiv m^2/(9H_0^2)$. This shows that there is an effective DE density

$$\rho_{\rm DE}(t) = \rho_0 \gamma Y(x)\,, \tag{114}$$

where $\rho_0 = 3H_0^2/(8\pi G)$. We can trade S_0 for Y, and rearrange the equations so that U and Y satisfy the coupled system of equations

$$Y'' + (3-\zeta)Y' - 3(1+\zeta)Y = 3U' - 3(1+\zeta)U\,, \tag{115}$$

$$U'' + (3+\zeta)U' = 6(2+\zeta)\,, \tag{116}$$

$$\zeta(x) \equiv \frac{h'}{h} = -\frac{3\Omega_M e^{-3x} + 4\Omega_R e^{-4x} - \gamma Y'}{2(\Omega_M e^{-3x} + \Omega_R e^{-4x} + \gamma Y)}\,. \tag{117}$$

The result of the numerical integration is shown in Fig. 1. In terms of the variable $x = \ln a$, radiation-matter equilibrium is at $x = x_{\rm eq} \simeq -8.1$, while the present epoch corresponds to $x = 0$. From the left panel of Fig. 1 we see that the effective DE vanishes in RD. This is a consequence of the fact that, in RD, $R = 0$, together with our choice of boundary conditions $U(t_*) = \dot{U}(t_*) = 0$ at some initial value t_* deep in RD. As a consequence, $\Box^{-1}R$ remains zero in an exact RD phase, and only begins to grow when it starts to feel the effect of non-relativistic matter. The evolution of the auxiliary field $U = -\Box^{-1}R$ is shown in the left panel of Fig. 2. We see however that, as we enter in the matter-dominated (MD) phase, the effective DE density start to grow, until it eventually dominates, as we see from the right panel of Fig. 1. The numerical value of $\Omega_{\rm DE}$ today can be fixed at any desired value, by choosing the parameter m of the nonlocal model (just as in ΛCDM one can chose Ω_Λ by fixing the value of the cosmological constant). In Fig. 1 m has been chosen so that, today, $\Omega_{\rm DE} \simeq 0.68$, i.e. $\Omega_M \simeq 0.32$. This is obtained by setting $\gamma \simeq 0.050$, which corresponds to $m \simeq 0.67 H_0$. Of course, the exact value of Ω_M, and therefore of m, will eventually be fixed by Bayesian parameter estimation within the model itself, as we will discuss below.

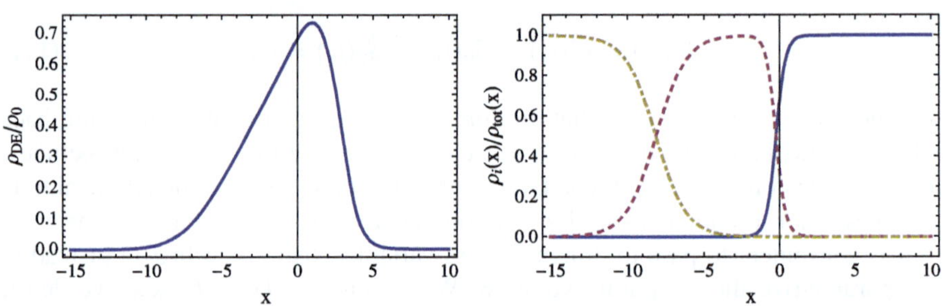

Fig. 1 *Left panel* the function $\rho_{\rm DE}(x)/\rho_0$, against $x = \ln a$, for the RT model (from [64]). *Right panel* the energy fractions $\Omega_i = \rho_i(x)/\rho_c(x)$ for i = radiation (*green*, *dot-dashed*) matter (*red*, *dashed*) and *dark* energy (*blue solid line*) (from [49])

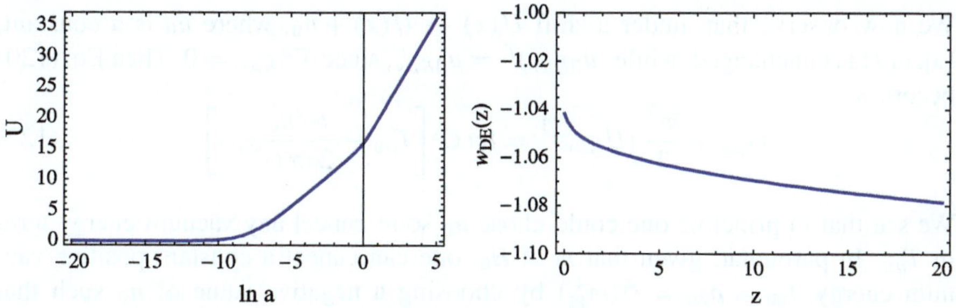

Fig. 2 *Left* the background evolution of the auxiliary field U, for the RT model. *Right* $w_{\rm DE}$ as a function of the redshift z, for the RT model (from [49])

We also define, as usual, the effective equation-of-state parameter of dark energy, $w_{\rm DE}$, from[13]

$$\dot{\rho}_{\rm DE} + 3(1 + w_{\rm DE})H\rho_{\rm DE} = 0\,. \tag{118}$$

Once fixed m so to obtain the required value of Ω_M, $\rho_{\rm DE}(x)$ is fixed, and therefore we get a pure prediction for the evolution of $w_{\rm DE}$ with time. The right panel of Fig. 2 shows the result, plotted as a function of redshift z. We observe that $w_{\rm DE}(z)$ is on the phantom side, i.e. $w_{\rm DE}(z) < -1$. This is a general consequence of Eq. (118), together with the fact that, in the RT model, $\rho_{\rm DE} > 0$, $\dot{\rho}_{\rm DE} > 0$, and $H > 0$, so $(1 + w_{\rm DE})$ must be negative. Near the present epoch we can compare the numerical evolution with the widely used fitting function [22, 63]

$$w_{\rm DE}(a) = w_0 + (1 - a)w_a\,, \tag{119}$$

(where $a = e^x$), and we get $w_0 \simeq -1.04$, $w_a \simeq -0.02$. These results are quite interesting, because they show that, at the level of background evolution, the nonlocal term generates an effective DE, which produces a self-accelerating solution with $w_{\rm DE}$ close to -1.

It is interesting to observe that, in terms of the field $U = -\Box^{-1}R$, Eq. (60) can be replaced by the system of equations

$$G_{\mu\nu} + \frac{m^2}{3}\left(U g_{\mu\nu}\right)^{\rm T} = 8\pi G\, T_{\mu\nu}\,, \tag{120}$$

$$\Box U = -R\,. \tag{121}$$

[13] The same expression for $w_{\rm DE}(x)$ can be obtained defining an effective DE pressure $p_{\rm DE}$ from the trace of the (i, j) component of the modified Einstein equation (105), and defining $w_{\rm DE}(x)$ from $p_{\rm DE} = w_{\rm DE}\rho_{\rm DE}$. The equivalence of the two definitions is a consequence of the fact that, because of the extraction of the transverse part in the RT model (and because of the general covariance of the action for the RR model), energy-momentum conservation is automatically ensured.

We now observe that, under a shift $U(x) \to U(x) + u_0$, where u_0 is a constant, Eq. (121) is unchanged, while $(u_0 g_{\mu\nu})^T = u_0 g_{\mu\nu}$, since $\nabla^\mu g_{\mu\nu} = 0$. Then Eq. (120) becomes

$$G_{\mu\nu} + \frac{m^2}{3}\left(U g_{\mu\nu}\right)^{\mathrm{T}} = 8\pi G\left[T_{\mu\nu} - \frac{m^2 u_0}{24\pi G} g_{\mu\nu}\right]. \tag{122}$$

We see that in principle one could chose u_0 so to cancel any vacuum energy term in $T_{\mu\nu}$. In particular, given that $m \simeq H_0$, one can cancel a constant positive vacuum energy $T_{00} = \rho_{\mathrm{vac}} = \mathcal{O}(m_{\mathrm{Pl}}^4)$ by choosing a negative value of u_0 such that $-u_0 = \mathcal{O}(m_{\mathrm{Pl}}^2/H_0^2) \sim 10^{120}$ (viceversa, choosing a positive value of u_0 amounts to introducing a positive cosmological constant). This observation is interesting, but unfortunately by itself is not a solution of the cosmological constant problem. We are simply trading the large value of the vacuum energy into a large value of the shift parameter in the transformation $U(x) \to U(x) + u_0$, and the question is now why the shifted field should have an initial condition $U(t_*) = 0$, or anyhow $U(t_*) = \mathcal{O}(1)$, rather than an astronomically large initial value.

The next point to be discussed is how the cosmological background evolution depends on the choice of initial conditions (112). To this purpose, let us consider first Eq. (116). In any given epoch, such as RD, MD, or e.g. an earlier inflationary de Sitter (dS) phase, the parameter ζ has an approximately constant value ζ_0, with $\zeta_0 = 0$ in dS, $\zeta_0 = -2$ in RD and $\zeta_0 = -3/2$ in MD. In the approximation of constant ζ Eq. (116) can be integrated analytically, and has the solution [64]

$$U(x) = \frac{6(2+\zeta_0)}{3+\zeta_0} x + u_0 + u_1 e^{-(3+\zeta_0)x}, \tag{123}$$

where the coefficients u_0, u_1 parametrize the general solution of the homogeneous equation $U'' + (3+\zeta_0)U = 0$. The constant u_0 corresponds to the reintroduction of a cosmological constant, as we have seen above. We will come back to its effect in Sect. 7.4. The other solution of the homogeneous equation, proportional to $e^{-(3+\zeta_0)x}$, is instead a decaying mode, in all cosmological phases. Thus, the solution with initial conditions $U(t_*) = \dot{U}(t_*) = 0$ has a marginally stable direction, corresponding to the possibility of reintroducing a cosmological constant, and a stable direction, i.e. is an attractor in the u_1 direction. Perturbing the initial conditions is equivalent to introducing a non-vanishing value of u_0 and u_1. We see that the introduction of u_0 will in general lead to differences in the cosmological evolution, which we will explore below, while u_1 corresponds to an irrelevant direction. In any case, it is reassuring that there is no growing mode in the solution of the homogeneous equation. Consider now Eq. (115). Plugging Eq. (123) into Eq. (115) and solving for $Y(x)$ we get [64]

$$\begin{aligned} Y(x) = &-\frac{2(2+\zeta_0)\zeta_0}{(3+\zeta_0)(1+\zeta_0)} + \frac{6(2+\zeta_0)}{3+\zeta_0} x + u_0 - \frac{6(2+\zeta_0)u_1}{2\zeta_0^2 + 3\zeta_0 - 3} e^{-(3+\zeta_0)x} \\ &+ a_1 e^{\alpha_+ x} + a_2 e^{\alpha_- x}, \end{aligned} \tag{124}$$

where

$$\alpha_{\pm} = \frac{1}{2}\left[-3 + \zeta_0 \pm \sqrt{21 + 6\zeta_0 + \zeta_0^2}\right]. \tag{125}$$

In particular, in dS there is a growing mode with $\alpha_+ = (-3 + \sqrt{21})/2 \simeq 0.79$. In RD both modes are decaying, and the mode that decays more slowly is the one with $\alpha_+ = (-5 + \sqrt{13})/2 \simeq -0.70$ while in MD again both modes are decaying, and $\alpha_+ = (-9 + \sqrt{57})/4 \simeq -0.36$. Thus, if we start the evolution in RD, in the space $\{u_0, u_1, a_1, a_2\}$ that parametrizes the initial conditions of the auxiliary fields, there is one marginally stable direction and three stable directions. However, if we start from an early inflationary era, there is a growing mode corresponding to the a_1 direction. Then Y will grow during dS (exponentially in x, so as a power of the scale factor), but will then decrease again during RD and MD. We will study the resulting evolution in Sect. 7.4, where we will see that even in this case a potentially viable background evolution emerges. In any case, it is important that in RD and MD there is no growing mode, otherwise the evolution would necessarily eventually lead us far from an acceptable FRW solution. This is indeed what happens in the model (58), where the homogeneous solutions associated to an auxiliary field are unstable both in RD and in MD (see Appendix. A of [49]), and is the reason why we have discarded that model.

7.1.2 The RR Model

Similar results are obtained for the RR model. Specializing to a FRW background, and using the dimensionless field $W(t) = H^2(t)S(t)$ instead of $S(t)$, Eqs. (80)–(83) become

$$h^2(x) = \Omega_M e^{-3x} + \Omega_R e^{-4x} + \gamma Y \tag{126}$$

$$U'' + (3 + \zeta)U' = 6(2 + \zeta), \tag{127}$$

$$W'' + 3(1 - \zeta)W' - 2(\zeta' + 3\zeta - \zeta^2)W = U, \tag{128}$$

where again $\gamma = m^2/(9H_0^2)$, $\zeta = h'/h$ and

$$Y \equiv \frac{1}{2}W'(6 - U') + W(3 - 6\zeta + \zeta U') + \frac{1}{4}U^2. \tag{129}$$

From this form of the equations we see that there is again an effective dark energy density, given by $\rho_{\rm DE} = \rho_0 \gamma Y$.

To actually perform the numerical integration of these equations, and also to study the perturbations, it can be more convenient to use a variable $V(t) = H_0^2 S(t)$ instead of $W(t) = H^2(t)S(t)$. Then Eqs. (126)–(128) are replaced by

$$h^2(x) = \frac{\Omega_M e^{-3x} + \Omega_R e^{-4x} + (\gamma/4)U^2}{1 + \gamma[-3V' - 3V + (1/2)V'U']}, \tag{130}$$

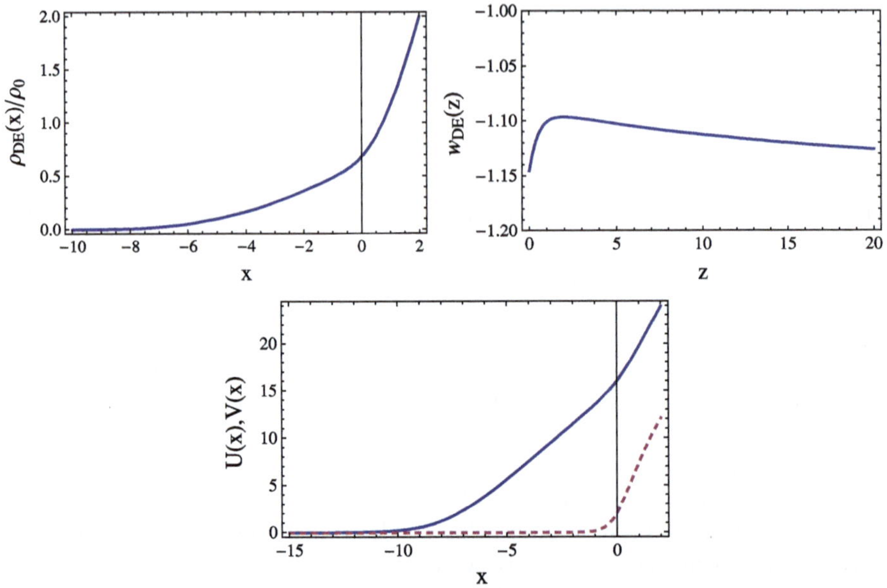

Fig. 3 *Upper left panel* the function $\rho_{\mathrm{DE}}(x)/\rho_0$ against $x = \ln a$, for the RR model. *Upper right panel* the function $w_{\mathrm{DE}}(z)$. *Lower panel* the background evolution of the auxiliary fields U (*blue solid line*) and V (*red dashed line*). From [37]

$$U'' + (3 + \zeta)U' = 6(2 + \zeta)\,, \tag{131}$$

$$V'' + (3 + \zeta)V' = h^{-2}U\,. \tag{132}$$

In Eqs. (131) and (132) appears $\zeta = h'/h$. In turn, h' can be computed explicitly from Eq. (130). The resulting expression contains V'' and U'', which can be eliminated using again Eqs. (131) and (132). This gives

$$\zeta = \frac{1}{2(1 - 3\gamma V)}\left[h^{-2}\Omega' + 3\gamma\left(h^{-2}U + U'V' - 4V'\right)\right], \tag{133}$$

where $\Omega(x) = \Omega_M e^{-3x} + \Omega_R e^{-4x}$. Then Eqs. (131) and (132), with h^2 given by Eq. (130) and ζ given by Eq. (133), provide a closed set of second order equations for V and U, whose numerical integration is straightforward.

The result of the numerical integration is shown in Fig. 3. Similarly to Eq. (112) for the RT model, we set initial conditions $U = U' = V = V' = 0$ at some initial time x_{in} deep in RD (we will see in Sect. 7.4.1 how the results depend on this choice). In this case we get $w_0 \simeq -1.14$, $w_a \simeq 0.08$ [67], so the RR model differs more from ΛCDM, compared to the RT model, at the level of background evolution. In the RR model, to obtain for instance a value $\Omega_M = 0.32$, i.e. $\Omega_{\mathrm{DE}} = 0.68$, we must fix $m \simeq 0.28 H_0$.

The dependence on the initial conditions can be studied as before. The equation for U is the same as in the RT model, so the homogeneous solution for U is again $u_0 + u_1 e^{-(3+\zeta_0)x}$. The homogeneous equation for V is the same as that for U, so similarly the homogeneous solution for V is $v_0 + v_1 e^{-(3+\zeta_0)x}$. In the early Universe we have $-2 \leq \zeta_0 \leq 0$ and all these terms are either constant or exponentially decreasing, which means that the solutions for both U and V are stable in MD, RD, as well as in a previous inflationary stage. From this point of view the RR model differs from the RT model which, as we have seen, has a growing mode during a dS phase. Note also the the constant u_0 now no longer has the simple interpretation of a cosmological constant term since, contrary to Eqs. (107), (83) is not invariant under $U \to U + u_0$.

7.2 *Cosmological Perturbations*

In order to asses the viability of these models, the next step is the study of their cosmological perturbations. This has been done in [37]. Let us considering first the scalar perturbations. We work in the Newtonian gauge, and write the metric perturbations as

$$ds^2 = -(1+2\Psi)dt^2 + a^2(t)(1+2\Phi)\delta_{ij}dx^i dx^j\,. \tag{134}$$

We then add the perturbations of the auxiliary fields, see below, we linearize the equations of motion and go in momentum space. We denote by k the comoving momenta, and define

$$\kappa \equiv k/k_{\rm eq}\,, \tag{135}$$

where $k_{\rm eq} = a_{\rm eq} H_{\rm eq}$ is the wavenumber of the mode that enters the horizon at matter-radiation equilibrium. To illustrate our numerical results, we use as reference values $\kappa = 0.1, 1, 5$. The mode with $\kappa = 5$ entered inside the horizon already during RD, while the mode $\kappa = 1$ reentered at matter-radiation equality. In contrast, the mode with $\kappa = 0.1$ was outside the horizon during RD and most of MD, and re-entered at $z \simeq 1.5$. Overall, these three values of k illustrate well the k dependence of the results, covering the range of k relevant to the regime of linear structure formation.

We summarize here the results for the RT and RR models, referring the reader to [37] for details and for the (rather long) explicit expression of the perturbation equations.

7.2.1 RT Model

In the RT model we expand the auxiliary fields as

$$U(t,\mathbf{x}) = \bar{U}(t) + \delta U(t,\mathbf{x})\,, \qquad S_\mu(t,\mathbf{x}) = \bar{S}_\mu(t) + \delta S_\mu(t,\mathbf{x})\,. \tag{136}$$

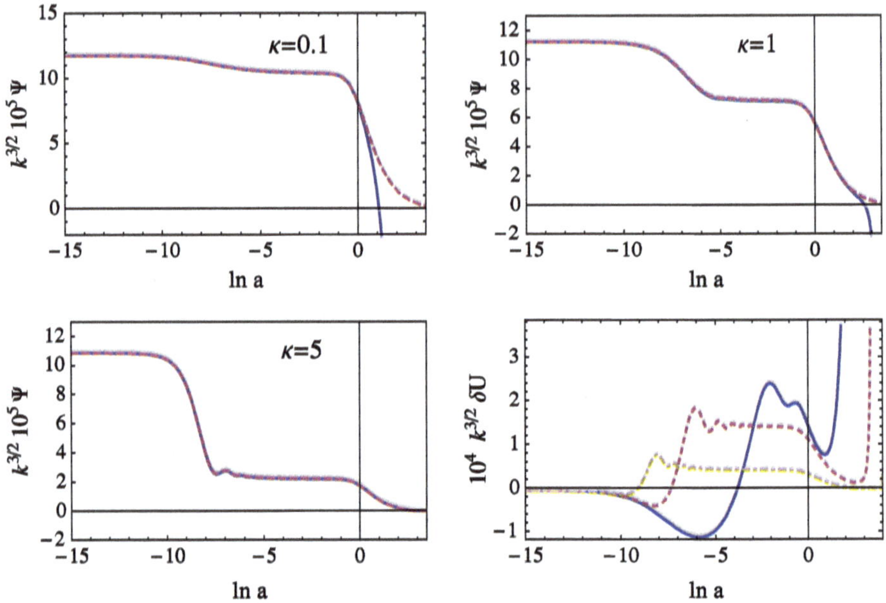

Fig. 4 $k^{3/2}\Psi(a;k)$ in the RT model (*blue solid line*) and in ΛCDM (*purple dashed line*), as a function of $x = \ln a(t)$, for $\kappa = 0.1$ (*left upper panel*), $\kappa = 1$ (*right upper panel*), $\kappa = 5$ (*lower left panel*). Observe that the quantity that we plot is $k^{3/2}\Psi(a;k)$ multiplied by a factor 10^5. *Lower right panel* the evolution of the perturbations δU for $\kappa = 0.1$ (*blue solid line*), $\kappa = 1$ (*purple, dashed*) and $\kappa = 5$ (*green, dot-dashed*)

In FRW, the background value $\bar{S}_i$ vanishes because there is no preferred spatial direction, but of course its perturbation δS_i is a dynamical variable. As with any vector, we can decompose it into a transverse and longitudinal part, $\delta S_i = \delta S_i^{\rm T} + \partial_i(\delta S)$ where $\partial_i(\delta S_i^{\rm T}) = 0$. Since we restrict here to scalar perturbations, we only retain δS, and write $\delta S_i = \partial_i(\delta S)$. Thus in this model the scalar perturbations are given by Ψ, Φ, δU, δS_0 and δS, see also [58, 73].

Figure 4 shows the time evolution of the Fourier modes of the Bardeen variable Ψ_k for our three reference values of κ (blue solid line) and compare with the corresponding result in ΛCDM (purple dashed line).[14] As customary, we actually plot $k^{3/2}\Psi_k$, whose square gives the variance of the field per unit logarithmic interval of momentum, according to

$$\langle \Psi^2(\mathbf{x})\rangle = \int \frac{d^3k}{(2\pi)^3}\langle |\Psi_{\mathbf{k}}|^2\rangle = \frac{1}{2\pi^2}\int_0^\infty \frac{dk}{k}\,\langle |k^{3/2}\Psi_k|^2\rangle\,, \tag{137}$$

[14] Figs. 4 and 5 have been obtained by Dirian *et al.* in the work leading to [37] although there, for reasons of space, we only published the corresponding figures relative to the RR model. Note also that the quantity plotted as Ψ in Fig. 6 of [37] was actually $-\Psi$.

where the bracket is the ensemble average over the initial conditions, that we take to be the standard adiabatic initial conditions. Note also that, if start the evolution choosing real initial conditions on Ψ_k, it remains real along the evolution.

We see from Fig. 4 that, up to the present time $x = 0$, the evolution of the perturbations is well-behaved, and very close to that of ΛCDM, even if in the cosmological future the perturbations will enter the nonlinear regime much earlier than for ΛCDM. In particular, the perturbation of the 'would-be' ghost field U, up to the present time, are small, with $k^{3/2}U_k \sim 10^{-4}$. Observe that in the cosmological future the perturbation becomes non-linear, both for Ψ_k and for δU_k, with the nonlinearity kicking in earlier for the lower-momentum modes.[15] This can be understood as follows. Any classical instability possibly induced by the nonlocal term will only develop on a timescale t such that mt is (much) larger than one. However, we have seen that, to reproduce the typical observed value of Ω_M, m is of order H_0, and in fact numerically smaller, with $m \simeq 0.28H_0$ for the RT model (see Sect. 7.3 for accurate Bayesian parameter estimation). Thus, instabilities induced by the nonlocal term, if present, only develop on a timescale larger or equal than to a few times H_0, and therefore in the cosmological future.

Beside following the cosmological evolution for the fundamental perturbation variables, such as $\Psi_k(x)$ (recall that $x \equiv \ln a(t)$ is our time-evolution variable, not to be confused with a spatial variable!), the behavior of the perturbations can also be conveniently described by some indicators of deviations from ΛCDM. Two useful quantities are the functions $\mu(x;k)$ and $\Sigma(x;k)$, defined by

$$\Psi = [1 + \mu(x;k)]\Psi_{\rm GR}\,, \tag{138}$$

$$\Psi - \Phi = [1 + \Sigma(x;k)](\Psi - \Phi)_{\rm GR}\,, \tag{139}$$

where the subscript 'GR' denotes the same quantities computed in GR, assuming a ΛCDM model with the same value of Ω_M as the modified gravity model. The advantage of using Ψ and $\Phi - \Psi$ as independent combinations is that the former enters in motion of non-relativistic particles, while the latter determines the light propagation. The numerical results for the RT model are shown in upper panels of Fig. 5. We see that, in this model, the deviations from ΛCDM are very tiny, of order of 1% at most, over the relevant wavenumbers and redshifts. In the forecast for experiments, $\mu(x;k)$ is often approximated as a function independent of k, with a power-like dependence on the scale factor,

$$\mu(a) = \mu_s a^s\,. \tag{140}$$

For the RT model we find that the scale-independent approximation is good, in the range of momenta relevant for structure formation, but the functional form (140) only

[15] Nevertheless, even the longest observable wavelength, which can be observed through their effect on the CMB, remain well linear up to the present epoch. We will see indeed from a full Boltzmann code analysis in Sect. 7.3 that the nonlocal models fit very well the CMB data.

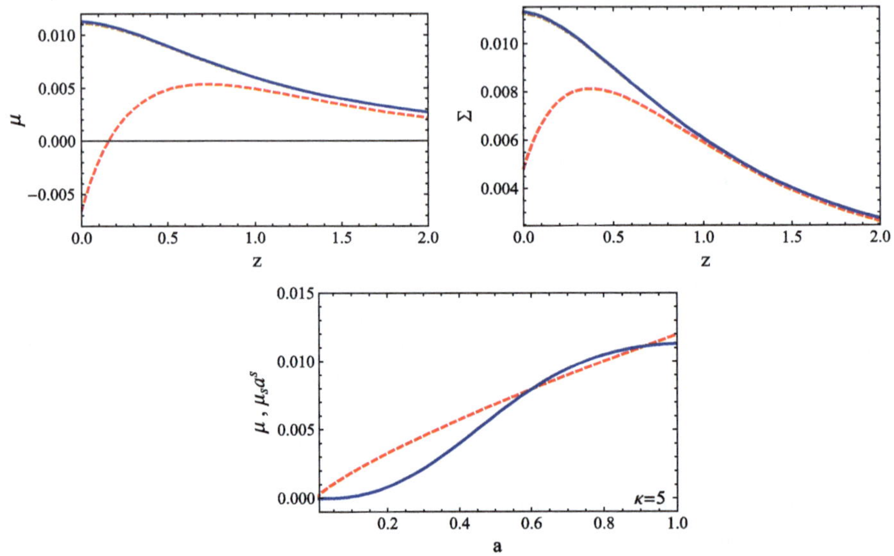

Fig. 5 *Upper left panel* $\mu(z;k)$, as a function of the redshift z, for $\kappa = 0.1$ (*red dashed*) $\kappa = 1$ (*brown dot-dashed*) and $\kappa = 5$ (*blue solid line*), for the RT model. The curves for $\kappa = 1$ and $\kappa = 5$ are almost indistinguishable on this scale. *Upper right panel* the same for $\Sigma(z;k)$. *Lower panel* $\mu(a;k)$, as a function of the scale factor a, for $\kappa = 5$ (*blue solid line*), for the RT model, compared to the function $\mu(a) = \mu_s a^s$ with $\mu_s = 0.012$ and $s = 0.8$ (*red dashed*)

catches the gross features of the a-dependence. The lower panel of Fig. 5 compares the function $\mu(a,k)$ computed numerically for $\kappa = 5$, with the function (140), setting $\mu_s = 0.012$ and $s = 0.8$.

Another useful indicator of deviations from GR is the effective Newton's constant, which is defined so that the Poisson equation for the Bardeen variable Φ takes the same for as in GR, with Newton's constant G replaced by a function $G_{\rm eff}(x;k)$. In the RT model, for modes inside the horizon, [37, 73],

$$\frac{G_{\rm eff}}{G} = 1 + \mathscr{O}\left(\frac{1}{\hat{k}^2}\right), \tag{141}$$

where $\hat{k} = k/(aH)$. This gives again the information that, for the RT model, deviations from ΛCDM in structure formations are quite tiny. We will see in more detail in Sect. 7.3 how the predictions of the model compare with that of ΛCDM for CMB,SNae, BAO and structure formation data.

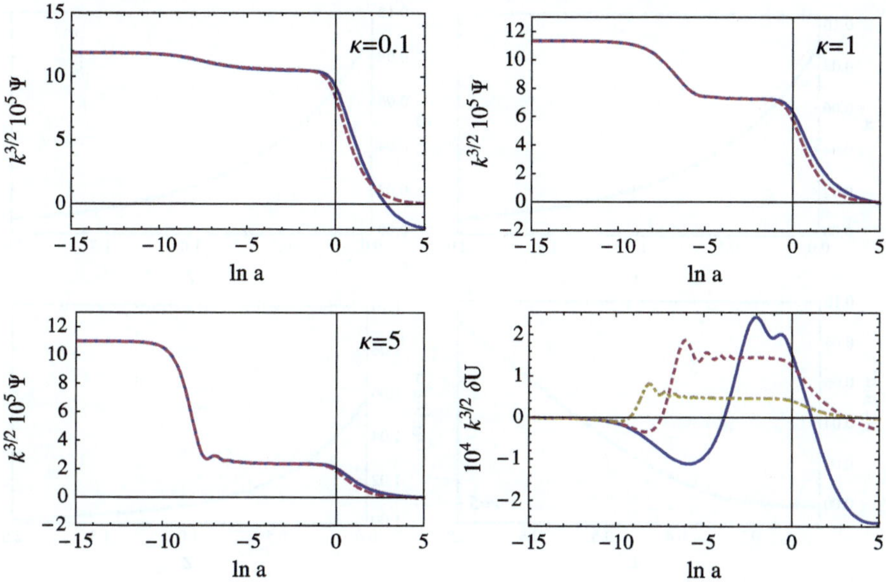

Fig. 6 $k^{3/2}\Psi(a;k)$ from the RR model (*blue solid line*) and from ΛCDM (*purple dashed line*), as a function of $x = \ln a(t)$, for $\kappa = 0.1$ (*left upper panel*), $\kappa = 1$ (*right upper panel*), $\kappa = 5$ (*left lower panel*). Observe that, on the vertical axis, we plot $10^5 k^{3/2}\Psi(a;k)$. Adapted from [37]. *Lower right panel* the evolution of the perturbations δU for $\kappa = 0.1$ (*blue solid line*), $\kappa = 1$ (*purple, dashed*) and $\kappa = 5$ (*green, dot-dashed*)

7.2.2 RR Model

In the RR model, in the study of perturbations we find convenient to use U and $V = H_0^2 S$ (rather than $W = H^2(t)S$). In the scalar sector we expand the metric as in Eq. (134) and the auxiliary fields as $U(t,\mathbf{x}) = \bar{U}(t) + \delta U(t,\mathbf{x})$, $V = \bar{V}(t) + \delta V(t,\mathbf{x})$. Thus, in this model the scalar perturbations are described by $\Psi, \Phi, \delta U$ and δV.

The results for the evolution of Ψ are shown in Fig. 6. We see that again the perturbations are well-behaved, and very close to ΛCDM. Compared to the RT model, the deviations from ΛCDM are somewhat larger, up to the present epoch. However, contrary to the RT model, they also stay relatively close to ΛCDM even in the cosmological future.

The functions μ and Σ are shown as functions of the redshift in the upper panels of Fig. 7, for our three reference value of the wavenumber. At a redshift such as $z = 0.5$, typical for the comparison with structure formation data, they are of order 5%, so again larger than in the RT model. For the RR model μ, as a function of the scale factor, is well reproduced by Eq. (140), with

$$\mu_s = 0.094\,, \qquad s = 2\,, \tag{142}$$

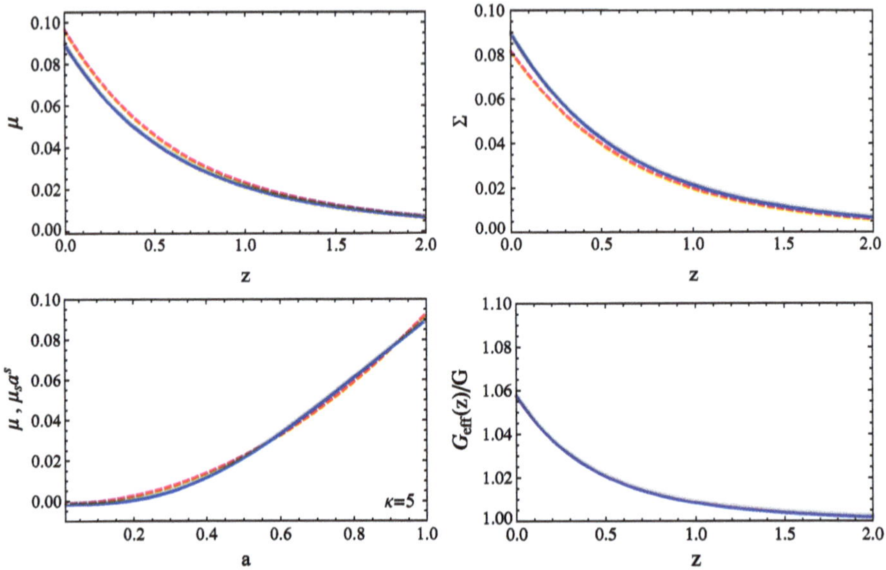

Fig. 7 *Upper left panel* $\mu(z; k)$, as a function of the redshift z for the RR model, for $\kappa = 0.1$ (*red dashed*) $\kappa = 1$ (*brown dot-dashed*) and $\kappa = 5$ (*blue solid line*). The curves for $\kappa = 1$ and $\kappa = 5$ are almost indistinguishable on this scale. *Upper right panel* the same for $\Sigma(z; k)$. *Lower left panel* $\mu(a; k)$, as a function of the scale factor a, for $\kappa = 5$ (*blue solid line*), for the RT model, compared to the function $\mu(a) = \mu_s a^s$ with $\mu_s = 0.094$ and $s = 2$ (*red dashed*). *Lower right panel* $G_{\rm eff}/G$ as a function of the redshift z, for sub-horizon modes, for the RR model. From [37]

see the lower panel of Fig. 7. By comparison, the forecast for EUCLID on the error $\sigma(\mu_s)$ over the parameter μ_s, for fixed cosmological parameters, is $\sigma(\mu_s) = 0.0046$ for $s = 1$ and $\sigma(\mu_s) = 0.014$ for $s = 3$ [83]. Thus (barring the effect of degeneracies with other cosmological parameters), we expect that the accuracy of EUCLID should be sufficient to test the prediction for μ_s from the RR model, and possibly also for the RT model.

Finally, the effective Newton's constant in the RR model, for sub-horizon scales, is given by

$$\frac{G_{\rm eff}(x; k)}{G} = \frac{1}{1 - 3\gamma \bar{V}(x)} \left[1 + \mathcal{O}\left(\frac{1}{\hat{k}^2}\right)\right]. \tag{143}$$

Thus in the sub-horizon limit, $G_{\rm eff}(x; k)$ becomes independent of k. However, contrary to the RT model, it retains a time dependence. The behavior of $G_{\rm eff}$ as a function of the redshift is shown in the lower right panel of Fig. 7.

Nonlinear structure formation has also been studied, for the RR model, using N-body simulations [11]. The result is that, in the high-mass tail of the distribution, massive dark matter haloes are slightly more abundant, by about 10% at

$M \sim 10^{14} M_{\odot}/h_0$. The halo density profile is also spatially more concentrated, by about 8% over a range of masses.[16]

Tensor perturbations have also been studied in [27, 39], for both the RR and RT models, and again their evolution is well behaved, and very close to that in ΛCDM.

7.3 Bayesian Parameter Estimation and Comparison with ΛCDM

The results of the previous sections show that the RR and RT nonlocal models give a viable cosmology at the background level, with an accelerated expansion obtained without the need of a cosmological constant. Furthermore, their cosmological perturbations are well-behaved and in the right ballpark for being consistent with the data, while still sufficiently different from ΛCDM to raise the hope that the models might be distinguishable with present or near-future observations. We can therefore go one step forward, and implement the cosmological perturbations in a Boltzmann code, and perform Bayesian parameter estimation. We can then compute the relevant chi-squares or Bayes factor, to see if these models can 'defy' ΛCDM, from the point of view of fitting the data. We should stress that this is a level of comparison with the data, and with ΛCDM, that none of the other infrared modifications of GR widely studied in the last few years has ever reached. The relevant analysis has been performed in [38], using the *Planck* 2013 data then available, together with supernovae and BAO data, and updated and extended in [39], using the *Planck* 2015 data.

In particular, in [39] we tested the nonlocal models against the *Planck* 2015 TT, TE, EE and lensing data from Cosmic Microwave Background (CMB), isotropic and anisotropic Baryonic Acoustic Oscillations (BAO) data, JLA supernovae, H_0 measurements and growth rate data, implementing the perturbation equations in a modified CLASS [16] code. As independent cosmological parameters we take the Hubble parameter today $H_0 = 100\ \mathrm{h\,km\,s^{-1}Mpc^{-1}}$, the physical baryon and cold dark matter density fractions today $\omega_b = \Omega_b h^2$ and $\omega_c = \Omega_c h^2$, respectively, the amplitude A_s and the spectral tilt n_s of primordial scalar perturbations and the reionization optical depth $\tau_{\rm re}$, so we have a 6-dimensional parameter space. For the neutrino masses we use the same values as in the *Planck* 2015 baseline analysis [1], i.e. two massless and a massive neutrinos, with $\sum_\nu m_\nu = 0.06$ eV, and we fix the effective number of neutrino species to $N_{\rm eff} = 3.046$.

Observe that, in the spatially flat case that we consider, in ΛCDM the dark energy density fraction Ω_Λ can be taken as a derived parameter, fixed in terms of the other parameters by the flatness condition. Similarly, in the nonlocal models m^2 can be taken as a derived parameter, fixed again by the flatness condition. Thus, not only

[16] The result of [11] were obtained using, for the RR model, the value of the cosmological parameters obtained in ΛCDM, before parameter estimation for the RR models was performed in [38, 39], see below. It would be interesting to repeat the analysis using the best-fit parameters of the RR model, and to extend it also to the RT model.

the nonlocal models have the same number of parameters as ΛCDM, but in fact the independent parameters can be chosen so that are exactly the same in the nonlocal models and in ΛCDM.

The results are shown in Table 1. On the left table we combine the *Planck* CMB data with JLA supernovae and with a rather complete set of BAO data, described in [39]. On the right table we also add a relatively large value of H_0, of the type suggested by local measurement. The most recent analysis of local measurements, which appeared after [39] was finished, gives $H_0 = 73.02 \pm 1.79$ [81]. In the last row we give the difference of χ^2, with respect to the model that has the lowest χ^2. Let us recall that, according to the standard Akaike or Bayesian information criteria, in the comparison between two models with the same number of parameters, a difference $|\Delta\chi^2| \leq 2$ implies statistically equivalence between the two models compared, while $|\Delta\chi^2| \gtrsim 2$ suggests "weak evidence", and $|\Delta\chi^2| \gtrsim 6$ indicates "strong evidence".[17]

Thus, for the case BAO+Planck+JLA, ΛCDM and the RT model are statistically equivalent, while the RR model is on the border of being strongly disfavored. Among the various parameter, a particularly interesting result concerns H_0, which in the nonlocal models is predicted to be higher than in ΛCDM. Thus, adding a high prior on H_0, of the type suggested by local measurements, goes in the direction of favoring the nonlocal models, as we see from the right table. In this case ΛCDM and the RT model are still statistically equivalent, although now with a slight preference for the RT model, while the RR model becomes only slightly disfavored with respect to the RR model, $\chi^2_{\rm RR} - \chi^2_{\rm RT} \simeq 2.8$, and statistically equivalent to ΛCDM, $\chi^2_{\rm RR} - \chi^2_{\Lambda{\rm CDM}} \simeq 1.4$.

In Table 1 we also give the derived values of $\gamma = m^2/(9H_0^2)$ for the nonlocal models. These central values for γ correspond to

$$m/H_0 \simeq 0.288 \qquad \text{(RR model)}\,, \tag{144}$$

$$m/H_0 \simeq 0.680 \qquad \text{(RT model)}\,. \tag{145}$$

From the values in the Table, in the case BAO+Planck+JLA, we find, for the total matter fraction $\Omega_M = (\omega_c + \omega_b)/h_0^2$, the mean values $\Omega_M = \{0.308, 0.300, 0.288\}$ for ΛCDM, the RT and RR models, respectively, and $h_0^2\Omega_M = \{0.141, 0.142, 0.143\}$, which is practically constant over the three models. Using BAO+Planck+JLA+($H_0 = 73.8$) these numbers change little, and become $\Omega_M = \{0.305, 0.298, 0.286\}$ for ΛCDM, the RT and the RR model, see [39] for full details, and plots of one and two-dimensional likelihoods. In particular, the left panel of Fig. 8 shows the two-dimensional likelihood in the plane (Ω_M, σ_8). We see that the nonlocal models

[17] The comparison of the χ^2 is not genuinely Bayesian. A more accurate method for comparing models, which is fully Bayesian, is based on Bayes factors. We checked in [39] that the results obtained from the computation of the Bayes factor are in full agreement with that obtained from the comparison of the χ^2.

Table 1 Parameter tables for ΛCDM and the nonlocal models. Beside the six parameters that we have chosen as our fundamental parameters, we give also the values of the derived quantities $z_{\rm re}$ (the redshift to reionization) and σ_8 (the variance of the linear matter power spectrum in a radius of 8 Mpc today). For the RR and RT models, among the derived parameters, we also give $\gamma = m^2/(9H_0^2)$. From [39]

Param	BAO+Planck+JLA			BAO+Planck+JLA+($H_0 = 73.8$)		
	ΛCDM	RT	RR	ΛCDM	RT	RR
$100\,\omega_b$	$2.228^{+0.014}_{-0.015}$	$2.223^{+0.014}_{-0.014}$	$2.213^{+0.014}_{-0.014}$	$2.233^{+0.014}_{-0.014}$	$2.226^{+0.014}_{-0.014}$	$2.217^{+0.014}_{-0.014}$
ω_c	$0.119^{+0.0011}_{-0.0011}$	$0.1197^{+0.0011}_{-0.00096}$	$0.121^{+0.001}_{-0.001}$	$0.1185^{+0.00097}_{-0.0011}$	$0.1194^{+0.001}_{-0.001}$	$0.1207^{+0.00096}_{-0.00097}$
H_0	$67.67^{+0.47}_{-0.5}$	$68.76^{+0.46}_{-0.51}$	$70.44^{+0.56}_{-0.56}$	$67.93^{+0.48}_{-0.43}$	$68.91^{+0.49}_{-0.5}$	$70.65^{+0.52}_{-0.54}$
$\ln(10^{10}A_s)$	$3.066^{+0.019}_{-0.026}$	$3.056^{+0.021}_{-0.023}$	$3.027^{+0.027}_{-0.023}$	$3.077^{+0.026}_{-0.019}$	$3.061^{+0.026}_{-0.022}$	$3.031^{+0.018}_{-0.022}$
n_s	$0.9656^{+0.0041}_{-0.0043}$	$0.9637^{+0.0039}_{-0.0041}$	$0.9601^{+0.004}_{-0.0039}$	$0.9671^{+0.0041}_{-0.0041}$	$0.9645^{+0.004}_{-0.0041}$	$0.9611^{+0.0038}_{-0.004}$
$\tau_{\rm re}$	$0.06678^{+0.011}_{-0.013}$	$0.0611^{+0.011}_{-0.013}$	$0.04516^{+0.014}_{-0.012}$	$0.07275^{+0.014}_{-0.01}$	$0.0641^{+0.013}_{-0.012}$	$0.04791^{+0.01}_{-0.011}$
$z_{\rm re}$	$8.893^{+1.1}_{-1.2}$	$8.359^{+1.2}_{-1.2}$	$6.707^{+1.7}_{-1.2}$	$9.435^{+1.3}_{-0.85}$	$8.636^{+1.3}_{-1.1}$	$7.02^{+1.1}_{-1.2}$
σ_8	$0.817^{+0.0076}_{-0.0095}$	$0.8283^{+0.0085}_{-0.0093}$	$0.8443^{+0.01}_{-0.0099}$	$0.8197^{+0.0096}_{-0.0075}$	$0.8298^{+0.0095}_{-0.0086}$	$0.8456^{+0.0081}_{-0.0088}$
γ	–	$5.15(4) \times 10^{-2}$	$9.21(7) \times 10^{-3}$	–	$5.17(4) \times 10^{-2}$	$9.24(7) \times 10^{-3}$
$\chi^2_{\rm min}$	13631.0	13631.6	13637.0	13637.5	13636.1	13638.9
$\Delta\chi^2_{\rm min}$	0	0.6	6.0	1.4	0	2.8

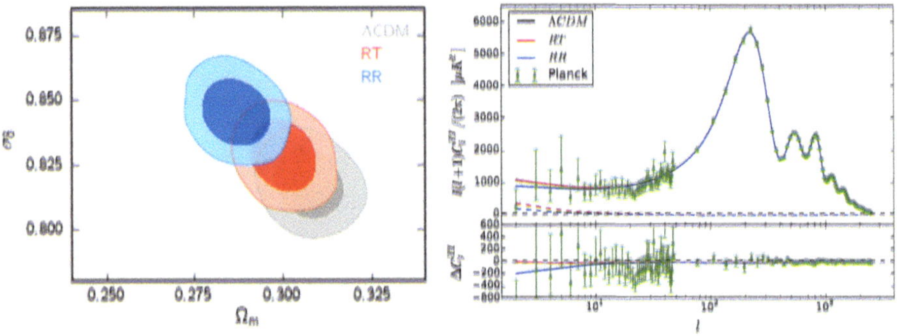

Fig. 8 *Left panel* $\sigma_8 - \Omega_m$ contour plot for Planck+BAO+JLA+($H_0 = 70.6$). *Right panel Upper plot* temperature power spectrum (*thick*), and the separate contribution from the late ISW contributions (*dashed*), for ΛCDM (*black*), RT (*red*) and RR (*blue*), using the best fit values of the parameters determined from BAO+JLA+Planck. The *black* and *red lines* are indistinguishable on this scale. The lower plot shows the residuals for ΛCDM and difference of RT (*red*) and RR (*blue*) with respect to ΛCDM. Data points are from Planck 2015 [1] (*green bars*). Error bars correspond to $\pm 1\sigma$ uncertainty. From [39]

predict slightly higher values of σ_8 and slightly lower values of Ω_M. The fit to the CMB temperature power spectrum, obtained with the data in Table 1, is shown in the right panel of Fig. 8.[18]

7.4 Extensions of the Minimal Models

The RR and RT models, as discussed above, are a sort of 'minimal models', that allow us to begin to explore, in a simple and predictive setting, the effect of nonlocal terms. However, even if the general philosophy of the approach should turn out to be correct, it is quite possible that the actual model that describes Nature will be more complicated. A richer phenomenology can indeed be obtained with some well-motivated extensions of these models, as we discuss in this section.

[18]We should also stress that the analysis in [38, 39] has been performed using, for the sum of the neutrino masses, the value of the *Planck* baseline analysis [1], $\sum_\nu m_\nu = 0.06$ eV, which is the smallest value consistent with neutrino oscillations. Increasing the neutrino masses lowers H_0. In ΛCDM this would increase the tension with local measurements, which is the main reason for choosing them in this way in the *Planck* baseline analysis. However, we have seen that the non-local models, and particularly the RR model, predict higher values of H_0, so they can accommodate larger neutrino masses without entering in tension with local measurements. A larger prior on neutrino masses would therefore favor the nonlocal models over ΛCDM. This possibility is currently being investigated [10].

7.4.1 Effect of a Previous Inflationary Era

The minimal models studied above are characterized by the fact that the initial conditions for the auxiliary fields and their derivatives are set to zero during RD. As we have discussed in Sect. 6, the choice of initial conditions on the auxiliary fields is part of the definition of the model, and different initial conditions define different nonlocal models. In principle, the correct prescription should come from the fundamental theory. We now consider the effect of more general initial conditions, in particular of the type that could be naturally generated by a previous phase of inflation.[19]

RT model. We consider first the effect of u_0 in the RT model [49]. From Eq. (123) we see that the most general initial condition of U amounts to a generic choice of the parameters u_0 and u_1, at some given initial time. The parameter u_1 is associated to a decaying mode, so the solution obtained with a nonzero value of u_1 is quickly attracted toward that with $u_1 = 0$. However, u_0 is a constant mode. We have seen in Eq. (122) that, in the RT model, the introduction of u_0 corresponds to adding back a cosmological constant term. From Eq. (122) we find that the corresponding value of the energy fraction associated to a cosmological constant, Ω_Λ, is given by $\Omega_\Lambda = \gamma u_0$. In the case $u_0 = 0$, for the RT model, $\gamma \simeq 5 \times 10^{-2}$, see Table 1. Then the effect of a non-vanishing u_0 will be small as long as $|u_0| \ll 20$. However larger values of u_0 can be naturally generated by a previous inflationary era. Indeed, we see from Eq. (123) that in a deSitter-like inflationary phase, where $\zeta_0 \simeq 0$, if we start the evolution at an initial time t_i at beginning an inflationary era and set $U(t_i) = \dot{U}(t_i) = 0$, we get, during inflation

$$U(x) = 4(x - x_i) + \frac{4}{3}\left(e^{-3(x-x_i)} - 1\right), \tag{146}$$

where $x_i = x(t_i)$. At the end of inflation, $x = x_f$, we therefore have

$$U(x_f) \simeq 4\Delta N, \tag{147}$$

where $\Delta N = x_f - x_i \gg 1$. Consider next the auxiliary field $Y(x)$. If we choose the initial conditions at the beginning of inflation so that the growing mode is not excited, i.e. $a_1 = 0$ in Eq. (124), at the end of inflation we also have $Y(x_f) \simeq 4\Delta N$. These values for $U(x_f)$ and $Y(x_f)$ can be taken as initial conditions for the subsequent evolution during RD. The corresponding results where shown in [49]. This choice of a_1 is however a form of tuning of the initial conditions on Y. Here we consider the most generic situation in which $a_1 \neq 0$. In this case during inflation Y will grow to a value of order $\exp\{0.79\Delta N\}$, where ΔN is the number of efolds and $\alpha_+ \simeq 0.79$ in a deSitter-like inflation. It will then decrease as $\exp\{-0.70x\}$ during the subsequent RD phase, see Eq. (125).

[19]We are assuming here that the effective nonlocal theory given by the RR or RT model is still valid at the large energy scales corresponding to primordial inflation. Whether this is the case can only be ascertained once one has a understood the mechanism that generates these nonlocal effective theories from a fundamental theory.

Despite the growth during inflation (exponential in x, so power-like in the scale factor a), the DE density associated to Y, $\rho_{DE} = \gamma Y \rho_0$, is still totally negligible in the inflationary phase, because $\rho_0 = \mathcal{O}(\text{meV}^4)$ is utterly negligible compared to the energy density during inflation. Thus, this growth of Y does not affect the dynamics at the inflationary epoch, nor in the subsequent RD era. Nevertheless, this large initial value at the end of inflation can produce a different behavior of Y near the present epoch, when the effective DE term $\gamma Y(x)$ becomes important.[20]

To be more quantitative let us recall that, if inflation takes place at a scale $M \equiv (\rho_{\text{infl}})^{1/4}$, the minimum number of efolds required to solve the flatness and horizon problems is given by

$$\Delta N \simeq 64 - \log \frac{10^{16}\,\text{GeV}}{M} . \tag{148}$$

The inflationary scale M can range from a maximum value of order $O(10^{16})$ GeV (otherwise, for larger values the effect of GWs produced during inflation would have already been detected in the CMB temperature anisotropies) to a minimum value around 1 TeV, in order not to spoil the predictions of the standard big-bang scenario. Assuming instantaneous reheating, the value of the scale factor a_* at which inflation ends and RD begins is given by $\rho_{\text{infl}} = \rho_{R,0}/a_*^4$, where $\rho_{R,0}$ is the present value of the radiation energy density, and as usual we have set the present value $a_0 = 1$. Plugging in the numerical values, for $x_* = \log a_*$ we find

$$x_* \simeq -65.9 + \log \frac{10^{16}\,\text{GeV}}{M} . \tag{149}$$

Recall also that RD ends and MD starts at $x = x_{\text{eq}} \simeq -8.1$. Thus, assuming that the number of efolds ΔN is the minimum necessary to solve the horizon and flatness problems, during RD (i.e. for $x_* < x < x_{\text{eq}}$) we have

$$\log[Y(x)] \simeq 0.79\Delta N - 0.70(x - x_*) , \tag{150}$$

where we used the fact that, during RD, $Y(x) \propto e^{-0.70x}$, see Eq. (125). In Fig. 9 we show the result for ρ_{DE} and w_{DE} obtained starting the evolution from a value $x_{\text{in}} = -15$ deep in RD, setting as initial conditions $U(x_{\text{in}}) = 4\Delta N$, $U'(x_{\text{in}}) = 0$, and with $Y(x_{\text{in}}) = \exp\{0.79\Delta N - 0.70(x_{\text{in}} - x_*)\}$ and $Y'(x_{\text{in}}) = -0.70 Y(x_{\text{in}})$, as determined by Eq. (150). We show the result for three different values of the inflationary scale M, and also show again, as a reference curve, the result for the minimal

[20]Two caveats are however necessary here. First, as already mentioned, we are assuming that the nonlocal models are valid in the early inflationary phase. Second, we are assuming that the large value of Y generated during inflation is still preserved by reheating. During reheating the energy density of the inflaton field is transferred to the radiation field. Since γY is just the DE energy density, it is in principle possible that even the energy density associated to Y is transferred to the radiation field, just as the inflaton energy density. In this case the evolution could resume at the beginning of RD with a small initial value of Y. Since, during RD, Y only has decaying modes, the solution would then be quickly attracted back to that obtained setting $Y(x_*) = 0$ at some x_* in RD.

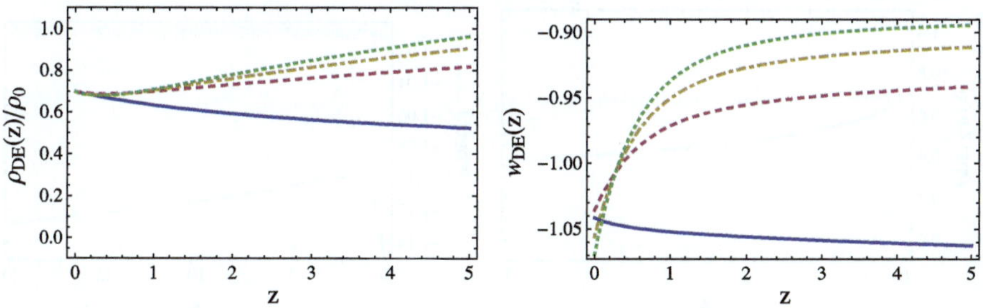

Fig. 9 *Left panel* $\rho_{\rm DE}/\rho_0$ for the RT model, shown against the redshift z, for the initial conditions on U and Y corresponding to the minimal model (i.e. $U = U' = Y = Y' = 0$ at an initial value $x_{\rm in} = -15$ in RD, *blue solid line*), and for the initial values of U, U', Y, Y' given by an inflationary phase with $M = 10^3$ GeV (*red dashed*), $M = 10^{10}$ GeV (*brown, dot-dashed*) and $M = 10^{16}$ GeV (*green, dotted*). In each case we adjust γ so to maintain fixed $\Omega_M = 0.30$, which gives $\gamma = 5.16 \times 10^{-2}$ for the minimal model, and $\gamma \simeq \{2.72 \times 10^{-3}, 1.04 \times 10^{-3}, 3.76 \times 10^{-4}\}$ for $M = \{10^3, 10^{10}, 10^{16}\}$ GeV, respectively. *Right panel* the corresponding results for $w_{\rm DE}$

RT model. We see that the results, already for the background evolution, are quantitatively different from the minimal case. Comparing with the observational limits of $w_{\rm DE}(z)$ from Fig. 5 of the *Planck* DE paper [2] we see that the predictions of these non-minimal nonlocal models for $w_{\rm DE}(z)$ are still consistent with the observational bounds, so even these models are observationally viable, at least at the level of background evolution. Observe that now, in the past, $w_{\rm DE}(z)$ is no longer phantom, since $\rho_{\rm DE}(x) = \gamma Y(x)$ now starts from a large initial value and, at the beginning, it decreases. Then, $w_{\rm DE}(z)$ crosses the phantom divide at $z \simeq 0.35$ (for $M = 10^3$ GeV), and $z \simeq 0.32$ (for $M = 10^{10}$ and $M = 10^{16}$ GeV). It is quite interesting to observe that, in the RT model, an early inflationary phase leaves an imprint on the equation of state of dark energy today, so that one could in principle infer the inflationary scale from a measurement of the function $w_{\rm DE}(z)$.

RR model. The situation in the RR model is different, because now the homogeneous solutions associated to the auxiliary fields U and V in Eqs. (131) and (132) only have constant or decreasing modes, in all cosmological epochs. In a deSitter epoch, setting $\zeta(x) = \zeta_0 = 0$, the solution of Eq. (131) is still given by Eq. (146), so again at the end of inflation $U(x_f) \simeq 4\Delta N$. Neglecting the second term in Eq. (146), we can set $U(x) \simeq 4(x - x_i)$ on the right-hand side of Eq. (132). Taking into account that during a deSitter inflationary phase $h(x)$ is constant, at a value $h_{\rm dS} = H_{\rm dS}/H_0$, the equation for $V(x)$ becomes

$$V'' + 3V' = \frac{4(x - x_i)}{h_{\rm dS}^2}. \tag{151}$$

If we start the evolution at an initial time x_i at beginning an inflationary era with initial conditions $V(x_i) = V'(x_i) = 0$ we get, during inflation,

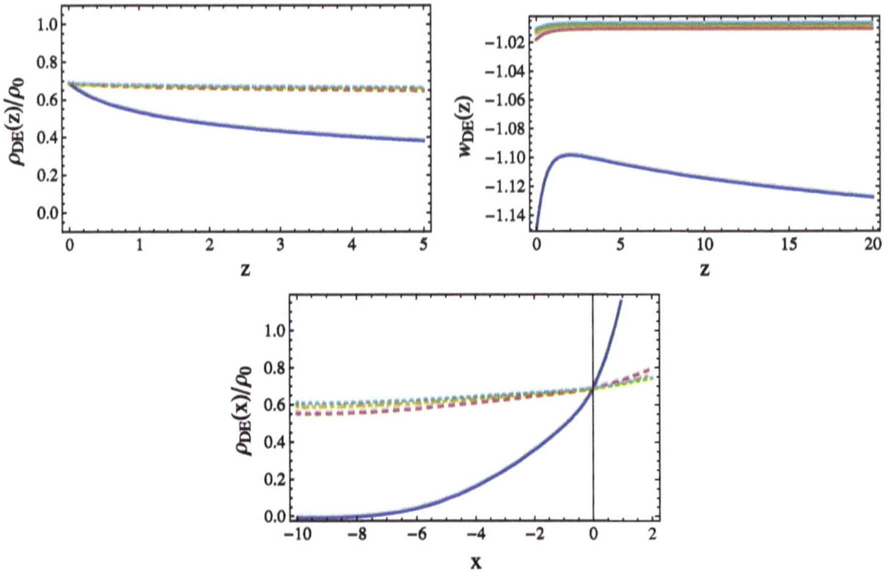

Fig. 10 *Upper left panel* $\rho_{\rm DE}(z)/\rho_0$ for the RR model, for the initial conditions on U and Y corresponding to the minimal model (i.e. $U = U' = V = V' = 0$ at an initial value $x_{\rm in} = -15$ in RD, *blue solid line*), and for the initial values of U given by an inflationary phase with $M = 10^3$ GeV (*red dashed*), $M = 10^{10}$ GeV (*brown, dot-dashed*) and $M = 10^{16}$ GeV (*green, dotted*). In each case we adjust γ so to maintain fixed $\Omega_M = 0.30$, which gives $\gamma = 9.12 \times 10^{-3}$ for the minimal model, and $\gamma \simeq \{1.18 \times 10^{-4}, 5.87 \times 10^{-5}, 3.73 \times 10^{-5}\}$ for $M = \{10^3, 10^{10}, 10^{16}\}$ GeV, respectively. *Upper right panel* the corresponding results for $w_{\rm DE}$. *Lower panel* the function $\rho_{\rm DE}(x)/\rho_0$ against $x = \ln a$

$$V(x) = \frac{2}{27h_{\rm dS}^2}\left[9(x-x_i)^2 - 6(x-x_i) + 2\left(1-e^{-3(x-x_i)}\right)\right]. \tag{152}$$

Then, at the end of inflation, $V(x_f) \simeq 2(\Delta N)^2/(3h_{\rm dS}^2)$. This value is totally negligible, since even for an inflationary scale as small as $M = 1$ TeV, $h_{\rm dS}^2 \sim 10^{15}$. Thus, as initial conditions for the subsequent evolution in RD, we can take $U(x_{\rm in}) = 4\Delta N$ and $V(x_{\rm in}) = 0$, at a value $x_{\rm in}$ deep in RD. Of course, one could take an initial value $V(x_{\rm in}) = \mathcal{O}(1)$, but this would not really affect the result. The point is that, for V, inflation does not generate a very large value at the beginning of RD.

The result is shown in Fig. 10 where, again, we express ΔN in terms of the inflationary scale using Eq. (148). We see that the RR model with a large initial value of u_0 gets closer and closer to ΛCDM. We find that $w_{\rm DE}(z=0)$ ranges from the value -1.017 for $M = 10^3$ GeV to the value -1.009 for $M = 10^{16}$ GeV, so the deviation with respect to ΛCDM are of order $(1-2)\%$. Observe also that in the cosmological future $\rho_{\rm DE}(x)$ continues to grow, although slowly, see the lower panel in Fig. 10.

From the point of view of the comparison with observations, a sensible strategy is therefore to start from the minimal RR model, since it predicts the largest deviations

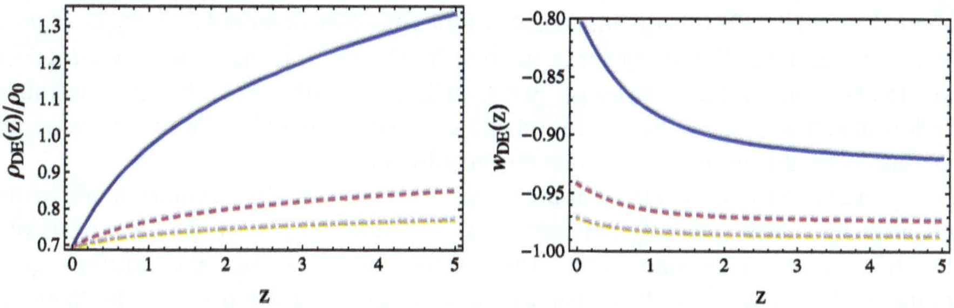

Fig. 11 *Left panel* $\rho_{\rm DE}(z)/\rho_0$ for the RR model, for the initial conditions $U = u_0$, $U' = V = V' = 0$ at an initial value $x_{\rm in} = -15$ in RD, with $u_0 = -30$ (*blue solid line*), -60 (*red dashed*) and -100 (*brown, dot-dashed*). The corresponding values of m/H_0 are $\{0.42, 0.12, 0.06\}$. These values corresponds to regime of the 'path B' solutions of [72]. *Right panel* the corresponding function $w_{\rm DE}(z)$

from ΛCDM and therefore can be more easily falsified (or verified). Indeed, already the next generation of experiments such as EUCLID should be able to discriminate clearly the minimal RR model from ΛCDM. However, one must keep in mind that the non-minimal model with a large value of u_0 is at least as well motivated physically as the 'minimal' model, but more difficult to distinguish from ΛCDM.

The RR model with a large value of u_0 is also conceptually interesting because it gives an example of a dynamical DE model that effectively generates a dark energy that, at least up to the present epoch, behaves almost like a cosmological constant, without however relying on a vacuum energy term, and therefore without suffering from the lack of technical naturalness associated to vacuum energy. Observe that these nonlocal models do not solve the coincidence problem, since in any case we must choose m of order H_0, just as in ΛCDM we must choose the cosmological constant Λ of order H_0^2. However, depending on the physical origin of the nonlocal term, the mass parameter m might not suffer from the problem of large radiative corrections that renders the cosmological constant technically unnatural.

Observe also that, just as in ΛCDM, the inflationary sector is a priori distinct from the sector that provides acceleration at the present epoch. Thus, one can in principle supplement the nonlocal models with any inflationary sector at high energy, adding an inflaton field with the desired inflaton potential, just as one does for ΛCDM. However in these nonlocal models, and particularly in the RR model, there is a very natural choice, which is to connect them to Starobinski inflation, since in a model where is already present a nonlocal term proportional to $R\Box^{-2}R$ is quite natural to also admit a local R^2 term. As first suggest in [26, 65], one can then consider a model of the form

$$S = \frac{m_{\rm Pl}^2}{2}\int d^4x\,\sqrt{-g}\left[R + \frac{1}{6M_{\rm S}^2}R\left(1 - \frac{\Lambda_{\rm S}^4}{\Box^2}\right)R\right], \tag{153}$$

where $M_S \simeq 10^{13}$ GeV is the mass scale of the Starobinski model and $\Lambda_S^4 = M_S^2 m^2$. As discussed in [26], at early times the non-local term is irrelevant and we recover the standard inflationary evolution, while at late times the local R^2 term becomes irrelevant and we recover the evolution of the non-local models, although with initial conditions on the auxiliary fields determined by the inflationary evolution.

A general study of the effect of the initial conditions on the auxiliary fields in the RR model has been recently performed in [72]. In particular, it has been observed that there is a critical value $\bar{u}_0 \simeq -14.82 + 0.67 \log \gamma$. For initial conditions $u_0 > \bar{u}_0$ the evolution is of the type that we have discussed above (denoted as 'path A' in [72]). For $u < \bar{u}_0$ a qualitatively different solution ('path B') appears. On this second branch, after the RD and MD epoch, there is again a DE dominated era, where however $w_{\rm DE}$ gets close to -1 but still remaining in the non-phantom region $w_{\rm DE} > -1$ (and, in the cosmological future, approaches asymptotically an unusual phase with $w_{\rm DE} = 1/3$, $\Omega_{\rm DE} \to -\infty$ and $\Omega_M \to +\infty$, see Fig. 4 of [72]). In Fig. 11 we show the evolution in the recent epoch for such a solution, for three different values of $u_0 = -30, -60, -100$. As we see from Eq. (129), the DE density in this case starts in RD from a non-vanishing value $\rho_{\rm DE}(x_{\rm in})/\rho_0 = (\gamma/4)u_0^2$. For instance, for $u_0 = -60$, requiring $\Omega_M = 0.3$ fixes $\gamma \simeq 0.00157$, so $\rho_{\rm DE}(x_{\rm in})/\rho_0 \simeq 1.4$. It then decreases smoothly up to the present epoch, where $\rho_{\rm DE}(x = 0)/\rho_0 \simeq 0.7$, resulting in a non-phantom behavior for $w_{\rm DE}(z)$.[21]

For sufficiently large values values of $-u_0$, this second branch is still cosmologically viable (while we see from the figure that, e.g., $u_0 = -30$ gives a value of $w_{\rm DE}(0)$ too far from -1 to be observationally viable), and has been compared to JLA supernovae in [72]. Observe however that a previous inflationary phase would rather generate the initial conditions corresponding to 'path A' solutions.

7.5 Exploring the Landscape of Nonlocal Models

The study of nonlocal infrared modifications of GR is a relatively recent research direction, and one needs some orientation as to which nonlocal models might be viable and which are not. At the present stage, the main reason for exploring variants of the models presented is not just to come out with one more nonlocal model that fits the data. Indeed, with the RT and RR models, both in their minimal and non-minimal forms discussed above, we already have a fair number of models to test

[21] In the RT model the situation is different. Indeed, in [49] it was found that cosmological solutions such that, today, $\rho_{\rm DE}(x = 0)/\rho_0$ is positive and equal to, say, 0.7, only exist for u_0 larger than a critical value $\bar{u}_0 \simeq -12$. Thus, again 'path A' solutions only exists only for u_0 larger than a critical value, but below this critical value there are no viable 'path B' solutions. The reason can be traced to the fact that in the RT model a non-vanishing initial value of u_0 corresponds to $\rho_{\rm DE}(x_{\rm in})/\rho_0 = \gamma u_0$, linear in u_0, while in the RR model corresponds to $\rho_{\rm DE}(x_{\rm in})/\rho_0 = (\gamma/4)u_0^2$. Thus, a negative value of u_0 in the RT model implies a negative initial value of $\rho_{\rm DE}(x_{\rm in})/\rho_0$, resulting in a qualitatively different evolution. In particular, for u_0 negative and sufficiently large, it becomes impossible to obtain $\rho_{\rm DE}(x = 0)/\rho_0$ positive and equal to 0.7 by the present epoch.

against the data. Rather, our main motivation at present is that identifying features of the nonlocal models that are viable might shed light on the underlying mechanism that generates their specific form of nonlocality from a fundamental local theory.

A first useful hint comes from the fact, remarked in Sect. 4.2, that at the level of models defined by equations of motions such as Eq. (58) or (60), models where $\Box^{-1}$ acts on a tensor such as $G_{\mu\nu}$ or $R_{\mu\nu}$ are not cosmologically viable, while models involving $\Box^{-1}R$, such as the RT model, are viable. A similar analysis can be performed for models defined directly at the level of the action. At quadratic order in the curvature, a basis for the curvature-square terms is given by $R^2_{\mu\nu\rho\sigma}$, $R^2_{\mu\nu}$ and R^2. Actually, for cosmological application it is convenient to trade the Riemann tensor $R_{\mu\nu\rho\sigma}$ for the Weyl tensor $C_{\rho\sigma\mu\nu}$. A natural generalization of the nonlocal action (61) is then given by

$$S_{\rm NL} = \frac{m_{\rm Pl}^2}{2}\int d^4x\sqrt{-g}\left[R - \mu_1 R\frac{1}{\Box^2}R - \mu_2 C^{\mu\nu\rho\sigma}\frac{1}{\Box^2}C_{\mu\nu\rho\sigma} - \mu_3 R^{\mu\nu}\frac{1}{\Box^2}R_{\mu\nu}\right], \tag{154}$$

where μ_1, μ_2 and μ_3 are parameters with dimension of squared mass. This extended model has been studied in [27], where it has been found that the term $R^{\mu\nu}\Box^{-2}R_{\mu\nu}$ is ruled out since it gives instabilities in the cosmological evolution at the background level. The Weyl-square term instead does not contribute to the background evolution, since the Weyl tensor vanishes in FRW, and it also has well-behaved scalar perturbations. However, its tensor perturbations are unstable [27], which again rules out this term.

These results indicate that models in which the nonlocality involves $\Box^{-1}$ applied on the Ricci scalar, such as the RR and RT model, play a special role. This is particularly interesting since, as we saw in Eq. (63), a term $R\Box^{-2}R$ has a specific physical meaning, i.e. it corresponds to a diff-invariant mass term for the conformal mode. The same holds for the RT model, since at linearized order over Minkowski it is the same as the RR model. This provides an interesting direction of investigation for understanding the physical origin of these nonlocal models, that we will pursue further in Sect. 8.

One can then further explore the landscape of nonlocal models, focusing on extensions of the RR model. Indeed, already the RT model can be considered as a nonlinear extension of the RR model, since the two models become the same when linearized over Minkowski. An action for the RT model would probably include further nonlinear terms beside $R\Box^{-2}R$, such as higher powers of the curvature associated to higher powers of $\Box^{-1}$. We have seen in Sect. 7.3 that the RT model appears to be the one that fits best the data, so it might be interesting to explore other physically-motivated nonlinear extensions of the RR model. In particular, in [26] we have explored two possibilities, that could be a sign of an underlying conformal symmetry, and that we briefly discuss next.

The Δ_4 model. A first option is to consider the model whose effective quantum action is

$$\Gamma_{\Delta_4} = \frac{m_{\rm Pl}^2}{2}\int d^4x\,\sqrt{-g}\left[R - \frac{m^2}{6}R\frac{1}{\Delta_4}R\right]. \tag{155}$$

where Δ_4 is the Paneitz operator (39). This operator depends only on the conformal structure of the metric, and we have seen that it appears in the nonlocal anomaly-induced effective action in four dimensions. In FRW the model can again be localized using two auxiliary fields U and V, so that the full system of equations reads [26]

$$h^2(x) = \frac{\Omega(x) + (\gamma/4)U^2}{1+\gamma[-3V' - 3V + (1/2)V'(U'+2U)]}, \tag{156}$$

$$U'' + (5+\zeta)U' + (6+2\zeta)U = 6(2+\zeta), \tag{157}$$

$$V'' + (1+\zeta)V' = h^{-2}U, \tag{158}$$

where as usual $\Omega(x) = \Omega_M e^{-3x} + \Omega_R e^{-4x}$. The effective DE density can then be read from $\rho_{\rm DE}(x)/\rho_0 = h^2(x) - \Omega(x)$. In the 'minimal' model with initial conditions $U(x_{\rm in}) = U'(x_{\rm in}) = V(x_{\rm in}) = V'(x_{\rm in}) = 0$ at some value $x_{\rm in}$ deep in RD, we find that the evolution leads to $w_{\rm DE}(z=0) \simeq -1.36$, too far away from -1 to be consistent with the observations. Also, contrary to the RR model, there is no constant homogeneous solution for U in RD and MD, because of a presence of a term proportional to U in Eq. (157). Rather, the homogeneous solutions are $U = e^{\alpha_\pm x}$ with $\alpha_+ = -2$ and $\alpha_- = -(3+\zeta_0)$, which are both negative in all three eras, and indeed whenever $\zeta_0 > -3$, which is always the case in the early Universe. Therefore, there is no 'non-minimal' model in this case. No large value for U or V is generated during inflation, and in any case even a large initial value at the end of inflation would decrease exponentially in RD, quickly approaching the solution of the minimal model. Therefore, this model is not cosmologically viable.

The conformal RR model. Another natural modification related to conformal symmetry would be to replace the $\Box$ operator in the RR model (or in the RT model) by the 'conformal d'Alembertian' $[-\Box + (1/6)R]$ [26], which again only depends on the conformal structure of space-time. We will call it the 'conformal RR model'. More generally, one can also study the model [70]

$$\Gamma_{\xi\rm RR} = \frac{m_{\rm Pl}^2}{2}\int d^4x\,\sqrt{-g}\left[R - \frac{m^2}{6}R\frac{1}{(-\Box+\xi R)^2}R\right], \tag{159}$$

with ξ generic, although only $\xi = 1/6$ is related to conformal invariance. Its study is a straightforward repetition of the analysis for the RR model. We can localize it by introducing two fields $U = (-\Box + \xi R)^{-1}R$ and $S = (-\Box + \xi R)^{-1}U$, and then Eqs. (130)–(132) become

$$h^2(x) = \frac{\Omega(x) + (\gamma/4)U^2}{1+\gamma[-3(V-\xi UV)' - 3(V-\xi UV) + (1/2)V'U']}, \tag{160}$$

$$U'' + (3+\zeta)U' + 6\xi(2+\zeta)U = 6(2+\zeta), \tag{161}$$

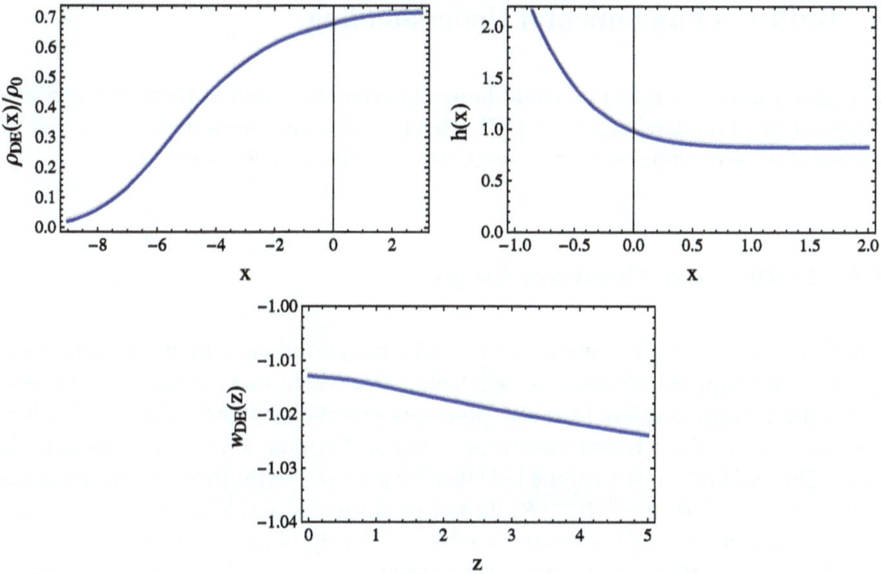

Fig. 12 *Upper left panel* the dark energy density $\rho_{\rm DE}(x)/\rho_0$ for the conformal RR model. *Upper right panel* the Hubble parameter $h(x)$. *Bottom panel* $w_{\rm DE}(z)$. From [26]

$$V'' + (3+\zeta)V' + 6\xi(2+\zeta)V = h^{-2}U\,, \tag{162}$$

where again $\zeta \equiv h'/h$. This models has some novel features compared to the $\xi = 0$ case [70]. Indeed, as we see from Fig. 12, the DE density goes asymptotically to a constant, and correspondingly also the Hubble parameter becomes constant, so the evolution approaches that of ΛCDM. This can also be easily understood analytically, observing that in a regime of constant (and non-vanishing) R, the operator $(-\Box + \xi R)^{-1}$ acting on R reduces to $(\xi R)^{-1}$. Then the nonlocal term in the action (159) reduces to a cosmological constant $\Lambda = m^2/(12\xi^2)$, leading to a de Sitter era with $H^2 = \Lambda/3 = m^2/(6\xi)^2$, i.e. $H = m/(6\xi)$. Similarly, from Eq. (161) we see that, asymptotically, $U \to 1/\xi$. Note that this solution only exists for $\xi \neq 0$. In particular, for the conformal RR model we have $\xi = 1/6$, so asymptotically $H \to m$ and $h \to 3\gamma^{1/2}$, in full agreement with the numerical result in Fig. 12.

As we see from the bottom panel in Fig. 12, for the physically more relevant case $\xi = 1/6$, $w_{\rm DE}(z)$ is very close to -1, for all redshifts of interest. Therefore, similarly to the non-minimal RR model discussed in Sect. 7.4.1, the conformal RR model is phenomenologically viable but more difficult to distinguish from ΛCDM, compared to the minimal RR model with $\xi = 0$.

8 Toward a Fundamental Understanding

The next question is how one could hope to derive the required form of the nonlocalities, from a fundamental local QFT. This is still largely work in progress, and we just mention here some relevant considerations, following Refs. [65, 66].

8.1 Perturbative Quantum Loops

The first idea that might come to mind is whether perturbative loop corrections can generate the required nonlocality. We have indeed seen that, among several other terms, the expansion in Eq. (16) also produces a term of the form $\mu^4 R\Box^{-2}R$, where μ is the mass of the relevant matter field (scalar, fermion or vector) running in the loops. One could then try to argue [24] that the previous terms in the expansion, such as $R\log(-\Box/\mu^2)R$ or $\mu^2 R\Box^{-1}R$, do not produce self-acceleration in the present cosmological epoch, and just retain the $\mu^4 R\Box^{-2}R$ in the hope to effectively reproduce the RR model. Unfortunately, it is easy to see that this idea does not work. Indeed, as we have seen in detail in Sect. 2, to obtain a nonlocal contribution we must be in the regime in which the particle is light with respect to the relevant scale, $|\Box/\mu^2| \gg 1$. In the cosmological context the typical curvature scale is given by the Hubble parameter, so at a given time t a particle of mass μ gives a nonlocal contribution only if $\mu^2 \lesssim H^2(t)$. In the opposite limit $\mu^2 \gg H^2(t)$ it rather gives the local contribution (18). Thus, to produce a nonlocal contribution at the present cosmological epoch, we need $\mu^2 \lesssim H_0^2$. Then, retaining only the Einstein–Hilbert term and the $\mu^4 R\Box^{-2}R$ term, we get an effective action of the form

$$\Gamma = \int d^4x\sqrt{-g}\left[\frac{m_{\rm Pl}^2}{2}R - R\frac{\mu^4}{\Box^2}R\right], \tag{163}$$

apart from a coefficient $\delta = \mathcal{O}(1)$ that we have reabsorbed in μ^4. Comparing with Eq. (61) we see that we indeed get the RR model, but with a value of the mass scale m given by

$$m \sim \frac{\mu^2}{m_{\rm Pl}}. \tag{164}$$

Since $\mu \lesssim H_0$, for m we get the ridiculously small value $m \lesssim H_0(H_0/m_{\rm Pl}) \sim 10^{-60}H_0$. To obtain a value of m of order H_0 we should rather use in Eq. (164) a value $\mu \sim (H_0 m_{\rm Pl})^{1/2}$, which is of the order of the meV (such as a neutrino!). However, in this case $\mu \gg H_0$, and for such a particle at the present epoch we are in the regime (18) where the form factors are local. Therefore we cannot obtain the RR model with a value $m \sim H_0$, as would be required for obtaining an interesting cosmological model. The essence of the problem is that, with perturbative loop corrections, the term $R\Box^{-2}R$ in Eq. (163) is unavoidably suppressed, with respect to the Einstein–Hilbert term, by a factor proportional $1/m_{\rm Pl}^2$.[22]

8.2 Dynamical Mass Generation for the Conformal Mode

The above considerations suggest to look for some non-perturbative mechanism that might generate dynamically the mass scale m [65]. An interesting hint, that follows from the exploration of the landscape of nonlocal models presented above, is that the models that are phenomenologically viable are only those, such as the RR and RT model, that have an interpretation in terms of a mass term for the conformal mode, as we saw in Eq. (63). Thus, a mechanism that would generate dynamically a mass for the conformal mode would automatically give the RR model, or one of its nonlinear extensions such as the RT model or the conformal RR model. Dynamical mass generation requires non-perturbative physics, in which case it emerges as a very natural consequence, as we know from experience with several two-dimensional models, as well as from QCD. As we discussed, an effective mass term for the gluon, given by the gauge-invariant but nonlocal expression (53), is naturally generated in QCD. The question is therefore whether some sector of gravity can become non-perturbative in the IR, in particular in spacetimes of cosmological relevance such as deSitter. Indeed, it is well know that in deSitter space large IR fluctuations can develop. This is true already in the purely gravitational sector, since the graviton propagator grows without bound at large distances, and in fact the fastest growing term comes from the conformal mode [7], although the whole subject of IR effects in deSitter is quite controversial (see e.g. [86] for a recent discussion and references).

Another promising direction for obtaining strong IR effects is given by the quantum dynamics of the conformal factor, which includes the effect of the anomaly-induced effective action. Indeed, the term $\sigma\bar{\Delta}_4\sigma$ in Eq. (42) can induce long-range correlations, and possibly a phase transition reminiscent of the BKT phase transition in two dimensions [4]. Further work is needed to put this picture on firmer ground.

Acknowledgements I am grateful to Giulia Cusin, Yves Dirian, Stefano Foffa, Maud Jaccard, Alex Kehagias, Nima Khosravi, Martin Kunz, Michele Mancarella, Ermis Mitsou and Valeria Pettorino for collaboration on different aspects of this project over the last few years. I also thank Ilya Shapiro for useful discussions. This work is supported by the Fonds National Suisse and by the SwissMAP National Center for Competence in Research.

[22] It has been pointed out in [72] that such a small value of m^2 could be compensated using a nonminimal model with a large value of $|u_0|$. This would however lead to a model indistinguishable from ΛCDM. Furthermore, with $m/H_0 \sim 10^{-60}$, the required value of u_0 would be huge. For instance, in the RT model $\Omega_\Lambda = \gamma u_0$. Since $\gamma \sim (m/H_0)^2$, this would require $u_0 \sim 10^{120}$. In the RR model, where the effective DE is quadratic in u_0, this would still require $u_0 \sim 10^{60}$. Observe that one should also tune the matter content so that the term $\mu^4/\Box^2$ in $k_W(-\Box/\mu^2)$ vanishes, since we have seen that this term induces unacceptable instabilities in the tensor sector.

References

1. P. Ade et al., Planck 2015 results XIII. Cosmological parameters (2015)
2. P. Ade et al., Planck 2015 results. XIV, Dark energy and modified gravity (2015)
3. L. Alberte, A.H. Chamseddine, V. Mukhanov, Massive gravity: resolving the puzzles. JHEP **1012**, 023 (2010). doi:10.1007/JHEP12(2010)023
4. I. Antoniadis, P.O. Mazur, E. Mottola, Criticality and scaling in 4-D quantum gravity. Phys. Lett. B **394**, 49–56 (1997). doi:10.1016/S0370-2693(96)01663-2
5. I. Antoniadis, P.O. Mazur, E. Mottola, Physical states of the quantum conformal factor. Phys. Rev. D **55**, 4770–4784 (1997). doi:10.1103/PhysRevD.55.4770
6. I. Antoniadis, P.O. Mazur, E. Mottola, Cosmological dark energy: prospects for a dynamical theory. New J. Phys. **9**, 11 (2007). doi:10.1088/1367-2630/9/1/011
7. I. Antoniadis, E. Mottola, Graviton fluctuations in de Sitter space. J. Math. Phys. **32**, 1037–1044 (1991). doi:10.1063/1.529381
8. I. Antoniadis, E. Mottola, 4-D quantum gravity in the conformal sector. Phys. Rev. D **45**, 2013–2025 (1992). doi:10.1103/PhysRevD.45.2013
9. Arkani-Hamed, N., Dimopoulos, S., Dvali, G., Gabadadze, G.: Nonlocal modification of gravity and the cosmological constant problem (2002)
10. A. Barreira, Y. Dirian, in preparation
11. A. Barreira, B. Li, W.A. Hellwing, C.M. Baugh, S. Pascoli, Nonlinear structure formation in Nonlocal Gravity. JCAP **1409**, 031 (2014). doi:10.1088/1475-7516/2014/09/031
12. A. Barvinsky, G. Vilkovisky, The generalized Schwinger-Dewitt technique in gauge theories and quantum gravity. Phys. Rep. **119**, 1 (1985). doi:10.1016/0370-1573(85)90148-6
13. A. Barvinsky, G. Vilkovisky, Beyond the Schwinger-Dewitt technique: converting loops into trees and in-in currents. Nucl. Phys. B **282**, 163 (1987). doi:10.1016/0550-3213(87)90681-X
14. A.O. Barvinsky, Serendipitous discoveries in nonlocal gravity theory. Phys. Rev. D **85**, 104,018 (2012). doi:10.1103/PhysRevD.85.104018
15. N. Birrell, P. Davies, in *Quantum Fields in Curved Space* (Cambridge University Press, Cambridge, 1982)
16. D. Blas, J. Lesgourgues, T. Tram, The cosmic linear anisotropy solving system (CLASS). Part II: approximation schemes. JCAP **7**, 034 (2011). doi:10.1088/1475-7516/2011/07/034
17. P. Boucaud, A. Le Yaouanc, J.P. Leroy, J. Micheli, O. Pene, J. Rodriguez-Quintero, Testing Landau gauge OPE on the lattice with a $\langle A^2 \rangle$ condensate. Phys. Rev. D **63**, 114,003 (2001). doi:10.1103/PhysRevD.63.114003
18. D. Boyanovsky, H.J. de Vega, R. Holman, D.S. Lee, A. Singh, Dissipation via particle production in scalar field theories. Phys. Rev. D **51**, 4419–4444 (1995). doi:10.1103/PhysRevD.51.4419
19. I.L. Buchbinder, S.D. Odintsov, I.L. Shapiro, *Effective Action in Quantum Gravity* (Institute of Physics, Bristol, 1992)
20. E. Calzetta, B. Hu, Closed time path functional formalism in curved space-time: application to cosmological back reaction problems. Phys. Rev. D **35**, 495 (1987). doi:10.1103/PhysRevD.35.495
21. M.A.L. Capri, D. Dudal, J.A. Gracey, V.E.R. Lemes, R.F. Sobreiro, S.P. Sorella, H. Verschelde, A Study of the gauge invariant, nonlocal mass operator Tr $\int d^4x\, F_{\mu\nu}(D^2)^{-1} F_{\mu\nu}$ in Yang-Mills theories. Phys. Rev. D **72**, 105,016 (2005). doi:10.1103/PhysRevD.72.105016
22. M. Chevallier, D. Polarski, Accelerating universes with scaling dark matter. Int. J. Mod. Phys. D **10**, 213–224 (2001). doi:10.1142/S0218271801000822
23. A. Codello, R.K. Jain, Covariant effective field theory of gravity I: formalism and curvature expansion (2015)
24. A. Codello, R.K. Jain, Covariant effective field theory of gravity II: cosmological implications (2015)
25. H. Collins, R. Holman, A. Ross, Effective field theory in time-dependent settings. JHEP **02**, 108 (2013). doi:10.1007/JHEP02(2013)108

26. G. Cusin, S. Foffa, M. Maggiore, M. Mancarella, Conformal symmetry and nonlinear extensions of nonlocal gravity. Phys. Rev. D **93**, 083,008 (2016). doi:10.1103/PhysRevD.93.083008
27. G. Cusin, S. Foffa, M. Maggiore, M. Mancarella, Nonlocal gravity with a Weyl-square term. Phys. Rev. D **93**, 043,006 (2016). doi:10.1103/PhysRevD.93.043006
28. D.A. Dalvit, F.D. Mazzitelli, Running coupling constants, Newtonian potential and nonlocalities in the effective action. Phys. Rev. D **50**, 1001–1009 (1994). doi:10.1103/PhysRevD.50.1001
29. F. David, Conformal field theories coupled to 2D gravity in the conformal gauge. Mod. Phys. Lett. A **3**, 1651 (1988). doi:10.1142/S0217732388001975
30. C. de Rham, G. Gabadadze, Generalization of the Fierz-Pauli action. Phys. Rev. D **82**, 044,020 (2010). doi:10.1103/PhysRevD.82.044020
31. C. de Rham, G. Gabadadze, A.J. Tolley, Resummation of massive gravity. Phys. Rev. Lett. **106**, 231,101 (2011). doi:10.1103/PhysRevLett.106.231101
32. S. Deser, Covariant decomposition and the gravitational Cauchy problem. Ann. Inst. Henri Poincare **7**, 149 (1967)
33. S. Deser, Closed form effective conformal anomaly actions in $D \gtrsim 4$. Phys. Lett. B **479**, 315–320 (2000). doi:10.1016/S0370-2693(00)00315-4
34. S. Deser, J. Trubatch, S. Trubatch, The massive spin-two field. Can. J. Phys. **44**, 1715 (1966)
35. S. Deser, R. Woodard, Nonlocal cosmology. Phys. Rev. Lett. **99**, 111,301 (2007). doi:10.1103/PhysRevLett.99.111301
36. S. Deser, R. Woodard, Observational viability and stability of nonlocal cosmology. JCAP **1311**, 036 (2013). doi:10.1088/1475-7516/2013/11/036
37. Y. Dirian, S. Foffa, N. Khosravi, M. Kunz, M. Maggiore, Cosmological perturbations and structure formation in nonlocal infrared modifications of general relativity. JCAP **1406**, 033 (2014). doi:10.1088/1475-7516/2014/06/033
38. Y. Dirian, S. Foffa, M. Kunz, M. Maggiore, V. Pettorino, Non-local gravity and comparison with observational datasets. JCAP **1504**, 044 (2015). doi:10.1088/1475-7516/2015/04/044
39. Y. Dirian, S. Foffa, M. Kunz, M. Maggiore, V. Pettorino, Non-local gravity and comparison with observational datasets. II. Updated results and Bayesian model comparison with ΛCDM. JCAP **1605**(05), 068 (2016). doi:10.1088/1475-7516/2016/05/068
40. J. Distler, H. Kawai, Conformal field theory and 2D quantum gravity. Nucl. Phys. B **321**, 509–527 (1989). doi:10.1016/0550-3213(89)90354-4
41. A. Dobado, A.L. Maroto, Particle production from nonlocal gravitational effective action. Phys. Rev. D **60**, 104,045 (1999). doi:10.1103/PhysRevD.60.104045
42. J.F. Donoghue, General relativity as an effective field theory: the leading quantum corrections. Phys. Rev. D **50**, 3874–3888 (1994). doi:10.1103/PhysRevD.50.3874
43. J.F. Donoghue, B.K. El-Menoufi, Non-local quantum effects in cosmology 1: quantum memory, non-local FLRW equations and singularity avoidance. Phys. Rev. D **89**, 104,062 (2014). doi:10.1103/PhysRevD.89.104062
44. J.F. Donoghue, B.K. El-Menoufi, Covariant non-local action for massless QED and the curvature expansion. JHEP **10**, 044 (2015). doi:10.1007/JHEP10(2015)044
45. D. Dudal, J.A. Gracey, S.P. Sorella, N. Vandersickel, H. Verschelde, A refinement of the Gribov-Zwanziger approach in the Landau gauge: infrared propagators in harmony with the lattice results. Phys. Rev. D **78**, 065,047 (2008). doi:10.1103/PhysRevD.78.065047
46. G. Dvali, Predictive power of strong coupling in theories with large distance modified gravity. New J. Phys. **8**, 326 (2006). doi:10.1088/1367-2630/8/12/326
47. P.G. Ferreira, A.L. Maroto, A few cosmological implications of tensor nonlocalities. Phys. Rev. D **88**, 123,502 (2013). doi:10.1103/PhysRevD.88.123502
48. S. Foffa, M. Maggiore, E. Mitsou, Apparent ghosts and spurious degrees of freedom in non-local theories. Phys. Lett. B **733**, 76–83 (2014). doi:10.1016/j.physletb.2014.04.024
49. S. Foffa, M. Maggiore, E. Mitsou, Cosmological dynamics and dark energy from non-local infrared modifications of gravity. Int. J. Mod. Phys. A **29**, 1450,116 (2014)
50. E.V. Gorbar, I.L. Shapiro, Renormalization group and decoupling in curved space. JHEP **0302**, 021 (2003). doi:10.1088/1126-6708/2003/02/021

51. E.V. Gorbar, I.L. Shapiro, Renormalization group and decoupling in curved space. 2. The Standard model and beyond. JHEP **0306**, 004 (2003). doi:10.1088/1126-6708/2003/06/004
52. S. Hassan, R.A. Rosen, Bimetric gravity from ghost-free massive gravity. JHEP **1202**, 126 (2012). doi:10.1007/JHEP02(2012)126
53. S. Hassan, R.A. Rosen, Resolving the ghost problem in non-linear massive gravity. Phys. Rev. Lett. **108**, 041,101 (2012). doi:10.1103/PhysRevLett.108.041101
54. M. Jaccard, M. Maggiore, E. Mitsou, A non-local theory of massive gravity. Phys. Rev. D **88**, 044,033 (2013)
55. M. Jaccard, M. Maggiore, E. Mitsou, Bardeen variables and hidden gauge symmetries in linearized massive gravity. Phys. Rev. D **87**, 044,017 (2013). doi:10.1103/PhysRevD.87.044017
56. S. Jhingan, S. Nojiri, S. Odintsov, M. Sami, I. Thongkool et al., Phantom and non-phantom dark energy: the cosmological relevance of non-locally corrected gravity. Phys. Lett. B **663**, 424–428 (2008). doi:10.1016/j.physletb.2008.04.054
57. R. Jordan, Effective field equations for expectation values. Phys. Rev. D **33**, 444 (1986). doi:10.1103/PhysRevD.33.444
58. A. Kehagias, M. Maggiore, Spherically symmetric static solutions in a non-local infrared modification of general relativity. JHEP **1408**, 029 (2014). doi:10.1007/JHEP08(2014)029
59. V.G. Knizhnik, A.M. Polyakov, A.B. Zamolodchikov, Fractal structure of 2D quantum gravity. Mod. Phys. Lett. A **3**, 819 (1988). doi:10.1142/S0217732388000982
60. T. Koivisto, Newtonian limit of nonlocal cosmology. Phys. Rev. D **78**, 123,505 (2008). doi:10.1103/PhysRevD.78.123505
61. T.S. Koivisto, Cosmology of modified (but second order) gravity. AIP Conf. Proc. **1206**, 79–96 (2010). doi:10.1063/1.3292516
62. N. Koshelev, Comments on scalar-tensor representation of nonlocally corrected gravity. Gravit. Cosmol. **15**, 220–223 (2009)
63. E.V. Linder, Exploring the expansion history of the universe. Phys. Rev. Lett. **90**, 091,301 (2003). doi:10.1103/PhysRevLett.90.091301
64. M. Maggiore, Phantom dark energy from nonlocal infrared modifications of general relativity. Phys. Rev. D **89**, 043,008 (2014)
65. M. Maggiore, Dark energy and dimensional transmutation in R^2 gravity (2015)
66. M. Maggiore, Perturbative loop corrections and nonlocal gravity. Phys. Rev. D **93**, 063,008 (2016). doi:10.1103/PhysRevD.93.063008
67. M. Maggiore, M. Mancarella, Non-local gravity and dark energy. Phys. Rev. D **90**, 023,005 (2014). doi:10.1103/PhysRevD.90.023005
68. A.V. Manohar, Effective field theories. Lect. Notes Phys. **479**, 311–362 (1997). doi:10.1007/BFb0104294
69. C.W. Misner, K. Thorne, J. Wheeler, *Gravitation* (Freeman, New York, 1973)
70. E. Mitsou, Aspects of infrared non-local modifications of general relativity. Ph.D. thesis, Geneva U. (2015). http://inspirehep.net/record/1360266/files/arXiv:1504.04050.pdf
71. V. Mukhanov, S. Winitzki, *Introduction to Quantum Effects in Gravity* (Cambridge University Press, Cambridge, 2007)
72. H. Nersisyan, Y. Akrami, L. Amendola, T.S. Koivisto, J. Rubio, Dynamical analysis of $R\Box^{-2}R$ cosmology: impact of initial conditions and constraints from supernovae (2016)
73. S. Nesseris, S. Tsujikawa, Cosmological perturbations and observational constraints on non-local massive gravity. Phys. Rev. D **90**, 024,070 (2014). doi:10.1103/PhysRevD.90.024070
74. S. Nojiri, S.D. Odintsov, Modified non-local-F(R) gravity as the key for the inflation and dark energy. Phys. Lett. B **659**, 821–826 (2008). doi:10.1016/j.physletb.2007.12.001
75. S. Perlmutter et al., Measurements of omega and lambda from 42 high redshift supernovae. Astrophys. J. **517**, 565–586 (1999). doi:10.1086/307221
76. J. Polchinski, A two-dimensional model for quantum gravity. Nucl. Phys. B **324**, 123–140 (1989). doi:10.1016/0550-3213(89)90184-3
77. J. Polchinski, *String Theory. Volume 1: An Introduction to the Bosonic String* (Cambridge University Press, Cambridge, 2007)

78. M. Porrati, Fully covariant van Dam-Veltman-Zakharov discontinuity, and absence thereof. Phys. Lett. B **534**, 209 (2002). doi:10.1016/S0370-2693(02)01656-8
79. R.J. Riegert, A nonlocal action for the trace anomaly. Phys. Lett. B **134**, 56–60 (1984). doi:10.1016/0370-2693(84)90983-3
80. A.G. Riess et al., Observational evidence from supernovae for an accelerating universe and a cosmological constant. Astron. J. **116**, 1009–1038 (1998). doi:10.1086/300499
81. Riess, A.G., et al.: A 2.4% Determination of the Local Value of the Hubble Constant (2016)
82. I.L. Shapiro, Effective action of vacuum: semiclassical approach. Class. Quant. Grav. **25**, 103,001 (2008). doi:10.1088/0264-9381/25/10/103001
83. Y. Song et al., Complementarity of weak lensing and peculiar velocity measurements in testing general relativity. Phys. Rev. **D84**, 083,523 (2011). doi:10.1103/PhysRevD.84.083523
84. N. Tsamis, R. Woodard, Nonperturbative models for the quantum gravitational back reaction on inflation. Ann. Phys. **267**, 145–192 (1998). doi:10.1006/aphy.1998.5816
85. R. Woodard, Nonlocal models of cosmic acceleration. Found. Phys. **44**, 213–233 (2014). doi:10.1007/s10701-014-9780-6
86. R.P. Woodard, Some inconvenient truths. JHEP **05**, 152 (2016). doi:10.1007/JHEP05(2016)152
87. J.J. York, Covariant decompositions of symmetric tensors in the theory of gravitation. Ann. Inst. Henri Poincare **21**, 319 (1974)

Emergence of Gravity and RG Flow

Ayan Mukhopadhyay

Abstract This is a tribute to Padmanabhan's works on the holographic principle which have consistently enunciated the profound philosophy that the classical equations of gravity themselves hold the key to understanding their holographic origin. I discuss how this can be realised by reformulating Einstein's equations in AdS as a non-perturbative RG flow that further leads to a new approach towards constructing strongly interacting QFTs. For a concrete demonstration, I focus on the hydrodynamic limit in which case this RG flow connects the AdS/CFT correspondence with the membrane paradigm.

1 A Route to Explore the Holographic Origin of Gravity

It is widely believed that *the holographic principle* holds the key to merging quantum and gravity together into a consistent framework. This principle broadly postulates that the gravitational dynamics in a given *volume* of spacetime can be described using degrees of freedom living at the *boundary* [1–4]. Thus gravity and at least one dimension of spacetime should be *both* emergent together from familiar quantum dynamics of many-body systems living on a holographic screen, whose embedding in the emergent spacetime should depend on the observer and the measurement process. A precise general statement of the holographic principle is still elusive although we do have a very concrete realisation in the form of AdS/CFT correspondence of string theory in which certain supergravity theories with stringy corrections in anti-de Sitter (AdS) space have been shown to have *dual* descriptions given by specific types of conformal Yang–Mills theories without gravity living at the boundary [5–7].

Thanu Padmanabhan (affectionately called Paddy by his collaborators and colleagues) has been one of the pioneers to first realise and emphasise that a general understanding of the holographic principle should start from a reformulation and reinterpretation of the classical gravity equations themselves. An enormously influ-

A. Mukhopadhyay (✉)
Institut Für Theoretische Physik, Technische Universität Wien,
Wiedner Hauptstr. 8-10, 1040 Vienna, Austria
e-mail: ayan@hep.itp.tuwien.ac.at

J.S. Bagla and S. Engineer (eds.), *Gravity and the Quantum*,
Fundamental Theories of Physics 187, DOI 10.1007/978-3-319-51700-1_17

ential volume of his publications has demonstrated that the classical equations of motion of gravity can be encoded via *surface terms* from which information about the bulk gravitational dynamics can be extracted [8–10]. Furthermore, the variation of these surface terms have been reinterpreted as differential thermodynamic identities in static spacetimes giving new insights into the general holographic origin of thermodynamic behaviour of classical black holes. My first research publication was in collaboration with him and it led to a general principle of construction of *holographic surface terms* for classical gravity which can encode classical bulk equations and reproduce the entropy of black holes when evaluated on-shell [11]. Generalising earlier work done by Paddy himself in the case of Einstein's gravity [8–10], together we were able to establish the existence of such surface-terms for Lanczos–Lovelock gravity in arbitrary dimensions. Looking back, it gives me a lot of pleasure that our work has continued to play a major role in subsequent works done by Paddy and his other collaborators in elucidating the holographic principle [12–14] (for a review see [15]), and even in taking steps towards understanding the holographic origin of dark energy [16].

My subsequent research (specially [17–20]) has also been directly influenced by the broad philosophy enunciated through Paddy's work that the classical gravity equations themselves hold the key to unravelling gravity's holographic origin. In the context of AdS/CFT correspondence, which is the most concrete example of holography, this question can be formulated in a precise way. Let us however state this question from a broader point of view by considering a class of gravitational spacetimes where one can naturally define a *spatial holographic direction* related with a *decreasing* energy scale. Such spacetimes include asymptotically anti-de Sitter (aAdS) spaces and those with horizons (such as black holes) where this holographic direction is the radial direction associated with a *wrap factor* or a *blackening function. If gravity is holographic, then the holographic radial direction should be related to a scale of precise kind of renormalisation group flow of the dual quantum system implying that holographic screens at constant values of this radial coordinate should contain complete information about a specific kind of coarse-grained description of the dual quantum system.* The broad question partly is, how do we make this statement precise and also how do we relate the freedom of choice of local coarse-graining for doing measurements in the dual quantum system to the emergence of diffeomorphism symmetry in the gravitational theory.

In the case of AdS/CFT correspondence, the precise microscopic QFT described by the data on the holographic screen at infinity (the boundary of the aAdS spacetime) is precisely known. Nevertheless, the precise general relation between the scale of the QFT and the emergent radial coordinate, meaning a correspondence between RG flow in the QFT and the radial evolution of data on holographic screens via gravitational dynamics is still unknown. As I will show below, an amazing lot can be known about this mysterious map between RG flow and radial evolution by appropriate reformulation of classical gravity equations themselves.

There is another aspect of the holographic origin of gravity which is also very enigmatic. Typically the map between classical gravity with a few fields and a dual QFT works only when the latter is strongly coupled [5–7]. This feature has revolutionised our understanding of strong coupling dynamics in quantum many-body systems. At strong coupling, the perturbative machinery of calculations with Feynman diagrams does not work and so far there is no better alternative to the holographic duality (whenever it is applicable) for calculating real-time quantum dynamics in presence of strong interactions. In order to calculate physical quantities via holography, one simply solves for the *asymptotic* data that lead to solutions in the dual gravity theory which are *free of naked singularities*. These lead to relations between the apriori independent leading (non-normalisable) and subleading (normalisable) modes of the gravitational fields near the boundary of AdS which satisfy two-derivative equations (such as Einstein's equations and the covariant Klein–Gordon equation). Each such field corresponds to an operator of the quantum theory. The non-normalisable modes correspond to the *sources* for local operators, and the normalisable modes correspond to the *expectation values* of the corresponding operators. Solving for the relations between these two that lead to dual geometries without naked singularities, we can obtain correlation functions, transport coefficients, etc. of the dual QFT. In fact, even if the Lagrangian description of the dual QFT is unknown, the dual gravity description gives us a concrete machinery to calculate all physical observables.

The enigmatic aspect is as follows. If the classical gravity equations can be reformulated as a RG flow, then this RG flow itself should know which microscopic UV data should lead to dual spacetimes in the theory of gravity without naked singularities. The RG flow is a first order evolution with the holographic radial direction moves towards the infrared of the dual QFT. The criterion for absence of naked singularities should be better obtained from the infrared behaviour of the RG flow as often in the ultraviolet the dual field theory can become weakly coupled so that the holographic classical gravitational description may no longer be valid [21]. Therefore, demanding appropriate requirements on the infrared holographic screen where the RG flow ends should ensure absence of naked singularities in dual gravity. Typically, this infrared holographic screen is the horizon. However, this infrared horizon screen should not be a fixed point of the RG flow so that the microscopic UV data can be recovered from the endpoint data by following back the first order scale evolution. *The question then is what is this infrared behaviour of RG flow that should be specified at the endpoint (the holographic screen coinciding with the horizon) which should lead us to the same UV data that is usually specified at the AdS boundary to ensure that the dual spacetimes do not have naked singularities.*

The data at the holographic horizon screen is expected to be very universal and characterised by a few parameters. As for example, although the microscopic UV data in the hydrodynamic limit consists of infinite number of transport coefficients which should be specified at the AdS boundary to obtain regular future horizons [22–24], the dynamics of the horizon is known to be characterised universally by a *non-relativistic incompressible Navier–Stokes fluid* with the shear viscosity being the only parameter as demonstrated via the membrane paradigm [25, 26]. Therefore, somehow the endpoint of RG flow that reformulates classical gravitational dynamics should be specified only by a *few* parameters which should determine the infinite number of physical observables of the dual QFT. A natural implication then is that *in any fixed number of dimensions only a class of gravitational theories (which may be constituted by finite or infinite number of higher derivative corrections to Einstein's equations) can be holographic.* Furthermore, *this can possibly be revealed by reformulating the classical equations of gravity in the form of RG flows, and then finding out when the absence of naked singularities in solutions of the gravitational theory can be translated into an appropriate criterion for the endpoint of the RG flow involving only a few infrared parameters.*

In the following section, I will describe how such a reformulation of classical gravity equations in AdS in the form of RG flows work and also how the infrared criterion for the RG flow can determine the microscopic UV data of the dual field theories. In Sect. 3, I will describe the construction of the RG flow in the field theory which will also define the latter in a constructive way in special limits. Special emphasis will be given on the hydrodynamic sector. I will conclude with an outlook.

2 Reformulating Gravity as a *Highly Efficient RG Flow*

The map between gravity in AdS and RG flow can be readily understood in the Fefferman–Graham coordinates which is well adapted for the description of the asymptotic behaviour of the spacetime metric and other gravitational fields from which the microscopic UV data of the dual field theory can be readily extracted. Therefore, we first describe how the map works in the Fefferman–Graham coordinates. The map can of course be expressed in any coordinate system and this will be related to the freedom of choosing the scale of observation in the dual field theory locally as we will show later. We will also consider pure Einstein's gravity for most of our discussion.

Any aAdS spacetime metric can be expressed in the Fefferman–Graham coordinates in the form:

$$ds^2 = \frac{l^2}{r^2}\left(dr^2 + g_{\mu\nu}(r,x)dx^\mu dx^\nu\right). \tag{1}$$

This coordinate system should be valid in a finite patch ending at the boundary which is at $r = 0$. Also l is called the AdS radius and in the holographic correspondence it provides units of measurement of bulk gravitational quantities which then corresponds to parameters and couplings of the dual field theory. The *boundary metric* $g^{(b)}_{\mu\nu}$ defined as:

$$g^{(b)}_{\mu\nu}(x) \equiv g_{\mu\nu}(r = 0, x) \tag{2}$$

is identified with the metric on which the dual field theory lives. For the sake of simplicity, unless stated otherwise we will assume that $g^{(b)}_{\mu\nu} = \eta_{\mu\nu}$ so that the dual field theory lives in flat Minkowski space. Our results of course can be generalised readily to any arbitrary curved boundary metric.

For later purposes, it is useful to define:

$$z^{\mu}{}_{\nu} \equiv g^{\mu\rho}\frac{\partial}{\partial r}g_{\rho\nu}. \tag{3}$$

Einstein's equations with a negative cosmological constant $\Lambda = -d(d-1)/(2l^2)$ in $(d+1)-$ dimensions in Fefferman–Graham coordinates can be written in the following form [18]:

$$\begin{aligned}\frac{\partial}{\partial r}z^{\mu}{}_{\nu} - \frac{d-1}{r}z^{\mu}{}_{\nu} + \mathrm{Tr}\, z\left(\frac{1}{2}z^{\mu}{}_{\nu} - \frac{1}{r}\delta^{\mu}{}_{\nu}\right) &= 2\, R^{\mu}{}_{\nu},\\ \nabla_{\mu}\left(z^{\mu}{}_{\nu} - \mathrm{Tr}\, z\delta^{\mu}{}_{\nu}\right) &= 0,\\ \frac{\partial}{\partial r}\mathrm{Tr}\, z - \frac{1}{r}\mathrm{Tr}\, z + \frac{1}{2}\mathrm{Tr}\, z^2 &= 0.\end{aligned} \tag{4}$$

Above, all indices have been lowered or raised with $g_{\mu\nu}$ or its inverse respectively. The first equation is the real dynamical equation and the latter ones are constraints that the data at that boundary $r = 0$ should satisfy. The radial dynamical evolution preserves the constraints, and therefore if they are satisfied at $r = 0$, they should be satisfied everywhere (for further details see [27]). It is to be noticed that in the above form the AdS radius l does not appear in the equations of motion.

Let us proceed now *without* assuming the traditional rules of AdS/CFT correspondence. We only assume that corresponding to any *solution* of $(d+1)-$ dimensional Einstein's equations with a negative cosmological constant that is free of naked singularities, there should exist a *state* in the dual $d-$dimensional field theory. The $(d+1)$ metric then should contain information about $\langle t_{\mu\nu}{}^{\infty}\rangle$, the expectation value of the microscopic energy-momentum tensor operator in the dual quantum state. For later convenience of analysing the hydrodynamic limit, we will consider $\langle t^{\mu}{}_{\nu}{}^{\infty}\rangle$ instead of $\langle t_{\mu\nu}{}^{\infty}\rangle$. When the boundary metric is $\eta_{\mu\nu}$, $\langle t^{\mu}{}_{\nu}{}^{\infty}\rangle$ should satisfy the Ward identities:

$$\partial_{\mu}\langle t^{\mu}{}_{\nu}{}^{\infty}\rangle = 0, \quad \mathrm{Tr}\,\langle t^{\infty}\rangle = 0 \tag{5}$$

The first one is the local conservation of the energy-momentum tensor and the latter comes from conformal invariance (we will later see why the dual quantum theory should have conformal invariance). The question is of course how to extract $\langle t^{\mu}{}_{\nu}{}^{\infty}\rangle$ from the dual spacetime metric. At this stage, it should be intuitively obvious that the microscopic Ward identities (5) should be related to the constraints of Einstein's equations (4).

We should now see the problem of identification of $\langle t^{\mu}{}_{\nu}{}^{\infty}\rangle$ from a broader perspective of connecting data on holographic screens at $r = constant$ with an appropriate RG flow in the dual QFT. Firstly, we identify the radial coordinate r with the inverse of the scale Λ of the dual quantum theory, i.e. we impose the relation

$$r = \Lambda^{-1}. \tag{6}$$

If the relation between r and Λ should be such that (i) it is state (i.e. solution) independent, and (ii) that the AdS radius l which has no direct interpretation in the dual QFT should not play a role in the mutual identifications, then the above is the only possibility given that $r = 0$ corresponds to the UV. Now on holographic screens at $r = constant$ we must identify the following pair of data $g_{\mu\nu}(\Lambda)$ and $\langle t^{\mu}{}_{\nu}(\Lambda)\rangle$. The effective metric $g_{\mu\nu}(\Lambda)$ can be seen as a generalised effective scale-dependent coupling or rather the source for the effective operator $\langle t^{\mu}{}_{\nu}(\Lambda)\rangle$. We should identify $g_{\mu\nu}(\Lambda)$ with $g_{\mu\nu}(r)$ that appears in the Fefferman–Graham metric (1) at $r = \Lambda^{-1}$ for reasons similar to those mentioned above. Firstly, as evident from (4), as a result of this identification the evolution equations for $g_{\mu\nu}(\Lambda)$ does not involve l which has no direct meaning in the dual QFT, and secondly the identification is also state (solution) independent. Furthermore, $g_{\mu\nu}(\Lambda)$ coincides then with the metric $\eta_{\mu\nu}$ on which the dual QFT lives at $\Lambda = \infty$. In usual perturbative RG flows, we do not talk about a background metric $g_{\mu\nu}(\Lambda)$ that evolves with the scale, however it makes perfect sense to do so in a special limit as explained below.

At this stage, we can introduce the notion of *highly efficient RG flow* [19, 20]. To understand this notion, it is first useful to classify operators in a QFT as *single-trace* and *multi-trace* operators. Single-trace operators are those which are gauge-invariant and which form the minimal set of generators of the algebra of all local gauge-invariant operators. All other gauge-invariant operators, which are multi-trace, are formed out of products and spacetime derivatives of the single-trace operators. It is these single-trace operators which are dual to gravitational fields in the holographic correspondence. The *large N limit* (where N is usually the rank of the gauge group in the QFT) is that in which the expectation values of the multi-trace operators *factorise* into those of the constituent single-trace operators. It is only in this limit that a QFT can have a holographic dual in the form of a classical gravity theory. Furthermore, when the QFT is strongly interacting, we expect there to be only few single-trace operators which have small scaling dimensions, because unless protected by symmetries there will be large quantum corrections to the anomalous dimensions at strong coupling. The remaining single-trace operators will decouple from the RG flow. The holographically dual classical gravity should then have only a few fields which are dual to the single-trace operators with small scaling dimensions.

Even in the large N and strong coupling limit, the single-trace operators can mix with multi-trace operators along the RG flow [28]. However, the RG flow can be thought of as a classical equations for scale evolution of single-trace operators in the sense that due to large N factorisation, the multi-trace operators can be readily replaced by the products of the constituents single-trace operators when their expectation values are evaluated in *any* state. It is expected that the gravitational theory can be truncated to pure gravity with a (negative) cosmological constant implying that there should be a consistent truncation of the dual RG flow equations to

$$\frac{\partial}{\partial \Lambda} t^{\mu}{}_{\nu}(\Lambda) = F^{\mu}{}_{\nu}[t^{\mu}{}_{\nu}(\Lambda), \Lambda], \tag{7}$$

with $F^{\mu}{}_{\nu}$ being non-linear in $t^{\mu}{}_{\nu}(\Lambda)$ so that it mixes with multi-trace operators built out of its products and derivatives along the RG flow.

In the strong interaction and large N limits, it is then useful to conceive a RG flow such that (despite $t^{\mu}{}_{\nu}(\Lambda)$ mixing with multi-trace operators constructed from its products and derivatives) at each scale there should exist an effective metric $g_{\mu\nu}(\Lambda)$ which is a non-linear functional of $t^{\mu}{}_{\nu}(\Lambda)$ and Λ, i.e. of the form

$$g_{\mu\nu}(\Lambda) = G_{\mu\nu}[t^{\mu}{}_{\nu}(\Lambda), \Lambda], \tag{8}$$

which is constructed in the fixed background metric $\eta_{\mu\nu}$ such that $t^{\mu}{}_{\nu}(\Lambda)$ preserves the form of the Ward identity

$$\nabla_{(\Lambda)\mu} t^{\mu}{}_{\nu}(\Lambda) = 0, \tag{9}$$

with $\nabla_{(\Lambda)}$ being the covariant derivative constructed from $g_{\mu\nu}(\Lambda)$. Therefore, an evolving metric $g_{\mu\nu}(\Lambda)$ which is a classical functional of $t^{\mu}{}_{\nu}(\Lambda)$ (in the sense mentioned before) emerges as a tool for defining an efficient RG flow which invokes an efficient mixing of single-trace operators with multi-trace operators such that the Ward identity for local energy and momentum conservation takes the same form at each scale. *This property of preservation of form of Ward identity for local conservation of energy-momentum constitutes the major ingredient for defining an highly efficient RG flow.* This definition is not complete as it does not tell us how such a RG flow can be constructed in the field theory – this will be described in the following section. Furthermore, we will also discuss the utility of such a RG flow in constructing strongly interacting large N field theories.

The major motivation of constructing a highly efficient RG flow is that it readily gives rise to a holographically dual classical gravity theory with full diffeomorphism invariance in one higher dimension due to the following theorem [19].

Theorem 1: Let us consider the $d-$dimensional scale evolution of $t^{\mu}{}_{\nu}(\Lambda)$ taking the schematic form (7) in a *fixed* background metric $g^{(b)}_{\mu\nu}$ such that there exists a background metric $g_{\mu\nu}(\Lambda)$ which is a functional of $t^{\mu}{}_{\nu}(\Lambda)$ and Λ in the same *fixed* background metric $g^{(b)}_{\mu\nu}$ as schematically represented by (8), and in which $t^{\mu}{}_{\nu}(\Lambda)$ satisfies the local conservation equation (9) at each Λ. Also let $g_{\mu\nu}(\Lambda)$ coincide with the fixed background metric $g^{(b)}_{\mu\nu}$ at $\Lambda = \infty$ so that $t^{\mu}{}_{\nu}{}^{\infty}$ satisfies $\nabla_{(b)\mu} t^{\mu}{}_{\nu}{}^{\infty} = 0$ with $\nabla_{(b)}$ being the covariant derivative constructed from $g^{(b)}_{\mu\nu}$.

We claim that as a consequence of the above assumptions, $g_{\mu\nu}(\Lambda)$ gives a $(d+1)-$dimensional bulk metric (1) in the Fefferman-Graham gauge with $r = \Lambda^{-1}$ such that it solves the equations of a *pure* $(d+1)-$classical gravity theory with *full* $(d+1)-$diffeomorphism invariance and a negative cosmological constant determined by the asymptotic curvature radius l. Also $g^{(b)}_{\mu\nu}$ is the boundary metric of this emergent asymptotically AdS spacetime.

This theorem ensures that a $(d+1)-$dimensional classical gravity with full diffeomorphism invariance can be rewritten as a *first order scale evolution* (7) of an effective energy-momentum tensor operator.

Let us now go back and see how Einstein's equation (4) can be reformulated into such a form as (7). Let us consider the background metric of the dual 4 dimensional field theory to be $\eta_{\mu\nu}$ where the following RG flow equation [19]

$$\begin{aligned}\frac{\partial t^{\mu}{}_{\nu}(\Lambda)}{\partial\Lambda} &= \frac{1}{\Lambda^3}\cdot\frac{1}{2}\Box t^{\mu}{}_{\nu}(\Lambda) - \frac{1}{\Lambda^5}\cdot\left(\frac{1}{4}\,\delta^{\mu}{}_{\nu}t^{\alpha}{}_{\beta}(\Lambda)t^{\beta}{}_{\alpha}(\Lambda) - \frac{7}{128}\,\Box^2 t^{\mu}{}_{\nu}(\Lambda)\right) - \\ &+\frac{1}{\Lambda^5}\log\Lambda\cdot\frac{1}{32}\cdot\Box^2 t^{\mu}{}_{\nu}(\Lambda) + \mathcal{O}\left(\frac{1}{\Lambda^7}\log\Lambda\right) \qquad (10)\end{aligned}$$

can be constructed. For the above RG flow, we can indeed construct the following unique $g_{\mu\nu}(\Lambda)$ as given by

$$\begin{aligned}g_{\mu\nu}(\Lambda) = \eta_{\mu\nu} &+ \frac{1}{\Lambda^4}\cdot\frac{1}{4}\eta_{\mu\alpha}t^{\alpha}{}_{\nu}(\Lambda) + \frac{1}{\Lambda^6}\cdot\frac{1}{24}\eta_{\mu\alpha}\Box t^{\alpha}{}_{\nu}(\Lambda) + \\ &+\frac{1}{\Lambda^8}\cdot\left(\frac{1}{32}\,\eta_{\mu\alpha}t^{\alpha}{}_{\rho}(\Lambda)t^{\rho}{}_{\nu}(\Lambda) - \frac{7}{384}\,\eta_{\mu\nu}t^{\alpha}{}_{\beta}(\Lambda)t^{\beta}{}_{\alpha}(\Lambda)\right. \\ &\left.+\frac{11}{1536}\,\eta_{\mu\alpha}\Box^2 t^{\alpha}{}_{\nu}(\Lambda)\right) + \\ &+\frac{1}{\Lambda^8}\log\Lambda\cdot\frac{1}{516}\cdot\eta_{\mu\alpha}\Box^2 t^{\alpha}{}_{\nu}(\Lambda) + \mathcal{O}\left(\frac{1}{\Lambda^{10}}\log\Lambda\right) \qquad (11)\end{aligned}$$

as a functional of $t^{\mu}{}_{\nu}(\Lambda)$ and Λ in the flat Minkowski space background such that when it is considered as an effective background metric, the scale-dependent Ward identity (9) is satisfied at each Λ (given that at $\Lambda = \infty$, the usual Ward identities (5) hold). Furthermore, the $5-$dimensional bulk metric (1) then satisfies Einstein's equations (4) with $r = \Lambda^{-1}$ and the cosmological constant set to $-6/l^2$. The *log* term in (10) is related to the conformal anomaly.

It is to be noted that the Ward identity (9) can also be recast as an effective operator equation, i.e. can be rewritten in a state-independent manner as an identity in flat Minkowski space $\eta_{\mu\nu}$. In the above example, (9) can be readily unpacked into

$$\begin{aligned}\partial_{\mu}t^{\mu}{}_{\nu}(\Lambda) &= \frac{1}{\Lambda^4}\cdot\left(\frac{1}{16}\partial_{\nu}\left(t^{\alpha}{}_{\beta}(\Lambda)t^{\beta}{}_{\alpha}(\Lambda)\right) - \frac{1}{8}t^{\mu}{}_{\nu}(\Lambda)\partial_{\mu}\,\mathrm{Tr}\,t(\Lambda)\right) + \\ &+\frac{1}{\Lambda^6}\cdot\left(\frac{1}{48}t^{\alpha}{}_{\beta}(\Lambda)\partial_{\nu}\Box t^{\beta}{}_{\alpha}(\Lambda) - \frac{1}{48}t^{\mu}{}_{\nu}(\Lambda)\partial_{\mu}\Box\,\mathrm{Tr}\,t(\Lambda)\right) + \\ &+\mathcal{O}\left(\frac{1}{\Lambda^8}\right). \qquad (12)\end{aligned}$$

We then explicitly see that the scale-dependent effective background $g_{\mu\nu}(\Lambda)$ as given by (11) serves to absorb the multi-trace contributions that spoil the usual Ward identity for local energy-momentum conservation. As a result, the Ward identity preserves its form (9) at each scale in the new scale-dependent background.

It should be immediately noted that although the RG flow (10) leads to the bulk metric in the Fefferman–Graham gauge, the underlying equations determining the latter should have underlying full diffeomorphism invariance. It can be readily argued that otherwise it is impossible that the RG flow (10) will be able to preserve a Ward identity of the form (9). In particular, absence of diffeomorphism invariance in the dual bulk theory that gives the evolution of $g_{\mu\nu}(\Lambda)$ will imply that there will be other propagating degrees of freedom in addition to $g_{\mu\nu}(\Lambda)$ in which case the Ward identity (9) should be modified.

The RG flow reformulation (10) of Einstein's equations has been demonstrated so far only in the asymptotic (i.e. UV) expansion. This series (10) has a *finite* radius of convergence related to the scale (radius) where the Fefferman–Graham coordinates has a coordinate singularity in the dual spacetime. In order to sum (10) to all orders in Λ^{-1}, we need to assume a specific form of the energy-momentum tensor such as the hydrodynamic form to be considered later. In the latter case, all orders in Λ^{-1} can be summed at any given order in derivative expansion. The radius of convergence is the scale corresponding to the location of the horizon at late time and is related to the final temperature.

The immediate question is how do we derive the RG flow reformulation of the classical gravity equations such as (10) corresponding to Einstein's equations. In order to answer this, it is sufficient to understand what does $t^{\mu}{}_{\nu}(\Lambda)$ correspond to in the dual gravitational theory. To do this a gauge-independent formulation of the map between RG flow and gravitational equations is helpful. We express the $(d+1)-$dimensional spacetime metric via ADM-like variables [29]:

$$\mathrm{d}s^2 = \alpha(r,x)\mathrm{d}r^2 + \gamma_{\mu\nu}(r,x)\,(\mathrm{d}x^\mu + \beta^\mu(r,x)\mathrm{d}r)\,(\mathrm{d}x^\nu + \beta^\nu(r,x)\mathrm{d}r)\,. \tag{13}$$

in which α is the analogue of the lapse function and β^μ is the analogue of the shift vector. Specifying conditions determining these amounts to gauge-fixing the diffeomorphism symmetry. For reasons (state independence and absence of explicit presence of l in the evolution equations) mentioned before, assuming that $r=0$ is the boundary the identification of Λ and $g_{\mu\nu}(\Lambda)$ should take the form [19]

$$r = \Lambda^{-1}, \quad g_{\mu\nu}(\Lambda = r^{-1}) = \frac{r^2}{l^2}\gamma_{\mu\nu}(r,x). \tag{14}$$

Note the above is not only true for Einstein's gravity but also for a general gravitational theory. In this case the form of $t^{\mu}{}_{\nu}(\Lambda)$ can also be fixed to a large extent by (i) requiring it to be state (solution) independent, (ii) demanding absence of explicit presence of l in its scale evolution, and (iii) requiring that it satisfies the Ward identity (9). In a general gravitational theory, these imply that $t^{\mu}{}_{\nu}(\Lambda)$ should take the form up to an overall multiplicative constant [19]:

$$t^{\mu}_{\ \nu}(\Lambda = r^{-1}) = \left(\frac{l}{r}\right)^{d} \cdot \left(T^{\mu}_{\ \nu}{}^{\rm ql} + T^{\mu}_{\ \nu}{}^{\rm ct}\right), \tag{15}$$

where $T^{\mu}_{\ \nu}{}^{\rm ql}$ is the quasi-local stress tensor that is conserved via equations of motion [30] and $T^{\mu}_{\ \nu}{}^{\rm ct}$ is a sum of gravitational counterterms built out of the Riemann curvature of $\gamma_{\mu\nu}$ and its covariant derivatives such that they satisfy (9) via Bianchi-type identities. Up to second order in derivatives, $T^{\mu}_{\ \nu}{}^{\rm ct}$ can be parametrised as:

$$T^{\mu}_{\ \nu}{}^{\rm ct} = -\frac{1}{8\pi G_N}\left[C_{(0)} \cdot \frac{1}{l} \cdot \delta^{\mu}_{\ \nu} + C_{(2)} \cdot l \cdot \left(R^{\mu}_{\ \nu}[\gamma] - \frac{1}{2}R[\gamma]\delta^{\mu}_{\ \nu}\right) + \cdots\right], \tag{16}$$

with $C_{(n)}$s being dimensionless constants that depend on the gravitational theory and G_N being the $(d+1)-$dimensional gravitational constant. Above, the indices have been lowered/raised by the induced metric $\gamma_{\mu\nu}$/its inverse. In the case of Einstein's gravity, $T^{\mu}_{\ \nu}{}^{\rm ql}$ is the Brown–York tensor:

$$T^{\mu}_{\ \nu}{}^{\rm ql} = -\frac{1}{8\pi G_N}\gamma^{\mu\rho}\left(K_{\rho\nu} - K\gamma_{\rho\nu}\right). \tag{17}$$

Here $K_{\mu\nu}$ is the extrinsic curvature of the hypersurface $r = constant$ given by

$$K_{\mu\nu} = -\frac{1}{2\alpha}\left(\frac{\partial\gamma_{\mu\nu}}{\partial r} - \nabla_{(\gamma)\mu}\beta_{\nu} - \nabla_{(\gamma)\nu}\beta_{\mu}\right), \tag{18}$$

with $\beta_{\rho} = \gamma_{\rho\mu}\beta^{\mu}$, and $K = K_{\mu\nu}\gamma^{\mu\nu}$. Therefore, in the Fefferman–Graham gauge, $t^{\mu}_{\ \nu}(\Lambda)$ should take the following form for Einstein's gravity:

$$\begin{aligned} t^{\mu}{}_{\nu}(\Lambda = r^{-1}) = \frac{l^{d-1}}{16\pi G_N}\Bigg[&\frac{1}{r^{d-1}} \cdot \left(z^{\mu}{}_{\nu} - ({\rm Tr}\, z)\,\delta^{\mu}{}_{\nu}\right) + 2 \cdot \frac{1}{r^{d}} \cdot \left(d - 1 - C_{(0)}\right) \cdot \delta^{\mu}{}_{\nu} - \\ &-2 \cdot \frac{1}{r^{d-2}} \cdot C_{(2)} \cdot \left(R^{\mu}{}_{\nu}[g] - \frac{1}{2}R[g]\delta^{\mu}{}_{\nu}\right) + \cdots\Bigg]. \end{aligned} \tag{19}$$

The overall multiplicative constant $l^{d-1}/(16\pi G_N)$ has been chosen by us above and cannot be fixed by the arguments presented before. This overall factor is actually proportional to N^2 of the dual field theory (as mentioned before l itself has no meaning in the dual QFT but the gravitational constant measured in units where $l = 1$ does have one). This overall factor can be fixed by identifying the temperature in the field theory in a thermal state to that of the Hawking temperature of the dual black hole. This however requires taking into account quantum effects. For later convenience, we rescale $t^{\mu}_{\ \nu}(\Lambda)$ by this overall factor $(16\pi G_N)/l^{d-1}$ so that N^2 is now absorbed in the definition of $t^{\mu}_{\ \nu}(\Lambda)$. There is still a genuine ambiguity in the definition of $t^{\mu}_{\ \nu}(\Lambda)$ which arises from the choice of the gravitational counterterm coefficients $C_{(n)}$s. Fixing this ambiguity leads us to a profound and surprising understanding of gravity itself as described below.

We first observe that the above ambiguity of choosing coefficients of gravitational counterterms has an immediate consequence for the map between gravity and RG flow. It implies that the equations of gravity can be reformulated into infinitely many RG flow equations of the form (7) for any choice of gauge fixing of bulk diffeomorphisms. Each of these formulations corresponds to a specific choice of gravitational counterterms $C_{(n)}$s. Furthermore, each such RG flow will require the existence of the same (unique) $g_{\mu\nu}(\Lambda)$ taking the schematic form (8) in which the effective Ward identity (9) will be satisfied, and which will lead to the same bulk metric that satisfies the dual diffeomorphism invariant gravitational equations with a specific gauge fixing.

It is of course desirable that at the UV fixed point, i.e. at $\Lambda = \infty$, $t^{\mu}{}_{\nu}{}^{\infty}$ is finite. This leads to fixing a finite number of leading counterterms, particularly [31–33]

$$C_{(0)} = d - 1, \quad C_{(2)} = -\frac{1}{d-2}, \quad \text{etc.} \tag{20}$$

It is interesting to note that $t^{\mu}{}_{\nu}{}^{\infty}$ is completely free of ambiguities when the boundary metric is $\eta_{\mu\nu}$, because all other counterterms, except a few leading terms, vanish in any asymptotically AdS space because of the enhancement of symmetries in the geometry in the asymptotic limit. We thus recover the result for $t^{\mu}{}_{\nu}{}^{\infty}$ as in the traditional AdS/CFT correspondence. This procedure is however unsatisfactory for two reasons. Firstly, we still have infinite ambiguities in the form of unfixed coefficients of the infinite number of gravitational counterterms which vanish asymptotically. Secondly, if we can genuinely rewrite gravity as RG flow, in the latter form it should be first order evolution so that we can either specify conditions at the UV or at the IR, but not at both places. It is more desirable that we restrict the IR as we need a sensible IR behaviour of the RG flow even in cases where the UV completion is unknown. This is specially relevant for finding holographic duals of theories like QCD where only the IR can be expected to be captured by a holographically dual classical gravity description at large N [21] – in the UV the emergent geometry can have a singularity implying the necessity of new degrees of freedom.

This ambiguity is fixed by the following theorem stated below [18–20].

Theorem 2: Up to an overall multiplicative constant for $t^{\mu}{}_{\nu}(\Lambda)$, there is a unique choice of the functional $F^{\mu}{}_{\nu}$ in (7) that reformulates a pure holographic classical gravity theory as RG flow such that the endpoint of the RG flow at $\Lambda = \Lambda_{\mathrm{IR}}$ can be converted to a fixed point in the hydrodynamic limit corresponding to *non-relativistic incompressible Navier-Stokes fluid* under the universal rescaling:

$$\Lambda_{\mathrm{IR}}^{-1} - \Lambda^{-1} = \xi \cdot \lambda^{-1} \quad t = \frac{\tau}{\xi}, \tag{21}$$

(corresponding to near horizon and long time behaviour of the dual gravitational dynamics) where ξ is taken to zero with λ and τ kept finite. This also corresponds to fixing the gravitational counterterms in (16) uniquely so that $t^{\mu}{}_{\nu}(\Lambda)$ is uniquely identified as a functional of the ADM variables in the dual pure gravitational theory. Even those counterterms which are necessary to cancel UV divergences are also determined by the prescribed IR behaviour.

Remarkably, the hydrodynamic limit can fix all the ambiguities of the RG flow which however has a state-independent formulation in terms of evolution of the operator $t^{\mu}{}_{\nu}(\Lambda)$ with the scale and which is valid even beyond this limit. Thus long wavelength perturbations of black holes unsurprisingly play a very fundamental role in understanding holographic correspondence as RG flow. We do not have a complete proof of this theorem, so actually it is still a conjecture. However very non-trivial calculations which will be sketched in the next section provide solid supporting verifications.

It is also important that the end point of the RG flow is not really fixed point although it becomes so after the rescaling (21) which has been first introduced in the context of gravitational dynamics in the hydrodynamic limit in the dual theory in [34]. As we will see in the next section, it implies that all physical parameters in $t^{\mu}{}_{\nu}(\Lambda)$ should satisfy appropriate bounds regrading how they behave at the endpoint. These bounds determine all integration constants of the first order RG flow and thus determine the UV values of physical observables. Remarkably, these UV values are exactly the same as those for which dual gravitational geometries are free of naked singularities. Since the hydrodynamic limit determines the RG flow uniquely, all physical observables beyond the hydrodynamic limit can also be obtained from the RG flow. *Therefore, not only that a holographic gravitational theory can be reformulated as a unique RG flow for every choice of gauge-fixing of diffeomorphism symmetry (up to an overall constant numerical factor for* $t^{\mu}{}_{\nu}(\Lambda)$*), the data which leads to regular horizons are also determined by this RG flow.* This IR criterion constitutes another crucial defining feature of a highly efficient RG flow as exemplified by (10) for Einstein's gravity.

Finally, we note that the choice of gauge fixing of the diffeomorphism symmetry is also encoded in the RG flow (which in cases other than the Fefferman–Graham gauge may contain auxiliary non-dynamical variables corresponding to the lapse function and the shift vector). This is due to the feature that any asymptotically AdS metric has a residual gauge symmetry which corresponds to conformal transformations for the dual theory at the boundary under which the dual theory must be invariant (up to quantum anomalies that are related to logarithmic terms necessary for regulating divergences of the on-shell gravitational action [31, 32]). Such diffeomorphisms which preserve the Fefferman–Graham gauge are called Penrose–Brown–Henneaux (PBH) transformations in the literature [35–37], and these can be readily generalised to other choices of gauge fixing [19]. These turn out to lead to automorphism symmetry of the dual RG flow equations (7) when they are formulated in a general fixed conformally flat background metric [19]. We have called this *lifted Weyl symmetry*. Deciphering this symmetry for a given highly efficient RG flow readily leads us to determine the corresponding gauge fixing in the dual gravity theory and thus also the choice of hypersurface foliation in the dual geometries used as holographic screens at various scales.

3 The Field Theory Perspective and the Hydrodynamic Limit

In the previous section, we have discussed reformulation of a holographic pure gravity theory as a highly efficient RG flow which can self-determine microscopic UV data by an appropriate IR criterion, and reproduce results of traditional holographic correspondence where these data are determined by explicitly solving the gravitational equations and demanding absence of naked singularities. In this section, following [20] we will show how such a RG flow can be constructed in the field theory and even define it constructively in the strong interaction and large N limits. We will illustrate the construction briefly in the hydrodynamic limit.

In the strong interaction and large N limits, a handful of single-trace operators (dual to the fields in the gravitational theory) can define at least some sectors of the full theory in the sense mentioned in the previous section. Instead of using the elementary fields to define the QFT, it then makes sense to use collective variables which are directly measurable and which parametrise the expectation values of these single-trace operators in all states. Such collective variables include the hydrodynamic variables and can be extended to include the shear-stress tensor and other non-hydrodynamic parameters also (see for instance [38–40]). At the very outset, it is clear that such an exercise of defining quantum operators via collective variables which parametrise their expectation values is futile except in the strong interaction and large N limits. Unless we are in the large N limit, the expectation values of the multi-trace operators do not factorise, therefore we need new collective variables for defining multi-trace operators. Also if we are not in the strong interaction limit, we will need to consider infinitely many single-trace operators. These will imply proliferation of the number of collective variables required to describe exact quantum dynamics.

The physical picture is as follows. Consider a set of microscopic single-trace operators O_I^∞ such as the energy-momentum tensor which can be parametrised by a set of collective variables X_A^∞ such as the hydrodynamic variables. Furthermore, the spacetime evolution of the expectation values $\langle O_I^\infty \rangle$ can be captured by equations of motions for the collective variables X_A^∞ such as the hydrodynamic equations with parameters η_M^∞ such as the transport coefficients. It is to be noted here that the hydrodynamics being mentioned here is not referring to any kind of coarse-graining, rather an asymptotic series involving perturbative derivative expansion (with infinite number of transport coefficients) which captures the dynamics near thermal equilibrium [41, 42]. Generally speaking, we can succinctly represent the quantum operators O_I^∞ through their expectation values $\langle O_I^\infty \rangle [X_A^\infty, \eta_M^\infty]$.

We can readily do an appropriate coarse-graining of our measurements of $\langle O_I^\infty \rangle$ and proceed to define $\langle O_I(\Lambda)\rangle$. The latter definition can be achieved via appropriate coarse-grained collective variables $X_A(\Lambda)$ which by construction follow similar equations as $X_A(\infty)$ but with new parameters $\eta_M(\Lambda)$. As in any RG flow, we expect that we need fewer parameters $\eta_M(\Lambda)$ to describe the spacetime evolution of $X_A(\Lambda)$ than the number of η_M^∞ we need to describe that of X_A^∞ to the same degree of approximation. In a highly efficient RG flow, we define the coarse-grained quantum operators $O_I(\Lambda)$ through their expectation values $\langle O_I(\Lambda)\rangle [X_A(\Lambda), \eta_M(\Lambda)]$ assuming that the coarse-grained operators are the same functionals of the coarse-grained collective variables at each scale (as in the UV) but with new scale-dependent parameters. Note that there is no explicit dependence on Λ in the functionals $\langle O_I(\Lambda)\rangle [X_A(\Lambda), \eta_M(\Lambda)]$.

In order to complete the construction we will need to define the constructive principles for coarse-graining that defines $X_A(\Lambda)$ which should follow similar equations at each scale but with new scale-dependent parameters $\eta_M(\Lambda)$. These three principles are listed below.

1. **High efficiency**: There should exist an appropriate background metric:
   $g_{\mu\nu}(\Lambda)[X_A(\Lambda), \eta_M(\Lambda), \Lambda]$
   and appropriate background sources:
   $J(\Lambda)[X_A(\Lambda), \eta_M(\Lambda), \Lambda]$
   at each Λ such that the Ward identity
   $$\nabla_{(\Lambda)\mu} t^\mu{}_\nu(\Lambda) = \sum{}' O_I(\Lambda) \nabla_{(\Lambda)\nu} J_I(\Lambda) \tag{22}$$
   is satisfied with $\nabla_{(\Lambda)}$ being the covariant derivative constructed from $g_{\mu\nu}(\Lambda)$ and $\sum'$ denoting summation over all effective single-trace operators except $t^\mu{}_\nu(\Lambda)$.
2. **Upliftability to operator dynamics**: The functionals $g_{\mu\nu}(\Lambda)[X_A(\Lambda), \eta_M(\Lambda), \Lambda]$ and $J(\Lambda)[X_A(\Lambda), \eta_M(\Lambda), \Lambda]$ can be uplifted to functionals of the single-trace operators. Therefore, they should assume the forms
   $g_{\mu\nu}(\Lambda)[O_I(\Lambda), \Lambda]$ and $J(\Lambda)[O_I(\Lambda), \Lambda]$
   so that the effective Ward identities (22) can be promoted to operator equations such as (12). As a consequence, it follows that the scale evolution equations for $O_I(\Lambda)$ such as (10) become state-independent equations involving single and multi-trace operators and Λ only, and thus without involving the collective variables explicitly.
3. **Good endpoint behaviour**: The IR end point of the RG flow where most of the parameters $\eta_M(\Lambda)$ blow up and some collective variables $X_A(\Lambda)$ become singular can be made regular under the universal rescaling (21). In the hydrodynamic limit, the endpoint should be converted to a fixed point corresponding to non-relativistic incompressible Navier–Stokes equations under the stated rescaling.

Our claim is that for every realisation of a highly efficient RG flow which satisfies the above three principles:

1. there corresponds a unique dual gravitational theory up to a choice of gauge-fixing of the bulk diffeomorphism symmetry that can have a dual holographic description as a strongly interacting large N QFT, and
2. there is unique set of UV data for (infinitely many) $\eta_M(\Lambda)$ which however can be resummed in the IR to a finite number of parameters characterising the dynamics at the endpoint (such as the shear viscosity of the infrared non-relativistic incompressible Navier–Stokes fluid), and also these UV data (such as the UV values of the infinitely many transport coefficients) are the same as those which determine the regularity of the future horizons in the dual gravitational theory corresponding to the RG flow.

The infrared end point typically corresponds to the location of the horizon at late time, and thus the highly efficient RG flow connects the AdS/CFT correspondence with the membrane paradigm. The highly efficient RG flow gives a constructive way to define strongly interacting large N QFTs by reformulating the holographic correspondence. The first two principles in our list defining highly efficient RG flows utilise the first theorem of reformulation of diffeomorphism invariant gravity and the third principle in our list utilises the second theorem discussed in the previous section. However, here our list of principles also presents a generalisation which is valid not only for the reconstruction of holographic pure gravity but also when the latter is coupled to a finite number of matter fields. The utility of highly efficient RG flow is actually deeper. It shows that all such QFTs and hence all holographic gravitational theories are determined by finite amount of data that governs the dynamics at the end point. Therefore, all holographic gravitational theories can be parametrised by a finite number of free parameters in any given dimension. How this parametrisation works has not been completely understood yet.

As an illustration, let us see how we construct highly efficient RG flows in the hydrodynamic limit [20]. Once again, let us revert back to the sector of states where $t^{\mu}{}_{\nu}(\Lambda)$ is the only single-trace operator with a non-vanishing expectation value for the sake of simplicity. The expectation value of $t^{\mu}{}_{\nu}(\Lambda)$ is parametrised by the (collective) hydrodynamic variables $u^{\mu}(\Lambda)$ and $T(\Lambda)$ which thus define the quantum operator. Furthermore, $u^{\mu}(\Lambda)$ can be assumed to satisfy Landau–Lifshitz definition in which case $u^{\mu}(\Lambda)$ is a timelike eigenvector of $t^{\mu}{}_{\nu}(\Lambda)$ with unit norm with respect to the background metric $g_{\mu\nu}(\Lambda)$ so that $u^{\mu}(\Lambda)g_{\mu\nu}(\Lambda)u^{\nu}(\Lambda) = -1$. The hydrodynamic variables $u^{\mu}(\Lambda)$ and $T(\Lambda)$ should satisfy hydrodynamic equations in the effective background $g_{\mu\nu}(\Lambda)$ with scale-dependent energy density $\varepsilon(\Lambda)$, pressure $P(\Lambda)$ and transport coefficients $\gamma^{(n,m)}(\Lambda)$, where n denotes the order in the derivative expansion (running from zero to infinity) and m lists the finite number of independent parameters at each order in the derivative expansion. At the first order in the derivative expansion, there are only two independent transport coefficients, namely the shear and the bulk viscosities.

The coarse-graining of $u^{\mu}(\Lambda)$ and $T(\Lambda)$ can be expressed both in integral or differential form. The latter form is more useful and is as shown below:

$$:\frac{\partial u^{\mu}(\Lambda)}{\partial \Lambda}: = a^{(0)}(\Lambda)u^{\mu}(\Lambda) + \sum_{n=1}^{\infty}\sum_{m=1}^{n_s} a_s^{(n,m)}(\Lambda)\,\mathscr{S}^{(n,m)}(\Lambda)\,u^{\mu}(\Lambda) +$$
$$+\sum_{n=1}^{\infty}\sum_{m=1}^{n_v} a_v^{(n,m)}(\Lambda)\,\mathscr{V}^{\mu\,(n,m)}(\Lambda)\,,$$
$$:\frac{\partial T(\Lambda)}{\partial \Lambda}: = b^{(0)}(\Lambda) + \sum_{n=1}^{\infty}\sum_{m=1}^{n_s} b_s^{(n,m)}(\Lambda)\mathscr{S}^{(n,m)}(\Lambda). \tag{23}$$

Above $\mathscr{S}^{(n,m)}$ denotes the independent hydrodynamic scalars that can be constructed from derivatives of $u^{\mu}(\Lambda)$ and $T(\Lambda)$ at the $n-th$ order in derivatives (with independent meaning that a linear sum of these scalars do not vanish using lower order equations of motion). When $n = 1$, there is only one such scalar, namely $(\partial \cdot u)$. Similarly, $\mathscr{V}^{\mu\,(n,m)}(\Lambda)$ denotes hydrodynamic vectors which are not parallel to $u^{\mu}(\Lambda)$ (as otherwise it can be expressed via a scalar multiplying $u^{\mu}(\Lambda)$). When $n = 1$, there is only one such vector, namely $(u(\Lambda)\cdot\partial)u^{\mu}(\Lambda)$. The symbols $:\cdots:$ stand for subtracting away non-hydrodynamic contributions. The coarse-graining actually arises from a truncation of the series (23) at a given order in the derivative expansion. So far this is the most general way to coarse-grain hydrodynamic variables which is consistent with the hydrodynamic limit.

Furthermore, we assume that the flow of the energy density, pressure and the transport coefficients take the form of ordinary differential equations:

$$\frac{\mathrm{d}\varepsilon(\Lambda)}{\mathrm{d}\Lambda} = K[\varepsilon(\Lambda), P(\Lambda), \Lambda],$$
$$\frac{\mathrm{d}P(\Lambda)}{\mathrm{d}\Lambda} = L[\varepsilon(\Lambda), P(\Lambda), \Lambda],$$
$$\frac{\mathrm{d}\gamma^{(n,m)}(\Lambda)}{\mathrm{d}\Lambda} = M^{(n,m)}[\varepsilon(\Lambda), P(\Lambda), \gamma^{(k\leq n,p)}(\Lambda), \Lambda], \tag{24}$$

in which the scale evolution of transport coefficients at n−th order in derivative expansion involves only those at the same or lower orders.

The mathematical problem of constructing highly efficient RG flows in the hydrodynamic limit now becomes well-defined. We simply need to solve for the parameters $a^{(0)}, b^{(0)}, a_s^{(n,m)}, a_v^{(n,m)}, b_s^{(n,m)}$ in (23) and the functionals K, L and $M^{(n,m)}$ appearing in (24) such that the three principles listed before are satisfied. Unfortunately, we do not yet know how this mathematical problem can be solved directly. Fortunately, there is a concrete algorithmic method [18] (developed using some results of [17, 27]) to reformulate the classical gravitational equations in the forms (23) and (24) which can be used to solve for these parameters indirectly so that we can satisfy the three principles and obtain all highly efficient RG flows.

The most subtle aspect of this procedure is in how we satisfy the third principle of good infrared behaviour. As discussed in the previous section, the reformulation of gravity as RG flow is subject to the ambiguities of undetermined counterterm coefficients. However there are a finite number of such terms at each order in the derivative expansion. The recipe is to proceed with these ambiguities which lead to unknown numerical constants in (23) and (24). In order for the endpoint to be governed by non-relativistic incompressible Navier–Stokes equations, $\varepsilon(\Lambda)$ must be finite at the endpoint Λ_{IR} where $\gamma^{(n,m)}(\Lambda)$ should satisfy bounds $\gamma^{(n,m)}(\Lambda) \leq (\Lambda - \Lambda_{\mathrm{IR}})^{-k(n,m)}$ with $k(n, m)$ being appropriate numerical constants which are independent of the RG flow or the dual gravitational theory [18]. It turns out that when we actually solve for $\gamma^{(n,m)}(\Lambda)$ the number of terms which diverge worse than the prescribed bounds are typically more than the number of integration constants available unless the counterterm coefficients which have been left undetermined so far are precisely chosen at each order in the derivative expansion. Setting these counterterm coefficients to such values, we can fix all integration constants of the RG flow and thus we can determine the UV values of all transport coefficients *uniquely*.

This procedure has been explicitly implemented for Einstein's gravity at zeroth, first and second orders in the derivative expansion. Remarkably, the UV values of the equations of state and the first and second order transport coefficients determined via this method matches exactly with the known values [22–24] which are required for the regularity of the future horizon.

We should understand how to solve for highly efficient RG flows independently without using the theorems for reformulation of dual gravitational theories so that we can classify all gravitational theories that are holographic and also where a finite number of IR parameters can determine all microscopic UV data in the dual theories.

4 Outlook

We have demonstrated that the reformulation of classical gravity as RG flow not only reveals how the holographic duality works but also gives us a deeper understanding of gravitational dynamics itself, in particular relating to what kind of data that determine the spacetime metric lead to absence of naked singularities.

An outstanding issue is to take another step to understand how to include quantum corrections in gravity while mapping it to a highly efficient RG flow whose notion also needs to be further generalised. In order to proceed, it should be useful to understand better how the three principles which define highly efficient RG flows themselves originate from a simpler and more holistic principle. Such a direction seems possible as there is evidence that classical gravity emerges from features of quantum entanglement in dual quantum systems [43]. In particular, it is known that classical minimal surfaces in dual geometries encode entanglement entropies in dual field theories [44]. It has also been argued elsewhere that efficient nonperturbative RG flows that coarse-grain quantum information efficiently such that they remove short range entanglement but preserve long range entanglement give rise to the holographic correspondence [45]. It is natural to speculate that when quantum gravity corrections are included the infrared end point for the dual RG flow is not characterised necessarily by local order parameters, but rather by non-local quantum order parameters related to patterns of global long range entanglement. This point of view also has a potential for defining quantum geometry in the emergent gravity theory.

I hold the point of view that a breakthrough in this direction is likely to come from a reformulation of classical gravity equations themselves which uses non-local geometric objects such as geodesics and minimal surfaces as the dynamical variables, and also which makes a tangible connection with the local RG flow perspective described in the present article. At present, how this can be realised seems a bit mysterious, however it is very likely that there are hidden treasures in classical gravity which are yet to discovered. It will not be surprising if the surface terms introduced by Paddy, and his novel variational principle (see [14] for instance) involving these surface terms which give classical gravitational equations in the bulk without using the metric as a dynamical variable, can shed some light in this direction.

Finally, I would like to mention that the reformulation of classical gravity equations as RG flows has also informed the development of a new approach for combining weak and strong coupling degrees of freedom of the quark-gluon plasma produced by heavy ion collisions self-consistently into a novel nonperturbative framework [46, 47]. Unravelling the holographic origin of gravity will surely revolutionise our understanding of nonperturbative aspects of quantum dynamics in the future.

Acknowledgements The research of A.M. is supported by a Lise-Meitner fellowship of the Austrian Science Fund (FWF), project no. M 1893-N27.

References

1. G. 't Hooft, The Holographic principle: opening lecture. Subnucl. Ser. **37**, 72–100 (2001). arXiv:hep-th/0003004
2. L. Susskind, The World as a hologram. J. Math. Phys. **36**, 6377–6396 (1995). arXiv:hep-th/9409089
3. R. Bousso, The Holographic principle. Rev. Mod. Phys. **74**, 825–874 (2002). arXiv:hep-th/0203101
4. J.M. Maldacena, TASI 2003 lectures on AdS/CFT, in *Progress in String Theory. Proceedings, Summer School, TASI 2003, Boulder, USA, 2–27 June 2003* (2003), pp. 155–203. arxiv:hep-th/0309246
5. J.M. Maldacena, The large N limit of superconformal field theories and supergravity. Int. J. Theor. Phys. **38**, 1133 (1999). arXiv:hep-th/9711200
6. E. Witten, Anti-de Sitter space and holography. Adv. Theor. Math. Phys. **2**, 253–291 (1998). arXiv:hep-th/9802150
7. S. Gubser, I.R. Klebanov, A.M. Polyakov, Gauge theory correlators from noncritical string theory. Phys. Lett. B **428**, 105–114 (1998). arXiv:hep-th/9802109
8. T. Padmanabhan, The Holography of gravity encoded in a relation between entropy, horizon area and action for gravity. Gen. Rel. Grav. **34**, 2029–2035 (2002). arXiv:gr-qc/0205090
9. T. Padmanabhan, Gravitational entropy of static space-times and microscopic density of states. Class. Quant. Grav. **21**, 4485–4494 (2004). arXiv:gr-qc/0308070
10. T. Padmanabhan, Holographic gravity and the surface term in the Einstein–Hilbert action. Braz. J. Phys. **35**, 362–372 (2005). arXiv:gr-qc/0412068
11. A. Mukhopadhyay, T. Padmanabhan, Holography of gravitational action functionals. Phys. Rev. D **74**, 124023 (2006). arXiv:hep-th/0608120

12. M. Akbar, R.-G. Cai, Thermodynamic behavior of field equations for f(R) gravity. Phys. Lett. B **648**, 243–248 (2007). arXiv:gr-qc/0612089
13. D. Kothawala, S. Sarkar, T. Padmanabhan, Einstein's equations as a thermodynamic identity: the cases of stationary axisymmetric horizons and evolving spherically symmetric horizons. Phys. Lett. B **652**, 338–342 (2007). arXiv:gr-qc/0701002
14. T. Padmanabhan, A. Paranjape, Entropy of null surfaces and dynamics of spacetime. Phys. Rev. D **75**, 064004 (2007). arXiv:gr-qc/0701003
15. T. Padmanabhan, Thermodynamical aspects of gravity: new insights. Reports Prog. Phys. **73**, 046901 (2010). arXiv:0911.5004
16. T. Padmanabhan, Dark energy and gravity. General Relativ. Gravit. **40**, 529–564 (2008). arXiv:0705.2533
17. S. Kuperstein, A. Mukhopadhyay, The unconditional RG flow of the relativistic holographic fluid. JHEP **11**, 130 (2011). arXiv:1105.4530
18. S. Kuperstein, A. Mukhopadhyay, Spacetime emergence via holographic RG flow from incompressible Navier–Stokes at the horizon. JHEP **11**, 086 (2013). arXiv:1307.1367
19. N. Behr, S. Kuperstein, A. Mukhopadhyay, Holography as a highly efficient renormalization group flow. I. Rephrasing gravity. Phys. Rev. D **94**(2), 026001 (2016). arXiv:1502.06619
20. N. Behr, A. Mukhopadhyay, Holography as a highly efficient renormalization group flow. II. An explicit construction. Phys. Rev. D **94**(2), 026002 (2016). arXiv:1512.09055
21. I.R. Klebanov, M.J. Strassler, Supergravity and a confining gauge theory: duality cascades and chi SB resolution of naked singularities. JHEP **08**, 052 (2000). arXiv:hep-th/0007191
22. G. Policastro, D.T. Son, A.O. Starinets, The shear viscosity of strongly coupled N=4 supersymmetric Yang–Mills plasma. Phys. Rev. Lett. **87**, 081601 (2001). arXiv:hep-th/0104066
23. R. Baier, P. Romatschke, D.T. Son, A.O. Starinets, M.A. Stephanov, Relativistic viscous hydrodynamics, conformal invariance, and holography. JHEP **04**, 100 (2008). arXiv:0712.2451
24. S. Bhattacharyya, V.E. Hubeny, S. Minwalla, M. Rangamani, Nonlinear fluid dynamics from gravity. JHEP **02**, 045 (2008). arXiv:0712.2456
25. T. Damour, Black hole eddy currents. Phys. Rev. D **18**, 3598–3604 (1978)
26. K.S. Thorne, R.H. Price, D.A. MacDonald, *Black Holes: The Membrane Paradigm* (Yale University Press, New Haven, 1986)
27. R.K. Gupta, A. Mukhopadhyay, On the universal hydrodynamics of strongly coupled CFTs with gravity duals. JHEP **03**, 067 (2009). arXiv:0810.4851
28. I. Heemskerk, J. Polchinski, Holographic and Wilsonian renormalization groups. JHEP **06**, 031 (2011). arXiv:1010.1264
29. R.L. Arnowitt, S. Deser, C.W. Misner, The dynamics of general relativity. General Rel. Grav. **40**, 1997–2027 (2008). arXiv:gr-qc/0405109
30. A. Balcerzak, M.P. Dabrowski, Generalized Israel junction conditions for a fourth-order brane world. Phys. Rev. D **77**, 023524 (2008). arXiv:0710.3670
31. M. Henningson, K. Skenderis, Holography and the Weyl anomaly. Fortsch. Phys. **48**, 125–128 (2000). arXiv:hep-th/9812032
32. V. Balasubramanian, P. Kraus, A stress tensor for anti-de sitter gravity. Commun. Math. Phys. **208**, 413–428 (1999). arXiv:hep-th/9902121
33. J. de Boer, E.P. Verlinde, H.L. Verlinde, On the holographic renormalization group. JHEP **08**, 003 (2000). arXiv:hep-th/9912012
34. I. Bredberg, C. Keeler, V. Lysov, A. Strominger, Wilsonian approach to fluid/gravity duality. JHEP **03**, 141 (2011). arXiv:1006.1902
35. R. Penrose, W. Rindler, *Spinors and Spacetime, Chap. 9*, vol. 2, Spinor and Twistor Methods in Space-Time Geometry (Cambridge University Press, Cambridge, 1986)
36. J.D. Brown, M. Henneaux, Central charges in the canonical realization of asymptotic symmetries: an example from three-dimensional gravity. Commun. Math. Phys. **104**, 207–226 (1986)
37. A. Schwimmer, S. Theisen, Diffeomorphisms, anomalies and the Fefferman–Graham ambiguity. JHEP **0008**, 032 (2000). arXiv:hep-th/0008082
38. R. Iyer and A. Mukhopadhyay, An AdS/CFT connection between Boltzmann and Einstein. Phys. Rev. D **81**, 086005 (2010). arXiv:0907.1156

39. R. Iyer, A. Mukhopadhyay, Homogeneous relaxation at strong coupling from gravity. Phys. Rev. D **84**, 126013 (2011). arXiv:1103.1814
40. M.P. Heller, R.A. Janik, M. Spaliński, P. Witaszczyk, Coupling hydrodynamics to nonequilibrium degrees of freedom in strongly interacting quark-gluon plasma. Phys. Rev. Lett. **113**(26), 261601 (2014). arXiv:1409.5087
41. M.P. Heller, R.A. Janik, P. Witaszczyk, Hydrodynamic gradient expansion in gauge theory plasmas. Phys. Rev. Lett. **110**(21), 211602 (2013). arXiv:1302.0697
42. G. Basar, G.V. Dunne, Hydrodynamics, resurgence, and transasymptotics. Phys. Rev. D **92**(12), 125011 (2015). arXiv:1509.05046
43. T. Faulkner, M. Guica, T. Hartman, R.C. Myers, M. Van Raamsdonk, Gravitation from entanglement in holographic CFTs. JHEP **03**, 051 (2014). arXiv:1312.7856
44. S. Ryu, T. Takayanagi, Holographic derivation of entanglement entropy from AdS/CFT. Phys. Rev. Lett. **96**,181602 (2006). arXiv:hep-th/0603001
45. B. Swingle, Entanglement renormalization and holography. Phys. Rev. D **86**, 065007 (2012)
46. E. Iancu, A. Mukhopadhyay, A semi-holographic model for heavy-ion collisions. JHEP **06**, 003 (2015). arXiv:1410.6448
47. A. Mukhopadhyay, F. Preis, A. Rebhan, S.A. Stricker, Semi-holography for heavy ion collisions: self-consistency and first numerical tests. JHEP **05**, 141 (2016). arXiv:1512.06445

Modelling Non-paradoxical Loss of Information in Black Hole Evaporation

Sujoy K. Modak and Daniel Sudarsky

Abstract We give general overview of a novel approach, recently developed by us, to address the issue black hole information paradox. This alternative viewpoint is based on theories involving modifications of standard quantum theory, known as "spontaneous dynamical state reduction" or "wave-function collapse models" which were historically developed to overcome the notorious foundational problems of quantum mechanics known as the "measurement problem". We show that these proposals, when appropriately adapted and refined for this context, provide a self-consistent picture where loss of information in the evaporation of black holes is no longer paradoxical.

1 Introduction

The black hole information problem [15] is one of the most debated and controversial problems of theoretical physics, and has been the focus of considerable attention from various theoretical viewpoints during the last four decades (see [18] for a pedagogic introduction). We in fact note Paddy's recent proposal [16, 17] connected to this issue. The main reason behind this activity is the fact that while according to the *unitary* evolution law of Quantum Mechanics (QM), all information about a quantum state at any time is encoded in the state at any other time, the process of thermal black hole

S.K. Modak (✉)
KEK Theory Center, High Energy Accelerator Research Organization (KEK), Tsukuba, Ibaraki 305-0801, Japan
e-mail: sujoy@post.kek.jp

S.K. Modak
Facultad de Ciencias - CUICBAS, Universidad de Colima, C.P. 28045 Colima, Mexico

D. Sudarsky
Instituto de Ciencias Nucleares, Universidad Nacional Autónoma de México, 04510 Mexico, DF, Mexico
e-mail: sudarsky@nucleares.unam.mx

J.S. Bagla and S. Engineer (eds.), *Gravity and the Quantum*,
Fundamental Theories of Physics 187, DOI 10.1007/978-3-319-51700-1_18

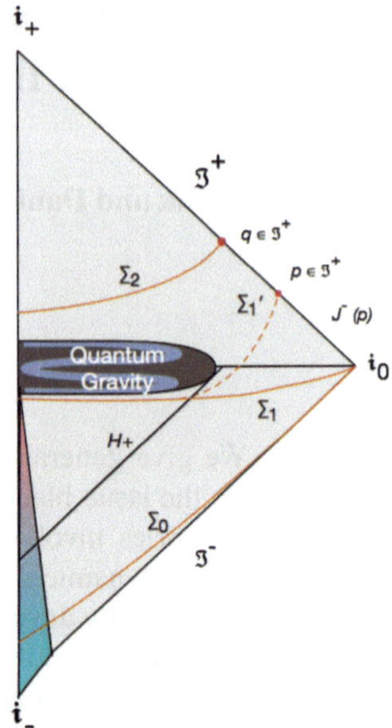

Fig. 1 Penrose diagram showing *black hole* formation and evaporation. The initial spacetime is Minkowskian at $\mathscr{I}^-$ and at the end of Hawking evaporation Quantum Gravity (QG) resolves the would be classical singularity making the final spacetime asymptotically flat at $\mathscr{I}^+$

evaporation via Hawing radiation poses a "threat" to such an expectation, leading to the so called "paradox".

More specifically, let us consider the formation (by gravitational collapse) of a black hole and its subsequent evaporation (by Hawking effect) as shown in Fig. 1. Let the initial state of the matter, defined on an initial Cauchy slice Σ_0, be characterized at the quantum level, by some pure, perhaps a coherent state. Under appropriate circumstances this matter collapses forming the black hole, and any quantum field, in this space-time will contribute to the Hawking radiation at late times. The radiation is characterized, in full, simply by the temperature and in particular it will be the same for any initial mass regardless the details of the initial quantum state. Thus unless some dramatic departure from the above picture takes place there would be no way to retro-dict the initial state from the former. The mapping from initial to final states would not be invertible and in particular it would fail to be represented in terms of a unitary operator.

The point is that, attempts to explain how the full state of the quantum fields *unitarily* related to the initial state would be encoded on late time hypersurfaces such as $\Sigma_2 \cup (J^-(q) \cap \mathscr{I}^+)$, have been *unsuccessful* until this date.

This, in turn, leads the majority of the physicists to believe that the fate of information and problem of unitarity in black hole evaporation is a *unique situation* where the unitarity of quantum evolution is in question. The recent finding of "firewall problem" [1] when considering a solution to the problem based on an approach known as based on "black hole complementarity" [32] is often presented as reflecting the tension between the *unitarity* of QM and *equivalence principle* of general relativity (GR).

We, however, need to be a bit more careful because the mapping implied by quantum theory is expected to be unitary only if it refers to states defined on complete Cauchy hypersurfaces. The point is that, while the initial state was characterized on Σ_0, which is a true Cauchy hypersurface, the late time hypersurfaces would fail to be Cauchy hypersurfaces due to the presence of a space-time singularity deep within the black hole. This singularity (or more precisely a surface arbitrarily close to it) can in fact be considered as a boundary of the space-time and thus as a necessary component of any true late time Cauchy hypersurface. However the singularity is generally expected to be a feature of the fact that we have not incorporated the aspects of quantum gravity which should cure this singularity removing the need to include the extra space-time boundary. It is then and only then that we face a truly problematic situation [22].

Therefore the real tension between quantum theory and general relativity in the context of black hole formation and evaporation arises only when we view that quantum gravity will remove the singularity and thus the need to include a spacetime boundary. It is then and only then that we face something that could be considered truly paradoxical. We could now ask ourselves, what would be the problem of adopting the position that, all processes involving black hole evaporation do in fact break the unitarity of quantum evolution? As we see it, the problem with that position would be that, as we just saw, we would be working in a context where we imagine having incorporated aspects of quantum gravity in the discussion. Having done that, it seems inevitable to view processes such as black hole formation and evaporation as part of a larger class of processes, after all, the black hole concept is essentially a global one. That is the notion of Black Hole is not one that could be considered as lying at the basic formulation of the theory, which is expected to be described in terms of some fundamentally local degrees of freedom, rather that the global notions such as event horizons, or trapped surfaces that should appear only as secondary and emerging entities. In fact we should expect that black hole creation and evaporation should appear in the theory occurring also as virtual process contributing to essentially all physical processes, thus raising the question of when precisely can we expect to have an exact unitary evolution law as dictated by standard quantum theory.

In fact the black hole information issue, motivated the analysis in [2] where it was argued that loss of unitary would be need to accompanied by unacceptably large violations of energy conservation or of causality. A subsequent study of the issue reveled however that those arguments were not very robust, and that such expectations could be radically modified [35].

These considerations open the door to considering the question of information loss in the context of possible of modified versions of quantum theory involving departure from unitary evolution at the fundamental level.

In fact it is fair to say that all the approaches that have been proposed so far for the recovery of information (and the full quantum state that is unitarily related to the initial one) have not been successful as they end up adding to more problematic aspects to the picture. A big motivation of this rather "one way traffic" is the adherence to the notion that failure of unitarity in black hole evaporation would completely invalidate QM. However, as we advocate here, the situation is not as simple, because the violation of unitarity is not only not unexpected in quantum context but rather a common occurrence in any situation, normally thought as involving the collapse of wave-function due to whatever reason (natural interaction or laboratory measurement). That is, we have further motivation to consider the issue at hand, in connection with the so called general measurement problem in quantum theory.

Therefore, in contrast with the established tradition we will discuss here an approach based on the exact opposite possibility, i.e. the *necessity of loss of information* during black hole evaporation just as in most ordinary situations involving the quantum regime, thus "dissolving" the paradox.

This "dissolution" comes of course at the cost of losing quantum mechanical unitarity and one might worry whether we would lose with it all the successes of standard QM. We furthermore note that from the foundational point of view regarding quantum theory, there have been various proposal of a modified version of quantum dynamics incorporating a *spontaneous collapse* of the wave-function to elevate QM from a *theory of measurement* to the *theory of reality* (see [27] for the terminology) in a manner that the subjective role of an observer becomes removed and one can treat QM objectively without introducing any extraneous notion of observer as an essential entity shaping reality. This collapse process is spontaneous and stochastic, and it is implemented in such a manner that the well established and experimentally successful predictions of quantum mechanics remain unaffected, while a gradual difference in the predictions appears as the quantum system's size approaches that of macroscopic object (for an account of ongoing experimental endeavor, see [4]). Thus, in building such type of theories, the aim is to resolve *the measurement problem* and eliminate various in-built two level descriptions of reality in Copenhagen interpretation; such as, micro/macro, classical/quantum, system/apparatus, system/observer, system/environment etc.

In this article, we will not discuss in detail any of these proposals (for that we refer the reader to the papers [5, 10–13, 23–29, 33, 34] as well as review articles [3, 4]), but we shall use some specific models and provide a concrete example offering a overview of the manner in which such models can deal with the information problem in black holes leading to a picture where the associated breakdown of unitarity is a part and parcel of the (modified) general quantum mechanical evolution.

2 Measurement Problem and Models of Wave-Function Collapse

According to the Copenhagen interpretation of quantum mechanics there are two distinct evolution rules for the quantum state/wave-function of a system. First, a continuous evolution as dictated by the Schrödinger equation and valid while the system is left alone and free from observations, and the second, a discontinuous and stochastic jump to one of the eigenstates (dictated by Born probability rule) of some self-adjoint operator in the Hilbert space once *measurement* by an external observer takes place. As characterized by R. Penrose, the first case is a unitary evolution or the U-process, while the second case is a reduction or the R-process, and *measurement* is a notion that separates these two processes. The problem is that within the standard view of QM, measurement does not have any kind of rigorous definition, nor is it clear when exactly it is performed during a evolution. In fact such vague and artificial division has been sharply criticized by Bell [7]

> ...If the theory is to apply to anything but highly idealized laboratory operations, are we not obliged to admit that more or less 'measurement-like' processes are going on more or less all the time, more or less everywhere? Do we not have jumping then all the time?

Also, it is an in-built aspect of the standard presentations, that without observer or some entity which is "measuring" the system, no specific outcome is presupposed in QM.

Within the community, working on the foundation of quantum mechanics, there are of course diverse viewpoints regarding the *measurement problem*. These include the Many World Interpretations, in its various forms, the Bohmian Mechanics program representing a reliance on nonlocal hidden variables, and the proposals for unifying the U and R processes, referred as the Dynamical Reduction Program (DRP), pioneered by Pearle, Ghirardi, Rimini and Weber.

The DRP first included a discrete process of collapse in the wave-function/quantum state, driven by an additional non-unitary and stochastic term modifying the Schrödinger evolution.

The basic idea in those proposals is that the evolution of systems with very small number of degrees of freedom is dominated by the standard part of the dynamics resulting in very small deviations from that predicted by standard theory, ensuring the reproduction of the stupendous success of quantum theory in high precision laboratory experiments whereas, the non-standard terms becomes dominant when a rather large number of degrees of freedom appear in a state representing a rather delocalized quantum superposition, thus ensuring the rapid collapse to one or the other of the classical looking components of Schrödinger cat states.

This feature ensures that when, what is normally called a measurement is performed, the system is driven to one or the other eigenstates of the apparatus' pointer's position simply because such pointer consists of a macroscopically large number of degrees of freedom. That is, as a result of the new general dynamical law the theory reproduces the standard predictions of quantum theory regarding the measurement of *the appropriate self-adjoint operator*. The first and is simplest successful model of this kind known as Ghirardi-Rimini-Weber (GRW) theory [11], which was later improved to deal with identical particles in a scheme that makes wave-function collapse a continuous process and known as the CSL theory [12, 26, 27]. Their recent advances in this direction have resulted in proposals for relativistic version of both type of theories [5, 28, 29, 33]. In this article we restrict ourselves to the non-relativistic framework of CSL theory to address the issue of black hole information and for the relativistic framework we refer the interested reader to our recent work [6].

2.1 CSL Theory: Non-relativistic Setting

The non-relativistic version of the CSL theory [12, 26, 27] is, at this point, much better explored than the relativistic counterpart and it is defined by following two equations: (i) A stochastically modified Schrödinger equation, whose solution is:

$$|\psi, t\rangle_w = \hat{\mathscr{T}} e^{-\int_0^t dt' \left[i\hat{H} + \frac{1}{4\lambda_0}[w(t') - 2\lambda_0 \hat{A}]^2 \right]} |\psi, 0\rangle, \tag{1}$$

where $\hat{\mathscr{T}}$ is the time-ordering operator, $w(t)$ is a random, white noise type classical function of time and its probability distribution is given by the second equation, (ii) the Probability Distribution (PD) rule:

$$PDw(t) \equiv {}_w\langle\psi, t|\psi, t\rangle_w \prod_{t_i=0}^{t} \frac{dw(t_i)}{\sqrt{2\pi\lambda_0/dt}}. \tag{2}$$

Thus the standard Schrödinger evolution and corresponding changes in the state corresponding to a "measurement" of the operator $\hat{A}$ are unified and the dynamics does not allow any superluminal signaling. In the non-relativistic limit, for a single particle, the proposal assumes that there is a spontaneous and continuous reduction characterized by $\hat{A} = \hat{\mathbf{X}}_\delta$, where $\hat{\mathbf{X}}_\delta$ is a suitably smeared position operator (with the smearing characterized by the scale δ). This smearing of the position operator is required to avoid an uncontrolled increase in energy associated with a point-like collapse event. The resultant theory can be applied to all situations without invoking any measurement device or observer. This framework can be easily extended to multiparticle system by choosing a set of operators representing each particle so that everything, including, the apparatuses are treated quantum mechanically. The final theory, thus seems to successfully address the measurement problem and completely overlook various two level descriptions in Copenhagen interpretation.

Since the final outcome of the collapse of an individual state vector is uncertain, it is useful to consider a collection of identical initial state and describe the evolution of an ensemble in the language of a density matrix. The CSL evolution of density matrix can be derived from a Lindblad type equation with a solution [12, 26, 27]

$$\rho(t) = \hat{\mathscr{T}} e^{-\int_0^t dt' \left[i(\underset{\rightarrow}{\hat{H}} - \underset{\leftarrow}{\hat{H}}) + \frac{\lambda}{2}[\underset{\rightarrow}{\hat{A}} - \underset{\leftarrow}{\hat{A}}]^2 \right]} \rho(0) \tag{3}$$

where the arrows mean the operators act on the left or right of $\rho(0)$.

3 Callan–Giddings–Harvey–Strominger (CGHS) Model

We choose the 2 dimensional version of black hole formation and evaporation, provided by the CGHS model [8], to exhibit an explicit realization of our proposal. The

Fig. 2 Penrose diagram for CGHS spacetime. Minkowskian and black hole regions are separated by a sharp gravitational collapse of null matter like photon

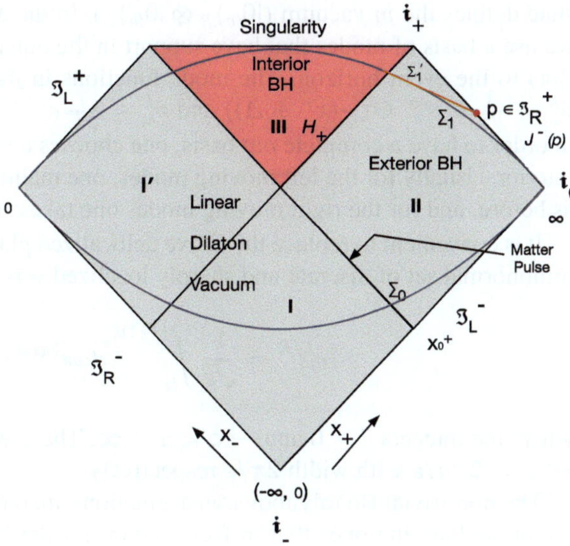

CGHS action is given by

$$S = \frac{1}{2\pi}\int d^2x\sqrt{-g}\left[e^{-2\phi}\left[R + 4(\nabla\phi)^2 + 4\Lambda^2\right] - \frac{1}{2}(\nabla f)^2\right]$$

where ϕ is the dilaton field, Λ^2 is a constant, and f is a real scalar field, representing matter. The Penrose diagram of CGHS model is shown in Fig. 2. Before the gravitational collapse ($x^+ < x_0^+$), the metric is Minkowskian, usually known as the dilaton vacuum (region I and I'), given by $ds^2 = -\frac{dx^+dx^-}{-\Lambda^2x^+x^-}$, whereas, at $x^+ > x_0^+$ it is represented by the black hole metric (region II, III)[1] $ds^2 = -\frac{dx^+dx^-}{\frac{M}{\Lambda}-\Lambda^2x^+(x^-+\Delta)}$. In regions I and I', natural Minkowskian coordinates are $y^+ = \frac{1}{\Lambda}\ln(\Lambda x^+)$, $y^- = \frac{1}{\Lambda}\ln(-\frac{x^-}{\Delta})$, with $-\infty < y^- < \infty$; $-\infty < y^+ < \frac{1}{\Lambda}\ln(\Lambda x_0^+)$. On the other hand, on the BH exterior (region II), where physical observers might exist, one has the coordinates $\sigma^+ = \frac{1}{\Lambda}\ln(\Lambda x^+) = y^+, \sigma^- = -\frac{1}{\Lambda}\ln(-\Lambda(x^- + \Delta))$ and the metric is $ds^2 = -\frac{d\sigma^+d\sigma^-}{1+(M/\Lambda)e^{\Lambda(\sigma^- - \sigma^+)}}$ with $-\infty < \sigma^- < \infty$ and $\sigma^+ > \sigma_0^+ = \frac{1}{\Lambda}\ln(\Lambda x_0^+)$. It is easy to check the asymptotic flatness of the black hole metric by introducing Schwarzschild like time t and space r coordinates [19, 20] using $\tanh(\Lambda t) = T/X$ and $-\frac{1}{\Lambda^2}(e^{2\Lambda r} - M/\Lambda) = T^2 - X^2$.

The quantum description of the field f can be made using two different natural bases. In the asymptotic past ($\mathscr{I}_L^- \cup \mathscr{I}_R^-$ or *in*) region, the basis mode functions are chosen to be: $u_\omega^R = \frac{1}{\sqrt{2\omega}}e^{-i\omega y^-}$ and $u_\omega^L = \frac{1}{\sqrt{2\omega}}e^{-i\omega y^+}$, with $\omega > 0$ (R and L indicate right and left moving modes respectively). The tensor product of respective vacuum

---

[1] More precisely, region I', although flat, is also part of the interior of the event horizon as nothing in that region can ever reach $\mathscr{I}_R^+$.

state defines the *in* vacuum ($|0_{in}\rangle_R \otimes |0_{in}\rangle_L$). In the asymptotic future (out region) we use a basis of modes that have support in the outside (exterior) and inside (interior) to the event horizon. The mode functions in the exterior to the horizon are: $v^R_\omega = \frac{1}{\sqrt{2\omega}} e^{-i\omega\sigma^-} \Theta(-(x^- + \Delta))$ and $v^L_\omega = \frac{1}{\sqrt{2\omega}} e^{-i\omega\sigma^+} \Theta(x^+ - x^+_0)$. Similarly, and in order to have a complete *out* basis, one chooses a set of modes for the black hole interior. Usually for the left moving modes, one maintains the same functional form as before, and for the right moving modes one takes: $\hat{v}^R_{\tilde{\omega}} = \frac{1}{\sqrt{2\tilde{\omega}}} e^{i\tilde{\omega}\sigma^-_{in}} \Theta(x^- + \Delta)$.

It is convenient to replace the above delocalized plane wave modes by a complete orthonormal set of discrete and sharply localized wave packets modes [9, 14],

$$v^{L/R}_{nj} = \frac{1}{\sqrt{\varepsilon}} \int_{j\varepsilon}^{(j+1)\varepsilon} d\omega e^{2\pi i \omega n/\varepsilon} v^{L/R}_\omega , \tag{4}$$

where the integers $j \geq 0$ and $-\infty < n < \infty$. These wave packets are peaked about $\sigma^{+/-} = 2\pi n/\varepsilon$ with width $2\pi/\varepsilon$ respectively.

The non-trivial Bogolyubov transformations are only relevant in the right moving sector, and are the ones that in fact account for the Hawking radiation. The initial state, corresponding to the vacuum for the right moving modes, and the left moving pulse (which leads to the formation of the black hole) $|\Psi_{in}\rangle = |0_{in}\rangle_R \otimes |Pulse\rangle_L$ can be expanded in the *out* basis:

$$N \sum_{F_{nj}} C_{F_{nj}} \left|F_{nj}\right\rangle^{ext} \otimes \left|F_{nj}\right\rangle^{int} \otimes |Pulse\rangle_L , \tag{5}$$

where the states $\left|F_{nj}\right\rangle$ are characterized by the finite occupation numbers $\{F_{nj}\}$ for each corresponding mode n, j; N is a normalization constant, and the coefficients $C_{F_{nj}}$'s are determined by the Bogolyubov transformations.

4 Gravitational Induced Collapse of Wave-Function and Loss of Information

As we have mentioned in the very beginning, the wave-function collapse model, as given by the CSL theory, needs to be adapted in order to be applicable to the problem at hand. One novel addition to the already developed CSL theory, is our hypothesis that *gravitational field enhances the rate of wave-function collapse*. This can be achieved by making the collapse rate as a function of the Weyl curvature scalar $W_{abcd}W^{abcd}$ as first suggested by Okon and Sudarsky [21]. That is, even in the absence of any measuring device/observer, in a spacetime region with enormously large curvature (such as inside the horizon and towards the center of a black hole), quantum superpositions are increasingly broken in a *stochastic* manner (provided by the CSL stochasticity) and produces similar effects as those caused by an external measurement usually considered in a laboratory context. It should be mentioned that such an effect of gravitation on quantum mechanics was strongly advocated by R. Penrose in several of his works (see, for instance, [30, 31]) and we consider those as a guiding path leading to our explicit demonstration.

We first note that in the multi-particle system, the CSL evolution (1) is generalized to

$$|\psi, t\rangle_{w_\alpha} = \hat{\mathcal{T}} e^{-\int_0^t dt' \left[i\hat{H} + \frac{1}{4\lambda_0} \sum_\alpha [w_\alpha(t') - 2\lambda_0 \hat{A}_\alpha]^2 \right]} |\psi, 0\rangle, \tag{6}$$

where α is an index labeling the set of particles. Our aim is to consider a CSL evolution (analogous to (6) applied to a field theory with the index α having a different meaning) of the initial state (5). Moreover we will treat CSL as an interaction term, so that the free Hamiltonian will be set to zero, i.e. $H = 0$ in (6). As we are using this equation in the context of QFT in curved spacetime, we need to choose a new operator, that must be constructed using the field operator and its derivatives. One such operator is the number operator (for right moving modes) defined in the interior Fock basis times the identity for the exterior Fock basis:

$$\hat{A}^\alpha = \hat{N}_{nj}^{int} \otimes \mathbb{I}^{ext} \tag{7}$$

for all n, j, where $\hat{N}_{nj}^{int} = \hat{N}_{nj}^{int(R)} \otimes \mathbb{I}^{int(L)}$ and $\mathbb{I}^{ext} = \mathbb{I}^{ext(L)} \otimes \mathbb{I}^{ext(R)}$. This ensures that the collapse will make the wave-packet to peak about particular values of n and j (which will be picked randomly depending on the specific realization of the noise $w_{nj}(t)$). In the standard CSL type evolution (6) it takes, strictly speaking, an infinite amount of time to fully collapse the wave-function to an eigenstate of the collapse operator due to the finite value of the collapse parameter λ_0. It is to be noted that the experimental bounds on λ_0 come from laboratory based experiments that are, of course, performed in a spacetime regions where curvature is negligible. Here we make a hypothesis that the collapse rate is in fact sensitive to the local curvature, so that a more general expression must have the following form

$$\lambda(W^2) = \lambda_0 \left(1 + (W^2/\mu^2)^\gamma\right) \tag{8}$$

where μ is an appropriate scale and $\gamma \geq 1$. One anticipated effect of this, is to generate an complete effective collapse of the quantum state, taking place in a finite time interval.

In the particular case of 2D models the Weyl scalar vanishes identically so instead, for this special case, we assume that λ is determined by the Ricci scalar. Thus, for this specific case (8) is replaced by,

$$\lambda(R) = \lambda_0 \left(1 + (R/\mu)^\gamma\right) \tag{9}$$

where, for the CGHS black hole $R = \frac{4M\Lambda}{M/\Lambda - \Lambda^2(T^2 - X^2)}$. The Kruskal time and space coordinates are respectively $T = \frac{x^+ + x^- + \Delta}{2}$ and $X = \frac{x^+ - x^- - \Delta}{2}$. Next we provide a brief account of calculation, further details can be found in [19, 20].

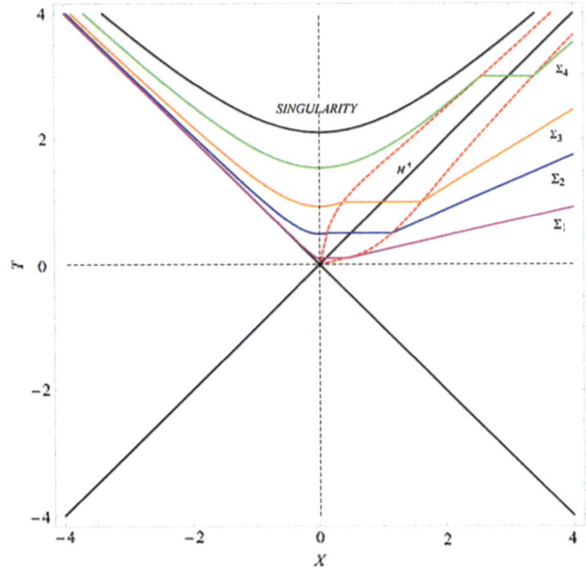

Fig. 3 Foliation of the CGHS spacetime by Cauchy slices. All the plots have fixed $M/\Lambda^3 = 4.42$, while $T = 0.1, 0.5, 1, 3$ (in *magenta*, *blue*, *orange* and *green*) as the joining $T = const.$ curves

Foliation of the spacetime: To implement CSL in the black hole model we need to foliate the spacetime with an appropriately defined spacelike/Cauchy slices, as plotted in Fig. 3, and use the time evolution from one slide to the other. As there is no natural notion of "time" in the relevant regions of these spacetimes, we take a convenient time parameter (τ) that will be used to characterize the quantum state evolution, according to the CSL dynamics.

We define the Cauchy slices to be Schwarzschild $r = const.$ inside the horizon and Schwarzschild $t = const.$ outside the horizon, and join them by surfaces with Kruskal $T = const.$. The intersection curves joining the family of Cauchy slices $r = const.$ and $T = const.$ at one end (inside horizon) is chosen to be $T_1(X) = \left(X^2 + \frac{M}{\Lambda^3}e^{-2\Lambda/\sqrt{X}}\right)^{1/2}$, whereas, at the other end (outside horizon) the intersection curve $T_2(X)$ can be found just by using a reflection about the event horizon $T = X$. We fix the value of the "time" parameter τ as the value of the coordinate T on the intersection of $r = const.$ with $X = 0$ (or T axis), so that the Ricci scalar is expressed as $R = \frac{4M\Lambda}{M/\Lambda - \Lambda^2\tau^2}$. It is now clear that R diverges for some finite value of $\tau = \tau_s = \frac{M^{1/2}}{\Lambda^{3/2}}$ corresponding to the divergence of R that characterizes the singularity.

Evolution of the quantum state: In standard CSL theory state is evolved according to the Eq. (6), which, in the present situation is subjected to the changing collapse parameter (9), that become a function of time parameter τ and, the collapse operators (7). The initial state for the right moving modes, traveling from $\mathscr{I}_R^-$ to $\mathscr{I}_R^+$ is denoted by the "in" vacuum for right moving modes, which can be expressed in the "out" basis according to (5) (we leave for the moment the "pulse" which is left moving and forms the black hole). The evolution equation of the state vector for right moving modes become

$$|\Psi,\tau\rangle_R = N\sum_F C_{F_{nj}} e^{-\int_o^\tau d\tau'[\frac{1}{4\lambda}\sum_{n,j}(w_{nj}-2\lambda F_{nj})^2]}|F_{nj}\rangle_R^{int}\otimes|F_{nj}\rangle_R^{ext} \tag{10}$$

where F_{nj} is the eigenvalue of the operator $\hat{N}_{nj}^{int}$ while acting on the state $|F\rangle^{int}$. As τ approaches τ_s the Cauchy slices tend to reach the spacetime singularity, and R diverges. This divergence in R makes the integral in (10) divergent, and thus the initial state collapses to a state with definite quantum numbers n, j, giving

$$\lim_{\tau\to\tau_s}|\Psi,\tau\rangle_R = NC_{F_{n_0j_0}}|F_{n_0j_0}\rangle_R^{int}\otimes|F_{n_0j_0}\rangle_R^{ext}, \tag{11}$$

on the hypersurface Σ_1 (Fig. 4) as it approaches the singularity. As the level of each mode's excitation depends on the realization of the stochastic value $w_{nj}(\tau_s)$, the final state after collapse, although remains pure, it is undetermined.

Evolution at the ensemble level as given by the density matrix: To account for the lack of predictability of the final state we consider a large collection of systems all prepared in the same initial state and use an ensemble description in terms of a density matrix. The evolution equation becomes

$$\rho_R(\tau)=N^2\sum_{F,G} e^{-\frac{\pi}{\Lambda}(E_F+E_G)}e^{-\sum_{nj}(F_{nj}-G_{nj})^2\int_{\tau_0}^{\tau}d\tau'\frac{\lambda(\tau')}{2}}|F\rangle_R^{int}\otimes|F\rangle_R^{ext}\langle G|_R^{int}\otimes\langle G|_R^{ext}. \tag{12}$$

Therefore, near the singularity (on Σ_1 in Fig. 4), as λ diverges in the exponential factor, the result is a diagonal density matrix of the form (omitting n, j from subscript for simplified notation and putting explicit expression for C_F):

$$\lim_{\tau\to\tau_s}\rho_R(\tau) = N^2\sum_F e^{-\frac{2\pi}{\Lambda}E_F}|F\rangle_R^{int}\otimes|F\rangle_R^{ext}\langle F|_R^{int}\otimes\langle F|_R^{ext}, \tag{13}$$

where $E_F = \sum_{nj}\omega_{nj}F_{nj}$ is the total energy of the final excited state.

The description of the state vector and density matrix is complete once we include the left moving matter pulse, so that

$$\lim_{\tau\to\tau_s}|\Psi,\tau\rangle_R = Ne^{-\frac{\pi}{\Lambda}E_{F0}}|F_0\rangle_R^{int}\otimes|F_0\rangle_R^{ext}\otimes|Pulse\rangle_L \tag{14}$$

$$\lim_{\tau\to\tau_s}\rho(\tau) = N^2\sum_F e^{-\frac{2\pi}{\Lambda}E_F}|F\rangle_R^{int}\otimes|F\rangle_R^{ext}\langle F|_R^{int}\otimes\langle F|_R^{ext}\otimes|Pulse\rangle_L\langle Pulse|_L. \tag{15}$$

where F_0 is understood as a specific particle excited state $F_{n_0j_0}$ and E_{F_0} is the energy of this state.

Quantum gravity (QG) and resolution of singularity: To pass from the hypersurface Σ_1 to Σ_f, in Fig. 4, one has to rely on a theory of QG which is likely to involve giving up the classical notion of "spacetime". In the absence of any fully workable theory of that kind, we make a few natural assumptions about QG theory, namely that – (i) it resolves the singularity and leads on the other side, to a regime describable

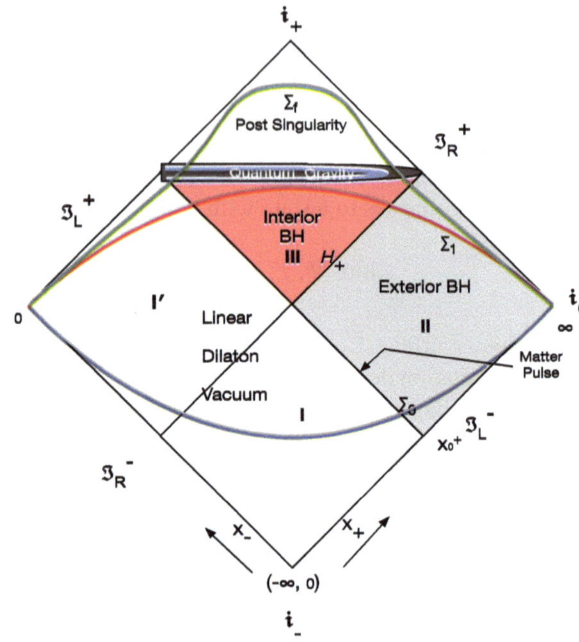

Fig. 4 Penrose diagram of CGHS spacetime with Quantum Gravity (QG) region

with standard classical notions of space-time, (ii) it does not lead to arbitrarily large violations of standard conservation laws such a energy conservation. If so, then QG makes following operation after combining the negative energy state $|F_0\rangle_R^{int}$ (which is complementary to Hawking radiation) with the positive matter $|Pulse\rangle$

$$|F_0\rangle_R^{int} \otimes |Pulse\rangle_L \to |p.s\rangle, \tag{16}$$

where $|p.s\rangle$ is a post-singularity quantum state with almost vanishing energy and residing as the complement of Hawking radiation near $\mathscr{I}_R^+$ on Σ_f. Then on the final hyper-surface Σ_f, the quantum state and the density matrix becomes

$$|\Psi\rangle_R = Ne^{-\frac{\pi}{\lambda}E_{F_0}} \otimes |F_0\rangle_R^{ext} \otimes |p.s\rangle \tag{17}$$

$$\rho = N^2 \sum_F e^{-\frac{2\pi}{\mu}E_F} |F\rangle_R^{ext} \langle F|_R^{ext} \otimes |p.s\rangle \langle p.s|\,. \tag{18}$$

$$= \rho_{thermal}^{ext} \otimes \mathbb{I}_{p.s.}^{ext} \tag{19}$$

The resulting picture, therefore, indicates that the final state on the Cauchy slice Σ_f, for an individual system is *pure, yet undetermined* while at the ensemble level it is *proper mixed state* as the density matrix is clearly thermal on the asymptotic regime times a state with a very low energy (and idealized to be vacuum) characterizing the remaining portion of Σ_f, (which is then taken to be a portion of flat spacetime). Thus the complete evolution is *non-unitary and information is lost*, mainly in the interior of the black hole, as a consequence of wave-function collapse. There is of course *nothing paradoxical* in this picture.

5 Discussions

We have put forward a novel proposal involving gravitational influenced wave-function collapse which we showed can account for the enormous loss of information in black hole evaporation which thus leads to a dissolution of the so called "paradox". On a broader perspective, this opens up a rather interesting possibility: that in the energy scale interpolating between, say, the current LHC (or the Standard Model) energy scale (about 10 TeV) and somewhere below the QG scale[2] (10^{16} TeV), there could be important effects describable in the context of *the standard model of particle physics adapted to the modified quantum field theory constructed on curved space-times with the additional feature of gravitational induced quantum state reduction, as provided, say, by one of the relativistic collapse proposals* [5, 28, 29, 33, 34]. Could it be, for instance, that the issue of the radiative corrections induced quantum instability of the Higgs potential are modified by the introduction of such modifications? Could we do with a scheme where supper-symmetry is not needed and have similar benefits arising, instead, from the effects of quantum collapse? We believe this line of inquire might offer interesting insights and modify the perspectives for physics beyond standard model, and perhaps the expectations for phenomenology of quantum gravity.

Acknowledgements We want to thank our collaborators E. Okon, L. Ortíz, I. Peña, D. Bedingham for their involvement and contribution towards this project. SKM is an International Research Fellow of Japan Society for the Promotion of Science. This research is partly funded by Grant-in-Aid to JSPS fellows (KAKENHI-PROJECT-15F15021). DS acknowledges partial financial support from DGAPA-UNAM project IG100316 and by CONACyT project 101712.

References

1. A. Almheiri, D. Marolf, J. Polchinski, J. Sully, Black holes: complementarity or firewalls? JHEP **1302**, 062 (2013)
2. T. Banks, L. Susskind, M.E. Peskin, Difficulties for the evolution of pure states into mixed states. Nucl. Phys. B **244**, 125 (1984)
3. A. Bassi, G.C. Ghirardi, Dynamical reduction models. Phys. Reports **379**, 257 (2003)
4. A. Bassi, K. Lochan, S. Satin, T.P. Singh, H. Ulbricht, Models of wave-function collapse, underlying theories, and experimental tests. Rev. Mod. Phys. **85**, 471 (2013)
5. D. Bedingham, Relativistic state reduction model. J. Phys. Conf. Ser. **306**, 012034 (2011)
6. D. Bedingham, S.K. Modak, D. Sudarsky, Relativistic collapse dynamics and black hole information loss. arXiv:1604.06537 [gr-qc]

[2] Up to the regime where we expect the validity of standard space-time notions as provided by General Relativity.

7. J. Bell, Against measurement (1990)
8. C.G. Callan, S.B. Giddings, J.A. Harvey, A. Strominger, Evanescent black holes. Phys. Rev. D **45**, R1005 (1992)
9. A. Fabbri, J. Navarro-Salas, *Modeling Black Hole Evaporation* (Imperial College Press, London, 2005)
10. G. Ghirardi, A. Rimini, T. Weber, A model for a unified quantum description of macroscopic and microscopic systems, in *Quantum Probability and Applications*, ed. by A.L. Accardi (Springer, Heidelberg, 1985), pp. 223–232
11. G. Ghirardi, A. Rimini, T. Weber, Unified dynamics for microscopic and macroscopic systems. Phys. Rev. D **34**, 470 (1986)
12. G. Ghirardi, P. Pearle, A. Rimini, Markov-processes in Hilbert-space and continuous spontaneouslocalization of systems of identical particles. Phys. Rev. A **42**, 7889 (1990)
13. G. Ghirardi, R. Grassi, P. Pearle, Relativistic dynamical reduction models: general framework and examples. Found. Phys. (J.S. Bell's 60th birthday issue) **20**, 1271 (1990)
14. S.B. Giddings, W.M. Nelson, Quantum emission from two-dimensional black holes. Phys. Rev. D **46**, 2486 (1992)
15. S.W. Hawking, Breakdown of predictability in gravitational collapse. Phys. Rev. D **14**, 2460 (1976)
16. K. Lochan, S. Chakraborty, T. Padmanabhan, Information retrieval from black holes. arXiv:1604.04987 [gr-qc]
17. K. Lochan, T. Padmanabhan, Extracting information about the initial state from the black hole radiation. Phys. Rev. Lett. **116**(5), 051301 (2016)
18. S.D. Mathur, The Information paradox: a pedagogical introduction. Classic. Quantum Gravity **26**, 224001 (2009)
19. S.K. Modak, L. Ortíz, I. Peña, D. Sudarsky, Non-paradoxical loss of information in black hole evaporation in a quantum collapse model. Phys. Rev. D **91**(12), 124009 (2015)
20. S.K. Modak, L. Ortíz, I. Peña, D. Sudarsky, Black hole evaporation: information loss but no paradox. Gen. Rel. Grav. **47**(10), 120 (2015)
21. E. Okon, D. Sudarsky, Benefits of objective collapse models for cosmology and quantum gravity. Found. Phys. **44**, 114–143 (2014)
22. E. Okon, D. Sudarsky, The black hole information paradox and the collapse of the wave function. Found. Phys. **45**, 461–470 (2015)
23. P. Pearle, Reduction of the state vector by a nonlinear Schrodinger equation. Phys. Rev. D **13**, 857 (1976)
24. P. Pearle, Towards explaining why events occur. Int. J. Theor. Phys. **18**, 489 (1979)
25. P. Pearle, Combining stochastic dynamical state-vector reduction with spontaneous localization. Phys. Rev. A **39**, 2277–2289 (1989)
26. P. Pearle, Collapse models. arXiv: quant-ph/9901077
27. P. Pearle, Collapse Miscellany. Contributed in 80th birthday Issue of Yakir Aharanov. arXiv:1209.5082 [quant-ph]
28. P. Pearle, Toward a relativistic theory of statevector reduction, in *Sixty-Two Years of Uncertainty*, ed. by A. Miller (Plenum, New York, 1990), pp. 193–214
29. P. Pearle, A relativistic dynamical collapse model. arXiv:1412.6723 [quant-ph]
30. R. Penrose, Newton, quantum theory and reality, in *Three Hundred Years of Gravitation*, ed. by S.W. Hawking, W. Israel (Cambridge University Press, New York, 1989), p. 17
31. R. Penrose, On gravity's role in quantum state reduction. General Relativ. Gravit. **28**, 581 (1996)
32. L. Susskind, L. Thorlacius, J. Uglum, The stretched horizon and black hole complementarity. Phys. Rev. D **48**, 3743 (1993)
33. R. Tumulka, A relativistic version of the Ghirardi–Rimini–Weber model. J. Stat. Phys. **125**, 821 (2006)
34. R. Tumulka, On spontaneous wave function collapse and quantum field theory. Proc. R. Soc. A **462**, 1897 (2006)
35. W.G. Unruh, R.M. Wald, On evolution laws taking pure states to mixed states in quantum field theory. Phys. Rev. D **52**, 2176–2182 (1995)

Relativistic Paths: A Feynman Problem

Jayant V. Narlikar

Abstract This paper describes the solution of a problem posed by R.P. Feynman in his lectures on path integrals in quantum mechanics. The problem discusses the motion of a relativistic electron. It is shown how the various familiar features of such an electron are captured by the path integrals of relativistic spinning particles.

1 Preamble

It is a pleasure to contribute this article to the proposed Paddy-60 volume. I had the privilege of being Padmanabhan's research guide as he embarked on his Ph.D., programme at the Tata Institute of Fundamental Research. At that stage I experienced how a teacher can learn from his student. In Paddy's case his strong points are an excellent understanding of the subject and ability to explain it to others. The present article arose out of an idea described in the book *Quantum Mechanics and Path Integrals* by R.P. Feynman and A. Hibbs. I trust that Paddy will appreciate the problem posed and the solution offered; more so since he used to play with path integrals in his early days of research.

2 Introduction

The path integral approach of Feynman [1] provides an elegant link between the classical and the quantum physics. This approach takes as its starting point, the classical action S describing the physical system. In general the state of the system can be described by a point in a suitably defined phase space. As the state changes,

This work in its original form was published in the Journal of the Indian Mathematical Society, **36**, 9, 1972.

J.V. Narlikar (✉)
Inter-University Centre for Astronomy and Astrophysics, Pune 411007, India
e-mail: jvn@iucaa.in

J.S. Bagla and S. Engineer (eds.), *Gravity and the Quantum*,
Fundamental Theories of Physics 187, DOI 10.1007/978-3-319-51700-1_19

the point moves along some path Γ. Although geometrically, any number of paths may be possible, classical physics requires the system to follow a unique path Γ_c, which satisfies the principle of stationary action:

$$\delta S = 0. \tag{1}$$

Here δS represents the change in S in going from Γ_c to another path in its neighbourhood. The uniqueness of Γ_c is responsible for the determinacy associated with classical physics.

In quantum physics (1) is replaced by a less definitive statement. Suppose the system is initially in a state described by point 1 and finally in state described by point 2. Consider all the geometrically possible paths from 1 to 2. Not all paths are equally important, however. For each path Γ we can compute the action functional $S(\Gamma)$. Now a quantum system is permitted to follow the path Γ, even though $\Gamma \neq \Gamma_c$. There is a well-defined probability amplitude that the system will follow the path Γ. Feynman gives a simple rule for computing the amplitude:

$$P(\Gamma) = (\text{Constant}) \exp\{iS(\Gamma)/\hbar\}, \tag{2}$$

where $\hbar$ is Planck's constant divided by 2π. The constant in (2) can be determined by normalizing probabilities. If we are only interested in the total amplitude that the system starts at 1 and end at 2, this is given by summing (2) over all permissible paths Γ. The final answer will depend on points 1 and 2 and may be written as $K(2, 1)$. Thus

$$K(2, 1) = \sum_{\Gamma} (\text{constant}) \exp(iS/\hbar) = \int \exp\{iS(\Gamma)/\hbar\} \mathcal{D}\Gamma, \tag{3}$$

where the summation over Γ is replaced by an integral since usually we are dealing with a continuum of uncountably many paths Γ. The constant factor has been absorbed in the measure of the path integral.[1]

The connection between the quantum and classical theories is now easy to establish. The latter follows from the former in the limit $\hbar \to 0$. When $\hbar \to 0$, then ratio $S/\hbar \to \infty$ in general, and the phase of the exponential exp $(iS/\hbar)$ changes rapidly even if S changes slowly from path to path. The path integral (3) may be approximated by the method of stationary phase, the significant contributions to $K(2, 1)$ coming from those paths for which $\delta S \approx 0$. In the limit $\hbar \to 0$ we finally arrive at a unique path Γ_c which satisfies (1).

This approach not only brings out the connection between the classical and the quantum physics, but it also throws light on why the principle of stationary action plays such an important part in the various branches of classical physics. The basic

[1]The definition of measure remains one of the difficult problems of path integral theory. Feynman was, however, able to arrive at important results without giving a precise general definition of measure.

idea described above had been qualitatively stated first by Dirac [2]. By giving it a quantitative form as in (3), Feynman was able to connect it up with the more conventional Schrödinger approach to quantum mechanics.

In spite of its advantages, and the many applications [cf. [3] for details], the path integral picture is not widely used in quantum theory. One reason is, that while it clarifies many of the conceptual difficulties of the non-relativistic theory, it is not so successful in describing the relativistic spin half particles. Since the concept of spin is lacking in the classical theory, and hence in the classical action, it is not obvious, how to define path amplitudes for spin half particles which satisfy the Dirac equation. In the present paper we shall discuss ideas which may lead to the solution of this formidable problem. To begin with, we shall consider a simplified problem, that of a free particle moving in one space + one time dimensions. In the rest of the paper we shall take $\hbar = 1$, and the velocity of light $c = 1$.

3 Motion in One Space + One-Time Dimensions

In their book *Quantum Mechanics and Path Integrals*, Feynman and Hibbs [3] discuss the motion of a Dirac particle in one space-like and one time-like dimension. Instead of giving a rule like (2) they give another which involves only those paths which are made of null segments. Briefly the rule may be described as follows.

Suppose the particle of mass m moves backward and forward in x-direction, starting at $x = 0$ at $t = 0$, and ending at $x = X$ at $t = T$, where $\mid X \mid \leq T$. Divide the interval $[0,T]$ into a large number n of small intervals of ε-duration, so that

$$n\varepsilon = T. \tag{4}$$

The particle is allowed to move only with the speed of light, so that if at the end of rth interval it is at X, then $\mid X_{r+1} - X_r \mid = \varepsilon$, for $1 \leq r < n$. Suppose in the entire interval $[0, T]$, the particle goes forward on n_1 occasions and backwards on n_2 occasions. Then

$$n_1 + n_2 = n = \frac{T}{\varepsilon}, n_1 - n_2 = \frac{X}{\varepsilon}. \tag{5}$$

A typical path, shown in Fig. 1, will therefore have null segments meeting in sharp corners. The amplitude for a path with R corners is given to be

$$(im\varepsilon)^R. \tag{6}$$

The propagator from [0, 0] to [X, T] is then obtained by summing over all paths an expression like (6). To avoid confusion we will use $\tilde{P}(\Gamma)$ for the 1 + 1 dimensional problem.

As mentioned above this rule appears to differ from that given in (2) in a radical way and seems to have an ad-hoc character about it. We shall, however, show how the connection between the two rules may be established, and how we can arrive at the Dirac propagator from this picture.

First, we note that just as $K(2, 1)$ in (3) is obtained from $P(\Gamma)$, we can relate $\tilde{P}(\Gamma)$ to $\tilde{K}(2,\ 1)$. This is done as follows. Divide the path Γ into a large number of small segments, denoting the intermediate points of division by $X_1, X_2, ..., X_{n-1}$, with X_0, X_n standing for the end points 1, 2. Consider the product

$$\prod_{r=1}^{n} \frac{1}{A_r} K(X_r, X_{r-1}), \tag{7}$$

where A_r represents some measure factor.[2] For suitably chosen A_r, the product (7) tends to $P(\Gamma)$ as $n \to \infty$, i.e. as the division becomes finer and finer. Thus we can build up $P(\Gamma)$ from a chain of K's. We can get back to $K(2,\ 1)$ by summing (7) over all paths, and using the property

$$K(X_{r+1}, X_{r_1}) = \int K(X_{r+1}, X_r) K(X_r, X_{r-1}) d^3 X_r, \tag{8}$$

where the integration is over the space coordinates x_r of X_r.

This method can be easily extended to the problem in question. We need $\tilde{K}(X_r, X_{r-1})$ for the case where X_r, X_{r-1} are close to each other. The $\tilde{K}(2, 1)$ in $1 + 1$ dimensions satisfies the inhomogeneous Dirac equation

$$\left[\gamma_4 \frac{\partial}{\partial t} - \gamma_a \frac{\partial}{\partial x} + im\right] \tilde{K} = \delta_2(x, t), \tag{9}$$

where, for convenience we have taken the coordinates of 1 at [0, 0] and of 2 at $[x, t]$. To solve (9) write

$$\tilde{K} = \left(\gamma_4 \frac{\partial}{\partial t} - \gamma_1 \frac{\partial}{\partial x} - im\right) \tilde{I}(x, t), \tag{10}$$

where

$$\frac{\partial^2 \tilde{I}}{\partial t^2} - \frac{\partial^2 \tilde{I}}{\partial x^2} + m^2 \tilde{I} = \delta(x)\delta(t). \tag{11}$$

In analogy with the non-relativistic case, we want a solution that vanishes for $t < 0$. This has been worked out in Appendix A. The result is

$$\tilde{I}(x, t) = \frac{1}{2}\theta(t)\theta(s^2) J_0(ms),\ \ s^2 = t^2 - x^2, \tag{12}$$

[2]This factor must have the dimensions (length)$^{-3}$ to make (7) dimensionless.

where θ is the Heaviside function and J_0 the Bessel function of order zero.

The $\tilde{K}(2, 1)$ obtained in this way satisfies a relation similar to (8):

$$\tilde{K}(X_{r+1}, X_{r-1}) = \int \tilde{K}(X_{r+1}, X_r)\gamma_4\tilde{K}(X_r, X_{r-1})d^3X_r \tag{13}$$

where the factor γ_4 is necessary to preserve spinor-covariance. For any path Γ, we can therefore form a product similar to that in (7), but with γ_4 appearing between successive factors.

In forming this product we need $\tilde{K}(X_r, X_{r-1})$ when the points are close together. Since the only length scale appearing in this problem is m^{-1}, we need the approximate form of $\tilde{K}$ for $| X_r - X_{r-1} | \ll m^{-1}$. We therefore look at $\tilde{K}$ given by (10) and (12) in the case where $0 < t = \varepsilon$ and $m\varepsilon \ll 1$. Since $s < \varepsilon$, we have $ms \ll 1$ and $J_0(ms) \approx 1$.

$$\begin{aligned}\tilde{K} &\approx \left(\gamma_4\frac{\partial}{\partial t} - \gamma_1\frac{\partial}{\partial x} - im\right)\left[\frac{1}{2}\theta(t+x) - \frac{1}{2}\theta(x-t)\right] \\ &= \frac{1}{2}(\gamma_4 - \gamma_1)\delta(t+x) + \frac{1}{2}(\gamma_4 + \gamma_1)\delta(t-x) - \frac{im}{2}[\theta(t+x) - \theta(x-t)]\end{aligned} \tag{14}$$

The two delta functions in (14) indicate that most of the amplitude is concentrated in the two directions $x = t = \varepsilon$, $x = -t = -\varepsilon$. To obtain the magnitude of this concentration we integrate K over $0 \leq x < \infty$ and over $-\infty < x \leq 0$ at $t = \varepsilon$. We get respectively

$$\tilde{P}_+ = \frac{1}{2}(\gamma_4 + \gamma_1) - \frac{im\varepsilon}{2}, \tilde{P}_- = \frac{1}{2}(\gamma_4 - \gamma_1) - \frac{im\varepsilon}{2}. \tag{15}$$

Although the $-im\varepsilon/2$ term really represents amplitude over $0 \leq x \leq \varepsilon$, we may lump it all at the end $x = \varepsilon$ and call $\tilde{P}_+$ as the amplitude along $x = +\varepsilon$. $\tilde{P}_-$ similarly represents amplitude along $x = -\varepsilon$. Since $m\varepsilon \ll 1$, the error involved is slight.

We now have passed from a continuous set of paths to a discrete set, as visualized in the beginning of this section. A typical path is made up of null segments like $x = \pm t$, and the amplitude along such a path is given by a series of factors $\tilde{P}_+$, $\tilde{P}_-$ with γ_4 in between. The following types of combinations would appear in a typical product:

$$\tilde{P}_+\gamma_4\tilde{P}_+, \tilde{P}_-\gamma_4\tilde{P}_-, \tilde{P}_+\gamma_4\tilde{P}_-, \tilde{P}_-\gamma_4\tilde{P}_+. \tag{16}$$

From (15) we get to order $(m\varepsilon)$,

$$\begin{gathered}\tilde{P}_+\gamma_4\tilde{P}_+ = \tilde{P}_+, \ \tilde{P}_-\gamma_4\tilde{P} = \tilde{P}_-, \\ \tilde{P}_+\gamma_4\tilde{P}_- = (-im\varepsilon)\gamma_4 P_-, \ \tilde{P}_-\gamma_4\tilde{P}_+ = (-im\varepsilon)\gamma_4\tilde{P}_+.\end{gathered} \tag{17}$$

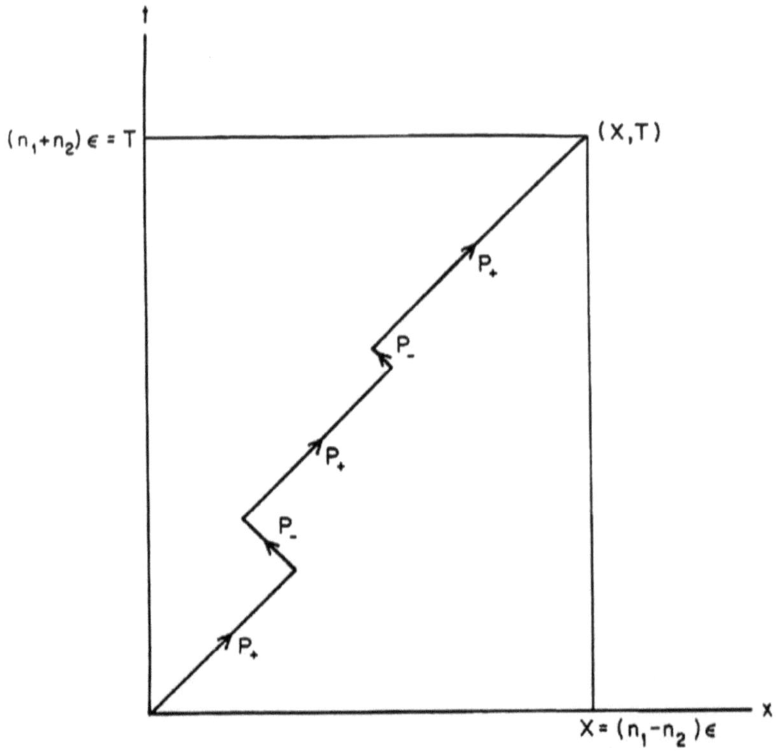

Fig. 1 Typical path of a particle moving backward and forward with the velocity of light.

Thus whenever two consecutive segments are in the same direction the amplitude is unaffected. When they are in opposite direction, we get a product $(-im\varepsilon)\gamma_4$). This explains why the rule given earlier in this section made use of paths with corners. The path described in Fig. 1 has the amplitude

$$\begin{aligned}\tilde{P}_+\gamma_4\tilde{P}_-\gamma_4\tilde{P}_+\gamma_4\tilde{P}_-\gamma_4\tilde{P}_+ &= (-im\varepsilon)\gamma_4\tilde{P}_-\gamma_4\tilde{P}_+\gamma_4\tilde{P}_-\gamma_4\tilde{P}_+\\ &= (-im\varepsilon)^2\gamma_4\cdot\gamma_4\tilde{P}_+\gamma_4\tilde{P}_-\gamma_4\tilde{P}_+\\ &= (-im\varepsilon)^3\gamma_4\tilde{P}_-\gamma_4\tilde{P}_+\\ &= (-im\varepsilon)^4\tilde{P}_+. \end{aligned}\tag{18}$$

Using the product rules (17) it is easy to see that the paths can be divided into four classes $\{\Gamma_{++}\}$, $\{\Gamma_{--}\}$, $\{\Gamma_{+-}\}$, $\{\Gamma_{-+}\}$. The amplitude for a Γ_{++} path begins with P_+ and ends with P_+. The others are similarly defined. The Γ_{++} and Γ_{--} paths have even number of corners whereas Γ_{+-}, Γ_{-+} paths have odd number of corners. To compute the total amplitude we need the total number of paths with a given number of corners.

Let $N_{++}(2R)$, $N_{--}(2R)$ respectively denote the number of Γ_{++} and Γ_{--} paths with $2R$ corners. Similarly let $N_{+-}(2R+1)$ and $N_{-+}(2R+1)$ denote the number of Γ_{+-} and Γ_{-+} paths with $2R+1$ corners. Then the total amplitude along paths with $R \gg 1$ corners is given by

$$\begin{aligned}\tilde{Q} &= \sum_{R\geq 1}\{N_{++}(2R)\tilde{P}_{+} + N_{--}(2R)\tilde{P}_{-}\}(-im\varepsilon)^{2R} \\ &+ \sum_{R\geq 0}\{N_{+-}(2R+1)\gamma_4\tilde{P}_{-} + N_{-+}(2R+1)\gamma_4\tilde{P}_{+}\}(-im\varepsilon)^{2R+1}.\end{aligned} \tag{19}$$

We are interested in $\tilde{Q}$ as $n \to \infty$, $\varepsilon \to 0$. However, to obtain the propagator from $[0,0]$ to $[X,T]$ we must divide Q by 2ε, since the above amplitude is distributed over an interval $\pm\varepsilon$ about $[X,T]$. Also, we must add the contribution from paths with *no* corners. We shall perform this calculation now. In the limit $n \to \infty$ we have

$$\begin{aligned}N_{++}(2R) &= \frac{(n_1-1)!(n_2-1)!}{(n_1-R)!(n_2-R+1)!R!(R-1)} \\ &\sim \frac{n_1^R n_2^{R-1}}{R!(R-1)!} = \frac{(T^2-X^2)^{R-1}(T+X)}{(2\varepsilon)^{2R-1}} \cdot \frac{1}{R!(R-1)!}, \\ N_{--}(2R) &\sim \frac{(T^2-X^2)^{R-1}(T-X)}{(2\varepsilon)^{2R-1}} \cdot \frac{1}{R!(R-1)!}, \\ N_{+-}(2R+1) &= \frac{n_1!n_2!}{(n_1-R)!(n_2-R)!R!R!} \\ &\sim \frac{n_1^R n_2^R}{R!R!} \sim \frac{(T^2-X^2)^R}{(2\varepsilon)^{2R}} \cdot \frac{1}{R!R!}, \\ N_{-+}(2R+1) &\sim \frac{(T^2-X^2)^R}{(2\varepsilon)^{2R}} \cdot \frac{1}{R!R!}.\end{aligned} \tag{20}$$

Using these approximations, and the following power series expansions for Bessel functions

$$J_0(\xi) = \sum_{R\geq 0} \frac{(-1)^R(\xi/2)^{2R}}{R!R!}, \quad J_0'(\xi) = \sum_{R\geq 1} \frac{\xi}{2} \cdot \frac{(-1)^R(\xi/2)^{2R-2}}{R!(R-1)!}, \tag{21}$$

we get

$$\tilde{Q} = \varepsilon\left[mJ_0'(mS) \cdot \frac{\gamma_4 T + \gamma_1 X}{S} - im\{J_0(mS) - 1\}\right], \tag{22}$$

where

$$S^2 = T^2 - X^2. \tag{23}$$

Equation (22) may be rewritten in the form

$$\tilde{Q} = 2\varepsilon\left(\gamma_4\frac{\delta}{\delta T} - \gamma_1\frac{\delta}{\delta X} - im\right)\left[\frac{1}{2}\{J_0(mS) - 1\}\right]. \tag{24}$$

Dividing by 2ε and adding (14) for $x = X$, $t = T$ as the contribution for paths with no corners, we get for $T > 0$

$$\tilde{K}[X, T; 0, 0] = \left(\gamma_4\frac{\delta}{\delta T} - \gamma_1\frac{\delta}{\delta X} - im\right)\left\{\frac{1}{2}j_0(mS)\theta(S^2)\right\}. \tag{25}$$

Thus we have a self-consistent picture in which the finite amplitude along a path can be built out of a series of infinitesimal propagators and the finite propagator is then obtained by summing the amplitude over all paths. The main difference between the relativistic and the non-relativistic case is that in the former case paths making a significant contribution to the amplitude are built out of null segments. In the latter case this is not so. *The relativistic picture is consistent with the fact that the eigenvalue of velocity of a Dirac particle is always* ± 1.

4 Motion in 3 + 1 Dimensions

The above picture can be generalized to 3+1 dimensions. Given a path Γ we define amplitude along it in terms of a chain of infinitesimal propagators. The propagator is given by the retarded solution of the inhomogeneous Dirac equation[3]

$$(\bar{\nabla}_2 + im)K(2, 1) = \delta_4(2, 1) \tag{26}$$

where $\bar{\nabla}_2$ is with respect to the coordinates of point 2. As in the 1+1 dimensional case we can write

$$K(2, 1) = (\bar{\nabla}_2 - im)I(2, 1) \tag{27}$$

where

$$(\Box_2 + m^2)I(2, 1) = \delta_4(2, 1). \tag{28}$$

The retarded solution for $I(2, 1)$ is

$$I(2, 1) = \frac{\theta(t_2 - t_1)}{2\pi}\left[\delta(S_{21}^2) - \frac{m}{2S_{21}}J_1(mS_{21})\theta(S_{21}^2)\right], \tag{29}$$

[3]For a vector A_i define $\bar{A}$ as $\gamma^i A_i$.

where S_{21}^2 is the square of the invariant distance between the coordinates $(\mathbf{x}_1, t_1)$, $(\mathbf{x}_2, t_2)$ of points 1 and 2. J_1 is the Bessel function of order 1.

The delta-function in (29) again emphasizes the importance of null directions. As in the 1+1 dimensional case we expect the 'important' paths to be made up of null segments. However, the null directions from a given point are not just two, but uncountably infinite. Hence it is not possible to look at the 3+1 case in terms of counting paths with corners. The main feature of the problems can, however, be described in terms of a perturbation expansion (cf. [4] for details). We write

$$K(2,1) = \sum_{n \geq 0} K^{(n)}(2,1), \tag{30}$$

where

$$\bar{\nabla} K^{(0)}(2,1) = \delta_4(2,1), \tag{31}$$

and for $n \geq 1$

$$\bar{\nabla} K^{(n)}(2,1) = -imK^{(n-1)}(2,1). \tag{32}$$

(31) can be solved in terms of the leading term of (29):

$$K^{(0)}(2,1) = \bar{\nabla} I^{(0)}(2,1), I^{(0)}(2,1) = \theta(t_2 - t_1)\frac{\delta(S_{21}^2)}{2\pi}. \tag{33}$$

A general $K^{(n)}(2,1), n \geq 1$ is given by

$$n = 2r : K^{(2r)} = (-im)^{2r}\bar{\nabla}_2 \int \dots \int I^{(0)}(2, P_r) I^{(0)}(P_r, P_{r-1}) \dots \\ \dots I^{(0)}(P_1, 1) d\tau_1 \dots d\tau_r, \tag{34}$$

$$n = 2r+1 : K^{(2r+1)} = (-im)^{2r+1} \int \dots \int I^{(0)}(2, P_r) I^{(0)}(P_r, P^{r-1}) \dots \\ \dots I^{(0)}(P_1, 1) d\tau_1 \dots d\tau_r, \tag{35}$$

where r is an integer. A typical $I^{(0)}$ describes propagation along a null segment, and (34), (35) represent summations over paths made up of null segments. An amplitude $(-im)^n$ is associated with the summation for $K^{(n)}$. This is the analogue of the $1+1$ dimensional case.

The propagator obtained so far is useful only when considered along with the hole theory. This is because it describes only forward propagation of particles of positive and negative energies. To describe electrons and positrons we need the Feynman propagator

$$K(2,1) = (\bar{\nabla} - im)I_+(2,1) = (\bar{\nabla} - im) \times \\ \times \left[\frac{\delta(S_{21}^2)}{4\pi} - \frac{m}{8\pi S_{21}} H_1^{(2)}(mS_{21}) \right], \tag{36}$$

where $H_1^{(2)}$ is the Hankel function of the second kind. It was shown in an earlier paper [5] how K_+ arises from $K(2, 1)$ when we take into account the electromagnetic interaction of electrons and positrons. We will therefore not go into those details here.

5 The Non-relativistic Approximation

The relativistic propagator described above has been obtained by summing amplitude over piecewise continuous paths made up of null segments. The non-relativistic propagator, on the other hand, is obtained by summing over all paths from 1 to 2 which always go forward in time. Also, the amplitude in the latter case is given by (2), *i.e.* by

$$P(\Gamma) = (\text{constant}) \exp\left[+i \int_{\Gamma} \frac{1}{2} m\dot{x}^2 dt \right] \tag{37}$$

for a free particle. It is therefore not clear how the non-relativistic case can be obtained from the relativistic one by a suitable approximation. In this section we show how this transition may be made.

First we obtain the non-relativistic forms for the propagators $K(2, 1)$ and $K_+(2, 1)$. For convenience we write

$$T = t_2 - t_1, \quad \mathbf{X} = \mathbf{x}_2 - \mathbf{x}_1, \quad X = \mid \mathbf{X} \mid . \tag{38}$$

The non-relativistic approximation is given by

$$T \gg X, mS_{21} \gg 1. \tag{39}$$

We therefore use the asymptotic formulae for J_1 and $H_1^{(2)}$:

$$J_1(mS_{21}) \sim \left(\frac{\pi m S_{21}}{2} \right)^{-1/2} \cos\left(mS_{21} - \frac{3\pi}{4} \right), \tag{40}$$

$$H_1^{(2)}(mS_{21}) \sim \left(\frac{\pi m S_{21}}{2} \right)^{-1/2} \exp\left\{ -i\left(mS_{21} - \frac{3\pi}{4} \right) \right\}, \tag{41}$$

and also use the approximations for S_{21}:

$$S_{21} \approx T, S_{21} \approx T - \frac{X^2}{2T} \tag{42}$$

respectively in the first and second factors of (40) and (41).

It is convenient to use the Dirac representation

$$\gamma_r = \begin{pmatrix} 1 & 0 \\ 0 & -1 \end{pmatrix}, \gamma = \begin{pmatrix} 0 & \sigma \\ -\sigma & 0 \end{pmatrix} \tag{43}$$

and write $K(2, 1)$, $K_+(2, 1)$ explicitly in matrix form. [The $\sigma = \{\sigma_1, \sigma_2, \sigma_3\}$ is the set of Pauli matrices.] We then have

$$K(2,1) \sim \left(\frac{m}{2\pi iT}\right)^{3/2} \times \begin{bmatrix} i^{-3/2}\exp\left(-mT + \frac{mX^2}{2T}\right)i & -\frac{\sigma\cdot\mathbf{x}}{T}\sin\left(mT - \frac{mX^2}{2T} - \frac{3\pi}{4}\right) \\ \frac{\sigma\cdot\mathbf{x}}{T}\sin\left(mT - \frac{mX^2}{2T} - \frac{3\pi}{4}\right) & -(-i)^{-3/2}\exp\left(mT - \frac{mX^2}{2T}\right)i \end{bmatrix} \tag{44}$$

and

$$K_+(2,1) \sim \begin{bmatrix} 1 & -\frac{\sigma\cdot\mathbf{x}}{2T} \\ \frac{\sigma\cdot\mathbf{x}}{2T} & 0 \end{bmatrix} \cdot \left(\frac{m}{2\pi iT}\right)^{3/2} \exp - i\left(mT - \frac{mX^2}{2T}\right), \tag{45}$$

for $T > 0$. $K_+(2, 1)$ for $T < 0$ can be written down similarly, while $K(2, 1)$ for $T < 0$ is zero.

The non-relativistic form of the Dirac equation separates the wave-function into a large part and a small part. The large part is propagated essentially by the top left-hand element of the propagator. This, we see, is

$$\left(\frac{m}{2\pi iT}\right)^{3/2} \exp - i\left(mT - \frac{mX^2}{2T}\right). \tag{46}$$

Equation (46) is just the non-relativistic propagator for Schroödinger equation.

We now give a rule for computing path amplitudes which leads to $K(2, 1)$ or $K_+(2, 1)$. The rule is obtained in the following way. The action for a relativistic particle is given by

$$S = -\int_\Gamma mds, \tag{47}$$

where

$$ds^2 = dt^2 - dx^2. \tag{48}$$

If we use (47) in (2), we will not arrive at a description of the Dirac particle - because (47) does not contain spin. To include spin we have to take the square root of (48) in the space of 4 × 4 matrices:

$$dt^2 - dx^2 = (\gamma_4 dt - \gamma \cdot d\mathbf{x})^2. \tag{49}$$

This is analogous to

$$\Box \equiv \bar{\nabla}^2 \tag{50}$$

which led to the Dirac equation. Thus (47) is replaced by

$$S = -\int_{\Gamma} m(\gamma_4 - \gamma \cdot \mathbf{x})dt. \tag{51}$$

In the non-relativistic approximation Γ will be an arbitrary path going forward in time.

However, the amplitude along Γ is not given by

$$P(\Gamma) = (\text{constant}) \cdot \exp\{-i \int m(\gamma_4 - \gamma \cdot \mathbf{x})dt\}. \tag{52}$$

The reason is that the right-hand side of (52) is independent of the path and depends only on end points. Such a prescription will not lead to a satisfactory quantum theory. Instead, we need a path-dependent amplitude. To achieve this we divide the path Γ into a large number of small segments and use (52) for each small segment. The $P(\Gamma)$ is then obtained by the ordered product of the amplitudes along the segments. If we write $\Gamma = \Gamma_1 + \Gamma_2 + \ldots + \Gamma_n$, and Γ_r, is a typical segment, then

$$P(\Gamma) = \prod_r P(\Gamma_r) = \prod (\text{constant}) \cdot \exp \cdot \{-i \int_{\Gamma r} m(\gamma_4 - \gamma \cdot x)dt\}. \tag{53}$$

Since for matrices A, B, the law

$$\exp\,(A + B) = (\exp A) \cdot (\exp B) \tag{54}$$

does not hold, the expression (53) as $n \to \infty$ is dependent on the particular path Γ.

It should be emphasized that the structure associated with this law is cruder than that discussed in the earlier section. The path Γ here can be approximated by another made up of null segments on a much finer scale. The amplitude along Γ is thus the sum of all such finer scale paths computed according to the last section. The above rule is at best an approximation that will be shown to work well.

We shall take Γ to be between $0 \leq t \leq T$ and let $P(t)$ denote the amplitude along the section of the path from 0 to t. Then it is easy to see that (53) corresponds to

$$\frac{dP}{dt} = -im(\gamma_4 - \gamma \cdot \dot{\mathbf{x}})P \tag{55}$$

where $\mathbf{x}(t)$ is the position of a typical point on Γ. The constant in (53) is taken to be unity.

It is easy to verify that (52) is not an integral of (55) unless x is constant. In that case the law (54) holds. We shall return to this special case later. We write P as a 4×4 matrix:

$$P = \begin{bmatrix} P_{11} \; P_{12} \\ P_{21} \; P_{22} \end{bmatrix} \tag{56}$$

where $P_{ij}(i, j = 1, 2)$ are 2×2 matrices. Using the representation (43) and writing $\mathbf{v} = \dot{\mathbf{x}}$, (55) takes the form

$$\dot{P}_{11} = -im(P_{11} - \sigma \cdot \mathbf{v}P_{21}), \dot{P}_{21} = im(P_{21} - \sigma \cdot \mathbf{v}P_{11}), \tag{57}$$

$$\dot{P}_{12} = -im(P_{12} - \sigma \cdot \mathbf{v}P_{22}), \dot{P}_{22} = im(P_{22} - \sigma \cdot \mathbf{v}P_{12}). \tag{58}$$

Initially we take $P_{11} = 1$. The initial values of P_{12}, P_{21}, P_{22} also need to be specified in order to complete the problem. It turns out that these are crucial in determining whether we finally arrive at (44) or (45). We shall settle this question at a later stage. To solve (57), put

$$\xi = P_{11}e^{imt}, \eta = P_{21}e^{-imt}. \tag{59}$$

Then we get

$$\dot{\xi} = ime^{2imt}(\sigma \cdot \mathbf{v})\dot{\eta}, = -ime^{2imt}(\sigma \cdot \mathbf{v})\xi. \tag{60}$$

Equation (60) can be solved in terms of the expansions

$$\xi = \sum_{n=0}^{\infty} \xi_{2n}, \eta = \sum_{n=0}^{\infty} \eta_{2n+1}, \tag{61}$$

where $\xi_0 = 1$ and for $n \geq 1$,

$$\dot{\xi}_{2n} = ime^{2imt}(\sigma \cdot \mathbf{v})\eta_{2n-1}, \tag{62}$$

$$\dot{\eta}_{2n+1} = -ime^{-2imt}(\sigma \cdot \mathbf{v})\xi_{2n-2}. \tag{63}$$

Given the initial conditions, and the function $v(t)$, we can solve (62), (63) by iteration.

In the non-relativistic approximation we have

$$|\mathbf{v}| \ll 1, |\dot{\mathbf{v}}| \ll m |\mathbf{v}|. \tag{64}$$

The first inequality implies that motion is slow compared with the speed of light. The second inequality means that the time scale over which velocity changes significantly is large compared to m^{-1}. The latter inequality suggests that the solution of (57) and (58) will be somewhat similar to that for $\mathbf{v}$ = constant. In this case the general solution of (55) is given by

$$P(t) = \exp\{-im(\gamma_4 t - \gamma\cdot\mathbf{x})\}.P_0 \tag{65}$$

where P_0 is an arbitrary matrix. Taking $P_0 = 1$ gives

$$P = \begin{Bmatrix} \cos mt\sqrt{(1-v^2)} - \frac{i\sin mt\sqrt{(1-v^2)}}{\sqrt{(1-v^2)}} & \frac{i\sigma\cdot\mathbf{v}}{\sqrt{(1-v^2)}}\sin mt\sqrt{(1-v^2)} \\ -\frac{i\sigma\cdot\mathbf{v}}{\sqrt{(1-v^2)}}\sin mt\sqrt{(1-v^2)} & \cos mt\sqrt{(1-v^2)} + \frac{i\sin mt\sqrt{(1-v^2)}}{\sqrt{(1-v^2)}} \end{Bmatrix} \tag{66}$$

However, this choice of initial conditions does not give either of K or K_+. As will be seen shortly, to obtain K we need

$$P_0 = \frac{\gamma_4 - \gamma\cdot\mathbf{v}}{\sqrt{(1-v^2)}}, \tag{67}$$

whereas to obtain K_+, P_0 is given by

$$P_0 = \frac{1}{2}\left\{1 + \frac{\gamma_4 - \gamma\cdot\mathbf{v}}{\sqrt{(1-v^2)}}\right\}. \tag{68}$$

We shall consider these two cases in detail.

In the first case (65) and (67) give, when $|v| \ll 1$,

$$P_{11} = \frac{\cos mt\sqrt{(1-v^2)}}{\sqrt{(1-v^2)}} - i\sin mt\sqrt{(1-v^2)} \sim \exp[-imt\sqrt{(1-v^2)}] \tag{69}$$

$$P_{21} = \frac{\sigma\cdot\mathbf{v}}{\sqrt{(1-v^2)}}\cos mt\sqrt{(1-v^2)} \sim \sigma\cdot\mathbf{v}\cos mt\sqrt{(1-v^2)}. \tag{70}$$

We now turn to the solution of (57) in the non-relativistic case. Guided by (69) and (70), but remembering that v now varies with t, we try

$$P_{11} \sim \exp - im \int_0^t \left\{1 - \frac{v^2(t_1)}{2}\right\} dt_1,$$

$$P_{21} \sim \sigma \cdot \mathbf{v}(t) \cos m \int_0^t \left[1 - \frac{v^2(t_1)}{2}\right] dt_1. \tag{71}$$

It is easy to verify that (57) is satisfied to within the non-relativistic approximation. P_{22}, P_{12} can be similarly obtained and we get

$$P(t) = \begin{Bmatrix} \exp - im \int_0^t \left\{1 - \frac{v^2(t_1)}{2}\right\} dt_1 & \sigma \cdot \mathbf{v}(t) \cos m \int_0^1 \left\{1 - \frac{v^2(t_1)}{2}\right\} dt_1 \\ \sigma \cdot \mathbf{v}(t) \cos m \int_0^t \left\{1 - \frac{v^2(t_1)}{2}\right\} dt_1 & - \exp im \int_0^t \left\{1 - \frac{v^2(t_1)}{2}\right\} dt_1 \end{Bmatrix}. \tag{72}$$

In the same way we can deal with the second case, and get

$$P_+(t) = \begin{Bmatrix} 1 & -\frac{1}{2}\sigma \cdot \mathbf{v}(t) \\ \frac{1}{2}\sigma \cdot \mathbf{v}(t) & 0 \end{Bmatrix} \cdot \exp - im \int_0^t \left\{1 - \frac{v^2(t_1)}{2}\right\} dt_1, t > 0. \tag{73}$$

Here we have written P_+ instead of P to distinguish between the two cases.

The propagator K or K_+ would now come out of summation of $P(T), P_+(T)$ over all paths from $(\mathbf{0}, 0)$ to $(\mathbf{X}, T)$. We already know (cf. [3] for details) that the path integral

$$\int \exp - im \int_0^T \left\{1 - \frac{\mathbf{v}^2(t_1)}{2}\right\} dt \mathscr{D}^3\mathbf{x}(t) = \left(\frac{m}{2\pi iT}\right)^{3/2} \exp \left\{\frac{imX^2}{2T} - imT\right\}. \tag{74}$$

The following results have been derived in Appendix B:

$$\int \mathbf{v}(T) \exp -im \int_0^T \{1 - \frac{\mathbf{v}^2(t_1)}{2}\Big\} dt \mathscr{D}^3\mathbf{x}(t)$$
$$= \frac{\mathbf{X}}{T}\left(\frac{m}{2\pi iT}\right)^{3/2} \exp \left\{\frac{imX^2}{2T} - im\,T\right\}, \tag{75}$$

and

$$\int \mathbf{v}(T) \cos m \int_0^T \{1 - \frac{\mathbf{v}^2(t_1)}{2}\Big\} dt \mathscr{D}^3\mathbf{X}(t)$$
$$= \left(\frac{m}{2\pi T}\right)^{3/2} \frac{\mathbf{X}}{T} \sin\left(mT - \frac{mX^2}{2T} - \frac{3\pi}{4}\right). \tag{76}$$

Using these results it is easy to see that

$$K(2, 1) = \int P \cdot \mathscr{D}^3\mathbf{X}(t), \tag{77}$$

$$K_+(2, 1) = \int P_+ \mathscr{D}^3\mathbf{X}(t). \tag{78}$$

We therefore see that the rule for path amplitude given in (53) leads to the correct propagator. To obtain $K(2, 1)$ we use (53) only for forward going paths, with P_0, the initial value given by (67). For $K_+(2, 1)$ we must use (53) for forward as well as backward going paths but with initial condition given by (68). In the non-relativistic case all paths are time like and no problems such as given by pair creation or annihilation are present. To deal with such problems, which arise frequently in electrodynamics, we must use the methods of [5].

6 Conclusion

To summarize, the motion of a Dirac particle may be looked at from two different ends. In the extreme relativistic limit, the velocities are comparable to the velocity of light c, and time scales short compared to m^{-1}. Here the motion is described by paths made of null segments, with many changes of direction occurring in time m^{-1}. The mass of the particle is responsible for changing the direction from one null segment to another. Because of any such changes, the motion of a forward going particle is time like over times large compared to m^{-1}.

In the non-relativistic limit we are concerned with this type of motion. Here paths are time like and do not change directions significantly over times of order m^{-1}. The amplitude in this case can be described by a relatively simple expression of the form $\exp(-im\bar{q})$ where q^i denotes a small section of the path (compared to m^{-1}). The amplitude along a finite section of the path is given by a product of such factors in the correct order. The sum of amplitudes over all paths leads to the non-relativistic limit of the Dirac or Feynman propagator, *including spin*.

It is instructive to show explicitly the part played by the second inequality of (64), in the non-relativistic limit. This is seen from the iterative solution described in the last section. To fix ideas we will take the initial condition given by $P_0 = 1$. This corresponds to $\eta_1 = 0$ at $t = 0$. Hence for $n \geq 1$

$$\xi_{2n}(t) = m^{2n} \int_0^t e^{2imt_1}\sigma \cdot \mathbf{v}(t_1)dt_1 \int_0^t e^{-2imt_2}\sigma \cdot \mathbf{v}(t_2)dt_2 \ldots$$
$$\ldots \int_0^{t_{2n-2}} e^{2imt_{2n-1}}\sigma \cdot \mathbf{v}(t_{2n-1})dt_{2n-1} \int_0^{t_{2n-1}} e^{-2imt_{2n}}\sigma \cdot \mathbf{v}(t_{2n})dt_{2n}, \tag{79}$$

$$\eta_{2n}(t) = -im^{2n-1} \int_0 e^{-2imt_1}\sigma \cdot \mathbf{v}(t_1)dt_1 \ldots \int_0^{t_{2n-2}} \sigma \cdot \mathbf{v}(t_{2n-1})e^{-2imt_{2n-1}}dt_{2n-1}. \tag{80}$$

Consider the last two integrals of (79) taken together. If we integrate by parts,

$$\int_0^{t_{2n-1}} e^{-2imt_{2n}}\sigma \cdot \mathbf{v}(t_{2n})dt_{2n} = \frac{i}{2m}\left\{e^{-2imt_{2n-1}}\sigma \cdot \mathbf{v}(t_{2n-1}) - \sigma \cdot \mathbf{v}(0)\right\}$$
$$- \frac{i}{2m}\int_0^{t_{2n-1}} e^{-2imt_{2n}}\sigma \cdot \dot{\mathbf{v}}(t_{2n})dt_{2n}. \tag{81}$$

In this the integral on the right-hand side is less important than the first term because of the second inequality of (64). Further, when we take the first erm of the right-hand side along with the integrand of the t_{2n-1} integral we get

$$\int_0^{t_{2n-2}} \frac{i}{2m}, \left\{v^2(t_{2n-1}) - e^{2imt_{2n-1}}\left[\sigma \cdot \mathbf{v}(t_{2n-1})\right]\left[\sigma \cdot (0)\right]\right\}dt_{2n-1}. \tag{82}$$

Again, because of the rapid oscillations of $e^{2imt_{2n-1}}$ the second term of the integrand makes very little contribution and we can approximate (82) by

$$\frac{i}{2m}\int_0^{t_{2n-2}} v^2(t_{2n-1})dt_{2n-1}. \tag{83}$$

Clearly, we can repeat this procedure for the subsequent integrals in $\xi_{2n}(t)$ and get, with redefinition of dummy variables,

$$\xi_{2n}(t) \sim m^{2n} \cdot \left(\frac{i}{2m}\right)^n \int_0^t v^2(t_1)dt_1 \int_0^{t_1} v^2(t_2)dt_2 \ldots \int_0^{t_{n-1}} v(t_n)dt_n$$
$$= \left(\frac{im}{2}\right)^n \cdot \frac{1}{n!}\left\{\int_0^t v^2(t)dt\right\}^n \tag{84}$$

so that

$$P_{11} = e^{-imt} \sum_{0}^{\infty} \xi_{2n}(t)$$
$$\sim \exp\left[- im \int_0^t \left\{ 1 - \frac{v^2(t_1)}{2} \right\} dt_1 \right]. \tag{85}$$

We can evaluate P_{12} similarly. With suitable account of initial conditions (72) and (73) can be obtained in this way.

The example of the relativistic spin-half particle discussed here illustrates the way Feynman's path integral technique can be applied to relativistic particles. The eigenvalue of velocity of a relativistic Dirac particle has the magnitude c. This result forms the basis of the path integral approach to describe such a particle. The reader familiar with the path integral approach to quantum mechanics will appreciate the problem solved here as another example of how the path integral technique is easy to visualise but difficult to implement.

Appendix A

To obtain the solution of (11) that vanishes for $t < 0$, put

$$I(x, t) = \int\int f(\omega, k) e^{ikx - i\omega t} \cdot \frac{d\omega}{2\pi} \cdot \frac{dk}{2\pi}, \tag{A.1}$$

where $-\infty < \omega < \infty$, $-\infty < k < \infty$.

Since

$$\delta(x)\delta(t) = \int_{-\infty}^{\infty} \int_{-\infty}^{\infty} e^{ikx - i\omega t} \frac{d\omega}{2\pi} \frac{dk}{2\pi}, \tag{A.2}$$

we get from (11),

$$f(\omega, k) = \frac{1}{m^2 + k^2 - \omega^2}. \tag{A.3}$$

Hence

$$I(x, t) = \int\int \frac{e^{ikx - i\omega t}}{m^2 + k^2 - \omega^2} \frac{d\omega}{2\pi} \frac{dk}{2\pi}. \tag{A.4}$$

To perform the ω integral we use contour integration in the complex x-plane. The poles are at $\omega = \pm\sqrt{(m^2 + k^2)}$. To arrive at a solution which vanishes for $t < 0$, we integrate parallel to real axis with $\omega = \omega_R + i\varepsilon$, where ω_R is the real part of ω, and $-\infty < \omega_R < \infty$. For $t < 0$ we can complete the contour by a semicircle at infinity in the upper half of the ω-plane. This contour has no poles and the integral along the semicircle vanishes. For $t > 0$ the contour is completed by a semicircle at infinity in the lower half of the ω-plane. This contour has poles with the residue

$$\frac{e^{ikx+i\surd(m^2+k^2)t}}{2\surd(m^2+k^2)} - \frac{e^{ikx-i\surd(m^2+k^2)t}}{2\surd(m^2+k^2)}. \tag{A.5}$$

The integral along the semicircle vanishes. Hence we get

$$\begin{aligned} I(x,t) &= \frac{-i}{4\pi}\int_{-\infty}^{\infty} \frac{e^{ikx}}{\sqrt{(m^2+k^2)}} \left\{ e^{i\sqrt{(m^2+k^2)}t} - e^{-i\sqrt{(m^2+k^2)}t} \right\} dk \\ &= \frac{1}{\pi}\int_0^{\infty} \frac{\cos kx \sin\surd(m^2+k^2)t}{\surd(m^2+k^2)} dk. \end{aligned} \tag{A.6}$$

We now consider two cases separately: (i) $t^2 > x^2$ and (ii) $t^2 < x^2$. In the first case put

$$t = s\cosh\theta, x = s\sinh\theta, k = m\sinh\alpha. \tag{A.7}$$

Then

$$\begin{aligned} I(x,t) &= \frac{1}{\pi}\int_0^{\infty} \cos\{ms\sinh\theta\sinh\alpha\}\sin\{ms\cosh\theta\cosh\alpha\}d\alpha \\ &= \frac{1}{2\pi}\int_0^{\infty} [\sin\{ms\cosh(\theta+\alpha)\} + \sin\{ms\cosh(\theta-\alpha)\}]d\alpha. \end{aligned} \tag{A.8}$$

In the two terms of the integrand, put $\theta + \alpha = u, \theta - \alpha = u$ respectively to get

$$\begin{aligned} I(x,t) &= \frac{1}{2\pi}\int_{\theta}^{\infty} \sin(ms\cosh u)du + \frac{1}{2\pi}\int_{-\infty}^{+\theta} \sin(ms\cosh u)du \\ &= \frac{1}{2\pi}\int_{-\infty}^{\infty} \sin(ms\cosh u)du \\ &= \frac{1}{2}J_0(ms). \end{aligned} \tag{A.9}$$

In the second case put

$$t = s\sinh\theta, x = s\cosh\theta, k = m\sinh\alpha. \tag{A.10}$$

Then we get, by proceeding as above

$$
\begin{aligned}
I(x,t) &= \frac{1}{\pi}\int_0^{\infty} \cos\{ms \sinh\alpha \cosh\theta)\} \sin\{ms \cosh\alpha \sinh\theta\} d\alpha \\
&= \frac{1}{2\pi}\int_0^{\infty} [\sin\{ms \sinh(\theta+\alpha)\} + \sin ms\{\sinh(\theta-\alpha)\}] d\alpha \\
&= \frac{1}{2\pi}\int_\theta^{\infty} \sin(ms \sinh u) du + \frac{1}{2\pi}\int_{-\infty}^{0} \sin(ms \sinh u) du \\
&= \frac{1}{2\pi}\int_{-\infty}^{\infty} \sin(ms \sinh u) du \\
&= 0.
\end{aligned}
$$

Hence the result follows:

$$
I(x,t) = \frac{1}{2} J_0(ms)\theta(s^2).\theta(t). \tag{A.11}
$$

Appendix B

Here we derive the results quoted in Eqs. (75) and (76) of the main text. Consider a typical path $\mathbf{X}(t)$ from (0, 0) to (x, T). For small ε, suppose $x(t)$ intersects the hyperplane $t = T - \varepsilon$ at $\mathbf{y}$ in

$$
\mathbf{y} = \mathbf{x}(T-\varepsilon). \tag{B.1}
$$

We can approximate $\mathbf{v}(T)$ by

$$
\mathbf{v}(T) \approx \frac{\mathbf{x}-\mathbf{y}}{\varepsilon} \tag{B.2}
$$

Since

$$
\exp im \int_0^T \frac{\mathbf{v}^2(t)}{2} dt = \left\{\exp im \int_0^{T-\varepsilon} \frac{\mathbf{v}^2(t)}{2} dt\right\} \cdot \left\{\exp im \int_{T-\varepsilon}^{T} \frac{\mathbf{v}^2(t)}{2} dt\right\}, \tag{B.3}
$$

we have

$$
\begin{aligned}
&\int \mathbf{x}(T) \exp\left\{-im\int_0^T \left[1 - \frac{\mathbf{v}^2(t)}{2}\right] dt\right\} \mathscr{D}^3\mathbf{x}(t) \\
&= e^{-imT} \int \left(\frac{\mathbf{x}-\mathbf{y}}{\varepsilon}\right) \frac{1}{A} \cdot \exp\left[\frac{im(\mathbf{x}-\mathbf{y})^2}{2\varepsilon}\right] d^3y \\
&\quad \times \exp \int \left\{+im \int_0^{T-\varepsilon} \frac{\mathbf{v}^2(t)}{2} dt\right\} \mathscr{D}^3\mathbf{x}(t).
\end{aligned} \tag{B.4}
$$

The path integral from 0 to $T - \varepsilon$ gives the nonrelativistic free particle propagator, and (B.4) becomes

$$e^{imT}\int \frac{\mathbf{x}-\mathbf{y}}{\varepsilon}\cdot\frac{1}{A}\cdot\exp\left[\frac{im(\mathbf{x}-\mathbf{y})^2}{2\varepsilon}\right]\cdot\left(\frac{m}{2\pi i(T-\varepsilon)}\right)^{3}/2 \\ \times\exp\left[\frac{imy^2}{2(T-\varepsilon)}\right]d^3\mathbf{y}. \qquad \text{(B.5)}$$

In (B.4) and (B.5) A is the measure constant, and is given by (*cf*. Feynman and Hibbs [3])

$$A = \left(\frac{2\pi i\varepsilon}{m}\right)^{3/2}. \qquad \text{(B.6)}$$

Substituting for A, we can evaluate (B.5) to get

$$\left(\frac{m}{2\pi iT}\right)^{3/2}\cdot\frac{\mathbf{x}}{T}\cdot d^{-imT+imx^2/(2T)} \qquad \text{(B.7)}$$

This is the same as the right-hand side of (75).

Equation (76) is the real part of (75) and hence we need the real part of (B.7). Writing $i = \exp\left(\frac{i\pi}{2}\right)$, we get the real part of (B.7) as

$$Re\left[\left(\frac{m}{2\pi iT}\right)^{3/2}\cdot\frac{\mathbf{x}}{T}e^{-imT+imx^2/(2T)}\right] \\ = \left(\frac{m}{2\pi T}\right)^{3/2}\cdot\frac{\mathbf{x}}{T}\cos\left(mT-\frac{m\mathbf{x}^2}{2T}+\frac{3\pi}{4}\right) \\ = \left(\frac{m}{2\pi T}\right)^{3/2}\cdot\frac{\mathbf{x}}{T}\sin\left(mT-\frac{m\mathbf{x}^2}{2T}-\frac{3\pi}{4}\right) \qquad \text{(B.8)}$$

This is the required result.

References

1. R.P. Feynman, Rev. Mod. Phys. **20**, 367 (1948)
2. P.A.M. Dirac, *The Principles of Quantum Mechanics*, 2nd edn. (The Clarenden Press, Oxford, 1935)
3. R.P. Feynman, A.R. Hibbs, *Quantum Mechanics and Path Integrals* (McGraw-Hill, New York, 1965)
4. F. Hoyle, J.V. Narlikar, Il Nuovo Cimento **7A**, 242 (1972)
5. F. Hoyle, J.V. Narlikar, Ann. Phys. (N.Y.), **62**, 44 (1971)

Einstein Equations from/as Thermodynamics of Spacetime

Krishnamohan Parattu

Abstract There are two results in the literature that seem closely related: Padmanabhan's interpretation of field equations near a null surface as a thermodynamic identity and Jacobson's derivation of Einstein equations from the Clausius relation of thermodynamics. I compare and contrast these two results in the framework of Gaussian null coordinates near a null surface.

Prologue

A STORY .....

In the court of Akbar Badshah ('Badshah', loosely translated, means 'emperor'), there was a musician called Tansen. He used to enthrall everyone in Akbar's Court with his superb performances. Once, after such a rendition, Akbar started praising him sky-high and said, "There can be no-one else in this world who can sing so well". Tansen disagreed, saying he knows of a hermit who lives in the jungle on the banks of Yamuna river who is far superior and that Tansen himself has learnt music from him for sometime. Akbar, who could not believe this, wanted to listen to this hermit in order to judge for himself. Since the hermit did not want any publicity, it was decided that Tansen will take Akbar near the place where the hermit lived and they should listen to his music without creating any disturbance.

They set out one day and reached the jungle near the river Yamuna, where, at a distance, they saw the hermit's hut. As the sun was setting on Yamuna, with all Nature at peace, the hermit came out his hut, sat on a rock facing the river and started singing. Akbar could immediately see that this was music of a completely different class which Tansen could never produce.

On their way back, Akbar queried, "Tansen, you say he taught you music; clearly, he has held back some techniques from you".

"No", said Tansen. "I know all the technical aspects of music he does."

"But, Tansen, then how do you account for such difference in quality?"

"It is simple. He sings for Yamuna while I sing for Badshah".

Source- Paddy's website

My first encounter- 2nd year Undergraduate

K. Parattu (✉)
IUCAA, Pune, India
e-mail: mailofkrishnamohan@gmail.com

K. Parattu
Perimeter Institute, Waterloo, Canada

J.S. Bagla and S. Engineer (eds.), *Gravity and the Quantum*,
Fundamental Theories of Physics 187, DOI 10.1007/978-3-319-51700-1_20

1 Introduction

I shall refer to Prof. Padmanabhan as Paddy, as he is popularly known, in this article.[1]

Emergent gravity paradigm has been Paddy's main research interest for the past decade [19, 21, 23]. The main motivation for this program is the observation that quantities that look and feel very much like thermodynamic quantities keep popping up in gravity all the time. Hence, it natural to ask "if gravity is the thermodynamic limit of the statistical mechanics of certain microscopic degrees of freedom ('atoms of space') [21]". One of the main results of this research program is the demonstration that Einstein's equations when projected on a null surface can be interpreted as a thermodynamic identity [1, 2, 10–12, 18]. As Paddy states, "If gravity is thermodynamic in nature, then the gravitational field equations must be expressible in a thermodynamic language [21]." The above results are the concrete realizations of this expectation. There is another result in the literature that smells very similar. Twenty years back, Jacobson showed that enforcing a thermodynamic identity on a local causal horizon is equivalent to enforcing Einstein's equations [8]. This work has also been followed up [3–5, 7, 15]. The question is often asked: What is the relation between these two approaches? In fact, this question was posed to me during my PhD defence! It is the purpose of this article to answer this question by comparing and contrasting the two results. Let me state at the outset that the two results are not the same result interpreted in two different ways, but arise from two different components of the Einstein equations at the null surface [1, 2]. In particular, the $T\delta S$ terms that appear in the two results are not the same. In Jacobson's case, the change in entropy is as you move along the null geodesics *on* the null surface. In Paddy's case, the change is as you move along the null geodesics *off* the null surface (along the auxiliary null vector).

I shall confine myself to Einstein's theory, and to four dimensions, in this article. For the comparison of the two approaches, I shall use Gaussian null coordinates (GNC) near a null surface [16, 17, 24].

The conventions used in this article are as follows: We use the metric signature $(-, +, +, +)$. The fundamental constants $G, \hbar$ and c have been set to unity. The Latin indices, $a, b, \ldots$, run over all space-time indices, and are hence summed over four values. Greek indices, $\alpha, \beta, \ldots$, are used when we specialize to indices corresponding to a codimension-1 surface, i.e. a 3-surface, and are summed over three values. Upper case Latin symbols, $A, B, \ldots$, are used for indices corresponding to two-dimensional hypersurfaces, leading to sums going over two values.

[1]This nickname ensured that PhD work was never far from my mind even while enjoying holidays in my home-state of Kerala, because of all the paddy fields around.

2 Comparing the Two Results in GNC

In order to compare Jacobson's and Paddy's approaches, we introduce a coordinate system adapted to a null surface. This coordinate system, named the Gaussian null coordinates (GNC), is quite general like the Gaussian normal coordinates [28] near a non-null surface. Hence, as far as we know, we are not imposing any restrictions on the null surfaces or the spacetimes by restricting to GNC. Further, the quantities that we will be referring to will be physical quantities that do not depend on the choice of coordinates. So, our results derived in the framework of GNC are general results valid around any null surface. The discussion of both Paddy's thermodynamic identity and the Raychaudhuri equation on the null surface which was used by Jacobson has been provided for the GNC metric in [1, 2]. For the Raychaudhuri equation, we shall take a slightly different route which makes it easier to compare with Jacobson's results while the results for Paddy's thermodynamic identity will be borrowed from the above two papers.

2.1 Gaussian Null Coordinates (GNC)

This coordinate system was introduced, as far as I know, by Moncrief and Isenberg [16]. The construction of these coordinates are also discussed in [6, 17, 24, 26]. We shall briefly detail the construction and note the essential properties of this coordinate system below.

In the case of Gaussian normal coordinates near a non-null surface, the construction proceeds by using geodesics normal to the surface. This won't work for the null case, since geodesics with tangent vectors along the surface normal, say ℓ^a, actually lie on the null surface. But this offers a unique direction on the null surface and the coordinate system on the surface can be set up adapted to this direction. To do this, choose any spacelike 2-surface on the null surface and assign coordinates (x^1, x^2) on that surface. Then, carry these coordinates along the null generators of the surface with some parameter u, not necessarily affine (see Appendix "Gaussian Null Coordinates with Affine Parametrization" for the case of affine parametrization), forming the third coordinate on the surface. To construct the coordinates in the region near the null surface, we introduce an auxiliary null vector k^a, satisfying $\ell_a k^a = -1$. Then, we carry the coordinates on the null surface along the null geodesics in the direction of k^a and take the fourth coordinate as an affine parameter $-r$ along these null geodesics, with $r = 0$ on the null surface. (Here, $-r$ has been used instead of r just to match with conventions we have been following.)

Then, the line element in GNC coordinates takes the following form:

$$ds^2 = -2r\alpha du^2 + 2dudr - 2r\beta_A dudx^A + q_{AB}dx^A dx^B \tag{1}$$

This line element contains six independent functions α, β_A and q_{AB}, all dependent on all the coordinates (u, r, x^A). The metric on the two-surface (i.e. $u =$ constant and $r =$ constant) is represented by q_{AB}. The surface $r = 0$ is the fiducial null surface while other $r =$ constant surfaces are not null in general. There are only 6 free functions in the metric, as can be expected when we use the 4 coordinate choices to restrict 10 components of the metric for a general spacetime.

We use the symbol s_a for the normal $\partial_a r$ to the $r =$ constant surfaces. This will be a null vector on the $r = 0$ null surface. In fact, s^a will go to the null vector ℓ^a, defined earlier, on the null surface. We introduce an auxiliary null vector k^a such that $k^a s_a = -1$ everywhere. The components of these quantities in (u, r, x^A) coordinates are as follows:

$$s_a = (0, 1, 0, 0)\,, \qquad s^a = \left(1, 2r\alpha + r^2\beta^2, r\beta^A\right) \tag{2a}$$

$$k_a = (-1, 0, 0, 0)\,, \qquad k^a = (0, -1, 0, 0) \tag{2b}$$

On the null surface, we introduce two spacelike vectors $\mathbf{e_A} = (\mathbf{e_1}, \mathbf{e_2})$ which satisfy $\ell_a e^a_A = k_a e^a_A = 0$. The four vectors

$$\left(\ell^a, k^a, e^a_1, , e^a_2\right) \tag{3}$$

form a basis near the null surface. Next we introduce the vector ξ^a:

$$\boldsymbol{\xi} = \frac{\partial}{\partial u} = (1, 0, 0, 0)\ . \tag{4}$$

which goes to ℓ^a on the null surface. This vector will be called the time development vector since it corresponds to the standard time direction (which is also a Killing direction) when Schwarzschild and Rindler metrics are written in GNC form (see Appendix B in [2]). Thus, we may take it as the time direction corresponding to the local Rindler observers in the local Rindler frame near the null surface. We have $\boldsymbol{\xi}^2 = -2r\alpha$. This is zero on the null surface, as expected since ξ^a goes to ℓ^a.

The vector $k^a = -\partial/\partial r$ is tangent to the ingoing null geodesic (ingoing since it points in the direction of decreasing r), which is affinely parametrized with affine parameter r. We denote λ_H to be the value of the affine parameter on the null surface. In the remaining discussions, we will work with λ defined through the following relation: $r = \lambda - \lambda_H$.

2.2 The Components of $F^a = G^a_b \xi^b$

In order to derive the thermodynamic identity, we focus on the vector $F^a = G^a_b \xi^b$. One way of seeing where this component comes from is to think in terms of the Noether current for gravity [1, 2] or in terms of the recently introduced gravitational momentum [1, 22]. Another way of thinking about this is to note that ξ^a goes to the null normal ℓ^a on the null surface and hence the various components of matter fluxes associated with the null surface are given by the various components of $G^a_b \xi^b$. We shall take projections of $G^a_b \xi^b$ along ℓ^a and k^a and show that these are the components used by Jacobson and Paddy respectively. The projection to the space spanned by e^a_A (see Eq. (3)) is obtained using the projector q^a_b. This gives rise to an equation of the form of the Navier–Stokes equation [1], but we shall not be discussing that result here.

2.2.1 Jacobson's Result from $G^a_b \xi^b \ell_a$ on the Null Surface

Jacobson's derivation of the Einstein equation proceeds by assuming the Clausius relation $\delta Q = T \delta S$, where the heat change δQ is taken as the matter flux across a null surface near a local equilibrium, T is the acceleration of an observer who perceives the local patch of the null surface as a local Rindler horizon and δS is the change in entropy which is proportional to the area change.

Consider the component

$$F^a \ell_a = G^a_b \xi^b \ell_a = G^{ab} \ell_a \ell_b = R^{ab} \ell_a \ell_b = 8\pi T^{ab} \ell_a \ell_b \,. \tag{5}$$

This projection actually picks up the component of F^a along k^a if you expand $F^a = A\ell^a + Bk^a + C^A e^a_A$ in the basis in Eq. (3). In GNC coordinates,

$$F^a \ell_a = R^{ab} \ell_a \ell_b = R^{rr} \,. \tag{6}$$

Evaluating this component of the Ricci tensor at the null surface $r = 0$, we obtain

$$\begin{aligned} R^{rr} &= \frac{\alpha}{\sqrt{q}} \partial_u \sqrt{q} - \partial_u^2 \ln \sqrt{q} + \frac{1}{4} \partial_u q_{AB} \partial_u q^{AB} \\ &= \alpha \partial_u \ln \sqrt{q} - \partial_u^2 \ln \sqrt{q} - \frac{1}{4} q^{AC} q^{BD} \partial_u q_{AB} \partial_u q_{CD} , \end{aligned} \tag{7}$$

where ∂_u denotes the operator $\partial / \partial u$. The induced metric on the null surface is given by $q_{ab} = g_{ab} + \ell_a k_b + k_a \ell_b$. Using q_{ab}, we can construct the second fundamental form on the null surface:

$$\Theta_{ab} = q_a^m q_b^n \nabla_m \ell_n \,. \tag{8}$$

Taking the trace of the second fundamental form, we get the expansion:

$$\Theta = g^{ab} \Theta_{ab} = q^{ab} \Theta_{ab} = q^{ab} \nabla_a \ell_b \,. \tag{9}$$

The second fundamental form and the expansion for the null surface at $r = 0$ in GNC are respectively

$$\Theta_{ab} = \frac{\partial_u q_{AB}}{2}; \quad \Theta = \frac{q^{AB}\partial_u q_{AB}}{2} = \partial_u \ln\sqrt{q} \tag{10}$$

Thus, we can write

$$F^a \ell_a = R^{rr} = \alpha\Theta - \partial_u\Theta - \Theta_{AB}\Theta^{AB} \tag{11}$$

$$= \alpha\Theta - \frac{1}{\sqrt{q}}\partial_u\left(\sqrt{q}\Theta\right) + \Theta^2 - \Theta_{AB}\Theta^{AB} \tag{12}$$

$$= \alpha\Theta - \frac{1}{\sqrt{q}}\partial_u\left(\sqrt{q}\Theta\right) - D, \tag{13}$$

where $D = \Theta_{AB}\Theta^{AB} - \Theta^2$ is identified as the dissipation corresponding to the null surface. This identification comes from the component $q^a_b F^b$ of F^a which can be written in a form similar to Navier–Stokes equations of fluid dynamics [1].

In order to compare with Jacobson's result, consider Eq. (11). We can write it in the form

$$\partial_u\Theta = \alpha\Theta - \Theta_{AB}\Theta^{AB} - R_{ab}\ell^a\ell^b \tag{14}$$

Once we decompose Θ_{ab} into its trace (expansion Θ), traceless symmetric part (shear σ_{ab}) and antisymmetric part (rotation ω_{ab}), this reduces to the form of the null Raychaudhuri equation [25] but with the first term being extra. This term appears because the null Raychaudhuri equation is usually defined with an affinely parametrized geodesic while the parameter u is not an affine parameter. If we derive the null Raychaudhuri equation without assuming affine parametrization, we can see that it is consistent with Eq. (14) [1].

But we shall follow the other route by keeping affine parametrization. Then, u has to be taken as an affine parameter in the construction of GNC. This would lead to the constraint that $\alpha = 0$ on the null surface $r = 0$ (see Appendix "Gaussian Null Coordinates with Affine Parametrization"). Enforcing this, Eq. (14) becomes

$$\partial_u\Theta = -\Theta_{AB}\Theta^{AB} - R_{ab}\ell^a\ell^b \tag{15}$$

This is the usual affinely parametrized null Raychaudhuri equation. In Jacobson's case, the $\Theta_{AB}\Theta^{AB}$ term is put to zero as a condition for local equilibrium. More explicitly, the rotation is zero since the null geodesics are taken to form the local horizon, while the expansion and shear are set to zero as a condition for equilibrium. If there is a given null surface with non-zero shear or expansion at a point, then these cannot be set to zero by choice of coordinates as they are geometrical quantities. But what is true is that given any point in spacetime and any null direction through that point, there is always a null surface through that point tangential to the null direction such that the expansion and shear are zero. To see this in GNC coordinates, first note that $\Theta_{ab} = \partial_u q_{ab}$. At the point P at which we want to construct the GNC coordinates on a null surface at local equilibrium, we first erect a local inertial frame (t, x, y, z). The x-axis is aligned so that the chosen null direction lies in the $x - t$

plane. If we now do a Rindler transformation, the metric can be put in the GNC form with the null surface coinciding with the local Rindler horizon with the chosen null direction along one of its generators. Since the transverse 2-metric is not touched in these transformations, they remain flat and we shall have $\partial_u q_{AB} = 0$ valid in the local inertial frame. Thus, given any point in spacetime and a null direction through it, one can always choose a null surface through the point tangential to the null direction such that $\Theta^{AB} = 0$; in other words, such that the shear, expansion and rotation are zero.

With such a choice of a null surface, Eq. (15) becomes

$$\partial_u \Theta = -R_{ab}\ell^a \ell^b , \tag{16}$$

which is precisely the form of the Raychaudhuri equation Jacobson integrated to get the geometric part needed to obtain the Einstein equation. Following Jacobson, we first integrate the above equation with the initial condition that $\Theta = 0$ at the point P at $u = 0$ to obtain

$$\Theta = -R_{ab}\ell^a \ell^b u . \tag{17}$$

Our convention is that u is increasing to the future. Jacobson's set-up involves imagining that the congruence was expanding a little to the past of P and then there was a matter flux through the congruence that provided just enough gravitational lensing to bring the expansion to zero at P. We shall integrate from the point P_0 with affine parameter value $-u_i$ to the past of P. The change in area of an infinitesimal cross-section around the chosen null generator is then given by

$$\delta A = \int_{-u_i}^{0} \Theta \sqrt{q} du d^2x = -\int_{-u_i}^{0} R_{ab}\ell^a \ell^b u \sqrt{q} du d^2x = \int_{0}^{u_i} R_{ab}\ell^a \ell^b \lambda \sqrt{q} d\lambda d^2x . \tag{18}$$

In the last step, we have changed the integration variable from u to $\lambda = -u$. Note that the change in area is positive if $R_{ab}\ell^a \ell^b > 0$, which is the condition that you obtain from the Einstein equations if the matter part satisfies the null energy condition. Finally, we assume that the entropy change that is to be associated with the null surface is proportional to its area change with some proportionality constant η:

$$\delta S = \eta \delta A = \eta \int_{0}^{u_i} R_{ab}\ell^a \ell^b \lambda \sqrt{q} d\lambda d^2x . \tag{19}$$

Next, we need an accelerating observer for whom the patch of null surface acts as a local Rindler horizon. Since the proper distance of an accelerated observer from the Rindler horizon is inversely proportional to the acceleration, we shall consider a highly accelerated observer so that the observer is very close to the horizon and we can be sure that our construction of the local inertial frame and local Rindler frame are valid. In fact, we may go to the light-like limit of the infinitely accelerated observer.

In order to find the suitable observer, we shall write the GNC form with affine parametrization for the Rindler metric. In flat Milnkowski space with coordinates (t, x, y, z), the observers moving along the integral curves of the generators of boosts along the x-direction are the ones who will observe the future part of the $x = t$ null plane as a Rindler horizon. These are observers moving along the integral curves of the vector $(x, t, 0, 0)$. We shall try to figure out which observers in our picture can be taken to correspond to these observers. First, we shall write down the metric in the following Rindler form [27]:

$$ds^2 = -2\kappa l dT^2 + \frac{dl^2}{2\kappa l} + dy^2 + dz^2 \,. \tag{20}$$

This form is obtained from the Minkowski metric by the coordinate transformation $x = \sqrt{2l/\kappa} \cosh \kappa t$ and $t = \sqrt{2l/\kappa} \sinh \kappa t$. Defining a new coordinate U by

$$U = T + \int \frac{dl}{2\kappa l}; \quad dT = dU - \frac{dl}{2\kappa l} \,, \tag{21}$$

we transform to

$$ds^2 = -2\kappa l dU^2 + 2dU dl + dy^2 + dz^2 \,. \tag{22}$$

We have got the metric in GNC form, but U is not an affine parameter. Looking at how the vector $\partial/\partial U$ should be scaled to make it affine, we can figure out that the transformation to coordinates

$$\lambda = \frac{e^{\kappa U}}{\kappa}; \quad s = \frac{l}{\kappa \lambda} \tag{23}$$

will do. This brings the metric in the form

$$ds^2 = 2d\lambda ds + dy^2 + dz^2 \,. \tag{24}$$

I am happy to say that we have rediscovered the flat metric in the double null form in a pleasantly roundabout way. This is now of the form of affinely parametrized GNC in Eq. (39). The observer can be then found to be the one moving along the integral curves of the vector $(\lambda, -s, 0)$.

Taking a cue from this, we shall look at the observers moving along the integral curves of $(u, -r, 0, 0)$ in the affine GNC metric of Eq. (39). So we anoint

$$\chi^a = (u, -r, 0, 0), \tag{25}$$

as the vector representing our observers. The normalized vector $\bar{\chi}^a$ will represent the four-velocity. Note that, on the null surface $r = 0$, we have

$$\chi^a = u \frac{\partial}{\partial u} = u \ell^a \,. \tag{26}$$

Writing $\bar{\chi}^a = N\chi^a$, the acceleration of the trajectories will be given by N and the corresponding acceleration temperature will be

$$T = \frac{N}{2\pi} \tag{27}$$

near the null surface.

The momentum four-vector associated to the matter flux by one such observer will be $NT^{ab}\chi_b$. The energy of the matter flux across the null surface as observed by this observer is the heat change

$$\delta Q = \int_{-u_i}^{0} NT^{ab}\chi_b \ell_a \sqrt{q} du d^2x = N \int_{-u_i}^{0} uT^{ab}\ell_b \ell_a \sqrt{q} du d^2x, \tag{28}$$

where we have used Eq. (26). Changing variables to $\lambda = -u$ as in Eq. (19), we obtain

$$\delta Q = -N \int_{0}^{u_i} \lambda T^{ab}\ell_b \ell_a \sqrt{q} d\lambda d^2x \,. \tag{29}$$

Now we have all the ingredients in place. Demanding $\delta Q = T\delta S$ and using Eqs. (19), (27) and (29), we obtain the condition

$$R_{ab}\ell^a \ell^b = -\frac{2\pi}{\eta} T_{ab}\ell^a \ell^b \,. \tag{30}$$

This is almost right, except for a pesky minus sign. If we obtain Einstein equation from the above relation, η will be set as negative and this would mean that area increase is entropy decrease from Eq. (19). We have been careful with signs around Eq. (19), so let us look back at Eq. (29). If we enforce the null energy condition $T^{ab}\ell_a \ell_b > 0$, the heat turns out to be negative. Since we require a positive heat change to correspond to the positive change in entropy, we redefine

$$\delta Q = N \int_{0}^{u_i} \lambda T^{ab}\ell_b \ell_a \sqrt{q} d\lambda d^2x \,. \tag{31}$$

This is in fact the correct definition. The original source of the extra minus sign was the fact that $\chi^a = u\partial/\partial u$ is past-directed in our region of integration since u is negative, and hence $-\chi^a$ should have been used.

Thus, we obtain the equation

$$R_{ab}\ell^a \ell^b = \frac{2\pi}{\eta} T_{ab}\ell^a \ell^b \,. \tag{32}$$

From here, we can follow Jacobson [8] to obtain the full Einstein equations.

2.2.2 Paddy's Result from $G^a_b \xi^b k_a$ on the Null Surface

Next, we shall discuss Paddy's approach where a certain component of the Einstein equations near a horizon takes a form similar to the first law of thermodynamics [1, 2, 10–12, 18]. In Paddy's approach the Einstein equations are not derived, but it is shown that a certain component of Einstein equations projected on a null surface has a thermodynamic interpretation. Note that this difference is superficial since even in Paddy's case one can derive the Einstein equations by starting from the thermodynamic identity backward and demanding that it holds for all null surfaces and even in Jacobson's case one can prove the Clausius relation starting from the Einstein equations.

To obtain Paddy's result, we look at another component of F^a. We shall take its projection along the auxiliary null vector k_a. Then, we obtain

$$F^a k_a = G^a_b \xi^b k_a = 8\pi T^a_b \xi^b k_a \; . \tag{33}$$

In this case, the work has already been done in [1, 2] and hence we just borrow the results. Working in GNC, the component $G^a_b \xi^b k_a = 8\pi T^a_b \xi^b k_a$ is interpreted and written in the form

$$\bar{F}\delta\bar{\lambda} = T\delta_{\bar{\lambda}} S - \delta_{\bar{\lambda}} E \; . \tag{34}$$

Here, F is the integral of $T^a_b \xi^b k_a$ over the null surface, interpreted as the force acting on the patch of the null surface and $\delta\lambda = \delta r$ is a small shift of the horizon in the direction of k^a. This is a small shift in the r-direction in GNC coordinates. Note that r is also an affine parameter.

On the RHS, $T = \alpha/2\pi$ is the acceleration temperature corresponding to the observers moving along the integral curves of ξ^a near $r = 0$ with the assumption that α is slowly varying in time. More precisely, we assume $\partial_u \alpha \ll \alpha^2$. The change in entropy δS is just the change in the 2-surface area, with appropriate factors, when the surface is shifted outward. The change in area is the integral of $\partial_r \sqrt{q}\delta r$ integrated over the patch of the null surface.

Finally, we have the change in energy. The quantity E here is given by the expression

$$\begin{aligned} E = & \frac{1}{16\pi} \int dr \int d^2x \, \sqrt{q} R^{(2)} - \frac{1}{8\pi} \int d^2x \partial_u \sqrt{q} \\ & - \frac{1}{16\pi} \int dr \int d^2x \sqrt{q} \left\{ \frac{1}{2} \beta_A \beta^A \right\} \; . \end{aligned} \tag{35}$$

Here, one term has been put to zero under the assumption that the u-constant, r-constant 2-surface on the null surface is closed. The identification of this quantity as the energy is due to the fact that it is able to reproduce the known expressions of energy in several known cases. For example, this reduces to the mass for the Schwarzschild metric.

With these identifications, Eq. (34) is of the form of the first law of thermodynamics.

3 Discussion

At first sight, Paddy's result and Jacobson's result look very similar. Jacobson's result says that one can derive Einstein equations from $\delta Q = T\delta S$ imposed on local Rindler horizons, which can be constructed along any null direction around any point in spacetime. Paddy's result is that a certain projection of the Einstein equations on a null surface can be interpreted to be of the form $\delta E = T\delta S - F\delta\lambda$. In thermodynamics, these relations will be called the Clausius relation and the first law of thermodynamics. In fact, one may be tempted to put both together and obtain the equation $\delta E = \delta Q - F\delta\lambda$. But closer scrutiny reveals that things are not that simple.

The apparent point of conflict is that both δQ in Jacobson's case and δE in Paddy's case are components of matter energy-momentum flux across the horizon. Since identification of which component is which physically is a little complicated near the null surface, I am sure many people must have thought that these are the same components. Once this is assumed, there appears to be a discrepancy with Paddy having an extra term compared to Jacobson's starting point, since the $T\delta S$ terms seem unambiguous.

I hope I have shed some light on this issue in this article, building up on work previously done in [1, 2]. Working in the framework of Gaussian null coordinates, one can see that

1. Paddy's result and Jacobson's result come from two different components of the Einstein tensor, and equivalently of the matter energy-momentum tensor, near the null surface.
2. The entropy change in the $T\delta S$ term is not the same in the two cases. In Paddy's case, the change is from the change in area along the null geodesics off the null surface, while the change in Jacobson's case is along the null geodesics on the null surface.

Another paper which compared the two approaches is [9]. It is not clear how the results there compare to the results stated here. The discussion in [9] had the general static metric introduced in [13, 14] as the reference metric although the final results are stated in tensorial form. It will be interesting to see how these results look when translated to Gaussian null coordinates.

Epilogue

> Theoretical physics is fun. Most of us indulge in it for the same reason a painter paints or a dancer dances...Occasionally, there are additional benefits like fame and glory and even practical uses; but most good theoretical physicists will agree that these are not the primary reasons why they are doing it. The fun in figuring out the solutions to Nature's brain teasers is a reward in itself.

Source- Paddy's book [20]

My first encounter- After PhD

Acknowledgements I am happy to dedicate this article to Paddy, who I was fortunate to have as my PhD supervisor, on the occasion of his 60th birthday as a tribute to his never-flagging enthusiasm for anything under the sun, his ocean-deep knowledge and his unmatched tenacity. I also acknowledge discussions with Sumanta Chakraborty, Bibhas Majhi and Dawood Kothawala on related topics in the past.

Appendix

Gaussian Null Coordinates with Affine Parametrization

The line element in GNC coordinates was given in Eq. (1) as

$$ds^2 = -2r\alpha du^2 + 2dudr - 2r\beta_A dudx^A + q_{AB}dx^A dx^B \, , \tag{36}$$

Here, α, β_A and q_{AB} are arbitrary functions. This was derived by taking u to be an arbitrary parameter along the null geodesics on the null surface $r = 0$ [24]. But suppose we now impose the condition that u is an affine parameter on the null surface. This is equivalent to the condition that $\xi^a \nabla_a \xi^b = 0$ at $r = 0$ for $\xi^a = \partial/\partial u$. For the above line element, we have

$$\xi^a \nabla_a \xi^b = \Gamma^b_{ac} \xi^a \xi^c = \Gamma^b_{uu} \, . \tag{37}$$

Equating this to zero, we get the following conditions at $r = 0$:

$$\partial_u g_{uu} = 0; \;\; \partial_r g_{uu} = 0; \;\; \partial_A g_{uu} = 0 \, . \tag{38}$$

Since $g_{uu} = -2r\alpha$, the first and third conditions are automatically satisfied, while the second condition implies $\alpha = 0$ at $r = 0$. This can be enforced by putting $\alpha = r\gamma$ where γ is an arbitrary function. Thus, the form of the GNC line element with affine parametrization is

$$ds^2 = -2r^2\gamma du^2 + 2dudr - 2r\beta_A dudx^A + q_{AB}dx^A dx^B \, . \tag{39}$$

Note that this form was indicated in [17] where the R_{uu} component was compared to the Raychaudhuri equation to claim that α has to be proportional to r. It is not clear from the text whether the author realized that this is necessary only if u is taken to be affine. But since affine parametrization can always be taken, it is true that the GNC metric can always be written in the above form.

References

1. S. Chakraborty, T. Padmanabhan, Thermodynamical interpretation of the geometrical variables associated with null surfaces. Phys. Rev. D **92**(10), 104,011 (2015). doi:10.1103/PhysRevD.92.104011
2. S. Chakraborty, K. Parattu, T. Padmanabhan, Gravitational field equations near an arbitrary null surface expressed as a thermodynamic identity. JHEP **10**, 097 (2015). doi:10.1007/JHEP10(2015)097
3. G. Chirco, S. Liberati, Non-equilibrium thermodynamics of spacetime: the role of gravitational dissipation. Phys. Rev. D **81**, 024,016 (2010). doi:10.1103/PhysRevD.81.024016
4. R. Dey, S. Liberati, A. Mohd, Higher derivative gravity: field equation as the equation of state (2016)
5. C. Eling, R. Guedens, T. Jacobson, Non-equilibrium thermodynamics of spacetime. Phys. Rev. Lett. **96**, 121,301 (2006). doi:10.1103/PhysRevLett.96.121301
6. H. Friedrich, I. Racz, R.M. Wald, On the rigidity theorem for space-times with a stationary event horizon or a compact Cauchy horizon. Commun. Math. Phys. **204**, 691–707 (1999). doi:10.1007/s002200050662
7. R. Guedens, T. Jacobson, S. Sarkar, Horizon entropy and higher curvature equations of state. Phys. Rev. D **85**, 064,017 (2012). doi:10.1103/PhysRevD.85.064017
8. T. Jacobson, Thermodynamics of space-time: the Einstein equation of state. Phys. Rev. Lett. **75**, 1260–1263 (1995). doi:10.1103/PhysRevLett.75.1260
9. D. Kothawala, The thermodynamic structure of Einstein tensor. Phys. Rev. D **83**, 024,026 (2011). doi:10.1103/PhysRevD.83.024026
10. D. Kothawala, Accelerated Observers, Thermal Entropy, and Spacetime Curvature (2016)
11. D. Kothawala, T. Padmanabhan, Thermodynamic structure of Lanczos-Lovelock field equations from near-horizon symmetries. Phys. Rev. D **79**, 104,020 (2009). doi:10.1103/PhysRevD.79.104020
12. D. Kothawala, S. Sarkar, T. Padmanabhan, Einstein's equations as a thermodynamic identity: the cases of stationary axisymmetric horizons and evolving spherically symmetric horizons. Phys. Lett. B **652**, 338–342 (2007). doi:10.1016/j.physletb.2007.07.021
13. A. Medved, D. Martin, M. Visser, Dirty black holes: space-time geometry and near horizon symmetries. Class. Quant. Grav. **21**, 3111–3126 (2004). doi:10.1088/0264-9381/21/13/003
14. A. Medved, D. Martin, M. Visser, Dirty black holes: symmetries at stationary nonstatic horizons. Phys. Rev. D **70**, 024,009 (2004). doi:10.1103/PhysRevD.70.024009
15. A. Mohd, S. Sarkar, Thermodynamics of local causal horizons. Phys. Rev. D **88**(2), 024,026 (2013). doi:10.1103/PhysRevD.88.024026
16. V. Moncrief, J. Isenberg, Symmetries of cosmological cauchy horizons. Commun. Math. Phys. **89**(3), 387–413 (1983). doi:10.1007/BF01214662
17. E.M. Morales, On a Second Law of Black Hole Mechanics in a Higher Derivative Theory of Gravity (2008). http://www.theorie.physik.uni-goettingen.de/forschung/qft/theses/dipl/Morfa-Morales.pdf
18. T. Padmanabhan, Classical and quantum thermodynamics of horizons in spherically symmetric space-times. Class. Quant. Grav. **19**, 5387–5408 (2002). doi:10.1088/0264-9381/19/21/306
19. T. Padmanabhan, Thermodynamical aspects of gravity: new insights. Rept. Prog. Phys. **73**, 046,901 (2010). doi:10.1088/0034-4885/73/4/046901
20. T. Padmanabhan, *Sleeping Beauties in Theoretical Physics*, vol. 895 (Springer, Berlin, 2015)
21. T. Padmanabhan, Exploring the nature of gravity, in *The Planck Scale II Wroclaw*, Poland, September 7–12, 2015 (2016). https://inspirehep.net/record/1419624/files/arXiv:1602.01474.pdf
22. T. Padmanabhan, Momentum density of spacetime and the gravitational dynamics. Gen. Rel. Grav. **48**(1), 4 (2016). doi:10.1007/s10714-015-1996-z
23. T. Padmanabhan, The atoms of space, gravity and the cosmological constant. Int. J. Mod. Phys. D **25**(07), 1630,020 (2016). doi:10.1142/S0218271816300202

24. K. Parattu, S. Chakraborty, B.R. Majhi, T. Padmanabhan, A boundary term for the gravitational action with null boundaries. Gen. Rel. Grav. **48**(7), 94 (2016). doi:10.1007/s10714-016-2093-7
25. E. Poisson, *A Relativist's Toolkit: The Mathematics of Black-Hole Mechanics*, 1st edn. (Cambridge University Press, Cambridge, 2007)
26. I. Racz, Stationary black holes as holographs. Class. Quant. Grav. **24**, 5541–5572 (2007). doi:10.1088/0264-9381/24/22/016
27. T. Padmanabhan, *Gravitation: Foundations and Frontiers* (Cambridge University Press, Cambridge, 2010)
28. R.M. Wald, *General Relativity*, 1st edn. (The University of Chicago Press, Chicago, 1984)

Thoughts on 50 Years in Astrophysics and Cosmology and on What Comes Next

Martin J. Rees

Abstract This paper presents a brief review of how some key developments in relativistic astrophysics and cosmology emerged over the last few decades - and offers some speculations on what issues will challenge the next generation.

1 Introduction

Anyone who reaches the age of 60 is likely to become interested in what happens to scientists as they grow older. There seem to be three destinies. First, and most common, is a diminishing focus on research - sometimes compensated by energetic efforts in other directions, sometimes just by a decline into torpor. A second pathway, followed by some of the greatest scientists, is an unwise and over-confident diversification into other fields. Those who follow this route are still 'doing science' - they want to understand the world and the cosmos but they no longer get satisfaction from researching in the traditional piecemeal way: they over-reach themselves, sometimes to the embarrassment of their admirers. (We can all think of some in this category). This trend is aggravated by the tendency for the eminent and elderly to be shielded from frank criticism. But there is a third way - the most admirable. This is to continue to do what one's competent at, accepting that there may be some new techniques that the young can assimilate more easily that the old, and that one can probably at best aspire to be on a plateau rather than scaling new heights. (There are of course some 'late flowering' exceptions. But whereas there are many composers whose last works are their greatest, there are few scientists for whom this is so.) The reason, I think is that composers can improve and deepen solely through 'internal development'; scientists, in contrast, need to absorb new concepts and new techniques if they want to stay at the frontier - and that's what gets harder as we get old.

The great physicist Lord Rayleigh was once asked, in his later years: 'Do scientists over 60 do more harm than good?' He responded 'No, provided that they stick to what they're good at, and don't criticize the work of younger scientists'. This precept was

M.J. Rees (✉)
Institute of Astronomy, Madingley Road, Cambridge CB3 0HA, UK
e-mail: mjr@ast.cam.ac.uk

J.S. Bagla and S. Engineer (eds.), *Gravity and the Quantum*,
Fundamental Theories of Physics 187, DOI 10.1007/978-3-319-51700-1_21

quoted by S Chandrasekhar - an eminent scientist whose sustained career certainly exemplified it to a high degree.

I'm confident that this third trajectory will be Paddy's too: that we can look forward to more original ideas, that his amazing intellectual energy will be undimmed, and there will be an continuing flow of books and general articles.

Many sciences - astronomy and cosmology certainly among them - advance decade by decade so that any practitioner can observe an 'arc of progress' during his or her career. So I'd like to offer some thoughts on what has happened in high energy astrophysics and cosmology during my own career, which started in the 1960s.

2 The Resurgence of Relativity: Some History

At that time, general relativity was already 50 years old. By then, relativists had developed beautiful mathematics and great insights but the topic was somewhat isolated from the mainstream of physics and astronomy. The gravitational effects governing ordinary stars and galaxies were weak enough to be adequately described by Newtonian theory - general relativity was no more than a tiny correction. And it wasn't well confirmed. The classic solar system tests confirmed Einstein with no better than 10% precision - and only in the post-Newtonian weak field limit.

The 1960s were transformational. On the theoretical side we had the Kerr Solution, the singularity theorems, the 'no hair' theorems, etc. But in parallel there were observational breakthroughs.

The first such breakthrough came in 1963 with the discovery of quasars – hyperluminous beacons in the centres of some galaxies, compact enough to vary within hours or days, but which vastly outshine their host galaxy. Quasars revealed that galaxies contained something more than stars and gas. That 'something' is a huge black hole lurking in their centres, though it took a further decade before that was properly appreciated (except by a few pioneers like Donald Lynden-Bell). I think a consensus would have developed faster if other lower-level forms of activity in galactic nuclei (e.g. in Seyfert Galaxies and radio galaxies) had already been more fully studied. Quasars, initially regarded as sui generis (whose redshifts, some thought, might have involved new physics) would then have been quickly recognized as extreme versions of a known phenomenon. Quasars are specially bright because, as we now recognize, they're energised by emission from magnetized gas swirling into a central black hole. These processes can generate the jets that inflate giant radio-emitting lobes that surround some galaxies.

Quasars were a key stimulus to the emergence of 'relativistic astrophysics'. But there were others. In particular, another surprise was the detection of neutron stars. One of the best-known objects in the sky is the Crab Nebula: the expanding debris from a supernova witnessed by Chinese astronomers in 1054 AD. It was a longtime puzzle what kept it shining, so blue and bright. The answer came when it was discovered that the innocuous seeming star in its centre was anything but normal. It was actually a neutron star spinning at 30 revs per second and emitting a wind of fast electrons that generated the blue light. In neutron stars relativistic effects are 10–20%. Not merely a tiny correction to Newton.

Some massive stars end their lives not as neutron stars but as black holes. But these holes are only detectable if they happen to be in interacting binaries, where they're lit up by accretion energy and emit strongly in the X-ray band. The fact that these X-ray sources are in binary systems means that their masses can be determined. Those that are more than 2 solar masses display irregular flickering. They're black holes. But those of lower mass vary periodically. They're spinning neutron stars - but energized by accretion, not by spindown as in pulsars.

The Italian/American space scientist Riccardo Giacconi has had a lifetime of achievement, including the discovery of stellar-mass black holes. But if history had been slightly different, he could have discovered neutron stars as well - finding them in binaries before radio astronomers found pulsars.

A sociological digression: In the West (and we are thinking back to an era when the Iron Curtain was almost impermeable) the inspirational gurus for relativistic astrophysics were John Wheeler in the US; and Roger Penrose and Dennis Sciama in England. In the Soviet Union, Yakov Zeldovich led a powerful group of theorists who progressed in parallel with those in the West. And there were distinguished individuals in Japan, India and elsewhere. Despite the impediments to travel, there were cordial and cooperative contacts among relativists - many of which I was able to observe close-up (though I was never a serious contributor to this subject).

The 1960s saw the first real advance in understanding black holes since the work of J Robert Oppenheimer and his co-workers in the late 1930s. They clarified what happens at $r = 2M$ in the Schwarzschild metric. (And its interesting to conjecture how much of the 1960s work Oppenheimer might have pre-empted if World War II hadn't broken out the very day his key paper with Snyder [1] appeared in Phys Rev).

A dead quasar – a quiescent massive black hole - lurks at the centre of most galaxies. Moreover there's a correlation between the mass of the hole and that of its host galaxy. The actual correlation is with the bulge (non-disc) component not the whole galaxy. Our own Galaxy harbours a modest-mass hole of around 4 million solar masses.

The holes grow by accretion of gas – a process that's surprisingly complicated. They also tidally disrupt and swallow entire stars. And as Blandford and Znajek [2] pointed out, energy can be electromagnetically extracted from a spinning hole and this is a specially effective way to produce the ubiquitous ultrarelativistic outflowing jets.

Are the holes spinning? We have some evidence. The swirling gas spiraling into them emits spectral lines in the x-ray band - smeared by Doppler and gravitational shifts in a fashion that depends on the metric. (The stable orbits extend closer to the hole, with higher binding energy, and larger red and blue shifts from Doppler and gravitational effects, when the hole is spinning in the same direction as the disc). There are some cases when the X-ray lines are so greatly broadened that the hole can't have a Schwarzschild metric and must have some spin. (See Reynolds [3] for a recent review). We don't yet know the distribution of spins. But the realization that all these objects are described by the Kerr metric - mass and spin, no other parameters – was something that hugely impressed Chandrasekhar, who wrote that "in my entire scientific life, the most shattering experience has been the realization that an exact solution of Einsteins equations provides the absolutely exact representation of untold numbers of massive black holes that populate the universe".

Any residual skepticism about the validity of Einstein's equations and the Kerr Metric was surely allayed in earlier 2016 when the LIGO detectors (with what was real 'beginner's luck') detected an event attributable to a merger of two holes, where the template of the event was excellently fit by computational models.

When two galaxies merge (as Andromeda and the Milky Way will in about 4 billion years) the black holes in the centre of each will spiral together forming a binary, which will shrink by emitting gravitational radiation and create a strong chirp when the two holes coalesce. Most galaxies have grown via a succession of past mergers. The consequent coalescences of these supermassive black holes would yield gravitational waves of much lower frequencies than LIGO can detect. These are the prime events to which LISA-type instruments in space would be sensitive.

3-D hydrodynamics incorporating MHD has allowed modelling of accretion flows, gamma ray bursts, and active galaxies. And breakthroughs in computing time-dependent space-times in general relativity have allowed calculations of the wave-form of gravitational radiation released by the collapse and merging of compact objects. The feedback on protogalaxies from stars and AGNs is still, however, modelled rather crudely. There has however been huge progress in simulating the emergence of cosmic structure, and setting galaxies in a cosmological context.

3 Probing the Early Universe

The standard Friedmann and Lemaitre models were known since the 1920s, but until the 60s astronomers had no instruments powerful enough to discriminate among them. Indeed the steady state model (where there was an infinite past) was still being debated. Quasars (especially those that were radio emitters too) transformed this. Because they're so luminous they allowed astronomers to probe back far enough along our past light cone to see how the universe was changing as it aged - and to infer that it was not in a steady state.

In the most distant reliably-known quasar [4], Lyman alpha is shifted from the UV almost into the infrared - wavelengths are stretched by 8.1. So we can infer that some galaxies had already assembled when the Universe was only about a tenth of its present age, and black holes, some of several billion solar masses had formed at their centres. But how much further back did the action actually start? When did the dark age end? To answer this, astronomers are looking for even more distant galaxies (they use lensing by clusters as nature's telescope). And they seek objects that may be exceptionally bright. For instance, gamma ray bursts outshine an entire galaxy by a factor of a million. Finding even one at say, a redshift of 15 would be a valuable clue.

But of course it's the cosmic microwave background – another great discovery of the 1960s – that's the crucial fossil of the earliest eras. Its thermal spectrum is the most compelling evidence for a big bang. Even by 1970 we had precise calculations of nucleosynthesis during the first 3 min. And now we famously have the angular fluctuations - most recently and precisely from the Planck spacecraft data. The slight ripples that induce these temperature fluctuations - detected via photons that have travelled freely since the Universe was 300,000 years old – subsequently enhanced their density contrast and condensed out into the first stars and galaxies.

Taking the CMB fluctuations as inputs, and including gravity and gas dynamics, computer simulations end up, after the 1000-fold expansion since the photons were last scattered, with properties that yield a good statistical fit to present structures and allow the cosmological parameters to be pinned down with a precision of a few percent.

We're definitely vindicated in extrapolating back to 1 s, because the calculated proportions of helium and deuterium produced (for a baryon density fitting other data) match beautifully with what's observed. Indeed we can probably be confident in extrapolation back to a nanosecond: that's when each particle had about 50 Gev of energy – an energy that can be achieved in the LHC - and the entire visible universe was squeezed to the size of our solar system.

But questions like"where did the fluctuations come from" and"why did the early universe contain the actual mix we observe of protons, photons and dark matter?" take us back to the even briefer instants when our universe was hugely more compressed still - into an ultra-high-energy domain where experiments offer no direct guide to the relevant physics.

For more than 30 years we've had the inflationary paradigm - seriously invoking an era when the Hubble radius was a billion times smaller than an atomic nucleus. It's argued that 'inflation' gives the best explanation of the flatness. The gaussianity of the fluctuation spectrum and the tilt in the amplitude are consistent with some inflationary models - and firming up the theory is a challenge for the future.

4 The Geometry and the Expansion

Let us now turn from the remote past to the long-range future forecast. In 1998 cosmologists had a big surprise. It was by then well known that the gravity of dark matter dominated that of ordinary stuff - but that together they contributed only about 30% of the critical density. This was thought to imply that we were in a universe whose expansion was slowing down, but not enough to eventually be halted. But, rather than slowly decelerating, the Hubble diagram of Type 1a supernovae famously revealed that the expansion was speeding up. Gravitational attraction was seemingly overwhelmed by a mysterious repulsive force latent in empty space resembling Einstein's lambda [5, 6].

And there was independent evidence supporting this. A straightforward low-density universe would have negative curvature. This can be tested because straightforward effects in the pre-recombination era lead to a peak in the fluctuation amplitude at a particular wavelength. The wavelength of this 'Doppler Peak' serves as a 'rigid rod' of known length. The angular scale of this feature in the CMB fluctuations depends on the curvature.

This peak was first detected by a balloon-borne experiment called Boomerang [7]. It is revealed with huge precision in data from the Planck spacecraft, on an angular scale that implies flatness.

For the universe to be 'flat', the missing 70% would need to be in some unclustered form. Moreover, this component, though dominant today, cannot have been dominant in the past, because it would have inhibited the growth of cosmic structure. It therefore has negative pressure (the 'PdV work' done during the expansion is negative.). And in the Friedman equations that implies acceleration. So even in the absence of the optical data on supernovae, the microwave background evidence for the 'Doppler Peak' would have allowed us to predicted cosmic acceleration.

Indeed if we'd just had the supernova Hubble diagram, some of us still wouldn't be convinced that a low-density universe with small deceleration could be ruled out. But the almost simultaneous CMB evidence for a 'flat' universe rules out this option. So together these very different measurements clinch the case. The fit between the fluctuation spectrum measured by the Planck spacecraft (in all spherical harmonics) and a 6-parameter model - and the realization that these fluctuations develop, under the action of gravity and gas dynamics, into galaxies and clusters with properties matching our actual cosmos – is an immense triumph. When the history of science in these decades is written, this will be one of the highlights - and I mean one of the highlights of all of science: up there with plate tectonics, the genome and only very few others.

An issue for physicists is the nature of the dark energy - is it time-independent, like Einstein's lambda, or was it different in the past? The data are consistent with constancy, but attempts to pin down the dependence are an important motive for surveys of high-redshift galaxies). Whereas the nature of dark matter may well be pinned down in a decade, dark energy won't be understood until we have a model for the graininess of space on the Planck scale - I'm not holding my breath for this - or some novel perspective of the kind that Paddy has pioneered.

I'd venture a gripe about the misuse of the famous 'pie diagram' that shows the constituents of space as roughly 70% dark energy, 25% dark matter, and 5% baryons (numbers that are pinned down more precisely by Planck data). This presentation gives a misleading perception of scientific priorities by making it seems that dark energy is overwhelmingly important - more so than dark matter. But from an astrophysical perspective the opposite is true. Dark matter is crucial: without taking it into account we can't understand the formation of structure, nor the present morphology of galaxies and clusters. In contrast, dark energy isn't crucial for any phenomena that interest astrophysicists. It has no significant role within galaxies and clusters. Moreover, since it depends less steeply on z than ordinary matter, its share of the 'pie' is very small beyond $z = 1$, so it played a minimal role in how cosmic structures emerged.

But despite its irrelevance for astronomy, dark energy may be the biggest fundamental challenge presented by the present day universe. That's why measuring its actual dependence on z is important (to test whether it really is independent of time, like Einstein's cosmological constant, or whether it is more complicated).

5 Beyond the Horizon

Another fundamental question is this: How large is physical reality? We can only observe a finite volume. The domain in causal contact with us is bounded by a horizon – a shell around us, delineating the distance light (if never scattered) could have travelled since the big bang. But that shell has no more physical significance than the circle that delineates your horizon if you're in the middle of the ocean. We'd expect far more galaxies beyond the horizon. There's no perceptible gradient across the visible universe - suggesting that similar conditions prevail over a domain that stretches thousands of times further. But that's just a minimum. If space stretched far enough, then all combinatorial possibilities would be repeated. Far beyond the horizon, we could all have avatars - and perhaps it would be some comfort that some of them might have made the right decision when we make a wrong one!

But even that immense volume may not be all that exists. 'Our' big bang may not be the only one. The physics of the inflation era is still isn't firm. But some of the options would lead to so-called 'eternal inflation' scenario, in which the aftermath of 'our' big bang could be just one island of space-time in an unbounded cosmic archipelago.

In scenarios like this, a challenge for 21st century physics is to answer two questions. First, are there many 'big bangs' rather than just one? Second - and this is even more interesting – if there are many, are they all governed by the same physics or not? Or is there a huge number of different vacuum states with different microphysics?

If the answer to this latter question is 'yes', there will still be underlying laws governing the multiverse - maybe a version of string theory. But what we've traditionally called the laws of nature will be just local bylaws in our cosmic patch. Even though it makes some physicists foam at the mouth, we then can't avoid the A-word - anthropics. Many domains could be still-born or sterile: the laws prevailing in them might not allow any kind of complexity. We therefore wouldn't expect to find ourselves in a typical universe - rather, we'd be in a typical member of the subset where an observer could evolve. It would then be important to explore the parameter-space for all universes (which requires having a believable and 'battle-tested' theory such as string theory) and also a way of putting a probability measure on each part of parameter space (which requires a firmly established model developed from, for instance, the 'eternal inflation' scenario).

I find that even those who are allergic to anthropic reasoning can be interested in exploring the consequences of changes in the laws and constants - different lambda, different G, different fluctuation amplitudes, etc. – if this is just presented as 'counterfactual physics' which helps to develop our intuition. This is analogous to the way some historians speculate on 'counterfactuals' such as what India might be like today if the Brits had never set foot there; and some biologists speculate on how our biosphere might have evolved if the dinosaurs hadn't been wiped out.

Some claim that unobservable entities aren't part of science. But few really think that. For instance, we know that galaxies disappear over the horizon as they accelerate away. But (unless we are in some special central position and the Universe has an 'edge' just beyond the present horizon) there will be some galaxies lying beyond our horizon - and if the cosmic acceleration continues they will remain beyond for ever. Not even the most conservative astronomer would deny that these never-observable galaxies are part of physical reality. These galaxies are part of the aftermath of our big bang. But why should they be accorded higher epistemological status than unobservable objects that are the aftermath of other big bangs?

[To offer an analogy: we can't observe the interior of black holes, but we believe what Einstein says about what happens there because his theory has gained credibility by agreeing with date in many contexts that we can observe. Likewise, if we had a model that described physics at the energies where inflation is postulated to have occurred, and if that model had been corroborated in other ways, then if it predicts multiple big bangs we should take that prediction seriously.]

If there's just one big bang, then we'd aspire to pin down why the numbers describing our Universe have the values we measure (the numbers in the 'standard model' of particle physics, plus those characterizing the geometry of the universe). But if there are many big bangs - eternal inflation, the landscape, and so forth – then physical reality is hugely grander than we'd have traditionally envisioned.

It could be that in 50 years we'll still be as flummoxed as we are today about the ultra-early universe. But maybe a theory of physics near the 'Planck energy' will by then have gained credibility. Maybe it will 'predict' a multiverse and in principle determines some of its properties – the probability measures of key parameters, the correlations between them etc.

Some don't like the multiverse; it means that we'll never have neat explanations for the fundamental numbers, which may in this grander perspective be just environmental accidents. This naturally disappoints ambitious theorists. But our preferences are irrelevant to the way physical reality actually is - so we should surely be open-minded.

Indeed there's an intellectual and aesthetic upside. If we're in a multiverse, it would imply a fourth and grandest Copernican revolution; we've had the Copernican revolution itself, then the realization that there are billions of planetary systems in our galaxy; then that there are billions of galaxies in our observable universe.

But we'd then realize that not merely is our observable domain a tiny fraction of the aftermath of our big bang, but our big bang is part of an infinite and unimaginably diverse ensemble.

About ten years ago I was on a panel at Stanford University where we were asked by someone in the audience how much we'd bet on the multiverse concept. I said that on the scale 'would you bet your goldfish, your dog, or your life?' I was nearly at the dog level. Andrei Linde, who had spent 25 years promoting 'eternal inflation', said he'd almost bet his life. Later, on being told this, Steven Weinberg said he'd happily bet Martin Rees' dog and Andrei Linde's life.

Andrei Linde, my dog, and I will all be dead before this is settled. But none of this should be dismissed as metaphysics. It's speculative science - exciting science. And it may be true.

6 Concluding Comments

Up till now progress in cosmology and high energy astrophysics has been owed 95% to advancing instruments and technology - less than 5% to armchair theory. I'd expect that balance to continue. But there is one big change, brought about by the fact that computer simulations have hugely advanced in power and expanded in range.

The 1960s were exhilarating for young astrophysicists. So much was new that the old guys didn't have a big head-start over the youngsters. But this is not an occasion for nostalgia. Indeed, I'm sure Paddy would want to emphasise that today is an equally good time for young researchers. The pace of advance has crescendoed rather than slackened; instrumentation and computer power have improved hugely; the frontiers are far more extensive. But in choosing a topic, those entering the subject shouldn't shoot for the most fundamental problem: they should multiply the importance of the problem by their perceived chances of making progress with it, and chose the one that maximizes that product.

Relativistic astrophysics is now 'mature', but other fields have opened up. Exo-planet research is only 20 years old, and serious work in astrobiology is really only starting. Some exo-planets may have biosphere – maybe some will harbour aliens who know all the answers already. And on that encouraging note I'll conclude by thanking Paddy for his friendship and stimulus over three decades, and expressing the hope that his amazing intellectual energy will be sustained for many more years.

References

1. J.R. Oppenheimer, J.R. Snyder, Phys. Rev. **56**, 455 (1939)
2. R.D. Blandford, R.L. Znajek, MNRAS **179**, 433 (1977)
3. C. Reynolds, Astronn. Nachr. **88**, 1 (2015)
4. D.J. Mortlock et al., Nature **474**, 616 (2011)
5. S. Perlmutter et al., Astrophys. J. **517**, 565 (1999)
6. A.D. Reiss et al., Astron. J. **116**, 1009 (1998)
7. P. De Bernadis et al., Nature **404**, 955 (2000)

Area Theorem: General Relativity and Beyond

Sudipta Sarkar

Abstract Gravity being the manifestation of the curvature of spacetime can create regions which are inaccessible to a class of observers. An example of such a region is the event horizon of black objects which acts as a one way causal boundary. The thermodynamics of space time horizons is believed to be a crucial input to understand the quantum dynamics of gravity. The basis of this thermodynamic analogy is the area theorem by Hawking which asserts that the area of the event horizon can not decrease in any classical process. The proof of the area theorem depends on both the validity of Einstein's equation as well as on the cosmic censorship hypothesis. A natural question in this regard could be to ask whether the thermodynamic properties of space time horizons can be generalized beyond general relativity? In this article, we will focus on the "area theorem" of black hole thermodynamics and discuss various possible generalization to higher curvature gravity. We will also discuss how the generalization of the area theorem beyond general relativity leads to various constraints in the couplings of higher curvature terms.

Over the last century, Einstein's ideas have taught us how the gravitational interaction affects the structure of spacetime. This dynamical fabric is the stage in which all other physical objects evolve and influence the way they propagate, sometimes in a very dramatic way, leading to regions that are causally inaccessible to any observer. The prototypical example of this is the event horizon of a black hole, the surface of no return for any infalling object. The event horizon is the absolute causal boundary in the space time and it is formally defined as the boundary of the closure of the causal past of future null infinity. Since the event horizon refers to the asymptotic structure of the space time, it is teleological by definition and therefore the construction of the event horizon requires future boundary conditions.

The event horizon of a black hole separates the space time into two causally disconnected regions. An outside observer has no access of the degrees of freedom inside the horizon and therefore she should be able to formulate the laws of (classical) physics, e.g. thermodynamics without probing the region inside the horizon. But, as

S. Sarkar (✉)
Indian Institute of Technology, Gandhinagar 382355, Gujarat, India
e-mail: sudiptas@iitgn.ac.in

J.S. Bagla and S. Engineer (eds.), *Gravity and the Quantum*,
Fundamental Theories of Physics 187, DOI 10.1007/978-3-319-51700-1_22

noticed by Bekenstein during the 70's, it is possible to lower the entropy of the outside universe by simply throwing an entropic object, say a coffee cup into the black hole. Such a process decreases the entropy of the outside world and leads to an apparent violation of the second law of thermodynamics. Bekenstein proposed a radical solution [1] of this problem by ascribing an entropy to the black hole itself proportional to the area of the event horizon. The main rationale for this idea was a fundamental result of black hole mechanics, namely the Hawking's area theorem [2].

The area theorem asserts that the area of the event horizon can not decrease when matter obeying null energy condition is thrown into the black hole. The area theorem is a direct consequence of the null Raychaudhuri equation for horizon generators (in an affine parametrization t) given by [3] ,

$$\frac{d\theta}{dt} = -\frac{\theta^2}{D-2} - \sigma_{ab}\sigma^{ab} - R_{ab}k^a k^b. \tag{1}$$

The Raychaudhuri equation controls the change of the expansion θ of the null generators k^a and if Einstein's equation along with null energy condition $T_{ab}k^a k^b > 0$ hold, then,

$$\frac{d\theta}{dt} < 0 \tag{2}$$

Note that this statement is true for all null surfaces in a spacetime which is a solution of Einstein's equation. Also, this is a statement on the second derivative of the area and what we now need is a boundary condition to constrain the change of the area. This is done by either appealing to Penrose's null completeness condition or by invoking cosmic censorship hypothesis [3, 4]. Such a condition can not be imposed on a generic null surface and therefore this singles out the preferred role of the event horizon. The conditions ensure that there is no caustic (i.e. $\theta \to -\infty$) in the future of the event horizon which implies that the expansion θ must be positive at every cross section of the event horizon and this leads to the area theorem. There are several possible extensions of area theorem to general causal horizons, for details of such generalizations, refer to [5] and references therein.

The Hawking area increase theorem ensures that the change of black hole entropy compensates the entropy of the object thrown into the horizon and saves the day for the second law. Later it was also found that all the four laws of black hole mechanics are, in fact, simply the ordinary laws of thermodynamics applied to a system with a black hole [6]. The fact that quantum effects make the black holes to radiate exactly with a thermal spectrum [7] eventually established that thermodynamic property of black holes is not mere an analogy but a deep indication of a fundamental feature of quantum gravity. The area theorem is then extended to semi classical gravity as the generalized second law (GSL) [8, 9] which says that the sum of horizon entropy and the entropy of matter fields outside can not decrease. It is the generalized second law which is the full statement of the second law of thermodynamics for a system containing a black hole.

As it is evident, the interpretation of the area of the event horizon as the entropy is crucially dependent on the validity of the Hawking's area theorem which in turn depends on the validity of the Einstein's field equations relating the focusing of null generators k^a of the horizon to the matter content via Raychaudhuri equation. But, it is quite natural to consider modified gravity theories which contains higher curvature terms in the action. As a typical example, consider the perturbative quantization of gravity which leads to a non renormalizable quantum theory and is confronted by uncontrollable infinities. If we treat such a non renormalizable theory as a low-energy effective field theory, adding new counter-terms and couplings at each new loop order, then the effective Lagrangian of gravity can be expressed as,

$$\mathscr{L} = \frac{1}{16\pi G}\left(R + \alpha\,\mathscr{O}(R^2) + \beta\,\mathscr{O}(R^3) + \cdots\right), \tag{3}$$

where $\alpha, \beta, \ldots$ are the new parameters in the theory with appropriate dimensions of length. At the level of the effective theory, all terms consistent with diffeomorphism invariance may appear, but from a phenomenological point of view, only a subset of terms which leads to a well behaved classical theory are obviously more desirable. The detailed structure of these terms will depend on the specifics of the underlying quantum gravity theory, nevertheless, one could adopt a bottom up approach and attempt to find a sub class of terms which retains the essential "good" features of general relativity (GR). In fact, such an approach could also be important to understand the dynamics of quantum gravity itself. So, if we turn on these higher curvature corrections, the field equation will get modified and the area theorem will not hold anymore. But, for specific higher curvature terms, we can still obtain exact black hole solutions as in case of GR [10, 11]. As a result, one could formulate the same problem with the second law for the outside universe and associate an entropy with the black hole horizon. Obviously such an entropy may not be proportional to the area of the horizon as the Hawking area theorem is no longer valid. In fact, following Bekenstein [1], it is possible to understand very easily why the black hole entropy in such a modified theory should not be proportional to the area. To see this, consider the simple case of spherical symmetry and assume that a set of identical particles with same mass m are collapsing in D dimensions to form a black hole of mass M. If each of these particles contains one bit of information (in whatever form, may be information about their internal states etc.), then the total loss of information due to the formation of the black hole will be M/m. Classically, this can be as low as possible, but quantum mechanically there is a bound on the mass of each constituent particle because we want the Compton wavelength of these particles to be less than the radius of the hole r_h. Then, the maximum loss of information will be Mr_h and this is a measure of the entropy of the hole. Note that, we have not used any information about the field equation yet. So, this is completely an off shell result. The field equation will provide a relationship between the mass M and the horizon radius.

Let us now consider the specific case of general relativity. If we solve the vacuum Einstein's equations for spherical symmetry, we obtain the usual Schwarzschild solution with $M \sim r_h^{D-3}$, and this lead to to black hole entropy being proportional to r_h^{D-2}, the area of the horizon.

Next comes the modified gravity, note that in GR, we could have guessed the relationship $M \sim r_h^{D-3}$ simply from dimensional ground. But, with higher curvature terms, we will have new dimensionful constants in our disposal and therefore there could be more complicated relationship between mass and horizon radius. For example, if we restrict ourselves up to only curvature square correction terms with a coupling constant α, we could have a relationship like $M \sim r_h^{D-3} + \alpha\, r_h^{D-5}$, and the second term can be regarded as the sub-leading correction to black hole entropy. So, this simple illustration shows how the presence of new dimensionful constants in modified gravity theories leads to the modification of the black hole entropy.

One could do much better than the above over-simplistic arguments and derive the first law of black hole thermodynamics for any diffeomorphism invariant theory of gravity provided the theory admits stationary black hole solutions with regular bifurcation surface [12, 13]. The entropy can be expressed as an integral of a local geometric quantity on the cross section of the event horizon and is identified with the Noether charge of the Killing isometry which generates the horizon. Although the derivation looks sufficiently general, there are several assumptions. The existence of a regular bifurcation surface is an important requirement. Also, we need to assume that the stationary event horizon is a Killing horizon of a Killing field. This is not automatically guaranteed beyond GR.

Even with these assumptions, there is one important limitation of the Noether charge algorithm. It does not uniquely fix the entropy expression beyond stationarity. One can always add ambiguity terms such that they vanish in the stationary case and hence can not be determined by the Wald's formalism [14, 15]. To be more precise, let us consider a stationary event horizon with null generators k^a in a spacetime which is a solution of a gravity theory with Lagrangian L. We consider another null normal to the horizon l^a which is normalized as $k^a l_a = -1$. Let θ_k and θ_l are the expansions associated with k^a and l^a and $\sigma_{ab}^{(i)}$ are the corresponding shears with $i = k,\ l$. Then, we define the Wald entropy as,

$$S_W = -2\pi \int_C \frac{\partial L}{\partial R_{abcd}} \varepsilon_{ab}\varepsilon_{cd} \sqrt{h}\, dA. \tag{4}$$

where $\varepsilon_{ab} = k_a l_b - k_b l_a$ is the binomal on the cross section of the horizon. This Wald entropy obeys the first law of horizon thermodynamics and coincides with the Noether charge of the Killing vector generating the horizon. But, due to the ambiguities in the Noether charge construction, the black hole entropy can always be expressed as,

$$S = -2\pi \int_C dA \left[\frac{\partial L}{\partial R_{abcd}} \varepsilon_{ab}\varepsilon_{cd} - p\, \theta_k \theta_l - q\, \sigma_k \sigma_l \right]. \tag{5}$$

Here we have used the notation $\sigma_{ab}^{(k)} \sigma^{(l)ab} = \sigma_k \sigma_l$. As can be seen from (5) that the ambiguous terms in the entropy formula involve equal number of k and l subscripts which follows from the fact that these are the only boost invariant combinations that can appear in the entropy functional. Also on a stationary slice the ambiguity terms vanish and then the expression coincides with the Wald formula (4). As a result, the First Law of black hole mechanics [13–15] doesn't fix the coefficients p and q.

It is expected that the features of GR will remain valid provided the contribution of these higher curvature terms are "small". As a result, if we demand the theory to be as consistence as GR, there has to be some constraints on these higher curvature coefficients. Such a constraint indeed may be obtained using the results of black hole thermodynamics by demanding that the intriguing relationship between black holes and thermodynamics survives with higher curvature corrections and the second law of black hole thermodynamics continues to hold. This may impose severe constraints on range of the higher curvature couplings $\alpha, \beta, \ldots$. In the absence of any experimental/observational test, it will be remarkable if only theoretical consistency requirements are enough to find such constraints. In fact, such bounds are so far only obtained in the context of AdS/CFT, from the consistency of the boundary gauge theory. But, any constraint from black hole thermodynamics will be far more general in nature and will be independent on any particular model of quantum gravity. If we want that any theory of quantum gravity in the low energy limit produces consistent corrections to GR in terms of higher curvature terms, the theory must reproduce these constraints.

But, if we hope to obtain any constraint on the higher curvature coupling from black hole thermodynamics using the second law, we first need to fix coefficients p and q etc. uniquely. To achieve this, we use the validity of the second law for linearized perturbations to the black hole.

Let us start with the equation of motion for a generic higher curvature metric theory of gravity which is of the form,

$$G_{ab} + H_{ab} = 8\pi T_{ab}\,, \tag{6}$$

where G_{ab} is the Einstein tensor coming from the Einstein-Hilbert part of the action and H_{ab} is the part coming from higher curvature terms–in theories with a cosmological constant there will also be an additional term proportional to the metric which we can absorb into G_{ab}. T_{ab} is the energy momentum tensor which we will assume to obey the Null Energy condition (NEC): $T_{ab}k^a k^b > 0$ for some null vector k^a. We will use the Raychaudhuri equation for null geodesic congruence which describes the evolution of the expansion along the horizon generating affine parameter t and the null generator of the horizon is $k^a = (\partial_t)^a$.

We write the entropy associated with the horizon as,

$$S = \frac{1}{4}\int_C (1+\rho)\, dA, \tag{7}$$

where $\rho(t)$ contains contribution from the higher curvature terms including ambiguities. In the stationary limit ρ will coincide with the Wald expression (4). Next we define the generalized expansion Θ as the rate of change of the entropy per unit area. The generalized expansion is related with the expansion of null generators as $\Theta = \theta_k + k^a \nabla_a \rho$. The evolution of Θ is can be written in a convenient form, $d\Theta/dt = -8\pi T_{kk} + E_{kk}$ where

$$E_{kk} = H_{kk} + \theta_k \frac{d\rho}{dt} - \rho R_{kk} \\ + k^a k^b \nabla_a \nabla_b \rho - \left(\frac{\theta_k^2}{D-2} + \sigma_k^2 \right) (1+\rho) \,. \tag{8}$$

Consider a situation when a stationary black hole is perturbed by some matter flux obeying NEC. The perturbation can be parametrized by some dimensionless parameter ε. Note that T_{kk} is linear ($\mathcal{O}(\varepsilon)$) in perturbation and so as θ_k, σ_k and $d\rho/dt$. We want to establish that the generalized expansion Θ is positive at every slice of the horizon provided the horizon reaches equilibrium in the future. We have already mentioned that T_{kk} is of order ε, so if E_{kk} is of higher order, i.e. $E_{kk} \sim \mathcal{O}(\varepsilon^2)$, then we obtain up to linear order $d\Theta/dt = -8\pi\, T_{kk} + \mathcal{O}(\varepsilon^2)$ which implies $d\Theta/dt < 0$, on every slice of the horizon. Since, we have already assumed that in the asymptotic future, the horizon again settles down to a stationary state, we must have $\Theta \to 0$ in the future. This will imply that Θ must be positive on every slice prior to the future and as a result the entropy given by (7) obeys a local increase law. Therefore, to establish the linearized second law, we only need to show that the linear order terms in E_{kk} exactly cancel each other. Interestingly, this alone will be enough to obtain the values of both the coefficients p and q introduced in the entropy functional.

Consider a particular example and start with the most general second order higher curvature theory of gravity in D dimensions. The action of such a theory can be expressed as,

$$S = \frac{1}{16\pi} \int d^D x \sqrt{-g} \left(R + \alpha\, R^2 + \beta\, R_{ab} R^{ab} + \gamma\, R_{abcd} R^{abcd} \right) \tag{9}$$

It is reasonable to assume that such a theory admits a stationary black hole solution as in the case of GR. We also expect to have a non stationary black hole solution with in-falling matter by perturbing this solution. Such a spacetime will be the counterpart of the Vaidya solution in general relativity for spherically symmetric case and can be expressed as,

$$ds^2 = -f(r,v)dv^2 + 2dvdr + r^2 d\Sigma_3^2 \tag{10}$$

Σ_3 can be any three dimensional space with positive, negative or zero curvature. We want to use this solution to investigate the issue of second law of black hole mechanics. Note that, the location of the event horizon $r = r(v)$ for this solution can be obtained by solving the equation $\dot{r} = dr(v)/dv = f(r,v)/2$ with appropriate boundary condition. Here, ($\dot{}$) and ($'$) denote respectively derivative with respect to v and r. The null generator of the event horizon is given by $k^a = \{1, f(r,v)/2, 0, \cdots, 0\}$

and the corresponding auxiliary null vector $l_a = \{-1, 0, 0, \cdots, 0\}$. The event horizon has nonzero expansion due to the perturbation caused by in falling matter. We will write the entropy associated with the horizon as in Eq. (5) with the ambiguity terms.

Note that, if we use the metric (10) with the choice of $d\Sigma_3^2$ to be a flat metric then all the shear term will vanish identically and q will remain undetermined. This happens because isometries of the metric on a horizon slice essentially coincides with that of a sphere. So we will break the symmetry by adding a cross term and the metric on the horizon slice takes the form,

$$ds^2 = -f(r, v)dv^2 + 2dvdr + r^2(dx^2 + dy^2 + dz^2 + \varepsilon_1 h(r, v)dxdy) \tag{11}$$

We will assume that this shear mode $h(r, v)$ will be balanced by some matter stress tensor still obeying the NEC. Also we do not require to find the explicit form of $h(r, v)$ for our analysis. We will calculate E_{kk} order by order in ε_1 and extract the coefficients of the linear order terms in ε from the evolution equation. Setting those terms to zero will satisfy the linearized second law and in the process p and q will be determined. Such a procedure immediately gives,

$$p = \beta + \frac{2\gamma}{D-2} \text{ and } q = \frac{2\gamma}{D-2} \tag{12}$$

This fixes all quadratic ambiguity terms in the horizon entropy. Note that there could be even higher order ambiguities but they do not contribute to the linearized second law. Using these values of coefficients, the entropy density for a general curvature square theory becomes [16],

$$\rho = \frac{1}{4} + \frac{1}{2}\left[\alpha R - \beta\left(R_{ab}k^a l^b - \frac{1}{2}\theta_k\theta_l\right) + \gamma \mathscr{R}\right], \tag{13}$$

where $\mathscr{R} = R_{abcd}\gamma^{ac}\gamma^{bd} - K_{ab(i)}K^{ab(i)} + K_{(i)}K^{(i)}$ is the *intrinsic* Riemann scalar on the horizon slice. On a non-equilibrium slice of the horizon this entropy expression differs from the result obtained from the Wald's formula in Eq. (4), but agrees with the holographic result for curvature square theories in [17–20] when the black hole is in an AdS space time. For the case general higher curvature theories, refer to [21].

It is indeed remarkable that the entropy for black holes in AdS space time which obeys linearized second law turns out to be related with the holographic entanglement entropy. It seems that somehow the validity of black hole thermodynamics is already encoded in the holographic principle; the holographic entanglement entropy satisfies the linearized second law while the Wald entropy does not.

Once we fix the linear ambiguity terms, the next obvious thing will be to go to the next order in perturbation and study the quantity E_{kk}. To simplify the calculation, we will discuss the case for spherically symmetric perturbations only and therefore shear is identically zero. Then, the evolution equation will be of the form,

$$\frac{d\Theta}{dt} = -8\pi T_{kk} - \zeta \theta_k^2, \tag{14}$$

Note that we must have $d\Theta/dt < 0$ to have a local entropy increase law. Also, we have $T_{kk} > 0$ by NEC. Now consider a situation where the stationary black hole is perturbed by some matter flux and we are examining the second law when the matter has already entered into the black hole. In that case, the above evolution equation does not have any contribution from matter stress energy tensor and the evolution will be driven solely by the θ_k^2 term. In such a situation, if we demand the entropy is increasing, we have to fix the sign of the coefficient of θ_k^2 term. We evaluate the coefficient in the stationary background and impose the condition that overall sign in front of θ_k^2 is negative. This will give us a bound on the parameters of the theory under consideration.

To compare with some related results, we now consider specific cases. First consider the case for Einstein Gauss-Bonnet (EGB) gravity for which $\beta = -4\alpha$ and $\gamma = \alpha$, with these values, the horizon entropy becomes,

$$S = \frac{1}{4} \int_C dA\,(1 + 2\gamma \mathscr{R}), \tag{15}$$

where $\mathscr{R}$ is the Ricci scalar intrinsic to the horizon. This is the well known expression for Jacobson-Myers entropy which satisfies the linearized second law [15, 22–24].

The expression of ζ is now given by,

$$\zeta = \frac{1}{D-2} + \frac{(D-4)\gamma}{(D-2)^2} \left[6\mathscr{R} - \frac{2\,(D-3)(D-2) f'(r,v)}{r(v)} \right]. \tag{16}$$

As discussed earlier, we will evaluate ζ for different stationary backgrounds and determine bounds on the coefficient γ from the constraint $\zeta > 0$. For the EGB gravity, we first consider the 5-dimensional spherically symmetric, asymptotically flat Boulware-Deser (BD) [10] black hole as the background, for which the horizon radius is related to the mass M as, $r_h^2 + 2\gamma = M$ and the existence of an event horizon demands $r_h^2 > 0$. Now, evaluating ζ for the above background at the horizon $r = r_h$, and imposing that $\zeta > 0$, we obtain the condition, $M > 2|\gamma|$ if $M > 0$. To understand this better, note that we require $M > 2\gamma$ to avoid the naked singularity of the black hole solution for $\gamma > 0$. Thus in this case for a spherically symmetric black hole, ζ will be positive and hence second law will be automatically satisfied. The condition of the validity of the second law is same as that for having a regular event horizon.

Also, for $\gamma > 0$, it is possible to make r_h as small as possible by tuning the mass M. But when γ is negative (a situation that appears to be disfavoured by string theory, see [10, 25] and references therein), r_h cannot be made arbitrarily small and it would suggest that these black holes cannot be formed continuously from a zero temperature set up. Notice that we could have reached the conclusion without the second law if M is considered to be positive–however, our current argument does not need to make this assumption. Due to this pathology, it would appear that the negative Gauss Bonnet coupling case would be ruled out in a theory with no cosmological constant.

The case for the 5-dimensional AdS black hole solution for EGB gravity with cosmological constant $\Lambda = -(D-1)(D-2)/2l^2$ as the background is more interesting. Now the horizon could be of planar, spherical or hyperbolic cross sections. We will first consider black brane solution with planar horizon. Then we obtain $\zeta = 1/(D-2)\,(1 - 2(D-1)\lambda_{GB})$ where we have introduced a rescaled coupling in D dimensions as $\lambda_{GB} l^2 = (D-3)(D-4)\gamma$. Again demanding positivity of ζ we get,

$$\lambda_{GB} < \frac{1}{2(D-1)}. \tag{17}$$

Remarkably, in $D = 5$ this coincides with the bound which has to be imposed to avoid instability in the sound channel analysis of quasi-normal modes of a black hole in EGB theory which is taken to be holographic dual of a conformal gauge theory. It was shown in [26] that when $\lambda_{GB} > 1/8$ the Schroedinger potential develops a well which can support unstable quasi normal modes in the sound channel. It is quite interesting to see that the second law knows about this instability. This bound on λ_{GB} produces a bound on η/s ratio and quite curiously that bound on η/s ratio in large D limit [27] tends to the Einstein value $\frac{1}{4\pi}$.

Another interesting case corresponds to the hyperbolic horizon. In this case, the intrinsic scalar is negative and if we also assume that $\gamma > 0$, then there is an obvious bound on the higher curvature coupling beyond which the entropy itself becomes negative and thereby looses any thermodynamic interpretation. Using the entropy expression in Eq. (15), this bound in general D dimension is found as $\lambda_{GB} < D(D-4)/4(D-2)^2$. If the analysis of the second law has any usefulness, it must provide a more stringent bound for the coupling γ and we will show that is indeed the case. Also, to analyze the case for hyperbolic horizons, we will only consider the so called zero mass limit. In the context of holographic entanglement entropy these topological black holes play an important role as shown in [28–30]. One can relate the entanglement entropy across a sphere to the thermal entropy in $R \times H^{D-2}$ geometry by a conformal transformation.

Now for holographic CFTs one has to evaluate the Wald entropy for these topological black holes as they are dual to the field theory placed on $R \times H^{D-2}$ to obtain the entanglement entropy across a spherical region at the boundary. In our context, imposing $\zeta > 0$, it turns out that the zero mass limit gives the most stringent bound on the coupling λ_{GB} given by,

$$\lambda_{GB} < \frac{9}{100}. \tag{18}$$

First note that this bound on λ_{GB} is independent of the dimensions. Also, comparing with the bound in Eq. (17) we can easily see that up to $D = 6$, the bound in Eq. (18) is strongest but from $D = 7$ onwards Eq. (17) is the strongest one. Next, in the five dimension, the bound in Eq. (18) quite curiously coincide with the tensor channel causality constraint [31–33]. For $D > 5$, this bound (18) from the second law will be stronger than the causality constraints.

In principle, it is possible to repeat these analysis for any higher curvature gravity theory to obtain similar bounds on the higher curvature couplings provided we have an exact stationary black hole solution as the background [34]. These bounds will be necessary if we demand that the second law of thermodynamics holds true for an observer outside the horizon. Any quantum theory of gravity which reproduces such higher curvature corrections and also aims to explain the microscopic origin of black hole entropy must satisfy these bounds. In fact, we can constrain various interesting gravity theories in 4 dimensions by our method. In 4 dimensions, our method is the only one to constrain these theories where the causality based analysis [35] is insufficient. For example, for critical gravity theories in $D = 4$ [36] analyzing black holes in AdS background we obtain the bound on the coupling (α_c), $-\frac{1}{2} \leq \alpha_c \leq \frac{1}{12}$. Also, for New Massive gravity in $D = 3$ [37, 38] we obtain the bound on couplings (σ) as, $-3 \leq \sigma \leq \frac{9}{25}$.[1]

Before discussing more conceptual issues, let us first summarize the main results: To obtain a generalization of the area theorem beyond general relativity, we need to fix the ambiguities in the definition of the Wald entropy for higher curvature gravity. We studied the linearized second law and fix such ambiguities uniquely. Next, we turned on the perturbation of higher order and obtained the evolution equation for the entropy. Using a spherically symmetric Vaidya type solution, we found that the requirement of entropy increase is that the quantity ζ in Eq. (16) must be positive. To obtain the bound on higher curvature coupling, we evaluate this ζ for various known static black hole solutions of the theory. For Einstein-Gauss Bonnet gravity, we get stringent bounds on the Gauss Bonnet couplings. Remarkably, some of these bounds coincide with the constraints on the coupling obtained by using the consistency of the boundary gauge theory. But, these bounds are independent of any quantum gravity model and should be a generic feature of the microscopic description of black holes.

Now, the important question which still left unanswered is the possible interpretation of these bounds from quantum gravity. In the context of fluid gravity duality, the area theorem of the black holes is related to the positivity of the divergence of the entropy current of the boundary theory [39]. It is reasonable that if these bounds are satisfied, a similar relationship also holds beyond GR and for the black holes in higher curvature theories. Then these bounds can be understood as the consistency condition for the applicability of the fluid gravity duality paradigm. Also, similar bounds are obtained from the causality and unitarity constraints of the boundary theory. It seems that the validity of the black hole thermodynamics in the bulk is indeed related with such conditions. In fact, if the black hole in a higher curvature gravity in the bulk does represent a thermal state in the boundary theory, it is possible that such

[1] The lower bound for both these two cases are coming from demanding the positivity of the entropy.

a connection is valid only for a finite range of higher curvature couplings. After all, the generic higher curvature terms introduce problematic features like perturbative ghosts etc. What is interesting is that even a well-formulated, ghost free higher curvature theory like Einstein Gauss Bonnet gravity also has these bounds. This may be directly related to some recent results on the causality constraints in these theories [35].

At the least, the existence of these bounds shows that a bottom up approach based on the study of theoretical consistency could also provide important clues about the necessary features expected from the quantum theory of gravity. The low energy approximation of any sensible quantum gravity theory should produce these higher curvature terms as the correction to GR. In the absence of any experimental or observational constraints, such bounds on the higher curvature couplings could be crucial to identify the correct approach of quantum gravity. The fact that AdS/CFT already reproduces some of these bounds is quite encouraging and can be considered as a strong support in favour of the importance of holographic principle in quantum gravity coming from the generalization of the Area theorem.

Acknowledgements The results discussed here are from the works [16, 34] done in collaboration with Srijit Bhattacharjee, Arpan Bhattacharyya, Aninda Sinha and Aron Wall. The research of SS is supported by the Department of Science and Technology, Government of India under the Fast Track Scheme for Young Scientists (YSS/2015/001346).

References

1. J.D. Bekenstein, Phys. Rev. D **7**, 2333 (1973). doi:10.1103/PhysRevD.7.2333
2. S.W. Hawking, Commun. Math. Phys. **25**, 152 (1972)
3. R.M. Wald, *General Relativity* (Univ. Pr, Chicago, USA, 1984)
4. S.W. Hawking, G.F.R. Ellis, *The Large Scale Structure of Space-Time* (Cambridge University Press, Cambridge, 1973)
5. T. Jacobson, R. Parentani, Found. Phys. **33**, 323 (2003). [arXiv:gr-qc/0302099]
6. J.M. Bardeen, B. Carter, S.W. Hawking, Commun. Math. Phys. **31**, 161–170 (1973)
7. S.W. Hawking, Commun. Math. Phys. **43**, 199 (1975) Erratum: [Commun. Math. Phys. **46**, 206 (1976)]. doi:10.1007/BF02345020
8. J.D. Bekenstein, Lett. Nuovo Cim. **4**, 737 (1972). doi:10.1007/BF02757029
9. J.D. Bekenstein, Phys. Rev. D **9**, 3292 (1974). doi:10.1103/PhysRevD.9.3292
10. D.G. Boulware, S. Deser, Phys. Rev. Lett. **55**, 2656 (1985)
11. R.C. Myers, J.Z. Simon, Phys. Rev. D **38**, 2434 (1988)
12. R.M. Wald, Phys. Rev. D **48**, 3427–3431 (1993). [gr-qc/9307038]
13. V. Iyer, R.M. Wald, Phys. Rev. D **50**, 846 (1994). [arXiv:gr-qc/9403028]
14. T. Jacobson, G. Kang, R.C. Myers, Phys. Rev. D **49**, 6587 (1994). [arXiv:gr-qc/9312023]
15. T. Jacobson, G. Kang, R.C. Myers, Phys. Rev. D **52**, 3518–3528 (1995). [gr-qc/9503020]
16. S. Bhattacharjee, S. Sarkar, A.C. Wall, Phys. Rev. D **92**(6), 064006 (2015). [arXiv:1504.04706 [gr-qc]]
17. D.V. Fursaev, A. Patrushev, S.N. Solodukhin, Distributional geometry of squashed cones. Phys. Rev. D **88**(4), 044054 (2013). [arXiv:1306.4000 [hep-th]]
18. S.N. Solodukhin, Entanglement entropy, conformal invariance and extrinsic geometry. Phys. Lett. B **665**, 305 (2008). [arXiv:0802.3117 [hep-th]]

19. J. Camps, Generalized entropy and higher derivative Gravity. JHEP **1403**, 070 (2014). [arXiv:1310.6659 [hep-th]]
20. X. Dong, Holographic entanglement entropy for general higher derivative gravity. JHEP **1401**, 044 (2014). [arXiv:1310.5713 [hep-th]]
21. A.C. Wall, arXiv:1504.08040 [gr-qc]
22. A. Chatterjee, S. Sarkar, Physical process first law and increase of horizon entropy for black holes in Einstein-Gauss-Bonnet gravity. Phys. Rev. Lett. **108**, 091301 (2012). [arXiv:1111.3021 [gr-qc]]
23. S. Kolekar, T. Padmanabhan, S. Sarkar, Entropy increase during physical processes for black holes in Lanczos-Lovelock gravity. Phys. Rev. D **86**, 021501 (2012). [arXiv:1201.2947 [gr-qc]]
24. S. Sarkar, A.C. Wall, Generalized second law at linear order for actions that are functions of Lovelock densities. Phys. Rev. D **88**, 044017 (2013). [arXiv:1306.1623 [gr-qc]]
25. A. Buchel, R.C. Myers, A. Sinha, Beyond eta/s = 1/4 pi. JHEP **0903**, 084 (2009). [arXiv:0812.2521 [hep-th]]
26. A. Buchel, R.C. Myers, Causality of holographic hydrodynamics. JHEP **0908**, 016 (2009). [arXiv:0906.2922 [hep-th]]
27. R. Emparan, D. Grumiller, K. Tanabe, Large-D gravity and low-D strings. Phys. Rev. Lett. **110**(25), 251102 (2013). [arXiv:1303.1995 [hep-th]]
28. R.C. Myers, A. Sinha, Phys. Rev. D **82**, 046006 (2010). [arXiv:1006.1263 [hep-th]]
29. R.C. Myers, A. Sinha, JHEP **1101**, 125 (2011). [arXiv:1011.5819 [hep-th]]
30. H. Casini, M. Huerta, R.C. Myers, JHEP **1105**, 036 (2011). [arXiv:1102.0440 [hep-th]]
31. D.M. Hofman, J. Maldacena, Conformal collider physics: energy and charge correlations. JHEP **0805**, 012 (2008). [arXiv:0803.1467 [hep-th]]
32. M. Brigante, H. Liu, R.C. Myers, S. Shenker, S. Yaida, The viscosity bound and causality violation. Phys. Rev. Lett. **100**, 191601 (2008). [arXiv:0802.3318 [hep-th]]
33. A. Buchel, J. Escobedo, R.C. Myers, M.F. Paulos, A. Sinha, M. Smolkin, Holographic GB gravity in arbitrary dimensions. JHEP **1003**, 111 (2010). [arXiv:0911.4257 [hep-th]]
34. S. Bhattacharjee, A. Bhattacharyya, S. Sarkar, A. Sinha, arXiv:1508.01658 [hep-th]
35. X.O. Camanho, J.D. Edelstein, J. Maldacena, A. Zhiboedov, Causality constraints on corrections to the graviton three-Point Coupling. arXiv:1407.5597 [hep-th]
36. H. Lu, C.N. Pope, Critical gravity in four dimensions. Phys. Rev. Lett. **106**, 181302 (2011). [arXiv:1101.1971 [hep-th]]
37. E.A. Bergshoeff, O. Hohm, P.K. Townsend, Massive gravity in three dimensions. Phys. Rev. Lett. **102**, 201301 (2009). [arXiv:0901.1766 [hep-th]]
38. D. Grumiller, O. Hohm, AdS(3)/LCFT(2): correlators in new massive gravity. Phys. Lett. B **686**, 264 (2010). [arXiv:0911.4274 [hep-th]]
39. S. Bhattacharyya, V.E. Hubeny, R. Loganayagam, G. Mandal, S. Minwalla, T. Morita, M. Rangamani, H.S. Reall, JHEP **0806**, 055 (2008). [arXiv:0803.2526 [hep-th]]

What Are the Atoms of the Space Time?

S. Shankaranarayanan

Abstract Equations of gravity when projected on space time horizons resemble Navier–Stokes equation of a fluid with a specific equation of state. Taking the view that the horizon fluid possesses some kind of physical reality beyond the formal mathematical similarity, we provide a statistical mechanical description for such fluids. We show that the model passes two crucial tests — obtaining the correct black hole entropy and negative bulk viscosity. We also show that the horizon fluid predicts the occurrence of mass gap thereby implying horizon-area quantization. We then give a brief sketch of how to go about identifying the atoms of space time.

Paddy

As I come to know *Paddy* more and more, I get reminded of the famous quote by Mr. Aldous Huxley: *The secret of genius is to carry the spirit of the child into old age, which means never losing your enthusiasm.* I must say that Paddy's enthusiasm for Physics has increased over the years, and his mind is young and creative. His enthusiasm has enabled to unearth new physics, and make original and fundamental contributions in Quantum Cosmology [1], Statistical mechanics of gravitating systems [2], Structure formation [3, 4], Dark energy [5–7] and emergent gravity [8–11]. His enthusiasm is quintessential for his prolific book writing. He now says that he is planning a bigger volume set spanning the whole of Physics. I wish him all the best and look forward to see more creative ideas from him in the future!

1 Introduction

In the modern viewpoint, general relativity (GR) and the standard model of particle physics (which is based on quantum field theory) are effective descriptions valid below a certain cutoff momentum scale. This implicitly assumes that the high-energy modes *decouple* from low-energy phenomena and that the low-energy influence of

S. Shankaranarayanan (✉)
School of Physics, Indian Institute of Science Education and Research Thiruvananthapuram, Trivandrum 695 016, India
e-mail: shanki@iisertvm.ac.in

J.S. Bagla and S. Engineer (eds.), *Gravity and the Quantum*,
Fundamental Theories of Physics 187, DOI 10.1007/978-3-319-51700-1_23

the high-energy modes is through the determination of coupling constants and mass of particles.

There are, however, two familiar settings where this separation breaks down and high-energy effects are, potentially, unmasked. These are the inflationary expansion of the universe and the black hole horizon. In the case of inflation, the exponential expansion causes a redshift from above to below the effective field theory cut-off. Hence, inflation provides us with the window of opportunity to observe high-energy — which may be inaccessible to terrestrial experiments — in the Cosmic Microwave Background. In the case of black holes, the low-energy outgoing modes evolve from modes with energies above the cut-off due to the redshift near the horizon [12–18]. This enables us to use black holes as theoretical laboratories to test models of quantum gravity.

At the heart of the difficulty of quantum gravity lies the different tenets quantum theory and general theory of relativity derive from. Whereas gravity is diffeomorphism invariant and has no preferred coordinate system for a space time, quantum theory requires the notion of a time-like Killing vector field to define particles; this implies that different observers have different particle interpretation. While general relativity describes dynamics of classical space time that has a causal structure, quantum theory deals with unitarily evolving states in a Hilbert space. Thus, the presence of gravity raises two fundamental questions: first, whether unitary evolution is preserved in curved space time; second, what is the right set of observables that are observer independent. The quantum nature of black hole evaporation, and in general, black hole thermodynamics may offer the key to answer these questions.

Many interesting features of gravity have arisen since the formal relation between the laws of thermodynamics and laws of black hole dynamics were found [10, 11, 14, 19]. Recent interest in the fluid-gravity correspondence can be considered as an extension of black hole thermodynamics, where charges are upgraded into local currents, and black hole entropy into a local entropy current [20–27]. While black hole mechanics highlights a disparity in the form of the black hole information problem, the fluid-gravity correspondence allows the possibility to connect macroscopic and microscopic physics through the study of the statistical properties of the fluid on the horizon of the black hole.

It's Deja Vu all over again! In late 1960s and early 1970s, the analogy between black holes and ordinary thermodynamics was treated a mathematical curiosity. However, only after Hawking's famous discovery of the evaporation of black holes, it was realized that the pairs of analogues between black holes and thermodynamics are indeed physically similar. Likewise, mathematical similarity between equations of General relativity near the black hole horizon and fluids is known for a long time [20–22]. In a series of papers, our group has shown that the horizon fluid is itself of physical interest, and that at least an effective theory describing this fluid as a condensate can be formulated [28–33]. In this review, we will discuss this programme in detail.

The rest of this review is organized as follows: The next section gives a brief review of black hole thermodynamics and the problems associated with that. Based on Ref. [33], in Sect. 3, we derive Damour–Navier Stokes equation for a generic $D-$

dimensional space time and obtain the first constraint equation for the horizon fluid. Based on Refs. [30, 33], in Sect. 4, we calculate the energy corresponding to the horizon fluid of a stationary, asymptotically flat space times. In Sect. 5, we describe the statistical description of the horizon fluid and using mean field theory, we obtain entropy of the horizon fluid [28, 29]. In Sect. 6, we develop the statistical mechanical description of the fluctuations of the horizon fluid, and using fluctuation-dissipation theory, we obtain the bulk viscosity of the horizon fluid. We also show how the matching of bulk viscosity with Damour's leads to quantization of horizon area. Finally, in Sect. 7, we discuss the implications of the results and attempt to answer the title of this article.

2 Black Hole Thermodynamics

One of the most remarkable features of black hole physics is the realization that black holes behave as thermodynamic systems and possess entropy and temperature. The pioneering works in the field of black hole thermodynamics started with Bekenstein [12, 13, 15, 17], who argued that the universal applicability of the second law of thermodynamics rests on the fact that a black hole must possess an entropy (S_{BH}) proportional to the area (A) of its horizon. The macroscopic properties of black holes were subsequently formalized by Bardeen, Carter and Hawking [14] as the four laws of black hole mechanics, in analogy with ordinary thermodynamics.

Hawking's demonstration of black hole thermal radiation [16, 18] paved the way to understand the physical significance of the temperature T_{H} (and hence the entropy-area proportionality). Hawking showed that quantum effects in the background of a body collapsing to a Schwarzschild black hole leads to the emission of a thermal radiation at a characteristic temperature:

$$T_{\text{H}} = \left(\frac{\hbar c}{k_{B}}\right)\frac{\kappa}{2\pi} = \left(\frac{\hbar c^{3}}{G k_{B}}\right)\frac{1}{8\pi M}, \tag{1}$$

where κ is the surface gravity, G is the Newton's constant in four dimensions, k_{B} is the Boltzmann constant, and M is the mass of the black hole. The factor of proportionality between temperature and surface gravity (and as such between entropy and area) gets fixed in Hawking's derivation [18], thus leading to the Bekenstein–Hawking area law:

$$S_{\text{BH}} = \left(\frac{k_{B}}{4}\right)\frac{A}{\ell_{\text{Pl}}^{2}}\ , \tag{2}$$

where $\ell_{\text{Pl}} = \sqrt{G\hbar/c^{3}}$ is the four dimensional Planck length.

Black-hole thermodynamics raises several important questions:

1. Unlike other thermodynamical systems, why is black hole entropy non-extensive? i.e. why S_{BH} is proportional to area and not volume?

2. Why is the black hole entropy large?
3. How S_{BH} concords with the standard view of the statistical origin? What are the black hole micro-states?

$$S \stackrel{?}{=} k_B \ln (\# \text{ of micro-states}) \tag{3}$$

These questions often seem related, which a correct theory of quantum gravity is expected to address. In the absence of a workable theory of quantum gravity, there have been several approaches which address one or several of the above questions. Most of the effort in the literature has been to understand the microscopic statistical mechanical origin of S_{BH} assuming that the black hole is in a (near) thermal equilibrium or not interacting with surroundings. However, black hole thermodynamics now has the *problem of Universality* [34]; at the leading order, several approaches using completely different microscopic degrees of freedom lead to Bekenstein–Hawking entropy [35]. Currently, it is not possible to identify which are the true degrees of freedom that are responsible for the black hole entropy [36]. Therefore other tests of reproducing black hole physics are key in distinguishing such models.

As we will show in the rest of this review, Fluid/gravity correspondence — projecting the Einstein equations onto the black hole horizon lead to Navier–Stokes style equation [20–27] — can provide a way to understand these black hole micro-states from the microscopic degrees of freedom of the horizon fluid. This is more interesting as we have a better understanding about the microscopic degrees of freedom of most fluid systems than gravity. More specifically, given a horizon fluid equation of state, it is possible to constrain the microscopic degrees of freedom and, hence, the *problem of Universality* encountered in black hole thermodynamics can be curtailed [28, 30, 37–39].

3 Fluid Gravity Correspondence in Arbitrary Dimensions

Historically, the Fluid-Gravity correspondence first received support from two closely related threads of work. Damour [20] and Thorne et al. [21, 22] showed that the black hole horizon can be described as a fluid. However, there exists a technical difference in the way the fluid gravity correspondence was deduced by these two groups, respectively. While, Damour [20] directly projected the Einstein's equations on a black hole horizon and showed that it gives rise to the Navier–Stokes equation of a 2 dimensional relativistic fluid that lives on the black hole horizon, Thorne et al. [21, 22] projected the equations of General Relativity on to a time-like hyper-surface close to the black hole horizon and obtained non-relativistic Navier–Stokes equation. In this review, the correspondence with the relativistic fluid would be more fruitful.

Padmanabhan obtained an entropy extremization principle to derive the Damour Navier Stokes (DNS) equation directly, that makes the hydrodynamical analogy with gravity self-contained [25]. In this section, we extend the analysis to D-dimensional space time [33]. As in Ref. [25], we will use small Latin letters for indices running over all D dimensions, Greek letters for $(D-1)$–dimensional null surface, and capital Latin for $(D-2)$ space-like dimensions.

Starting with the Einstein equations,

$$R_{ab} - \frac{1}{2}g_{ab} + \Lambda g_{ab} = 8\pi T_{ab}, \tag{4}$$

consider a null surface, which is therefore traced out by null geodesics. These are described by the geodesic equation (in this case with a non-affine parameterization),

$$l^a \nabla_a l_b = \kappa l_b. \tag{5}$$

We can choose coordinates such that

$$l = \partial_t + v^A \partial_A; \qquad l^a = (1, v^A, 0). \tag{6}$$

We can also, for convenience, construct another null vector, such that $k \cdot l = -1$.

The metric on the $(D-2)$–dimensional surface, denoted by q_{AB}, for which

$$ds^2 = q_{AB}(dx^A - v^A dt)(dx^B - v^B dt), \tag{7}$$

$$q_{ab} = g_{ab} + l_a k_a + l_b k_a. \tag{8}$$

It can explicitly be seen that

$$q_{ab} l^b = q_{ab} k^b = 0. \tag{9}$$

The Damour–Navier–Stokes equation is a consequence of the contracted Codazzi equation formed with l and q_{AB},

$$\begin{aligned} R_{mn} l^m q^n_a &= \left(\frac{1}{2}g_{ab} - \Lambda g_{ab} + 8\pi T_{ab}\right) l^m q^n_a \\ &= 8\pi T_{ab} l^m q^n_a. \end{aligned} \tag{10}$$

Here the last equality is through the relation of (9). Note that the cosmological terms drops out as

$$\Lambda g_{ab} l^m q^n_a = \Lambda(q_{mn} - l_m k_n - l_n k_m) l^m q^n_a = 0. \tag{11}$$

Re-write the LHS as

$$R_{mA} l^m = R_{\mu A} = \nabla_\mu(\nabla_A l^\mu) - \partial_A(\nabla_\mu l^\mu). \tag{12}$$

We expand both terms of the RHS in terms of expansion, shear and the velocity v_A.

Define

$$\chi^\beta_\alpha = \nabla_\alpha l^\beta \tag{13}$$

and

$$\omega_\alpha = \chi^0_\alpha, \tag{14}$$

which is the energy-momentum vector of the horizon fluid
As in Eq. (21) of [25] we expand out the second term of the RHS,

$$\partial_A(\nabla_\mu l^\mu) = \nabla_A l^A + \nabla_0 l^0 = \theta + \omega_A v^A + \omega_0 = \theta + \kappa, \tag{15}$$

and is independent of D. The other term of Eq. (12),

$$\nabla_\mu(\nabla_A l^\mu) = \nabla_\mu \chi^\mu_A, \tag{16}$$

can be evaluated by taking a frame where we neglect Christoffel symbols, so that

$$\nabla_\mu \chi^\mu_A = \partial_\mu \chi^\mu_A = \partial_0 \omega_A + \partial_B \chi^B_A. \tag{17}$$

Now use the fact that

$$\chi_{AB} = \Theta_{AB} + \omega_A v_B, \tag{18}$$

where we can split Θ_{AB} into trace and traceless parts

$$\Theta_{AB} = \sigma_{AB} + \left(\frac{1}{D-2}\right)\theta\delta^A_B. \tag{19}$$

Note the dimensional dependent prefactor, ensuring $\mathrm{tr}(\Theta) = \theta$.
Putting all this together into Eq. (12), we have,

$$R_{mA} l^m = (\partial_0 + v^B \partial_B)\omega_A - \partial_A(\kappa + \theta) + \partial_B \left(\sigma^B_A + \frac{1}{D-2}\theta\delta^B_A\right), \tag{20}$$

so that

$$8\pi T_{mA} l^m = (\partial_0 + v^B \partial_B)\omega_A + \partial_B \sigma^B_A - \partial_A \kappa - \frac{D-3}{D-2}\partial_A \theta \tag{21}$$

or equivalently

$$\frac{D\Pi_A}{dt} = -\frac{\partial}{\partial x^A}\left(\frac{\kappa}{8\pi}\right) + 2\frac{1}{16\pi}\sigma^B_{A|B} - \left(\frac{D-3}{D-2}\right)\frac{1}{8\pi}\frac{\partial\theta}{\partial x^A} - l^a T_{aA}. \tag{22}$$

We thus arrive at Navier–Stokes equation with pressure:

$$P = \frac{\kappa}{8\pi} = k_B \frac{T}{4}, \tag{23}$$

shear viscosity

$$\eta = \frac{1}{16\pi}, \tag{24}$$

and bulk viscosity

$$\xi = -\left(\frac{D-3}{D-2}\right)\frac{1}{8\pi}. \tag{25}$$

We would like to stress the following points regarding this result:

1. The above result relating fluid's pressure and horizon's surface gravity holds true for a general null surface. In the case of Schwarzschild black hole and in Schwarzschild radial coordinates, Damour–Navier–Stokes equation leads to the following simple equation $p_{,i} = 0$. Where i labels the angular coordinates. This is a trivial way to express the zero-th law of black hole thermodynamics, which says that the black hole temperature (or equivalently the horizon surface gravity) is constant across the horizon. For fluid, this means that the pressure is constant across the fluid.
2. Damour–Navier–Stokes equation in D−dimensional space time leads to the first constraint Eq. (23) relating the Pressure and temperature. As we will see in the next section, the horizon fluid also satisfies another constraint and this makes the horizon fluid an unique fluid system.
3. For the 4-dimensional Schwarzschild black hole the constraint (23) can be somewhat naively derived from the definition of pressure using the black hole entropy,

$$P = \frac{\partial S_{\rm BH}}{\partial A} \cdot T_H = \frac{T_H}{4}. \tag{26}$$

   Equation (26) boils down to the fact that for Schwarzschild black hole:

$$dM = P \cdot dA = P \cdot dV. \tag{27}$$

4. The horizon fluid also has another peculiar feature, its bulk viscosity is negative.

4 Horizon Fluid Energy

To understand thermodynamic relations for the horizon fluids, we must first define the appropriate notion of the fluid energy. There are several definitions of energy in general relativity [40–42]. Here we want a quasi-local notion to associate to the black hole horizon. Various competing definitions exist, but our task is simplified by the fact that we have a naturally defined fluid energy density on horizon

$$\omega_0 \equiv \nabla_0 l^0 = \kappa \tag{28}$$

where the equality follows from (15), noting that as the $v_A\omega_B$ is antisymmetric its trace is automatically zero.

The total energy of the horizon is now simply

$$\int \kappa \mathrm{d}A \tag{29}$$

where both κ and A are functions of all the black hole variables (M, Q, a). One can quickly see that this gives the expected M in the $D-$dimensional Schwarzschild.

In fact, from Eqs. (12.5.33)–(12.5.37) in [42], one can see that this is exactly

$$E = \int \kappa \mathrm{d}A = \oint_S \mathrm{d}S_{\mu\nu}\nabla^\mu l^\nu; \qquad l = \xi + \Omega\psi \tag{30}$$

when Ω is the rotational velocity at the horizon, making l the standard combination considered when evaluating, e.g. the surface gravity of black holes. This corresponds to a quasi-local mass frequently used in the literature [43, 44], evaluated on the horizon. This is almost equal to the well-known Komar mass, the difference being the replacement $\xi \to l$. Physically this corresponds to a quasi-local energy in the co-rotating frame [45], clearly the appropriate choice for the fluid that co-rotates with the horizon, as ours does.

The black hole is an odd system in many ways, as P, T, E and A are not independent. Instead they obey (23) and an extra constraint equation. The form of this equation varies depending on the class of black holes. We derive this relation for the D-dimensional Schwarzschild black hole and the (4D) Kerr-Newman black hole.

4.1 D-Dimensional Schwarzschild

The D-dimensional Schwarzschild, also known as a the Schwarzschild-Tangherlini black hole [46], has the form

$$\mathrm{d}s^2 = -\left(1 - \left(\frac{r_H}{r}\right)^{D-3}\right)\mathrm{d}t^2 + \frac{\mathrm{d}r^2}{1 - \left(\frac{r_H}{r}\right)^{D-3}} + r^2\mathrm{d}\Omega^2_{D-2}, \tag{31}$$

where

$$\Omega_{D-2} = \frac{2\pi^{\frac{D-1}{2}}}{\Gamma\left(\frac{D-1}{2}\right)}. \tag{32}$$

The horizon "area" is now given by

$$A = \Omega_{D-2} r_H^{D-2}. \tag{33}$$

and the standard temperature is

$$T = \frac{(D-3)}{4\pi k_B r_H}. \tag{34}$$

Here the horizon radius is related to the Komar (and equivalently the ADM) mass of the black hole by

$$r_H = \left(\frac{16\pi M.}{(D-2)\Omega_{D-2}^2} \right)^{\frac{1}{D-3}}. \tag{35}$$

This M is the energy of the horizon fluid, so (33)–(35) combine to give a constraint equation

$$E = \left(\frac{D-2}{D-3} \right) \frac{A}{4} k_B T. \tag{36}$$

4.2 Kerr–Newman

The charged, rotating black hole,

$$ds^2 = -\left[dt - a\sin^2\theta d\phi\right]\frac{\Delta}{\rho^2} + \left[(r^2 + a^2)d\phi - adt\right]^2 \frac{\sin^2\theta}{\rho^2} - \left[\frac{dr^2}{\Delta} + d\theta^2\right]\rho^2, \tag{37}$$

where

$$\rho^2 = r^2 + a^2\cos^2\theta; \quad \Delta = r^2 - 2Mr + Q^2 + a^2. \tag{38}$$

has inner and outer horizons given by

$$r_\pm = M + \sqrt{M^2 - Q^2 - a^2} \tag{39}$$

We will only be concerned by the physically relevant outer horizon.

The horizon area and black hole temperature can be easily seen to be

$$A = 4\pi(r_+^2 + a^2); \qquad T = \frac{1}{2\pi k_B} \frac{r_+ - r_-}{2(r_+^2 + a^2)}. \tag{40}$$

And the energy on horizon is [44]

$$E = \frac{r_+ - r_-}{2}. \tag{41}$$

Combining these results we can see that

$$E = \frac{A}{2} k_B T. \tag{42}$$

Note that this is identical to the expression for the D-dimensional black hole. These constraint equations play a crucial role in the derivation of the entropy and the bulk viscosity. These point that the constraint Eqs. (23), (36), (42) imply that the physical mechanism that drive the horizon fluid from an initial configuration, say, P_1, T_1, E_1, A_1 to final configuration P_2, T_2, E_2, A_2 can not be arbitrary and, hence, the fluctuations of these macroscopic quantities are also constrained.

5 Modeling the Horizon Fluid

As mentioned above, the horizon fluid corresponding to any stationary black hole is an odd system; the macroscopic parameters P, T, E and A (volume) are not independent. Combining Eqs. (23) and (42) one gets

$$P = k_{_B} \frac{T}{4} = \frac{E}{2\,A}.$$

This is an equation of state of a 2-D ideal massless relativistic gas [47]. The equation of state provides a possibility to identify a microscopic model for generic asymptotically flat 4-dimensional black hole space times using the fluid-gravity correspondence. For a microscopic theory defined by a Hamiltonian, thermodynamic quantities are typically a function of three independent parameters (E, N, V), where E is system's energy, N is the number of particles and V is the volume. However, the parameter space of the fluid corresponding to these black holes is one-dimensional and can be expressed by a single parameter $V = A$ (or E.) Therefore, if a microscopic model is supposed to reproduce the properties of the black holes, one needs to constrain the space of states of the fluid to only one dimension. This means from the point of view of the *relevant* states of the system one need not treat the variables E, N, V as independent, but they must fulfill two constrains: $E = E(V)$ and $N = N(V)$.

5.1 Ideal Gas Model of Horizon Fluid

The energy levels of a free non-relativistic particles living on a sphere is [48, 49]:

$$\varepsilon_\ell^{nr} = \frac{\ell(\ell+1) + \alpha^2}{2R^2} \tag{43}$$

where R is a radius of the sphere and α^2 is some constant. (In a strict sense there are differences between [48, 49] on whether the constant is allowed to be non-zero. However any energy quantum spectrum can be always shifted by a constant, which in best case is fixed by gravity considerations, such as these, so physically we stick to the spectrum given by Eq. (43) with *arbitrary* α.) The energy levels correspond to the Laplacian on the sphere and therefore have the degeneracy of the spherical harmonics, given as $g_\ell = 2\ell + 1$.

As one can see from the Hamiltonian used in [48], we can obtain spectrum of a massless relativistic scalar particle by a simple transformation:

$$\varepsilon_\ell^r = \sqrt{2\varepsilon^{nr}} = \frac{\sqrt{\ell(\ell+1)+\alpha^2}}{\sqrt{2}\cdot R} = \sqrt{\frac{4\pi\{\ell(\ell+1)+\alpha^2\}}{V}}. \tag{44}$$

This can be (through our constrains) expressed as a simple function of temperature

$$\varepsilon_\ell^r = \sqrt{\{\ell(\ell+1)+\alpha^2\}}\cdot T = \tilde{\varepsilon}_\ell \cdot T, \tag{45}$$

with

$$\tilde{\varepsilon}_\ell = \sqrt{\ell(\ell+1)+\alpha^2}.$$

($\tilde{\varepsilon}_\ell$ is independent on the black hole parameters.)

Immediate question is: What does the spectrum (45) physically imply? [28]. For a simple harmonic oscillator[1] the equipartition law ceases to hold for very low temperatures, when the temperature is of a comparable value to the discrete spacing between the energy levels. Furthermore, for temperatures of a value comparable to the spacing between the energy levels the average particle's energy is very close to the ground state energy. From Eq. (45) one observes that, for the horizon fluid, the temperature always is of a comparable value to the spacing between the energy levels, which means the equipartition law is not applicable.

This implies that the free gas is supposed to have mean particle's energy very near the ground state. In such case we can use the spectrum (45) to see that the mean particle's energy must be:

$$\bar{E} = \gamma \cdot k_B T,$$

where $\gamma \geq |\alpha|$ and γ being approximately of the same order as $|\alpha|$. Since

$$E = N\bar{E} = N\gamma \cdot k_B T \equiv (8\pi k_B T)^{-1},$$

this implies

$$N = \frac{1}{8\pi\gamma T^2} = \frac{A}{2\gamma}.$$

Therefore

$$A = 2\gamma \cdot N. \tag{46}$$

[1]The spacing between the energy levels from Eq. (44) approaches for large ℓ harmonic oscillator's energy levels, for small energy levels the difference between the levels is larger than in the case of oscillator.

Equation (46) is remarkable as it gives Bekenstein's [13] quantization of the black hole horizon area (since N is by definition a natural number). It gives a completely new and independent insight into Bekenstein's result. Furthermore, the insight does not rely on the quantum theory, only on the fluid interpretation of gravity. (The constant γ is here arbitrary, but can be fixed to obtain the most popular form of Bekenstein type of spectrum as $\gamma = 4\pi$. Since it will be shown that $\gamma = |\alpha|$, it can be demonstrated that this fixing gives wave-length of the particle in the ground state equal to the circumference of the black hole horizon.) As we will show later, the same quantization condition will be required to derive transport coefficients from the fluctuation-dissipation theorem.

5.2 Horizon Fluid as a Bose Einstein Condensate

The interesting feature of the above analysis is that constrains (23), (42) relation between the Black-Hole area and energy, lead to a ground state populated by the Bose particles. This strongly suggests the occurrence of a phase transition and Bose Einstein condensation (BEC).

The other hint comes from the fact that the specific heat for the Schwarzschild goes as $(T - T_c)^{-1}$ [50, 51] and diverges as T goes to T_c. This suggests that the system might be near a critical point. Such a system can be thought of near the critical point of a second order phase transition and can be described by Mean Field theory. While the Mean Field Theory cannot give the correct value of the critical exponent, it can naturally account for the power law divergence of the specific heat of a black hole horizon fluid [50–52].

With this insight, the following assumptions can be made [29]:

1. There is a temperature T_c (critical temperature), at which, all the N microscopic DOF on the horizon form a condensate.[2]
2. The system always remains close to the critical point, where the phase transition takes place.

The above assumptions immediate leads us to the total energy of the fluid of N particles [28]

$$E \propto N\varepsilon \propto N/r_H = N\alpha k_B T \tag{47}$$

where α is constant. The above energy should satisfy the constraint between energy E, A and T. Following (36) and (42), we have

$$E = A k_B T/\gamma, \tag{48}$$

[2]By microscopic DOF we refer to fluid degrees of freedom. These are effective DOF and not the fundamental Planck scale DOF of the theory as we consider long wavelength description of the horizon.

γ is dimension dependent constant. From (47), we get,

$$N = E/(\alpha k_B T) = A/(\gamma\alpha). \tag{49}$$

In the next section we use this as the key ingredient to model the horizon fluid within the mean field theory [29].

5.3 *Mean Field Theory of the Horizon Fluid*

The above discussions point to the fact that we can model the horizon fluid system using mean field theory [53–55]. The order parameter for a collection of particles, which forms a condensate at a certain transition temperature, is the wave function for the state (ψ) whose modulus is equal to the number density of particles ρ, i.e. $\psi \propto \sqrt{\rho}$.

Assuming for simplicity, the black hole horizon fluid system to be homogeneous. Defining the order parameter (η) of the homogeneous fluid as [29]

$$\eta = \sqrt{KN} \qquad \text{K a positive constant,} \tag{50}$$

Of course, now, the physical significance of the order parameter could no longer be supplied directly from the microscopic model.

Following Landau–Lifshitz [53] (Sect. 143), we can write down the Mean field theory in terms of the Thermodynamic potential, Φ, where the independent variables are T and the chemical potential μ, i.e., $\Phi = -P\,A$. It is important to note that using the constraint (23) and take the derivative w. r. t T one may obtain the entropy. However, as mentioned earlier, horizon fluid is highly constrained; A and T are related. The mean field construction using the order parameter η uses these constraints to obtain $a(P)$ and $B(P)$. Expanding Φ about T_c, we have,

$$\Phi = \Phi_0 + a(P)(T - T_c)\eta^2 + B(P)\eta^4; \tag{51}$$

where $a(P)$ and $B(P)$ are unknown phenomenological functions. Using (23), (50) and the constraint (49), we get,

$$-TA = 4\left[\Phi_0 + \kappa a(P)(T - T_c)\frac{A}{2\alpha} + \kappa^2 B(P)(\frac{A}{2\alpha})^2\right] \tag{52}$$

Matching the coefficients of A on both sides, we have,

$$a = -\frac{\alpha}{2\kappa}, \tag{53}$$

which shows that a is a negative number. Using (53), we also get the second mapping constraint,

$$\Phi_0(P, T) + \frac{1}{4} T_c A + \kappa^2 B(P) \left(\frac{A}{2\alpha} \right)^2 = 0. \tag{54}$$

It is important to note that $\eta = 0$ is the symmetric phase and $\eta \neq 0$ is the asymmetric phase [53]. In our case, since $a < 0$, this implies that the system is in the symmetric phase for $T < T_c$ and asymmetric phase for $T > T_c$.

In the asymmetric phase, the order parameter η has the value for which, the thermodynamic Potential is minimum. The minimization of the Thermodynamic potential with respect to η gives the condition

$$\eta^2 = \frac{a(T_c - T)}{2B} \Rightarrow \kappa N = \frac{a(T - T_c)}{2B} \tag{55}$$

Using (49) and (53), this can be expressed as

$$\frac{(T - T_c)}{2B} = \frac{\kappa^2 A}{\alpha^2}. \tag{56}$$

Denoting the entropy of the system in the symmetric and asymmetric phase by S_0 and $S_0 + \Delta S$, respectively, we have

$$\Delta S = -\frac{\partial \Phi}{\partial T} = \frac{a^2}{2B}(T - T_c). \tag{57}$$

From (53) and (57), we get,

$$\Delta S = \frac{A}{4}. \tag{58}$$

This is the first test of the horizon fluid picture and validates the modeling of the horizon fluid as a critical system. This has the following interesting implications:

1. This is a generic result for any $D-$dimensional stationary (spherical or axi-symmetric) black hole in General relativity.
2. One of the key results from black hole thermodynamics is that black holes in General Relativity have an entropy $S = k_B A/4$. This is a key test for modeling the horizon fluid, which has proven difficult for several prior models [37–39]; the fact that this holds true for a large class of black holes is encouraging for the success of our model of horizon fluid.

In the next section, we put the model to test by studying the fluctuations of the number of the particles in horizon fluid.

6 Fluctuations of the Horizon Fluid

It is well known that fluids can be described by two sets of parameters, namely susceptibilities (thermodynamic derivatives) and transport coefficients. While the first set of parameters correspond to changes in the local variables; the other set involve fluxes of thermodynamic quantities [56–58]. Obviously, these parameters can not be determined within fluid mechanics, however, can be derived using the theory of fluctuations that relate susceptibility/transport coefficient to autocorrelation function of a dynamical variable [56–59].

Statistical fluctuations of normal fluids is well-understood [60]. As mentioned earlier, horizon fluid is an unusual fluid. Unlike normal fluids, it is one parameter system, whose energy, pressure, temperature, volume (area in this case) are not independent of each other. Recently, we have explicitly show that it is possible to construct a theory of the Fluctuations in the horizon fluid in analogy with normal fluids [32]. The interesting feature of this approach is that it is general and is independent of the details of the horizon fluid.

6.1 Theory of Fluctuations for the Horizon Fluid: Broad Outline

The basic strategy that we adopt for the fluctuations to evaluate the transport coefficient of the horizon fluid are as follows [32]:

1. The fluctuations about the equilibrium position of the system are about the minimum value of the potential Φ (51). This implies that the leading order variation in the potential would be of second order in that variable.
2. Any dynamical process that the event horizon of a black hole undergoes corresponds to the horizon fluid system moving from an initial non-equilibrium state towards an equilibrium state. However, so long as the evolution of the event horizon follows from a classical theory of Gravity, which is what we shall look at here, the system is not far from equilibrium in the corresponding fluid picture.
   The above assumption imposes a restriction on the kind of dynamical processes we consider on the Gravity side. In the fluid side, we are looking at processes that increase the entropy of the system. However, on the Gravity side, it is the area of the event horizon that is proportional to the entropy [13, 18, 35] and even in a classical theory, it does not always increase. In the classical theory, in order for the area of the event horizon to increase, some additional conditions have to be imposed [22], like the condition that the black hole is strongly asymptotically predictable [35, 42].
3. Write down a Langevin equation that governs the transport processes to be considered [31].
4. Assume that the random fluctuations that the system undergoes are of a much shorter time-scale than the time period over which the system goes over from

one state to another due to external influence. These latter processes can also be viewed as fluctuations that the system undergoes; only they are of a much longer time-scale. This makes the probability distributions for the random short-time fluctuations and the long time ones independent of each other. It is possible to compute the correlation functions that contain all the information about the system from the assumptions that the random fluctuations obey the principle of Equipartition and both the probability distributions are given by Maxwell-Boltzmann. Another necessary input is the black hole constraints, which reflects the fact that Horizon fluid is actually a one parameter system.

5. The Transport coefficients for the horizon fluid can then be computed using a method given by Kubo [57] either in real space or in frequency space [32].

6.2 *Teleological Condition and Bulk Viscosity*

Using the theory of fluctuations [57], transport coefficients can be related to auto-correlation function of number density fluctuations (δN). Following Kubo [57], the coefficient of bulk viscosity is given by:

$$\zeta = \left(\frac{1}{n}\right)\frac{1}{Ak_BT}\int_{-\infty}^{\infty} dt \sum_a \sum_b \langle J^{aa}(0) J^{bb}(t)\rangle, \tag{59}$$

where, $n = Tr(\delta_{ab})$, a, b run from $1, \ldots (D-2)$ and the current is

$$J^{ab} = \delta_{ab}\,\delta(PV) = V\delta P\,\delta_{ab}. \tag{60}$$

To calculate the current, we need to know how the horizon fluid responds to the external influence. Interestingly, for the event horizon, the response to any external influence is anti-causal. In particular, if matter- energy falls through the event horizon, then the area of the event horizon increases till the matter-energy passes through the horizon. This is not unphysical as the event horizon of a black hole is defined globally in the presence of the future light-like infinity [22].

Due to this unusual property of the horizon, the horizon fluid also exhibits anti-causal response, i.e. the response of the horizon takes place before the external influence occurs [22]. Since from the fluid point of view, the system is initially out of equilibrium and slowly moves towards the final state in equilibrium, it follows that the external influence brings the system to equilibrium, so that there is no further evolution of the system from the state it is in. This is referred to as the teleological nature of horizon [22].

For a class of systems, it has been shown in the literature that if the system exhibits anti-causal transport process, then the anti causal transport coefficients have an opposite sign to their causal counterparts [61]. For normal fluids, external influence drives the system out of equilibrium. For the horizon fluid, it is the reverse; the system moves towards equilibrium in anticipation of the external influence like infusion of energy into the fluid. This is the anti-causal response of the horizon fluid.

Rewriting Eq. (60) as an entropic force — moving the system back to equilibrium — which, for the horizon fluid, takes the form:

$$F_{\mathrm{Th}} = P A \, \delta A \, \theta(-t) \tag{61}$$

where $\theta(t)$ is the theta function, and enforces the anti-casual, teleological nature of the horizon (for a detailed discussion, see [32]). The bulk viscosity can therefore be rewritten as

$$\begin{aligned} \zeta &= \frac{1}{A k_B T} \int_{-\infty}^{\infty} dt \, \langle F_{Th}(t) F_{Th}(0) \rangle \\ &= \frac{P^2}{A k_B T} \int_{-\infty}^{\infty} dt \, \langle \delta A(t) \delta A(0) \rangle \, \theta(-t) \\ &= \frac{(\alpha\gamma)^2 k_B T}{16A} \int_{-\infty}^{\infty} dt \, \langle \delta N(t) \delta N(0) \rangle \, \theta(-t). \end{aligned} \tag{62}$$

where the last equality comes from (23) and (47). Since our interest lies in the long wavelength (fluid) limit, we may evaluate the viscosity from the linear response of the horizon fluid [57]:

$$\zeta = \lim_{\varepsilon \to 0} \Im \left[\frac{(\alpha\gamma)^2 k_B T \left\langle \delta N^2(0) \right\rangle}{16A} \int_{-\infty}^{\infty} dt \exp[i(\omega - i\varepsilon)] \theta(-t) \right] \tag{63}$$

leading to

$$\zeta = -\frac{(\alpha\gamma)^2 k_B T}{16A} \frac{\left\langle \delta N^2(0) \right\rangle}{\omega}. \tag{64}$$

where ω corresponds to the lowest energy mode of the fluctuations that the horizon fluid support and sustain. Assuming that the fluctuations satisfy Maxwell-Boltzmann statistics, we get [33]

$$\left\langle \delta N^2(0) \right\rangle = \frac{4\,A}{(\gamma\,\alpha)^2}, \tag{65}$$

and the bulk viscosity simplifies to

$$\zeta = -\frac{k_B\,T}{4\,\omega}. \tag{66}$$

The above result points to something important regarding which we would like to stress the following:

1. Bulk viscosity is independent of the constants α, γ; these can only be determined with the knowledge of the microscopic theory. This is consistent as the fluid

description does not require complete knowledge of the microscopic degrees of freedom.

2. Bulk viscosity depends on the horizon fluid temperature and the lowest energy mode of the fluctuations. To go about determining the lowest energy mode, it may be important to get a physical insight. Let us consider fluctuations that cause a change in the horizon fluid area from A to $A + dA$. For the horizon fluid corresponding to Schwarzschild black hole, the change in area is only due to the change in the quasi-local energy. In Ref. [32] this was done by picking the largest wavelength to be the circumference of the black hole. However, for generic black holes like Kerr-Neumann, the change in the area can also be due to the change in the electrostatic energy or quasi-local energy $\kappa \times dA$ [42]. Using the fact that the lowest energy modes are adiabatic (slowly evolving), and that the energy and area are strongly constrained, the minimum energy change can be related to a minimum change in area through (47) and (49):

$$\Delta E = (\Delta N)\alpha T \propto (\Delta A)T, \tag{67}$$

and, thus, the minimum energy mode for any stationary black hole is

$$\omega = \Delta E = \left(\frac{D-2}{D-3}\right) \Delta A \, \frac{T}{4}. \tag{68}$$

The above result shows the existence of the lowest energy mode of fluctuation (δN) of the horizon fluid relating to the *minimum change* or *quantization* of the horizon area.

6.3 Implications

It is important to note that starting from a minimal model of the horizon fluid, while the entropy of the horizon fluid does not put any additional condition, however, matching the viscosity provides a condition on the area spectrum.

The area quantization has long been a key result of both horizon thermodynamics and various quantum gravity proposals. The first attempt goes back to Bekenstein [13], who uses the fact that particles entering the black hole cannot have zero size, to find the minimum area increase when one is absorbed by a Kerr-Newman black hole. This result is very general [28], and in D-dimensions leads to

$$\Delta A_{\rm min} = 8\pi \ell_P^{D-2}, \tag{69}$$

where we have briefly re-instituted an explicit Planck length, ℓ_P for clarity. Substituting Bekenstein's minimum area in (68), we get,

$$\omega = 2\pi \left(\frac{D-2}{D-3}\right) k_B\, T. \tag{70}$$

Substituting the above form of minimum energy mode in (66), we get

$$\zeta = -\left(\frac{D-2}{D-3}\right)\frac{1}{8\pi}. \tag{71}$$

It is important to note that this is the expression for bulk viscosity for all asymptotically flat space times in all dimensions and matches exactly the expression from the DNS equation (22). This is a *non-trivial* result, using fluctuation-dissipation [57] and the fact that the horizon is anti-causal [32], it is not possible to obtain the result without invoking area quantization.

So far, it has not been necessary to place a value on α. However, if we take $\Delta N_{\min} = 1$, from Eq. (69), we have

$$\alpha = 8\pi, \tag{72}$$

corresponding to [13]. Naturally, there have been several alternative proposals from counting black hole micro-states or Bohr's principle with quasi-normal modes [62, 63], and arguments from Loop quantum Gravity [64], for various values of α. Note that all these proposal give answers of the same order of magnitude (see [65]).

7 What Are the Atoms of Space Time?

The flow chart below provides a bird's-eye view of the fluid-gravity programme [28–33] that is described in this article.

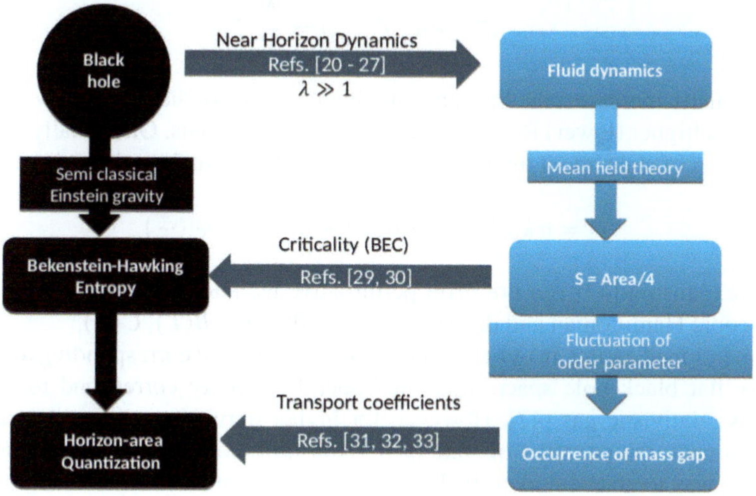

The point of view taken by us in this research programme is that the horizon fluid possesses some kind of physical reality beyond the formal similarity. As the reader can notice, the key advantage of this strategy is that, for most part, the analysis is independent of the details of the underlying microscopic theory and hence the conclusions about the negative specific heat, occurrence of mass gap and hence horizon-area quantization are independent of the model of quantum gravity. In our research programme, any horizon fluid satisfying the two constraints (23, 36), leads to an entropy that is identical to Bekenstein–Hawking entropy and bulk viscosity that is identical to that of Damour.

Mean Field Theory ignores the fluctuations of the order parameter. Using a general, yet, simple model, we showed that the Mean field theory could lead to Bekenstein–Hawking entropy. This strongly suggests that any approach that predicts Bekenstein–Hawking entropy should be treated at the same level as a Mean Field model in Condensed matter systems that implicitly ignore fluctuations. Taking the fluctuations about the mean value to be Gaussian, we have shown that it is possible to deduce the dynamical first law of thermodynamics for Black holes. It also possible to interpret Raychaudhury equation for the null geodesic equation for the null congruences on the black hole horizon as a Transport equation.

The bulk viscosity, found via the fluctuation-dissipation theorem, is matched with the expected value by the selection of a lowest energy mode. The needed value is found to correspond exactly to a minimum area change first proposed by Bekenstein [13]. We see that another macroscopic quantity that requires some microscopic information to recover.

All these results lead us to the following question: *What are the atoms of space time?* The answer to this question lies in mapping *cluster expansion technique* to the horizon fluid [47]. If the gas is made up of molecules that interact, however if it is dilute (number density $n \ll 1$) that the rate of occurrence of interactions is rare; it is then possible to approximate the gas equation of state to be an ideal gas:

$$P = n\, k_{_B}\, T \qquad n = \frac{N}{V}$$

As the density increases, the Pressure should deviate from the ideal gas equation of state and is higher (lower) for repulsive (attractive) potentials. One usually expands, the equation of state about the low density ($n \ll 1$) limit and can be written as

$$P = n\, k_{_B}\, T \left(1 + n\, B(T) + n^2\, C(T) + \cdots\right)$$

Cluster expansion [47] is a powerful perturbative technique where one starts from microscopic Hamiltonian and determine the coefficients $B(T)$, $C(T)$.

As mentioned earlier, in our view point, the horizon fluid corresponding to asymptotically flat black hole space times in General relativity correspond to an ideal massless relativistic gas [47]. However, for higher derivative gravity theories like Gauss-Bonnet, the equation of state of horizon fluid corresponds to non-ideal gas [30]. One way to go about identify the atoms of space time is to systematically determine the coefficients $B(T)$, $C(T)$ order by order from an interacting Hamiltonian corresponding to the horizon fluid. This is currently underway and I hope to say something more concrete soon.

Acknowledgements I thank Jozef Skákala, Swastik Bhattacharya, Jose Luis López, Bethan Cropp for collaboration and extensive discussions which have made my understanding clear. I also thank *Paddy* for discussions on several issues. The work is supported by DST-Max Planck Partner Group on Cosmology and Gravity; part of the work was supported by Ramanujan Fellowship of DST, India.

References

1. J. Narlikar, T. Padmanabhan, *Gravity, Gauge Theories and Quantum Cosmology* (D Reidel Pub Co, 1986)
2. T. Padmanabhan, Phys. Rept. **188**, 285 (1990). doi:10.1016/0370-1573(90)90051-3
3. T. Padmanabhan, *Structure Formation in the Universe*, vol. 1 (Cambridge University Press, Cambridge, 1993)
4. T. Padmanabhan, *Theoretical Astrophysics,* vol. 1 and 3 (Cambridge University Press, Cambridge, 2002)
5. T. Padmanabhan, Phys. Rept. **380**, 235 (2003). doi:10.1016/S0370-1573(03)00120-0
6. T. Padmanabhan, General relativity and gravitation **40**, 529 (2008). doi:10.1007/s10714-007-0555-7
7. T. Padmanabhan, Research in astronomy and astrophysics **12**, 891 (2012). doi:10.1088/1674-4527/12/8/003
8. T. Padmanabhan, Phys. Rept. **406**, 49 (2005). doi:10.1016/j.physrep.2004.10.003
9. T. Padmanabhan, A.I.P. Conf, Proc. **939**, 114 (2007). doi:10.1063/1.2803795
10. T. Padmanabhan, Rep. Prog. Phys. **73**(4), 046901 (2010). doi:10.1088/0034-4885/73/4/046901
11. T. Padmanabhan, Entropy **17**, 7420 (2015). doi:10.3390/e17117420
12. J.D. Bekenstein, Nuovo Cimento Lett **4**, 737 (1972)
13. J.D. Bekenstein, Phys. Rev. D **7**, 2333 (1973). doi:10.1103/PhysRevD.7.2333
14. J.M. Bardeen, B. Carter, S.W. Hawking, Commun. Math. Phys. **31**, 161 (1973). doi:10.1007/BF01645742
15. J.D. Bekenstein, Phys. Rev. D **9**, 3292 (1974). doi:10.1103/PhysRevD.9.3292
16. S.W. Hawking, Nature **248**, 30 (1974). doi:10.1038/248030a0Letter
17. J.D. Bekenstein, Phys. Rev. D **11**, 2072 (1975). doi:10.1103/PhysRevD.11.2072
18. S.W. Hawking, Commun. Math. Phys. **43**, 199 (1975)
19. T. Jacobson, Phys. Rev. Lett. **75**, 1260 (1995). doi:10.1103/PhysRevLett.75.1260
20. T. Damour, in *Marcel Grossmann Meeting: General Relativity*, ed. by R. Ruffini (1982)
21. R.H. Price, K.S. Thorne, Phys. Rev. D. **33**, 915 (1986). doi:10.1103/PhysRevD.33.915
22. K.S. Thorne, R.H. Price, D.A. Macdonald, *The Membrane Paradigm*, vol. 19 (Yale University Press, New Haven, 1986)
23. G. Policastro, D.T. Son, A.O. Starinets, J. High Energy Phys. **9**, 043 (2002). doi:10.1088/1126-6708/2002/09/043
24. S. Bhattacharyya, S. Minwalla, V.E. Hubeny, M. Rangamani, J. High Energy Phys. **2**, 045 (2008). doi:10.1088/1126-6708/2008/02/045
25. T. Padmanabhan, Phys. Rev. D. **83**(4), 044048 (2011). doi:10.1103/PhysRevD.83.044048
26. S. Kolekar, T. Padmanabhan, Phys. Rev. D **D85**, 024004 (2012). doi:10.1103/PhysRevD.85.024004
27. I. Bredberg, C. Keeler, V. Lysov, A. Strominger, J. High Energy Phys. **2012**(7) (2012). doi:10.1007/jhep07(2012)146

28. J. Skakala, S. Shankaranarayanan, Int. J. Mod. Phys. D **25**, 1650047 (2016). doi:10.1142/S0218271816500474
29. S. Bhattacharya, S. Shankaranarayanan, arXiv: 1411.7830 (2014)
30. J.L. López, S. Bhattacharya, S. Shankaranarayanan, Phys. Rev. D **94**, 024029 (2016). doi:10.1103/PhysRevD.94.024029
31. S. Bhattacharya, S. Shankaranarayanan, Int. J. Mod. Phys. D **24**(12), 1544005 (2015). doi:10.1142/S0218271815440058
32. S. Bhattacharya, S. Shankaranarayanan, Phys. Rev. D **93**(6), 064030 (2016). doi:10.1103/PhysRevD.93.064030
33. B. Cropp, S. Bhattacharya, S. Shankaranarayanan, (2016). doi:http://dx.doi.org/10.1103/PhysRevD.95.024006
34. S. Carlip, Lect. Notes Phys. **769**, 89 (2009). doi:10.1007/978-3-540-88460-6-3
35. R.M. Wald, Liv. Rev. Rela. **4**, 6 (2001). doi:10.12942/lrr-2001-6
36. T. Jacobson, R. Parentani, Found. Phys. **33**(2), 323 (2003). doi:10.1023/A:1023785123428
37. G. Chapline, E. Hohlfeld, R.B. Laughlin, D.I. Santiago, Int. J. Mod. Phys. A **18**, 3587 (2003). doi:10.1142/S0217751X03016380
38. G. Dvali, C. Gomez, Phys. Lett. B **716**(1), 240 (2012). doi:http://dx.doi.org/10.1016/j.physletb.2012.08.019
39. G. Dvali, C. Gomez, Eur. Phys. J. C **74**, 2752 (2014). doi:10.1140/epjc/s10052-014-2752-3
40. R. Beig, Phys. Lett. A **69**(3), 153 (1978). doi:10.1016/0375-9601(78)90198-6
41. L.B. Szabados, Living reviews in relativity **7**(4) (2004). doi:10.1007/lrr-2004-4
42. R.M. Wald, *General Relativity* (University of Chicago press, Chicago, 1984)
43. R. Kulkarni, V. Chellathurai, N. Dadhich, Class. Quant. Grav. **5**(11), 1443 (1988). doi:10.1088/0264-9381/5/11/007
44. V. Chellathurai, N. Dadhich, Class. Quant. Grav. **7**(3), 361 (1990). doi:10.1088/0264-9381/7/3/010
45. T. Dray, R. Kulkarni, C.A. Manogue, Gen. Relat. Grav. **24**(12), 1255 (1992). doi:10.1007/BF02418212
46. F.R. Tangherlini, Il Nuovo Cimento (1955–1965) **27**(3), 636 (1963). doi:10.1007/BF02784569
47. K. Huang, *Statistical Mechanics* (Wiley, New York, c1987, 2010)
48. H. Kleinert, S.V. Shabanov, Phys. Lett. A **232**, 327 (1997). doi:10.1016/S0375-9601(97)00410-6
49. S.T. Hong, W. Tae, Kim, Y.J. Park, Mod. Phys. Lett. A **15**(31), 1915 (2000)
50. C.O. Lousto, Phys. Rev. D **51**, 1733 (1995). doi:10.1103/PhysRevD.51.1733
51. C.O. Lousto, Gen. Relat. Grav. **27**, 121 (1995). doi:10.1007/BF02107952
52. P.C.W. Davies, Proc. Roy. Soc. Lond. A **353**(1675), 499 (1977). doi:10.1098/rspa.1977.0047
53. L. Landau, E.M. Lifshitz, *Course of Theoretical Physics*, Vol. 5, Statistical physics, Pt. 1 (Pergamon Press, London, 1980)
54. H.E. Stanley, *Introduction to Phase Transitions and Critical Phenomena*, vol. 1 (Oxford University Press, Oxford, 1971)
55. L.P. Kadanoff, *Statistical Physics: Statics, Dynamics and Renormalization*, vol. 1 (World Scientific, Singapore, c2000., 2010)
56. L.P. Kadanoff, P.C. Martin, Ann. Phys. **24**, 419 (1963). doi:10.1016/0003-4916(63)90078-2
57. R. Kubo, Rep. Progr. Phys. **29**, 255 (1966). doi:10.1088/0034-4885/29/1/306
58. R. Zwanzig, Ann. Rev. Phys. Chem. **16**(1), 67 (1965). doi:10.1146/annurev.pc.16.100165.000435
59. J.W. Dufty, Phys. Rev. **176**, 398 (1968). doi:10.1103/PhysRev.176.398
60. D.J. Evans, G.P. Morriss, *Statistical Mechanics of Nonequilibrium Liquids*, vol. 1 (Academic Press, London, c1990., 2010)
61. D.J. Evans, D.J. Searles, Phys. Rev. E **53**, 5808 (1996). doi:10.1103/PhysRevE.53.5808
62. J.D. Bekenstein, V.F. Mukhanov, Phys. Lett. B **360**, 7 (1995). doi:10.1016/0370-2693(95)01148-J
63. S. Hod, Phys. Rev. Lett. **81**, 4293 (1998)
64. O. Dreyer, Phys. Rev. Lett. **90**, 081301 (2003)
65. J. Skakala, S. Shankaranarayanan, Phys. Rev. D **89**, 044019 (2014). doi:10.1103/PhysRevD.89.044019

From Quantum to Classical in the Sky

Suprit Singh

Abstract Inflation has by-far set itself as one of the prime ideas in the current cosmological models that seemingly has an answer for every observed phenomenon in cosmology. More importantly, it serves as a bridge between the early quantum fluctuations and the present-day classical structures. Although the transition from quantum to classical is still not completely understood till date, there are two assumptions made in the inflationary paradigm in this regard: (i) the modes (metric perturbations or fluctuations) behave classically once they are well outside the Hubble radius and, (ii) once they become classical they stay classical and hence can be described by standard perturbation theory after they re-enter the Hubble radius. We critically examine these assumptions for the tensor modes of (linear) metric perturbations in a toy three stage universe with (i) inflation, (ii) radiation-dominated and (iii) late-time accelerated phases. The quantum-to-classical transition for these modes is evident from the evolution of Wigner function in phase space and its peaking on the classical trajectory. However, a better approach to quantify the degree of classicality and study its evolution was given by Mahajan and Padmanabhan [1] (Mahajan and Padmanabhan, Gen. Rel. Grav. 40, 661, 2008; Gen. Rel. Grav. 40, 709, 2008) using a *classicality parameter* constructed from the parameters of the Wigner function. We study the evolution of the classicality parameter across the three phases and it turns out that the first assumption holds true, there is emergence of classicality on Hubble exit, however the latter assumption of "once classical, always classical" seems to lie on a shaky ground.

If there has been a paradigm shift in cosmology, we can (with a wide consensus) say that it was the inclusion of inflation – a period of an exponential growth at the 'inception' of our universe – even though in an ad-hoc manner to do away with the shortcomings of the hot big bang model. These were old time issues such as the flatness and horizon problems and the choice of initial conditions. However, inflation really came as *deus ex machina* to explain many other significant features along with taking care of the above problems in its assorted bag of tricks. It, specifically, gave a way to

S. Singh (✉)
Department of Physics & Astrophysics, University of Delhi, New Delhi 110 007, India
e-mail: ssingh2@physics.du.ac.in

J.S. Bagla and S. Engineer (eds.), *Gravity and the Quantum*,
Fundamental Theories of Physics 187, DOI 10.1007/978-3-319-51700-1_24

unite the two scales[1] which I will take as the *quantum* and *classical* in our universe for our interests. In inflationary cosmology – the large scale structures, anisotropies of the cosmic microwave background, primordial magnetic fields and primordial gravitational waves all have their origins in the *quantum* fluctuations that reigned during that epoch [2]. These fluctuations are the metric perturbations on the otherwise smooth, homogenous and isotropic background, sourced by the fluctuations of an inflaton field in the simplest of models. Inflation, essentially, sets everything up in a vacuum state defined well back in the past and hence these fluctuations are *quantized* and evolve very much like a quantum field on a curved background.[2] That is, an infinite set of harmonic oscillator modes in Fourier space evolving in competition with the comoving Hubble length scale which decreases during the exponential expansion. Inflation, then makes the following two assumptions in this regard,

(i) the modes behave classically once they are outside the Hubble radius and,
(ii) once they become classical, they stay classical and hence can be described by standard perturbation theory after they re-enter the Hubble radius.

That is, after the end of inflation, these fluctuations are taken as classical, gaussian random fields on re-entry in the radiation-dominated phase which grow due to gravitational instability. Now, the details of how the quantum-to-classical transition occurs is one of the fundamental questions that stands open in the inflationary paradigm of cosmology. Various mechanisms such as decoherence etc. are invoked to explain the transition, but it is still not completely well understood. Also, without any clear quantitative measure for the classicality of the fluctuations, one has to deal with a "two-level" description of fluctuations being either fully classical or fully quantum mechanical while the reality may lie on the middle ground. Finally, it is imperative to test the very foundations of inflation and the *quantum origin* of seeds of cosmic structures. Could it be by designing a cosmological Bell-type experiment [3]? Or, recovering the quantum information from present-day observations of the sky [4]. Inflation, for all we know, is like modelling a black-box subcircuit using equivalent model circuits à la Thévenin and Norton, that can give similar observations.[3] The actual reality could be far different from the constructed models. These questions are essential to be answered if we have to put inflation on *fundamenta incocussa.*

The first question in the quest is, "How to quantify the degree of classicality?". This is usually determined from the peaking of the Wigner function on the classical trajectory. However, relying on the Wigner function alone can lead to ambiguities. This is seen, for example, in Schwinger effect where there is pair production under strong external electric field. The Wigner function in this case is uncorrelated and

[1]Usually, this means that it bridges the two scales in the high energy physics which are about 0.2 eV and 10^{12} GeV or more.

[2]Note that the perturbations do backreact on the geometry and also mix up on non-linear scales which makes them complicated over test quantum fields.

[3]Models of inflation have parameters such as the scalar spectral index (n_s), amplitude (A_s) of scalar power spectrum, tensor-to-scalar ratio (r), non-gaussianity (f_{NL}) etc. which are tested against observations.

concentrated on the classical trajectory both at very early times (when the field is in the in vacuum) and at late times (when particle number has reached an asymptotically constant limit). Thus, the peaking of Wigner function alone cannot be sufficient criterion for the emergence of classicality. For this, Mahajan and Padmanabhan [1] proposed using a *classicality parameter* (a measure of phase space correlations) which they showed works fairly well and fits the common intuition of classicality using various examples. We employed the same construction for a test scalar field in the setting of a three stage universe [5]. The results were a bit surprising regarding the two assumptions in inflation we mentioned before and the subject of the present article is to *emphasize* that the surprising results hold even for the primordial gravitational waves, i.e., the tensor modes of metric perturbations.

The idea is the following: We consider a three stage background universe[4] that comprises of an (i) inflationary de Sitter stage, (ii) radiation-dominated stage and finally, (iii) Cosmological constant (Λ) dominated late-time accelerated stage. The transitions points maintain the continuity of scale factor and its derivative (the hubble parameter). We shall use the scale factor as the time parameter since it is monotonic. Hence, the three stages can be expressed in terms of the comoving Hubble radius as,

$$L(a) = \begin{cases} (Ha)^{-1} & a \le e^{1/2} \\ (a/He) & e^{1/2} \le a \le (e/\epsilon)^{1/2} \\ (a\epsilon H)^{-1} & a \ge (e/\epsilon)^{1/2} \end{cases} \tag{1}$$

for the inflationary de Sitter ($H_{\rm inf} = H$), radiation-dominated and late-time de Sitter ($H_\Lambda = \epsilon H$) stages respectively. These are just 45° lines in the logarithmic plot (see Fig. 1). The three stages form a rich terrain and a natural sandwich of two characteristic length scales,[5] $L_{max} = 1/(H\epsilon^{1/2}e^{1/2})$ and $L_{min} = 1/(He^{1/2})$. Then, any length scale within this band has three transition points where it goes super-Hubble, sub-Hubble and finally super-Hubble again. The modes characterized by the length scales larger than L_{max} exit the Hubble radius in the first stage and remain super-Hubble after that while the modes with wavelengths smaller than L_{min} remain sub-Hubble and exit the Hubble radius only in the third stage. This relative behaviour of different modes with respect to the comoving Hubble radius leads to difference in the evolution of classicality for each of them. We assume simplest model of inflation and the details are not relevant for the present analysis. But a very brief idea of background evolution in the simplest single-field, slow-roll inflationary model is given in a separate digressive box for completeness (see Ref. [6] for the detailed physics of inflation).

---

[4]This is quite close to our real universe since matter domination lasted for a very small duration compared to the radiation-dominated stage.

[5]These are actually $L(a)$ evaluated at the transition points, which themselves, are obtained using the continuity of $L(a)$ across the phases.

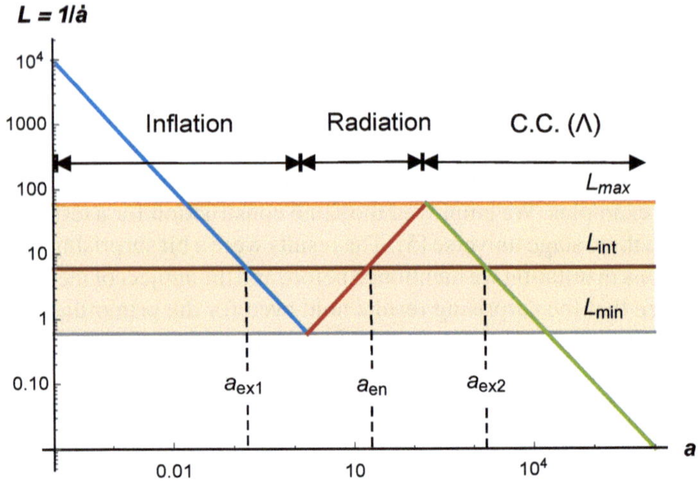

Fig. 1 The Cosmic Sandwich. A (logarithmic) plot of comoving Hubble radius, $L = (da/dt)^{-1}$ where t is the usual comoving time with the scale factor a for $\epsilon = H_\Lambda/H_{\rm inf} = 10^{-4}$ showing the inflationary phase (decreasing), radiation-dominated phase (increasing) and in late-time cosmological constant dominated (de Sitter) phase (decreasing) where the lines have slopes: ± 1. We have a natural band of two characteristic length scales, L_{max} and L_{min} such that any length scale within has three transition points where it goes super-Hubble, sub-Hubble and finally super-Hubble again

Inflation in a *Mustard* Seed

In the simplest single-field models of inflation, the action is just

$$S = \frac{1}{2}\int d^4x\sqrt{-g}\left[R - (\nabla\psi)^2 - 2V(\psi)\right] \quad \text{(Box1.1)}$$

where ψ is the scalar inflaton field which is minimally coupled to gravity and we have set the units in which $\hbar = c = 8\pi G_N = 1$. The homogenous *ansatz* is

$$ds^2 = -dt^2 + a^2(t)dx_i dx_i\ ; \qquad a(t) = \exp\left(\int dt\ H(t)\right), \quad \text{(Box1.2)}$$

where we have expressed the scale factor in terms of a quasi-constant Hubble parameter, $H(t)$. The variation leads to the constraint and the dynamical equations:

$$
\begin{aligned}
3H^2 = \rho_\psi &= \frac{1}{2}\dot{\psi}^2 + V, \\
\dot{H} &= -\frac{1}{2}\dot{\psi}^2, \\
0 &= \ddot{\psi} + 3H\dot{\psi} + dV/d\psi;
\end{aligned}
\tag{Box1.3}
$$

of which only two are independent. A successful inflation requires the universe to expand by, at least, around 60 e-foldings for a sufficiently long time and eventually end into a radiation dominated phase. This implies that inflation is a quasi-de Sitter stage with parameters

$$
\epsilon = -\frac{\dot{H}}{H^2} = \frac{3\dot{\psi}^2}{2\rho_\psi}, \quad \delta = \epsilon - \frac{\dot{\epsilon}}{2H\epsilon} = -\frac{\ddot{\psi}}{H\dot{\psi}}, \quad \eta = 2\epsilon - \delta \tag{Box1.4}
$$

satisfying $\{\epsilon, |\delta|, |\eta|\} \ll 1$ and $\mathcal{O}(\epsilon^2, \delta^2, \epsilon\delta) \ll \epsilon$ referred to as the slow roll conditions. The dynamics at the leading order is then dictated by,

$$
3H^2 \simeq V; \quad 3H\dot{\psi} \simeq -dV/d\psi \tag{Box1.5}
$$

which, given a potential, can be solved for the scale factor and the scalar field in the slow roll limit.

What we need are the fluctuations around the classical solution. We can decompose the metric fluctuations or perturbations on the background Friedmann solution into scalar, vector and tensor modes. These do not mix at the linear level owing to the symmetries of the background and can be studied independently of each other. We shall choose to work with the tensor modes alone in this article.[6] This is for simplicity, since at linear order, the tensor perturbations are gauge-invariant and cause no backreaction to the inflationary background. Also, the tensor modes carry on without much distortion through the reheating phase and leave their imprints on the cosmic microwave background. Therefore, we begin with the quadratic action,

$$
S_2 = \frac{1}{8}\int \mathbf{dx}\, d\tau\; a^2\left[(h'_{ij})^2 - (\nabla h_{ij})^2\right], \tag{2}
$$

which we get by expanding of the Einsten–Hilbert action (Box 1.1) to second order, using the equations of motion and restricting only to the tensor perturbations. Note that we have now shifted to using the conformal time,

[6]The scalar mode of perturbations in a suitable gauge requires a more careful study by incorporating backreaction and the effects after re-entry in the Hubble radius [7].

$$\tau \equiv \int \frac{dt}{a(t)} \tag{3}$$

and prime denotes derivative with respect to this time. The tensor mode action Eq. (2) is same the action for a massless scalar field up to a normalization factor. Thus, we can consider the dynamics of a massless scalar field in our three stage background neglecting the two states of polarizations so that we have,

$$S_h = \frac{1}{2}\int \mathrm{d}^4x\,\sqrt{-g}\;\partial_a h\;\partial^a h = \frac{1}{2}\int \mathbf{dx}\;\mathrm{d}\tau\;a^2\left[h'^2 - (\nabla h)^2\right]. \tag{4}$$

Due to the translational invariance of the Friedmann metric, we can decompose the field into independent Fourier modes and re-express the action as

$$S_h = \frac{1}{2}\int \mathrm{d}\mathbf{k}\int \mathrm{d}\tau\, a^2\left(h'^2_{\mathbf{k}} - k^2\,h^2_{\mathbf{k}}\right), \tag{5}$$

where $k = |\mathbf{k}|$. We have, thus, reduced the task to tackling the dynamics of decoupled, non-relativistic harmonic oscillators with time-dependent mass and frequencies. We quantize the system in the Schrödinger picture. The (time-dependent) Schrödinger equation for the system,

$$i\frac{\partial}{\partial\tau}\Psi(h_{\mathbf{k}},\tau) = \left(-\frac{1}{2a^2(\tau)}\frac{\partial^2}{\partial h^2_{\mathbf{k}}} + \frac{1}{2}a^2(\tau)\;k^2\;h^2_{\mathbf{k}}\right)\Psi(h_{\mathbf{k}},\tau) \tag{6}$$

admits time-dependent, form-invariant, Gaussian states with vanishing mean given by:

$$\Psi(h_{\mathbf{k}},\tau) = N\exp\left[-\alpha_k(\tau)h^2_{\mathbf{k}}\right] = N\exp\left[-\frac{a^2(\tau)k}{2}\left(\frac{1-z_k}{1+z_k}\right)h^2_{\mathbf{k}}\right]. \tag{7}$$

The time evolution of the wave function is captured in the functions $\alpha_k(\tau)$ or $z_k(\tau)$ which can be seen to satisfy the equations:

$$\alpha'_k = \frac{2\alpha_k^2}{a^2} - \frac{1}{2}a^2k, \tag{8}$$

$$z'_k + 2\,i\,k\,z_k + \left(\frac{a'}{a}\right)(z_k^2 - 1) = 0 \tag{9}$$

on substitution. These are non-linear first order Riccati-type equations and rather difficult to handle. But, we can introduce another function $\mu_k(\tau)$, defined through $\alpha_k = -(i\,a^2/2)(\mu'_k/\mu_k)$ which gives an equivalent easy-to-solve and familiar[7] second-order linear differential equation,

[7] In a remarkable co-incidence, the equation for μ_k is exactly the same as the field equation for $h_{\mathbf{k}}$ which we would get by directly varying the action in Eq. (5).

$$\mu_k'' + 2\left(\frac{a'}{a}\right)\mu_k' + k^2\mu_k = 0, \tag{10}$$

The function z_k, referred to as the *excitation* parameter, is a measure of departure from adiabatic evolution and is related to μ_k by

$$z_k = \left(\frac{k\,\mu_k + i\mu_k'}{k\,\mu_k - i\mu_k'}\right). \tag{11}$$

Thus, it suffices to solve for μ_k given the boundary conditions to determine the evolution of the system. Now the question is, how to *quantify* the degree of classicality of such a state? We track this using a correlation function which we refer to as the *classicality* parameter. It is a measure of the *phase space* correlations of a system defined as

$$\mathcal{C} \equiv \frac{\langle pq\rangle}{\sqrt{\langle p^2\rangle\,\langle q^2\rangle}}. \tag{12}$$

We can evaluate this quantity for our system using the Wigner function,

$$\mathcal{W}(\phi_{\mathbf{k}}, \pi_{\mathbf{k}}) = \frac{1}{\pi}\exp\left[-\frac{\phi_{\mathbf{k}}^2}{\sigma_k^2} - \sigma_k^2\left(\pi_{\mathbf{k}} - \mathcal{J}_k\,\phi_{\mathbf{k}}\right)^2\right],$$

defined in the $\phi_{\mathbf{k}} - \pi_{\mathbf{k}}$ phase space of the oscillator for the state in Eq. (7) to get,

$$\mathcal{C}_k = \frac{\mathcal{J}_k\sigma_k^2}{\sqrt{1 + (\mathcal{J}_k\sigma_k^2)^2}}; \quad \mathcal{J}_k\sigma_k^2 = \frac{2\,\mathrm{Im}(z_k)}{1 - |z_k|^2}. \tag{13}$$

So, when the Wigner distribution is an uncorrelated product of gaussians in $\phi_{\mathbf{k}}$ and $\pi_{\mathbf{k}}$, i.e., $\mathcal{J}_k = 0$, which is the case for the ground state, $\mathcal{C}_k = 0$ implying a pure quantum state. Otherwise, $|\mathcal{C}_k| \leq 1$ with a correlated Wigner function and hence implies deviation from the quantum nature towards classical behaviour which is maximal at the extremities.

All that is required now is to solve Eq. (10) for μ_k with appropriate initial conditions. We choose the standard Bunch-Davies vacuum initial condition during inflation and the mode functions in the subsequent stages are stitched together by demanding the continuity of μ_k and its derivative. All other quantities can be obtained once μ_k is known throughout the history of the universe. It is to be noted that, in reality, we have $\epsilon = \mathcal{O}\left(10^{-53}\right)$ which is a very, very small number and the only feasible study is through appropriate approximations in the analytical computation (see Ref. [5] for details). Here we present only the numerically computed results taking $\epsilon = 0.0001$ for visual and conceptual clarity. We show the evolution of parameters of the Wigner function in Figs. 2, 3 and 4 and the snapshots depicting the evolution of Wigner function in phase space in Fig. 5.

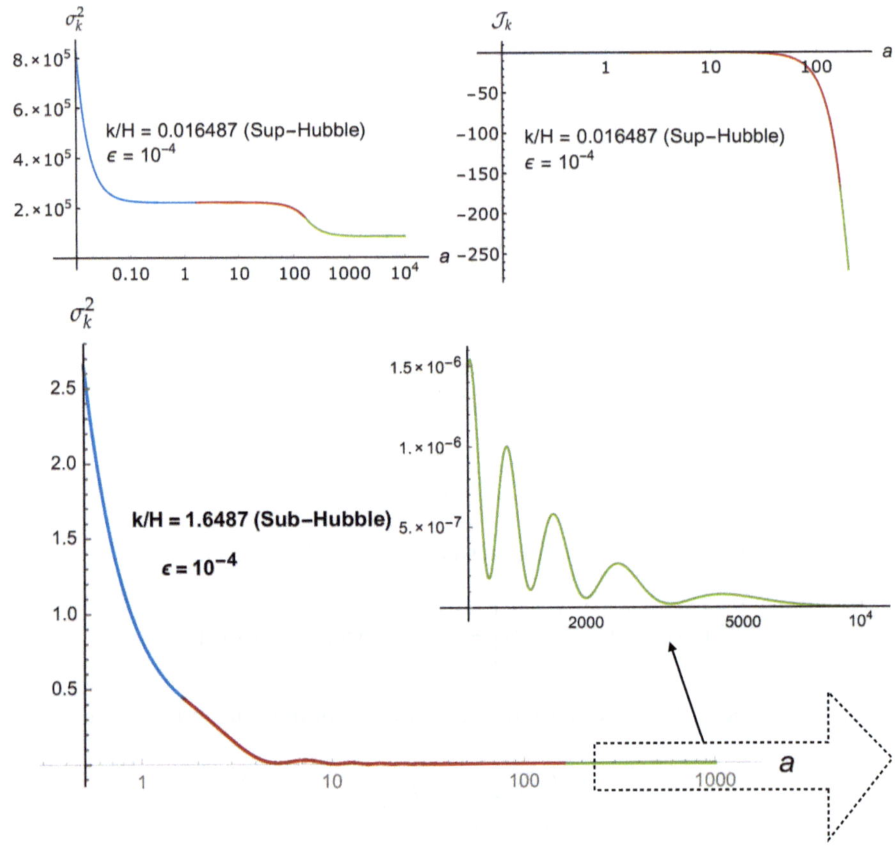

Fig. 2 Evolution of the parameters σ_k^2 and $\mathcal{J}_k$ of the Wigner function with the scale factor for $\epsilon = 0.0001$ for super- and sub-Hubble modes. The color scheme is as follows: early inflationary phase (*blue*), radiation dominated phase (*red*) and late-time de Sitter phase (*green*)

It is evident from the plots of the functions σ_k^2 and $\mathcal{J}_k$, the Wigner function starts uncorrelated with $\sigma_k^2 \to \infty$ and $\mathcal{J}_k \to 0$ at early times for all the modes when they are at sub-Hubble scales. The Wigner function is peaked highly on the ϕ_k-axis (the first plot in Fig. 5). Further on σ_k decreases sharply and $\mathcal{J}_k$ increases to negative values implying increasing anti-correlation. The Wigner function now starts to spreads out in the direction of momentum axis and lessens its spread on the position axes. The dynamics after this is strictly governed by the competition of the mode with the Hubble radius. The modes that exit the Hubble radius and go super-Hubble see a saturation of σ_k^2 and a monotonic increase of $\mathcal{J}_k$ to negative infinity. The Wigner function peaks on the momentum axis implying classical behaviour. For the modes which remain sub-Hubble until the final third stage, the limits $\sigma_k^2 \to 0$ and $\mathcal{J}_k \to -\infty$ are reached non-trivially with oscillations in between during the radiation-dominated epoch.

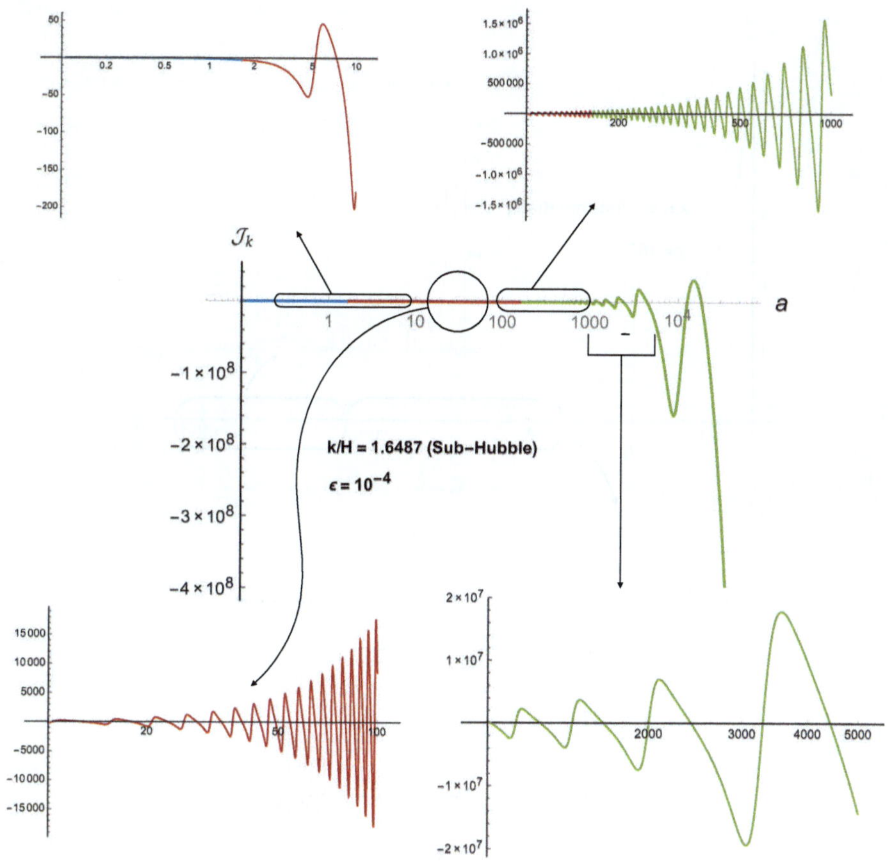

Fig. 3 Evolution of the parameter $\mathcal{J}_k$ of the Wigner function with the scale factor for $\epsilon = 0.0001$ for the sub-Hubble mode along with zoomed-in plots showing oscillations at different scales

The most interesting to us are the intermediate modes that exit the Hubble radius during inflation and then re-enter in the radiation-dominated epoch. The plots in Fig. 4 show a slight saddle of saturation as the mode exits the Hubble radius but then on re-entry the oscillations set in for both σ_k and $\mathcal{J}_k$ which damp to zero and increase to negative infinity respectively. The corresponding evolution of Wigner function in the phase space is also highly non-trivial. On towards Hubble-exit, the plots for $a = 0.05$, 0.1, 0.6 and 1 shows the swift peaking of Wigner function on momentum axes which persists for $a = 2$ when the mode is super-Hubble even though the Universe has made a transition to the second stage. However after re-entry, due to the oscillations, the Wigner function even though peaked on momentum axis, keeps on twisting and turning until the exit in the third stage to finally settle in the configuration shown for $a = 50, 000$. But clearly, the notion and quantifying the degree of *classicality* is not so easy relying on the Wigner function alone. Its

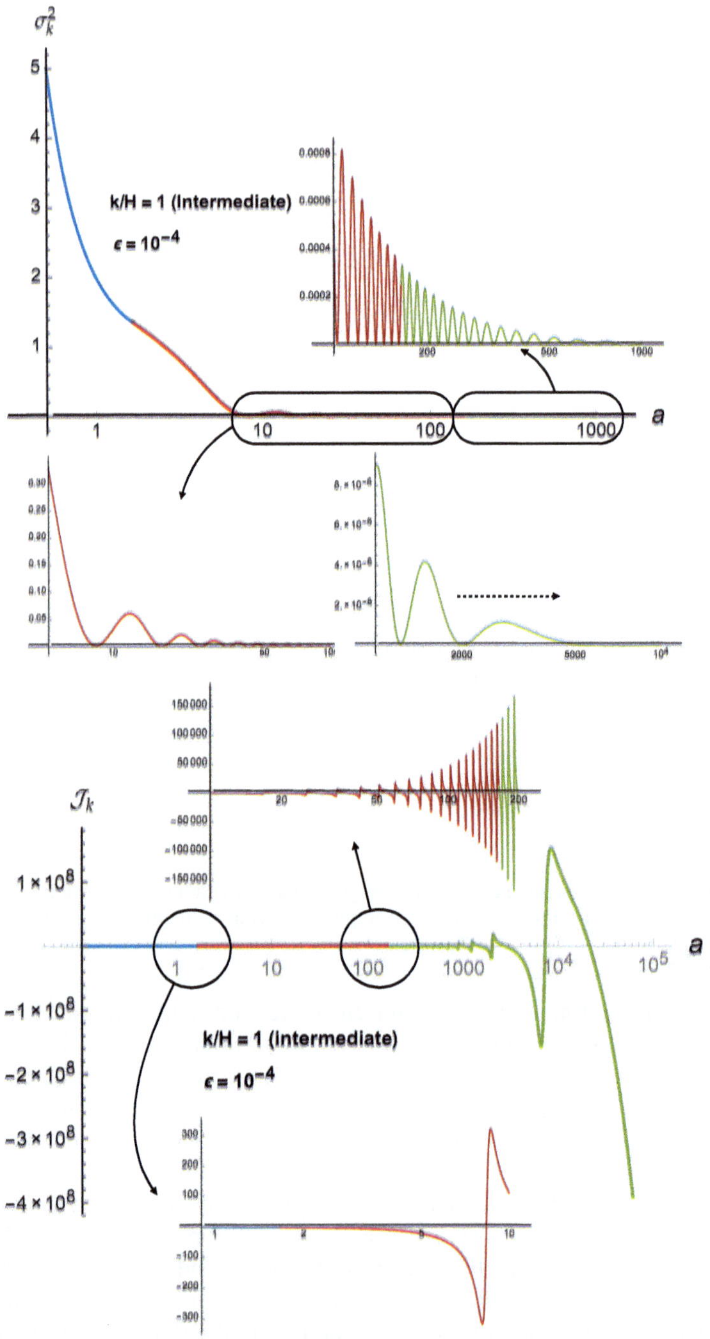

Fig. 4 Evolution of the parameters σ_k^2 (*above*) and $\mathcal{J}_k$ (*below*) of the Wigner function with the scale factor for an intermediate mode and $\epsilon = 10^{-4}$ along with zoomed-in plots at different times

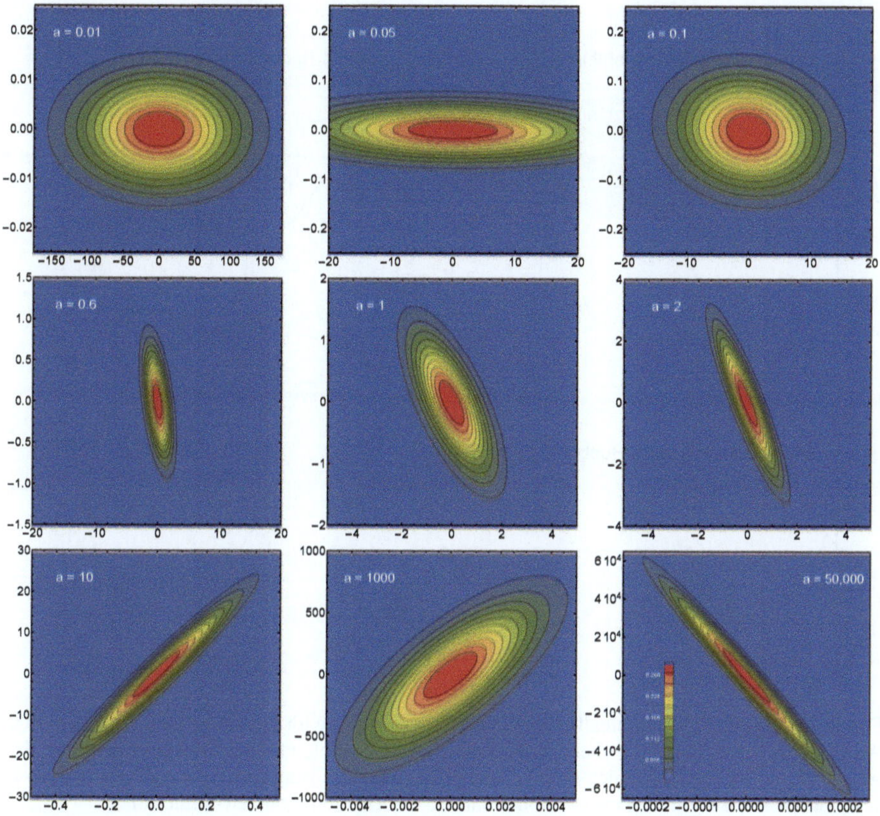

Fig. 5 Snapshots depicting the evolution of the Wigner function in the $\phi_k - \pi_k$ phase space for the intermediate mode ($k/H = 1$) and $\epsilon = 0.0001$

parameters lie in large ranges, for example, σ_k^2 is very large at early-times and $\mathcal{J}_k$ goes to negative infinity at late-times with oscillatory behaviour in between, leading to a very complicated evolution of the Wigner function in the phase space. This is where the classicality parameter helps being restricted to a finite domain $[-1, 1]$ by construction and as shown in Fig. 6 for different k/H values. We see the following evident features:

- The classicality parameter $\mathcal{C}_k$ starts from zero in the beginning of inflationary phase (indicating quantum origin) and after the Hubble exit whether it is in the early or late-time de Sitter stage, $\mathcal{C}_k \to -1$ and modes behave classically.
- For any mode with $k = k_{int}$ which lies within the $[k_{min}, k_{max}]$ band, we have a classical description near the end of the inflationary phase as $\mathcal{C}_k \to -1$ but as the universe makes a transition to radiation phase, it starts oscillating. These oscillations last all through the radiation phase and the late-time de Sitter phase till the mode exits the Hubble radius and then one finds $\mathcal{C}_k$ saturates at -1. Thus

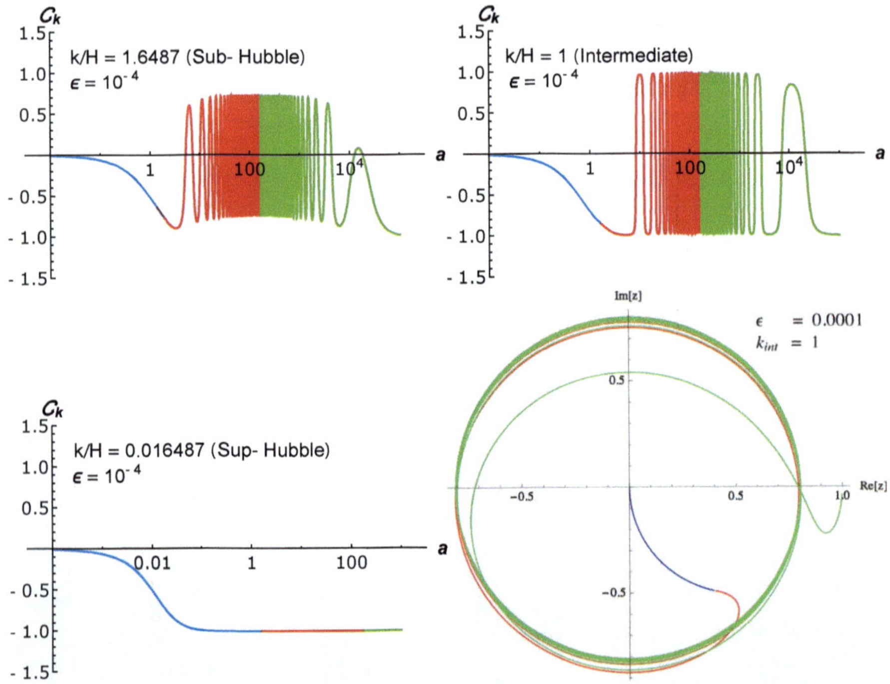

Fig. 6 Evolution of the classicality parameter $\mathcal{C}_k$ and the z_k with the scale factor for $\epsilon = 0.0001$. We clearly see that $|\mathcal{C}_k| \rightarrow 1$ whenever a mode exits the Hubble radius, however on re-entry, there are significant oscillations which seem to imply the dynamics is away from a completely classical description. The evolution is clearly non-adiabatic evident from non-zero values of z_k

in between, when the mode is sub-Hubble, the oscillations imply that the system is away from a *completely* classical description.

- The sub-Hubble mode does not reach -1 in the first two phases but once it exits the Hubble radius in the late-time de Sitter phase it reaches that value. On the other hand, the super-Hubble mode, once it exits from the initial de Sitter phase always remains super-Hubble and saturates with $\mathcal{C}_k \rightarrow -1$.
- This non-trivial behaviour of classicality parameter has a reason which is evident from the plot of z_k in Fig. 6. The function z_k is a complex quantity and its non-zero value measures departure from the adiabatic evolution of modes with respect to the background. There is a delay in the change of course of z_k and the background evolution. The system persists in the previous dynamics even though the Universe has made a transition to the next stage. Further, the rotor-like behaviour of z_k is what causes oscillations in the Classicality parameter. The rotations start in the radiation-dominated phase and persist in the late-time phase until the mode exits the Hubble radius.

Conclusively then, we have, for the two aforementioned *assumptions* of inflation,

✓ the emergence of classicality on Hubble exit checked,
✗ but "once classical, always classical" does not seem to hold.

What does it mean for the present-day observations that assume a classical-only description of the modes that re-enter the Hubble radius? Certainly, we require a re-think and re-check of the standard procedures. Even possible effects on the non-gaussianity parameter $f_{\rm NL}$ because of the non-classical behavior of perturbations after re entry. However, it doesn't stop here and brings more questions to the table such as: How will the above conclusion fare with the back-reaction included in and for the scalar perturbations? How does it compare with other constructs that describe the quantum-to-classical transition such as the *quantum discord* [8]? What does it mean to have a quantum nature intertwined with the classical and can this quantum nature have any imprints in the cosmic microwave background which may be extracted [9]?

As much as it is important to test inflation to its roots, it is a poetic quest to truly understand our *quantum origins* if it indeed turned from quantum to classical in the sky!

Acknowledgements The research of S.S. is supported by Dr. D.S. Kothari Postdoctoral fellowship from University Grants Commission, Govt. of India. I also thank Sujoy and Paddy for the collaboration leading to Ref. [5].

References

1. G. Mahajan, T. Padmanabhan, Gen. Rel. Grav. **40**, 661 (2008); G. Mahajan, T. Padmanabhan, Gen. Rel. Grav. **40**, 709 (2008)
2. A.R. Liddle, D.H. Lyth, *Cosmological Inflation and Large-Scale Structure* (Cambridge University Press, Cambridge, 1999)
3. J. Maldacena, arXiv:1508.01082
4. J. Martin, V. Vennin, Phys. Rev. D **93**, 023505 (2016)
5. S. Singh, S.K. Modak, T. Padmanabhan, Phys. Rev. D **88**, 125020 (2013)
6. D. Baumann, *TASI Lectures on Inflation*, arXiv:0907.5424
7. S. Singh, *Work in Progress*
8. H. Olliver, W.H. Zurek, Phys. Rev. Lett. **88**, 017901 (2002)
9. E. Lim, Phys. Rev. D **91**, 083522 (2015)

Classical and Quantum: A Conflict of Interest

T.P. Singh

Abstract We highlight three conflicts between quantum theory and classical general relativity, which make it implausible that a quantum theory of gravity can be arrived at by quantising classical gravity. These conflicts are: quantum nonlocality and space-time structure; the problem of time in quantum theory; and the quantum measurement problem. We explain how these three aspects bear on each other, and how they point towards an underlying noncommutative geometry of space-time.

This article is warmly dedicated to my Ph.D. supervisor Thanu Padmanabhan, on the happy occasion of his sixtieth birthday. One of the most important things I learnt from Paddy was to look for one's own questions and one's own answers, instead of necessarily accepting someone else's line of thought. I hope this lesson is reflected in the ideas presented in this article. In particular, Paddy himself might not agree with some or many of these ideas, and in that sense the lesson has probably been learnt well!

1 Some Limitations of Quantum Theory

Quantum theory is extraordinarily successful, and is not contradicted by any experiment. This is true for its non-relativistic version, as well as for relativistic quantum mechanics, and for quantum field theory. However, its successes should not blind us to the limitations of its theoretical structure, as we understand it today. First and foremost though, it is important to remember, and not often emphasized, that quantum mechanics has not been tested in all parts of the parameter space that are in principle accessible in table-top laboratory experiments. We have in mind tests of quantum linear superposition [Schrodinger cat states] for mesoscopic objects. The largest objects for which the superposition principle has been tested have a mass of

T.P. Singh
Tata Institute of Fundamental Research,
Homi Bhabha Road, Mumbai 400005, India
e-mail: tpsingh@tifr.res.in

J.S. Bagla and S. Engineer (eds.), *Gravity and the Quantum*,
Fundamental Theories of Physics 187, DOI 10.1007/978-3-319-51700-1_25

about 10^5 a.m.u. and the smallest objects which are known to behave classically have a mass of about a microgram [i.e. about 10^{18} a.m.u.]. In between, there is a technologically challenging range of some thirteen orders of magnitude, where there are no experimental tests of the superposition principle, although significant progress is now taking place since the last few years. [We note that macroscopic superpositions of internal states as in superconductors and Bose–Einstein condensates do not negate the previous statement. More on this later.] In this untested intermediate range, maybe there is a quantum-to-classical transition which can be explained by environmental decoherence and the many-worlds interpretation, or maybe by Bohmian mechanics. Alternatively, it maybe the case that there is a new dynamics such as spontaneous collapse, to which quantum and classical mechanics are approximations, and whose effects become significant in this intermediate regime, and which is responsible for the quantum-to-classical transition. To believe that quantum mechanics will definitely not be violated in this yet untested regime is akin to believing, if one were in the nineteenth century, that Newton mechanics will not be violated at high speeds or for small objects, even though the theory was not then tested at high speeds or for small objects. Of course with hindsight we know that such faith in Newtonian mechanics was misplaced, and accordingly we should reserve our judgement about quantum mechanics as well, until these thirteen orders of magnitude have been covered by experiments.

Quantum mechanics is generally taught to students as a 'final' theory, with rarely a mention of the unsatisfying aspects of its theoretical construction. Many physicists painfully 'unlearn' the theory in their later years, and realise the extreme peculiarity of the structure of the theory. The strangest aspect is the extreme dependence of the theory on its own classical limit, for its very construction and interpretation. One starts from the classical [Lagrangian or Hamiltonian] dynamics of the theory for the chosen degrees of freedom, and one must know the classical action and the Poisson brackets. Then the peculiar procedure 'quantize' is invoked: configuration variables and their canonical momenta are raised to the level of operators, and Poisson brackets are replaced by ad hoc quantum commutation relations. It works perfectly, but one is left wondering if the construction is fundamental: one should have been able to write down the principles of quantum theory ab initio, and derive classical mechanics from them, rather than the other way round.

The dependence on classical limit continues when one faces the task of interpreting the results of experiments on quantum systems, giving rise to the infamous quantum measurement problem [1]. There is a need for a so-called classical measuring apparatus: an object which is not found in superposition of position states, so that classical pointer states [which define the outcome of a measurement] can be defined. But then we are faced with tough questions. How large should an object be before it can be called classical? Quantum mechanics is silent about this. And the classical apparatus which quantum mechanics so much depends on for its interpretation, is something whose classical properties [in particular, the absence of position superposition of pointer states] should have been derived from quantum mechanics, rather than assuming its existence a priori, as if it had nothing to do with quantum theory per se.

It is well-known of course that things get more difficult from this point on. The evolution of the state of the quantum system is described by the Schrödinger equation: this evolution is deterministic and linear. The process of measurement by the classical apparatus breaks both linear superposition and determinism. Although there is no randomness in the initial conditions for the Schrödinger evolution, the outcomes of the measurement are random and probabilistic. This is an unparalleled situation in physics: probabilities without random initial conditions. The fact that probabilities arise during measurement, implies that something has to give. It means that either the probabilities are not real but only apparent, or that there is an aspect of randomness in the dynamics, or in the initial conditions, which is not evident in the Schrödinger equation.

Not only is there a dependence on its own classical limit, but there is also a dependence of quantum theory on external spacetime structure. We emphasize two aspects of this: one which suggests a possible conflict with special relativity, and the other which strongly suggests that the present formulation of the theory should possess an equivalent, but a more fundamental, formulation. The first of these has to do with the EPR paradox and non-local quantum correlations, which suggest that quantum events influence each other outside the light cone. One possible implication of this is that wave-function collapse in quantum theory is simply not compatible with the spacetime structure dictated by special relativity, and in order to describe collapse satisfactorily one perhaps needs to introduce a new 'quantum' structure of spacetime.

The second aspect, rarely emphasized, has to do with the fact that the time that appears in quantum theory is part of a classical spacetime geometry, which geometry is produced by classical macroscopic objects. But these classical objects are in turn a limiting case of quantum theory! Once again, the dependence of the theory on its own limit is evident. Clearly, there then ought to exist an equivalent reformulation of quantum theory which does not refer to a classical time.

We thus see that there are at least three different ways in which quantum mechanics depends on its own classical limit, or on classical spacetime structure. These give rise to the quantum measurement problem, the problem of quantum nonlocality, and the problem of time in quantum theory. In the next three sections we briefly review some developments which address these problems, and their inter-relationship. In the last section we discuss what these problems and their possible resolutions imply for a future quantum theory of gravity.

2 The Quantum Measurement Problem

Modern approaches to addressing the measurement problem broadly fall into three classes. The first is to say that collapse of the wave function is only an apparent process, and in reality no collapse ever takes place This is the essence of the many worlds interpretation. There is no need to modify or reformulate quantum theory. The second is to say that there is randomness in the initial conditions, but the evolution by itself is deterministic. This is Bohmian mechanics - a mathematical reformulation of quantum mechanics. The third is to say that there is randomness in the dynamics, and the deterministic Schrödinger evolution is only an approximation to the random dynamics. This is the essence of collapse models.

According to the many worlds interpretation, the evolution is deterministic Schrödinger evolution through and through, and upon a measurement the universe, including the observer, 'splits' into many branches, with a given branch possessing only one out of the various possible outcomes. The other branches contain, respectively, one or the other outcomes. The different branches do not interfere with each other, presumably because of decoherence. [There is a vast literature on decoherence, including the experiments and models by [2–4], books by [5–7] and the seminal papers [8–10] and reviews [11–15].] The collapse of the wave function is only apparent, not real, and there is no need to modify quantum mechanics. The hard part about many worlds is to understand where the probabilities come from? If the evolution is always deterministic Schrödinger evolution, then why do the outcomes obey the Born probability rule? Various explanations have been put forward, but they do not appear convincing enough [16–27].

Bohmian mechanics is a neat and precise reformulation of quantum theory, where additional equations of motion are introduced for the positions of particles. The wave function, which satisfies the Schrödinger equation, also enters in the equation of motion of particles. The theory is a deterministic theory of particles in motion. Randomness enters in a classical sense, via random initial conditions, chosen such that the outcomes of experiments obey the Born rule. Bohmian mechanics, as well as many worlds, make the same predictions as quantum theory, and they would be falsified if collapse models, which predict departures from quantum theory, are experimentally verified [28–35].

Collapse models, first developed in the eighties, propose a stochastic, nonlinear modification of the Schrödinger equation, and introduce the new feature that collapse of the wave function is a spontaneous process, not having anything to do per se with the act of measurement [36–40]. There is no longer any need for the vaguely defined measuring apparatus, nor an artificial divide between a 'quantum system' and a 'classical apparatus'. The nonlinear modification breaks linear superposition, while its stochastic nature ensures that the outcome of the broken superposition is random. The structure of the modifying terms is chosen in such a way that the random outcomes are realised according to the Born probability rule. The theory introduces two new constants of nature, a rate constant λ which determines the rate of collapse, and a critical length r_c to which the collapsed wave function is confined. The rate constant has been assigned an ad hoc value of $10^{-17}\,\mathrm{s}^{-1}$ for a nucleon - this means that the wave function of a nucleon undergoes spontaneous collapse once every 10^{17} s. Understandably then, the nonlinear modification is completely negligible for the nucleon and it behaves perfectly quantum mechanically, obeying the Schrödinger equation. However, for a particle of mass m, the rate constant is assumed to be $(m/m_N)\lambda$, where m_N is the nucleon mass, and hence the rate constant scales with mass. For macroscopic objects, the wave function collapses extremely rapidly; this explains the classical nature of macroscopic objects, and in particular it explains why pointer position states are classical.

Collapse models thus also provide a natural solution to the measurement problem. Before a quantum system interacts with the measuring apparatus, its microscopic nature ensures that the rate constant is very small, and the superpositions are long lived. Upon its interaction with the so-called measuring apparatus [which is macroscopic] their entangled state represents a macroscopic superposition, which involves the superposition of pointer position states. This state is extremely short lived, according to the model, and very quickly 'collapses' to one of the outcomes, while obeying the Born rule.

These models propose that there is a new stochastic dynamics, to which quantum mechanics is the microscopic approximation, and classical mechanics is the macroscopic approximation. The stochastic effect is negligible in the microscopic limit. On the other hand it is extremely prominent in the macro limit, so that quantum evolution effectively appears like classical evolution on trajectories which obey Newtonian dynamics. The quantum to classical transition is naturally explained, and there is no longer any need for a measuring apparatus, to explain the results of measurements.

The most interesting thing about collapse models is not that they are necessarily correct, but rather that they are experimentally testable and that in principle they make predictions which are different from those of quantum mechanics. In the micro regime, the rate constant is so small that the models are indistinguishable from Schrödinger evolution and hence make essentially the same experimental predictions as quantum mechanics. In the macro regime the predictions are the same as that of classical mechanics. It is in the in-between mesoscopic regime - the thirteen orders of magnitude alluded to at the beginning of the article - that the model predictions markedly differ from that of quantum mechanics. The principle effect is that in this range the lifetime of a quantum superposition is neither too large nor too small, but in a range suitable for experimental detection. Thus if a mesoscopic object, having a mass of say a billion a.m.u., is prepared in a superposed state by passing it through a diffraction grating, then according to collapse models this superposition will decay before the particle reaches the detecting screen, and hence no interference pattern will be seen. If this happens, it of course violates quantum mechanics, and is evidence for collapse models. Experiments of this nature form the subject of matter wave interferometry, and they have played a very important role in constraining collapse models and putting bounds on the rate constant λ [41]. A great technological challenge is to eliminate'impurities' such as ambient radiation and gas which cause environmental decoherence, and mask and mimic the loss of superposition caused by collapse models. The largest objects for which superposition has been verified through interferometry have a mass of about 10^5 a.m.u. and this puts an upper bound on λ of about $10^{-5}\,\mathrm{s}^{-1}$ [42].

A different class of experiments which are becoming important in testing and constraining collapse models have to do with a side effect of these models. Namely, the stochastic process which introduces randomness in the dynamics also causes stochastic heating of the affected quantum particle, and hence a very tiny violation of energy momentum conservation [43]. The fact that such a violation has not been observed in laboratory experiments and in astronomical observations puts powerful bounds on λ, the strongest current bound being that $\lambda < 10^{-8}\,s^{-1}$ [44]. Various new experiments have been proposed to test the effects of stochastic heating [45–48]. Eventually, in order to verify or rule out collapse models, experiments must push this bound all the way down to 10^{-17}, below which value collapse models may not be able to solve the measurement problem, and other explanations such as many worlds and Bohmian mechanics would start to appear more favorable.

Collapse models do indeed have some limitations, which call for their better theoretical understanding. The models are purely phenomenological in nature, having been proposed with the express purpose of solving the quantum measurement problem. The mathematical structure of the stochastic nonlinearity is designed so as to give rise to the Born probability rule. In that sense the models do not predict or prove the Born rule; rather they have the Born rule built into them. Thus the question as to what is the fundamental origin of the probabilities still remains unanswered. [The same is true of the many worlds picture, and of Bohmian mechanics as well.] We really do not know what is the cause of this randomness. Why should there be in nature this stochastic noise field which these models employ?

Two ideas which bear on this question in a serious way deserve mention. One is that this stochasticity has to do with gravity and spacetime structure. Gravitational fields are produced by macroscopic bodies, and the latter obey the uncertainty principle of quantum mechanics. It seems plausible [though not fool-proof] that this introduces an uncertainty in the produced gravitational field, and hence fluctuations in the spacetime geometry. This might be the source of randomness sought for by collapse models. It is then natural to ask how these fluctuations in the geometry affect the motion of a quantum particle which obeys Schrödinger evolution? Various model studies have shown that spacetime fluctuations produce gravitationally induced decoherence of the wave function, with the effect becoming more prominent as the mass of the quantum particle is increased [49–81]. While these results are very encouraging, they do not yet provide a collapse model. Gravity can cause decoherence, but it is not yet clear how (if at all) it causes collapse of the wave function (selection of one of the various outcomes) and how it explains the Born probability rule. The conceptual status of gravity in such models is also not very clear: is gravity classical, quantum, semiclassical, or something else? Nonetheless, since we know that gravity exists, it is very promising to investigate if it is the source of the nonlinear stochasticity in collapse models.

The second idea for a fundamental origin of collapse models is to consider if quantum theory is an approximation to a deeper underlying theory, and if the nonlinear stochastic modification arises as a higher order correction to the leading approximation. That quantum theory should perhaps be formulated differently, starting from some fundamental principles, is already indicated by the extreme dependence of the current formulation of the theory on its own classical limit. This is the essence of the theory of Trace Dynamics [TD], developed by Adler and collaborators [82–85]. TD is the classical dynamics of matrices q_r whose elements can either be odd grade [fermionic sector F] or even grade [bosonic sector B] elements of Grassmann numbers. The Lagrangian in this dynamics is defined as the trace of a polynomial function of the matrices and their time derivatives. Lagrangian and Hamiltonian dynamics can then be developed in the conventional manner. In TD, the matrix-valued configuration variables q_r and their conjugate momenta p_r all obey arbitrary commutation relations amongst each other. However, as a consequence of a global unitary invariance of the dynamics there occurs in TD an important conserved charge, known as the Adler–Millard charge

$$\tilde{C} = \sum_B [q_r, p_r] - \sum_F \{q_r, p_r\} \tag{1}$$

whose existence is central to the subsequent development of the theory.

Assuming that one is not examining the dynamics exactly, one develops an equilibrium statistical thermodynamics for the classical dynamics described by TD. If one considers a sufficiently large system [of many, many particles, each particle being a matrix, as if there were a gas of matrices], the 'system point' can in the long run be assumed to scan all of phase space. The phase space probability distribution achieves equilibrium [i.e. a uniform distribution over phase space]. The equilibrium distribution can be determined by maximising the entropy, as is done in statistical mechanics. The equipartition of the Adler–Millard charge leads to certain Ward identities, which in turn lead to the important result that thermal averages of canonical variables obey quantum dynamics and quantum commutation relations. In particular, the emergent q operators commute with each other, and so do the p operators. This is how quantum theory is seen as an emergent phenomenon. The quantum state satisfying the Schrödinger picture is recovered as usual, by implementing a transition from the Heisenberg picture to the Schrödinger picture. TD is a classical deterministic theory, and time evolution of the matrices is described in the standard way, with respect to a flat Minkowski spacetime background. However, TD is *not* a hidden variable theory, because the matrix variables exist at a distinctly different underlying level, as compared to the quantum theoretical degrees of freedom, with the latter arising only upon statistical coarse-graining, in the conventional sense of statistical mechanics. Hence the arguments of Bell's theorem against local hidden variable theories do not apply to TD.

Furthermore, if one considers the inevitable statistical fluctuations of the Adler–Millard charge about equilibrium, this leads to a collapse model type modification of the nonrelativistic Schrödinger equation. These fluctuations are the sought for source of randomness. One does not understand TD well enough to uniquely predict the modified theory. In particular one still does not have a proof of the origin of Born probability rule in TD, but TD is perhaps the only theory to date, apart from gravity, which provides a fundamental explanation for randomness, by way of the statistical fluctuations. The collapse models, which are highly successful phenomenologically, are one possible modification admitted by TD. The modification, ignorable for microscopic objects but significant for large objects, solves the quantum measurement problem and leads to emergent classical behavior in macroscopic systems. The fluctuations of the conserved charge about its equilibrium value carry crucial information about the arbitrary commutation relations amongst the configuration variables and their momenta in the underlying TD.

Coming back to collapse models, another of their limitations is that they are non-relativistic. Various attempts to construct relativistic collapse models face difficulties, a feature shared also by Bohmian mechanics. Perhaps this is an indicator that collapse may not be compatible with special relativity, especially in the light of quantum non-locality related issues which we discuss in a subsequent section below.

We take this occasion to mention that macroscopic quantum states such as super-conductors and Bose–Einstein condensates, which are made from superposition of internal degrees of freedom, do not invalidate collapse models. The constraints on the rate constant λ from such systems are rather weak.

We conclude this section by noting that important theoretical and experimental progress is currently being made on the quantum measurement problem, and on removing this aspect of the dependence of the theory on its classical limit. We can expect some exciting developments in this problem in the coming decade or so.

3 The Problem of Time in Quantum Theory

The time in quantum theory is part of a classical spacetime geometry, which geometry is produced by macroscopic bodies, which in turn are a limiting case of quantum objects, whose evolution is described with respect to this very time! It is evident that in order to avoid this self-reference there ought to exist an equivalent reformulation of quantum theory, which does not refer to classical time. This problem is no less severe than the measurement problem, but somehow it gets far less attention, if any at all.

In searching for such a reformulation we are guided by the assumption that such a reformulation should also throw light on the quantum measurement problem. After all both these problems arise from the dependence of quantum theory on its classical limit, and a common explanation is not implausible. We are also motivated by the fact that Trace Dynamics already seeks to obtain quantum theory, and its stochastic nonlinear modification, from underlying deeper principles, albeit while retaining the classical structure of spacetime. From our point of view however, as expressed above, the dependence of quantum theory on classical time seems to be a limitation, and we have made preliminary attempts to extend TD to remove the dependence on classical time. This is still work in progress and we summarize below what has been understood so far [86–88].

To achieve a formulation of quantum theory without classical time, we first generalized Trace Dynamics so as to make space-time coordinates also into operators. Associated with every degree of freedom there now are coordinate operators $(\hat{t}, \hat{\mathbf{x}})$ with arbitrary commutation relations amongst them. From these we construct a Lorentz invariant line-element $d\hat{s}^2$, and we define the important notion of Trace time s as follows:

$$ds^2 = Trd\hat{s}^2 \equiv Tr[d\hat{t}^2 - d\hat{x}^2 - d\hat{y}^2 - d\hat{z}^2] \tag{2}$$

A Poincaré invariant dynamics is constructed, in analogy with ordinary special relativity, and in analogy with TD, but with the difference that evolution is now defined with respect to trace time s. The theory, as before, admits a conserved Adler–Millard charge, and the degrees of freedom now involve bosonic and fermionic components of space-time operators as well. Because the space-time operators have arbitrary commutation relations, there is now no point structure or light-cone structure, nor a notion of causality, although the line-element is Lorentz invariant.

From this generalized TD, we constructed its equilibrium statistical thermodynamics, as before. The equipartition of the Adler–Millard charge results in the emergence of a generalized quantum dynamics [GQD] in which evolution is with respect to the trace time s, and the thermally averaged space-time operators $(\hat{t}, \hat{\mathbf{x}})$ are now a subset of the configuration variables of the system. It is significant that these averaged operators commute with each other. This is the originally sought after reformulation of quantum theory which does not refer to classical time. In the non-relativistic limit we recover the generalized Schrödinger equation

$$i\hbar\frac{d\Psi(s)}{ds} = H\Psi(s) \tag{3}$$

To go beyond special relativity, one must invoke an operator structure for the spacetime metric. Here the program runs into difficulties. It has been argued by Adler that the metric must retain its classical [non-operator] structure in TD. If a way can be found around this, we expect the development to proceed along the following lines.

To demonstrate the equivalence of the reformulation [GQD] with standard quantum theory, one must first explain how the classical Universe, with its classical matter fields and ordinary space-time, emerges from the GQD in the macroscopic approximation. Like in TD, one would next allow for inclusion of stochastic fluctuations of the Adler–Millard charge, in the Ward identity. This should result in a non-linear stochastic Schrödinger equation, but now with important additional consequences. One considers the situation where matter starts to form macroscopic clumps (as for example in the very early universe, right after the Big Bang). These stochastic fluctuations become increasingly significant as the number of degrees of freedom in the clumping system increases. As in collapse models, these fluctuations result in macroscopic objects being localized, but now not only in space, but in time as well! This means that the time operator associated with every object becomes classical (i.e. it takes the form: a c-number times a unit matrix).

The localization of macroscopic objects is thus accompanied by the emergence of a classical space-time. This is in accordance with the Einstein hole argument: classical matter fields and the metric they produce are required to give physical meaning to the point structure of spacetime. If, and only if, the Universe is dominated by macroscopic objects, as is the case in today's Universe, can one also talk of the existence of a classical space-time. When this happens, the trace proper time s can be identified with classical proper time. After the Universe reaches this classical state, it sustains this state, because of the continuous action of stochastic fluctuations on macroscopic objects, thereby simultaneously achieving the existence of a classical space-time geometry. Since the underlying generalized TD is Lorentz invariant, the emergent classical space-time is locally Lorentz invariant too. However there is a key difference: unlike in the underlying theory, now the light-cone structure, and causality, are emergent features, because the space-time coordinates have become c-numbers now.

Irrespective of this pre-existing classical spacetime background, a microscopic system in the laboratory is described at a fundamental level in terms of its own non-commutative space-time (2), via the generalized TD associated with it. Subsequent to coarse-graining, this results in the system's GQD (3) with its trace time. If we assume that stochastic fluctuations can be ignored, this GQD has commuting $\hat{t}$ and $\hat{\mathbf{x}}$ operators. These, because of their commutativity, can be mapped to the c-number t and $\mathbf{x}$ coordinates of the pre-existing classical universe, and trace time can then be mapped to ordinary proper time. This is hence a mapping to ordinary special relativity, and one recovers standard relativistic quantum mechanics in this way, as well as its non-relativistic limit. If this program can be fully implemented, it will establish as to how standard quantum theory is recovered from the reformulation which does not depend on classical time.

Thus in our scenario the problem of time and the problem of measurement are related to each other. If one starts from a formulation of quantum theory which does not have classical time, then, in order to recover classical time and spacetime geometry from it, one must also recover from this formulation the macroscopic limit of matter fields. This is because classical geometry and classical matter fields go hand in hand. And to recover the classical limit for macroscopic objects is the same thing as solving the measurement problem. Because the latter problem can be restated as: why are macroscopic objects not found in superposition of position states? The measurement problem is a subset of the larger question: how does the classical structure of spacetime and matter emerge from an underlying quantum theory of spacetime and matter?

4 Quantum Non-locality and Space-Time Structure

The essence of the EPR paradox is that measurement on one part of a quantum system instantaneously influences another part of the same (correlated) quantum system, even if the two sub-systems are space-like separated. To Einstein, this suggested that quantum theory is incomplete. However, experimental measurements on entangled quantum states indeed demonstrate non-local correlations and indeed suggest the existence of an acausal action at a distance across space-like separated regions. This has been confirmed by increasingly precise loophole free tests of violation of Bell's inequalities by quantum systems. Although such correlations cannot be used for superluminal signaling, the acausal nature of the influence suggests the possibility of a conflict with special relativity and Lorentz covariance. This so-called spooky action at a distance has been debated extensively, but numerous investigations over decades have not provided a satisfactory resolution of the issue. On the other hand, a remarkable experiment shows that even if one assumes that the influence travels causally in a hypothetical privileged frame of reference, its speed would have be at least four orders of magnitude greater than the speed of light. That in itself could lead to problems with special relativity. Furthermore, we have seen above that attempts to construct relativistic versions of collapse models run into difficulties. To us this is possibly a signal that wave function collapse is not compatible with classical spacetime structure [light-cones, and causality].

A possible resolution might come from the underlying noncommutative structure of spacetime that we have proposed, and which was discussed above, in the context of the problem of time. It may well be that trying to describe collapse from the viewpoint of ordinary spacetime is not the right thing to do, when one goes over from the absolute Newton time of non-relativistic quantum mechanics, to the relative time of special relativity [89]. Collapse is perceived as instantaneous in terms of ordinary time, but there is nothing to say that this is the correct time to use. We have to pay heed that this classical time is external to quantum theory.

Let us go back to the Generalized Quantum Dynamics [GQD] where evolution of the quantum system is described with respect to trace time s. Before the measurement takes place, the stochastic fluctuations of the Adler–Millard charge can be neglected for the quantum system [since it is microscopic], and as we observed above, its GQD can be mapped to standard quantum theory. However, when the measurement is done, the collapse inducing stochastic fluctuations in the space-time operators $\hat{t}$, $\hat{\mathbf{x}}$ associated with the quantum system become significant. These operators now carry information about the arbitrary commutation relations of the underlying generalized TD and they no longer commute with each other. This implies that they cannot be mapped to the ordinary space-time coordinates of special relativity. Here, simultaneity can only be defined with respect to the trace time s, and there is no special relativistic theory of wave function collapse. In this picture, collapse and the so-called non-local quantum correlation takes place only in the non-commutative space-time (2), which lacks point structure, lacks light-cone structure, and is also devoid of the notion of distance. Therefore one can only say that collapse takes place at a particular trace time, which is Lorentz invariant, and it is not physically meaningful to talk of an influence that has travelled, nor should one call the correlation non-local. In this picture the wave function does not know distance - it just is. We once again see that getting rid of classical spacetime from quantum theory removes another one of its peculiarity, the so-called spooky action at a distance.

If, as is conventionally done, one tries to view and describe the measurement on the entangled quantum state from the view-point of the Minkowski space-time of special relativity, the process inevitably appears acausal and non-local. However, such a description should not be considered valid, because there is no map from the fluctuating and noncommuting $\hat{t}$, $\hat{\mathbf{x}}$ operators to the commuting t and $\mathbf{x}$ coordinates of ordinary special relativity. No such map exists in the non-relativistic case either. However, in the non-relativistic case, because there is an absolute time, it becomes possible to model the fluctuations as a stochastic field on a given space-time background, as is done in collapse models, and collapse is instantaneous in this absolute time; however it does not violate causality.

We see that while on the one hand the problem of time is related to the measurement problem, on the other hand, the resolution of the time problem can alleviate the mysterious nature of quantum non-locality. It will be interesting to investigate if one can make an experimental proposal to verify if noncommutative spacetime is indeed the way to understand the spooky action at a distance.

Undoubtedly, much more work needs to be done, to put the ideas of the present and the previous section on a firm footing.

5 Implications for a Quantum Theory of Gravity?

The three problems that we have discussed here could all be called a conflict between quantum theory and general relativity. The measurement problem has to do with the classical [as opposed to quantum] nature of macroscopic objects. These objects are intimately tied up with spacetime geometry through the laws of general relativity. To the extent that quantum theory does not explain the properties of macroscopic objects, it maybe said to be in conflict with general relativity. The problem of time is a direct conflict of course, because quantum objects do not produce a classical spacetime geometry. And also, quantum non-locality does not seem consistent with classical spacetime structure.

Given all this, should we aim to construct a quantum theory of gravity by 'quantizing' classical general relativity? It seems rather unnatural to do so. It is a fine thing to quantize other fundamental forces, because they take spacetime structure as given, and because they do not face the kind of conflict that gravity faces with quantum theory. By quantizing general relativity, we seem to violate the rules of the game. There is this classical spacetime structure whose existence is pre-assumed while writing down the quantum rules: how can these rules then be applied to that very structure? It does not seem a logical thing to do, and there is no guarantee that the correct quantum theory of gravity will emerge in this way.

Rather, we see pressing reasons - measurement problem, time problem, non-locality - which suggest the need to modify both quantum theory and spacetime structure, when one starts trying to resolve the conflict between classical and quantum. We should not quantize gravity; rather there is an underlying theory - perhaps a combination of noncommutative geometry and Trace Dynamics as suggested here - or something else, from which both quantum theory and gravitation are emergent. Gravitation emerges in the full classical limit, when both matter and gravity are treated classically. Quantum theory emerges, upon coarse graining the underlying theory, when only the gravity sector is treated classically. It maybe that this underlying theory is arrived at by demanding that physical laws be covariant under general coordinate transformations of non-commuting coordinates, thus bringing together the element of general covariance from relativity, and the element of noncommutativity from quantum theory. Given the nonlinearity of gravitation, it seems rather unlikely that the principle of quantum linear superposition can survive such a union!

Acknowledgements It is my pleasure to thank my collaborators on these topics - Angelo Bassi, Hendrik Ulbricht, Saikat Ghosh, Shreya Banerjee, Srimanta Banerjee, Sayantani Bera, Suratna Das, Sandro Donadi, Suman Ghosh, Kinjalk Lochan, Seema Satin, Priyanka Giri, Navya Gupta, Bhawna Motwani, Ravi Mohan and Anushrut Sharma.

References

1. J.A. Wheeler, W.H. Zurek, *Quantum Theory and Measurement* (Princeton University Press, New Jersey, 1983)
2. M. Brune, E. Hagley, J. Dreyer, X. Maitre, A. Maali, C. Wunderlich, J.M. Raimonde, S. Haroche, Phys. Rev. Lett. **77**, 4887 (1996)
3. R.A. Harris, L. Stodolsky, J. Chem. Phys. **74**, 2145 (1981)
4. S. Gerlich, L. Hackermüller, K. Hornberger, A. Stibor, H. Ulbricht, M. Gring, F. Goldfarb, T. Savas, M. Muri, M. Mayor, M. Arndt, Nat. Phys. **3**, 711 (2007)
5. E. Joos, H. Zeh, C. Kiefer, D. Giulini, J. Kupsch, I.O. Stamatescu, *Decoherence and the Appearance of a Classical World in Quantum Theory* (Springer, Berlin, 2003)
6. M. Schlosshauer, *Decoherence and the Quantum-to-Classical Transition* (Springer, Heidelberg, 2007)
7. H.P. Breuer, F. Petruccione, *Relativistic Quantum Measurement and Decoherence* (Springer, Heidelberg, 2000)
8. H.D. Zeh, Found. Phys. **1**, 69 (1970)
9. E. Joos, H. Zeh, Z. Phys. **59**, 223 (1985)
10. A.O. Caldeira, A.J. Leggett, Phys. Rev. Lett. **46**, 211 (1981)
11. M. Schlosshauer, Rev. Mod. Phys. **76**, 1267 (2005)
12. B. Vacchini, K. Hornberger, Phys. Rep. **478**, 71 (2009)
13. W.H. Zurek, Phys. Today **44**, 36 (1991)
14. W.H. Zurek, Rev. Mod. Phys. **75**, 715 (2003)
15. G. Bacciagaluppi, (2007), http://stanford.library.usyd.edu.au/entries/qm-decoherence/
16. H. Everett III, Rev. Mod. Phys. **29**, 454 (1957). doi:10.1103/RevModPhys.29.454
17. D.S. DeWitt, N. Graham, *The Many-Worlds Interpretation of Quantum Mechanics* (Princeton University Press, Princeton, 1973)
18. M. Tegmark, Nature **448**, 23 (2007)
19. A. Kent, Int. J. Mod. Phys. A **5**, 1745 (1990)
20. D. Wallace, Stud. Hist. Philos. Mod. Phys. **34**, 415 (2003)
21. D. Deutsch, *The Fabric of Reality: The Science of Parallel Universes and its Implications* (Penguin Books, New York, 1998)
22. L. Vaidman, (2002). http://plato.stanford.edu/entries/qm-manyworlds/
23. S.D.H. Hsu, (2011). arXiv:1110.0549
24. S. Saunders, J. Barrett, A. Kent, D. Wallace (eds.), *Many Worlds? Everett, Quantum Theory, and Reality* (Oxford University Press, New York, 2010)
25. H. Putnam, British J. Philos. Sci. **56**, 615 (2005)
26. J.A. Barrett, P. Byrne (eds.), *The Everett Interpretation of Quantum Mechanics: Collected Works 1955–1980* (Princeton University Press, Princeton, 2012)
27. G. Bacciagaluppi, (2001). http://philsci-archive.pitt.edu/504/2/cracow.pdf
28. D. Bohm, Phys. Rev. **85**, 166 (1952)
29. D. Bohm, Phys. Rev. **85**, 180 (1952)
30. J. Bub, *Interpreting the Quantum World* (Cambridge University Press, Cambridge, 1997)

31. D. Dürr, S. Goldstein, N. Zanghì, J. Stat. Phys. **67**, 843 (1992)
32. P.R. Holland, *The Quantum Theory of Motion: An Account of the de Broglie-Bohm Causal Interpretation of Quantum Mechanics* (Cambridge University Press, Cambridge, 1993)
33. D. Bohm, B.J. Hiley, *The Undivided Universe: An Ontological Interpretation of Quantum Theory* (Routledge, New York, 1995)
34. D. Dürr, S. Teufel, *Bohmian Mechanics* (Springer, Heidelberg, 2009)
35. D. Dürr, S. Goldstein, *Quantum Physics Without Quantum Philosophy* (Springer, Heidelberg, 2012)
36. P. Pearle, Phys. Rev. D **13**, 857 (1976)
37. G.C. Ghirardi, A. Rimini, T. Weber, Phys. Rev. D **34**, 470 (1986)
38. G.C. Ghirardi, P. Pearle, A. Rimini, Phys. Rev. A **42**, 78 (1990)
39. P. Pearle, *Open Systems and Measurement in Relativistic Quantum Theory*, vol. 526, Lecture Notes in Physics (Springer, Berlin, 1999)
40. A. Bassi, G.C. Ghirardi, Phys. Rep. **379**, 257 (2003)
41. K. Hornberger, S. Gerlich, P. Haslinger, S. Nimmrichter, M. Arndt, Rev. Mod. Phys. **84**, 157 (2012)
42. A. Bassi, K. Lochan, S. Satin, T.P. Singh, H. Ulbricht, Rev. Mod. Phys. **85**, 471 (2013)
43. S.L. Adler, J. Phys. A **40**, 2935 (2007)
44. A. Vinante, M. Bahrami, A. Bassi, O. Usenko, G. Wijts, T.H. Oosterkamp, Phys. Rev. Lett. **116**, 090402 (2016)
45. B. Collett, P. Pearle, Found. Phys. **33**, 1495 (2003)
46. S. Bera, B. Motwani, T.P. Singh, H. Ulbricht, Sci. Rep. **5**, 7664 (2015)
47. M. Bahrami, M. Paternostro, A. Bassi, H. Ulbricht, Phys. Rev. Lett. **112**, 210404 (2014)
48. D. Goldwater, M. Paternostro, P.F. Barker, (2015). arXiv:1506.08782
49. F. Karolyhazy, Nuovo Cimento **42A**, 390 (1966)
50. F. Karolyhazy, A. Frenkel, B. Lukács, in *Quantum Concepts in Space and Time*, ed. by R. Penrose, C.J. Isham (Clarendon, Oxford, 1986)
51. F. Karolyhazy, Magy. Fiz. Foly. **22**, 23 (1974)
52. F. Karolyhazy, in *Sixty-two Years of Uncertainty*, ed. by A. Miller (Plenum, New York, 1990)
53. F. Karolyhazy, in *Fundamental Problems of Quantum Physics*, ed. by M. Ferrero, A. van der Merwe (Kluwer Academic Publishers, Netherlands, 1995)
54. F. Karolyhazy, A. Frenkel, B. Lukacs, in *Physics as Natural Philosophy*, ed. by A. Shimony, H. Feshbach (MIT Press, Cambridge, 1982)
55. A. Frenkel, in *Quantum Mechanics - A Half Century Later*, ed. by J.L. Lopes, M. Paty (Reidel, Dordrecht, 1977)
56. A. Frenkel, Found. Phys. **20**, 159 (1990)
57. A. Frenkel, in *Fundamental Problems of Quantum Physics*, ed. by M. Ferrero, A. van de Merwe (Kluwer Academic Publishers, Netherlands, 1995)
58. A. Frenkel, Found. Phys. **32**, 751 (2002)
59. A. Frenkel, in *Experimental Metaphysics*, vol. 1, ed. by A. Shimony, R.S. Cohen, M. Horne, J. Stachel (Kluwer Academic Publishers, Boston, 1997), p. 39
60. L. Diósi, Phys. Lett. A **120**, 377 (1987)
61. L. Diósi, J. Phys. A **40**(12), 2989 (2007)
62. L. Diósi, Phys. Rev. A **40**, 1165 (1989)
63. L. Diósi, B. Lukacs, Ann. Phys. **44**, 488 (1987)
64. R. Penrose, Gen. Rel. Grav. **28**, 581 (1996)
65. R. Penrose, Phil. Trans. R. Soc. Lond. A **356**, 1927 (1998)
66. R. Penrose, *Mathematical Physics 2000* (Imperial College Press, London, 2000)
67. L. Diósi, Phys. Lett. A **105A**, 199 (1984)
68. D.H. Bernstein, E. Giladi, K.W.R. Jones, Mod. Phys. Lett. A **13**, 2327 (1998)
69. D. Giulini, A. Großardt, Class. Quant. Grav. **28**, 195026 (2011)
70. R. Harrison, I. Moroz, K.P. Tod, Nonlinearity **16**, 101 (2003)
71. R. Harrison, I. Moroz, K.P. Tod, Class. Quantum Gravity **15**, 2733 (1998)
72. R. Ruffini, S. Bonazzola, Phys. Rev. **187**, 1767 (1969)

73. D. Giulini, A. Großardt, Class. Quant. Grav. **29**, 215010 (2012)
74. D. Giulini, A. Großardt, Class. Quant. Grav. **30**, 155018 (2013)
75. B.L. Hu, (2014). arXiv:1402.6584
76. C. Anastapoulos, B.L. Hu, New J. Phys. **16**, 085007 (2014)
77. M. Bahrami, A. Grossardt, S. Donadi, A. Bassi, New J. Phys. **16**, 11 (2014)
78. S. Colin, T. Durt, R. Willox, (2014). arXiv:1402.5653
79. S. Bera, S. Donadi, K. Lochan, T.P. Singh, Found. Phys. **45**, 1537 (2015)
80. S. Bera, R. Mohan, T.P. Singh, Phys. Rev. D **92**, 025054 (2015)
81. S. Bera, P. Giri, T.P. Singh, (2016). arXiv:1608.02102
82. S.L. Adler, Nucl. Phys. B **415**, 195 (1994)
83. S.L. Adler, *Quantum Theory as an Emergent Phenomenon* (Cambridge University Press, Cambridge, 2004)
84. S.L. Adler, A.C. Millard, Nucl. Phys. B **473**, 199 (1996)
85. S.L. Adler, J. Phys. A **39**, 1397 (2006)
86. T.P. Singh, in *Re-thinking Time at the Interface of Physics and Philosophy: The Forgotten Present*, ed. by T. Filk, A. von Muller, (Springer, Heidelberg, 2015). arXiv:1210.81110
87. K. Lochan, T.P. Singh, Phys. Lett. A **375**, 3747 (2011)
88. K. Lochan, S. Satin, T.P. Singh, Found. Phys. **42**, 1556 (2012)
89. S. Banerjee, S. Bera, T.P. Singh, Int. J. Mod Phys. **25**, 1644005 (2016)

Four Principles for Quantum Gravity

Lee Smolin

Abstract Four principles are proposed to underlie the quantum theory of gravity. We show that these suffice to recover the Einstein equations. We also suggest that MOND results from a modification of the classical equivalence principle, due to quantum gravity effects.

1 Introduction

This paper is dedicated to Thanu Padmanabhan on his 60th birthday, as it reflects longstanding concerns and insights of his [1–5]. *With thanks for his vibrant originality and dedication to the search for the fundamental principles of nature.*

Albert Einstein [6] taught us to distinguish between *principle theories* and *constructive theories.* The latter are descriptions of particular phenomena, fields or particles that constitute nature. These are specified in terms of dynamical equations of motion that the constituents obey. Principle theories are different: they give us universal principles that all physical phenomena must obey, whatever fields or particles constitute nature. The paradigmatic example Einstein used is the laws of thermodynamics. Einstein used the distinction to argue that special relativity is a principle theory. He used this to distinguish special relativity from its rivals, principally the Lorentz theory of the electromagnetic aether, which he argued is a constructive theory. The lesson is that when one can manage to encompass a phenomena in terms of a principle theory, that that will likely be superior to a constructive theory.

Once we have a principle theory, we can use it to frame and constrain candidate constructive theories.

In this contribution I propose we frame quantum gravity as a principle theory. Loop quantum gravity, string theory, causal sets, CDT, etc. can all be seen as constructive theories that tell us what quantum spacetime might be "made of". These may contain

L. Smolin (✉)
Perimeter Institute for Theoretical Physics, 31 Caroline Street North,
Waterloo, ON N2J 2Y5, Canada
e-mail: lsmolin@perimeterinstitute.ca

J.S. Bagla and S. Engineer (eds.), *Gravity and the Quantum*,
Fundamental Theories of Physics 187, DOI 10.1007/978-3-319-51700-1_26

some elements of the truth about what constitutes quantum spacetime. But I would like to suggest an alternative road in which we first seek principles.

These principles should have some non-trivial consequences. First of all, they should reproduce what we know already. In particular, the field equations of general relativity should emerge in a suitably defined classical or coarse grained limit. In addition, they should entail novel phenomena or provide an unexpected explication of known phenomena.

In this paper, I propose four principles and show that together they do entail that general relativity emerges as a coarse grained approximation. We also get new insights into the physics at small cosmological constant, Λ, both negative and positive. In the case of $\Lambda < 0$ we find evidence that aspects of the AdS/CFT correspondence are related to these principles. This is not new, but involves an interpretation of some existing results. But in the opposite case of small, positive, cosmological constant, I find a surprise: a tentative argument that the principle of equivalence, and hence gravity, is modified at low accelerations, compared to

$$a_* = c^2\sqrt{\frac{\Lambda}{3}}. \tag{1}$$

As I discuss in Sect. 5, this may be related to MOND.

Here are the four principles:

1. **The principle of absolute causality and relative locality** *A quantum spacetime consists of a set of events whose fundamental properties include causality (i.e. causal relations), energy and momentum* [7–11].[1] *Classical spacetime is emergent, as is locality. Locality is also relative to the positions of observers and the energy and other properties of the probes they employ to measure and tracks distant causal processes* [14, 15].
2. **Correspondence principle** *Classical spacetimes emerge in a suitable limit, by coarse graining the causal structure of sufficiently large quantum spacetimes.*
3. **The weak holographic principle**: *The area of a surface, which is defined by the causal structure as the boundary of a subsystem, is a measure of the channel capacity of that surface to serve as a channel of information in and out of that subsystem* [16].
4. **The quantum equivalence principle**: *The observers who, in the absence of curvature and a cosmological constant, see the vacuum to be a maximal entropy thermal state, are those that in the classical limit are uniformly accelerating. Hence those observers who see the vacuum to have zero temperature must be inertial* [17].

The first is a strengthing of relative locality as originally proposed in [14, 15].[2] We should stress that the correspondence principle guarantees the existence of a

[1]The notion of causal sets was introduced by [7]. Causal sets built out of intrinsic structures was developed by [11–13].

[2]A different proposal for relative locality is in [18].

classical manifold and metric, g_{ab}, but does not require that that metric satisfy the Einstein equations or any other dynamical law.

The third principle is not new [16, 19, 20], but it will be important to propose a form of it suitable for a fundamental theory of quantum gravity. The last is a development of a principle proposed in [17], and will require some explanation.

I should note that it may be attractive to think that the holographic principle is a consequence of the others, particularly the quantum equivalence principle. But there is a simple reason to think the holographic hypothesis and the quantum equivalence principle are independent. This is that they introduce independent constants. The holographic hypothesis introduces a quantum of area, A_p, to define surface entropy. The quantum equivalence principle introduces $\hbar$, in the guise of a boost temperature (with dimensions of the boost Hamiltonian, which are action.) The ratio gives us Newton's constant, $G = \frac{A_p}{\hbar}$. We will see below how they work together to give us Einstein gravity.

The quantum equivalence principle does more than deepen the classical equivalence principle, it puts limits on the older idea. These occur in the presence of a positive cosmological constant, which arises because there is no observer that sees the vacuum to have zero temperature. I argue in Sect. 5 that this leads to a violation of the equality of gravitational and inertial mass. This violation is essentially a renormalization group effect, hence it depends on temperature. But because of the quantum equivalence principle, this temperature dependence transmutes into an acceleration dependence. This, I briefly show in Sect. 5, can explain MOND [21, 22], and hence obviates the need for dark matter to explain the galaxy rotation curves.[3]

Once we have a set of principles from which the semiclassical limit follows, we can seek to show that various constructive theories realize the principles. For example, a key problem with background independent approaches to quantum spacetime histories, such as spin foam models, causal sets and causal dynamical triangulations, is showing that they have a good classical limit. This problem has two aspects. First, we have to show that there is a classical limit described in terms of an emergent classical spacetime. Then one has to show that the metric of this spacetime satisfies the Einstein equations. Rather than approach these questions directly, the principles I propose guarantee these outcomes, the first directly as a result of the correspondence principle, the second as a consequence of them all.

An objection that might be made of our strategy is that loop quantum gravity (or string theory, or CDE) is already a principle theory, whose principles are merely those of general relativity and quantum field theory. That is partly true, but it is important to mention that along the way from principles to the physics there are, in each case, technical choices that need to be made. These choices make the theory at least partly constructive. We are interested in a different question which is whether there are any new principles acting in the world which govern quantum gravitational phenomena.

---

[3]This does not yet address the need for dark matter on scales of clusters and large scale structure. It is possible that these are explained by dark matter while MOND explains the galaxy rotation curves.

I begin introducing the new principles in Sect. 2, The aim of Sect. 3 is to sketch the recovery of general relativity in a suitable limit. Section 4 contains a short discussion of how the AdS/CFT correspondence is related to the quantum equivalence principle proposed here. In Sect. 5, I sketch a route to deriving the phenomenology of MOND from the quantum equivalence principle, after which we conclude.

2 Four Principles for Quantum Gravity

I now propose four principles for quantum gravity. In each case there is a short statement of the principle, followed by a more detailed explication.

2.1 The Principle of Absolute Causality and Relative Locality

The principle of absolute causality and relative locality. *Causal relations are fundamental to nature as are energy and momentum, whose flow follows those relations. Classical spacetime is emergent, as is locality. Locality is also relative to the positions of observers and the energy and other properties of the probes they employ to measure and tracks distant causal processes* [14, 15].

Explication: *This first principle tells us what the theory is to be about. We want to specify that, just as in general relativity, a quantum spacetime can be understood to be about events and their causal relations. The principle also asserts the primacy of causality and of energy and momentum over spacetime and locality.*

A quantum spacetime consists of a set of events whose fundamental properties include causality (i.e. causal relations), energy and momentum. [7–11].[4] *Further, for every causal diamond $\mathcal{CD}(e,f)$ in the quantum spacetime, we posit that there corresponds a Hilbert space, $\mathcal{H}(e,f)$ which records quantum information measurable on the waist, $\mathcal{B}(e,f)$.*

There are a variety of structures which exist such as causal pasts, causal futures, causal diamonds and their boundaries.

We briefly recall their definitions.

- Causal diamonds on causal sets
  We are given two distinct events $e < f$ in a causal set. We define immediately the causal past of f, denoted $\mathcal{P}(f)$, consisting of d such that $d < f$. The immediate causal past of f is the subset of $\mathcal{P}(f)$ consisting of those events reached from f by one causal link into the past, and is denoted, $\mathcal{IP}(f)$. Similarly we define the causal future, and the immediate causal future, of e, denoted respectively by $\mathcal{F}(e)$ and $\mathcal{IF}(e)$. The causal diamond is

[4]The notion of causal sets was introduced by [7]. Causal sets built out of intrinsic structures was developed by [11–13].

$$\mathcal{CD}(e,f) = \mathcal{F}(e) \cap \mathcal{P}(f) \tag{2}$$

The boundary of the past of f, denoted $\partial\mathcal{P}(f)$ consists of elements of $\mathcal{P}(f)$ some of whose immediate futures are not containned in $\mathcal{P}(f)$. Similarly, we define the boundary of the future of e, denoted $\partial\mathcal{F}(e)$. The *waist* of the causal diamond, $\mathcal{CD}(e,f)$, is their intersection.

$$\mathcal{W}(e,f) = \partial\mathcal{F}(e) \cap \partial\mathcal{P}(f) \tag{3}$$

Quantum mechanics is fundamentally about one subsystem of nature probing the rest.[5] The most elementary act of observation a subsystem of the universe can make is to send a probe out into the world at one event and receive a response back at a future event. The act of probing the world is represented by a causal diamond, making them primary structures. Furthermore, the geometry of momentum space is more fundamental than the geometry of spacetime.

2.2 The Correspondence Principle

The correspondence principle: *Classical spacetimes emerge in a suitable limit, by coarse graining the causal structure of sufficiently large quantum spacetimes.*

Explication: *Consider a quantum spacetime, $\mathcal{Q}$, a subset of whose causal diamonds have large spacetime volumes (In Planck units.) To it there corresponds a classical metric spacetime, $(\mathcal{M}, g_{ab})$ such that for every causal diamond of $\mathcal{Q}$ whose spacetime volume $\mathcal{V}(e,f)$ is sufficiently large, there is a corresponding causal diamond, $\tilde{\mathcal{CD}}(e,f)$, in $(\mathcal{M}, g_{ab})$, whose spacetime volume and waist area coincide.*

Hence, among the observables in $\mathcal{H}(e,f)$ is the area of its waist, $\hat{A}[\mathcal{B}]$, the spacetime volume, $\mathcal{V}(e,f)$, and the spacetime curvature scalar averaged over the causal diamond, given by $<\mathcal{R}>_(e,f)$. Moreover, we require that there is an action of the Lorentz group on $\mathcal{H}(e,f)$, denoted $\mathcal{L}$.. The correspondence allows us to pull back $\mathcal{L}$ to the action of $SL(2,C)$ on functions on S^2.

The waist of the classical causal diamond is an S^2. There is a mapping from functions, f, on the S^2 into states $\psi(f)$ in $\mathcal{H}(e,f)$.

We call the pair, quantum and classical, a *paired causal diamond.*

Note that in the context of a constructive approach to quantum gravity, such as causal sets, *CDT* or spin foam models, this is a result we seek to demonstrate. Whether a given constructive quantum gravity theory satisfies the correspondence principle is then a test for adequacy of a quantum gravity theory. Thus, this correspondence principle plays the same role as the correspondence principle did in the development of the quantum theory, it is a criterion for adequacy, which the Schroedinger quantum mechanics passed by virtue of Ehrenfest's theorem.

[5] This idea is developed in relational quantum theory [23–25] and relative locality [14, 15].

We should also expect that there are, as in quantum physics, limits to the correspondence principle that arise from novel phenomena. In the present case these could arise from relative locality as well as from the ultra-low acceleration *MOND* regime. It is interesting that both point to modifications of general relativity at large distances.

2.3 The Weak Holographic Principle

The weak holographic principle: *The area of a surface, which is defined by the causal structure as the boundary of a subsystem, is a measure of the channel capacity of that surface, to serve as a channel of information in and out of that subsystem* [16].

Explication: *In a quantum causal structure we can define special surfaces as the intersection of causal past of one event with the causal future of another. The area of these surfaces are a measure of the channel capacity of these surfaces. In more detail, let us consider a quantum causal diamond,* $\mathcal{CD}(e,f)$*, to which there corresponds a classical causal diamond* $\tilde{\mathcal{CD}}(e,f)$*. A consequence of the metric geometry is that the waist or corner,* $\tilde{\mathcal{B}}$ *of the classical causal diamond is a space like* S^2*. By the correspondence to the quantum causal structure,* $\tilde{\mathcal{B}}$ *inherits the Hilbert space,* $\mathcal{H}(e,f)$ *and its observables. These include the area* $\hat{\mathcal{A}}[\mathcal{B}]$*. The Hilbert space* $\mathcal{H}(e,f)$ *represents the information that the observer represented by a causal diamond may obtain about the world by means of probes that measure information coming into its waist.*

$$S[\mathcal{B}] = \frac{\mathcal{A}[\mathcal{B}]}{A_p} \tag{4}$$

We will find below that $A_p = 4\hbar G$*, but it is introduced as an independent constant with units of area.*

2.4 The Quantum Equivalence Principle

The quantum equivalence principle: *The observers who, in the absence of curvature and a cosmological constant, see the vacuum to be a maximal entropy thermal state, are those that in the classical limit are uniformly accelerating. Hence those observers who see the vacuum to have zero temperature must be inertial* [17].

Explication: *Worked out in detail, the principle has two parts. The first part is purely quantum, the second develops the correspondence between the quantum and classical aspects of a paired causal diamond.*

The first part specifies the existence of dimensionally reduced thermal states corresponding to uniformly accelerated observers, boosted to the infinite momentum frame.

Consider an observer in a quantum spacetime who is uniformly accelerated to an arbitrarily high boost. The observer sees physics to be conformally invariant, dimensionally reduced by the elimination of the longitudinal direction, and thermal in the sense that it is described by the maximally entangled state generated by the boost Hamiltonian.

To every quantum causal diamond, in the quantum causal structure, and every boost generator, $\hat{K}_z$ *in* $\mathcal{L}$ *we associate a thermal state, with maximal entanglement entropy, in a* $2+1$ *(generally a* $(d-1)+1$*) dimensional conformal field theory, which lives on the waist, This is given by* [26]

$$\rho_U = e^{-\frac{2\pi}{\hbar} H_{Boost}} \tag{5}$$

where H_{Boost} *is the positive definite boost Hamiltonian, corresponding to* $\hat{K}_z$ *which acts on degrees of freedom on the waist. Note that the boost Hamiltonian must be dimensionless. The dimensionless Unruh temperature is* [27],

$$T_U = \frac{\hbar}{2\pi} \tag{6}$$

The second part requires that when the classical spacetime emerges, the locally accelerated trajectories correspond to observers that describe the vacuum as a thermal state with maximal entanglement entropy, as was posited in [17].

Specifically, it requires that *in the cases in which* $\mathcal{CD}(e,f)$ *is a paired causal diamond, there corresponds to it a causal diamond in a classical spacetime. The classical metric* g_{ab} *has a radius of curvature, R* [28]. *The state (5) corresponds to what an observer in the dual classical spacetime, uniformly boosted to the infinite momentum frame, might see. More specifically, in the limit where the spacetime volume is large in Planck units, but small compared to* R^4*, the metric has an approximate boost killing field. Then,* ρ_U *describes the quantum state of the causal diamond as seen by an observer moving with respect to an approximate boost killing field of the metric* g_{ab}.

2.4.1 Motivation

In the classical equivalence principle we make use of the notion of a freely falling, or inertial, frame, which the Lorerntzian geometry gives us before hand. We then define a uniformly accelerated frame as one which is accelerated relative to an inertial frame.

But the notion of an inertial frame does not appear naturally in the fundamentals of quantum gravity. The reason is that it depends on a limit of weak coupling to define. An inertial frame is one that *"has no forces on it."* Except, of course, that it must interact with its surroundings if it is to serve as an observer. Such an observer can only make sense when defined in a weak coupling limit. However, such a limit is antithetical to being deep in the quantum gravity regime, where all the degrees of freedom are interacting with each other.

Thus, in the quantum version we start with the accelerated elevator and later take a weak coupling limit where we transform to the freely falling frames.

What we want, then, is a quantum notion of a uniformly accelerating frame. The result of Unruh suggests that this should correspond to a thermal state. Notice that uniform acceleration implies an unlimited boost, corresponding to a uniform acceleration carried out for an unlimited time. In this time, an observer is accelerated to a boost, γ, relative to its initial motion, which is arbitrarily large. Thus we need a notion of an observer corresponding to what high energy physicists call the *infinite momentum frame* [29]. The quantum equivalence principle thus is going to assert that the limit of a uniformly accelerating observer is related to the infinite momentum frame.

We recall that in the infinite momentum frame the longitudinal coordinate is length contracted to an arbitrarily small interval, so that spheres are quashed down into pancakes. Thus a uniformly accelerating observer should see a world reduced by the elimination of the longitudinal direction.

Note that the limit of infinite boost, when $\gamma \to \infty$, can be seen as the limit of a renormalization group transformation. Asking that the quantum physics have a limit is then the same as asking that there be a fixed point of the renormalization group, i.e. a scale invariant theory. So the limit theory has to be a conformal field theory.

We can say this a different way. Greenberger [30] explains why, strictly speaking, the equivalence principle cannot hold for a particle of finite mass. The fact that the Compton wavelength $\lambda_C = \frac{\hbar}{m}$ is finite means that how a massive particle falls in a gravitational field will depend on its mass. But this does not rule out applying the equivalence principle to massless particles. If we generalize this to QFT we would say that any mass scale in a QFT is an impediment to the satisfaction of the equivalence principle, hence the quantum equivalence principle describes what a uniformly boosted observer sees in terms of a conformal field theory.

Moreover, the conformal field theory we need is one whose degrees of freedom can be attributed to the waist, which is a two dimensional sphere. Hence the longitudinal direction disappears. This is a well known characteristic of physics in the infinite momentum frame. If an object of longitudinal size, r is length contracted by a γ so large that $\frac{r}{\gamma} < l_{Pl}$ then it no longer is describable in terms of extension in a classical geometry. Either it disappears because it becomes part of quantum geometry at the Planck scale (if the lorentz symmetry is unmodified) or it disappears because it gets squeezed down to l_{Pl} as a limit in a deformed version of lorentz symmetry.

We note that this disappearance of a dimension in a short distance of high energy limit may be connected to the phenomena of dimensional reduction observed in several approaches to quantum gravity [31, 32].

2.4.2 Comments on the Quantum Equivalence Principle

We can make some simple comments on this new proposal:

- **NOTE 0**: Equation (6) is where $\hbar$ is introduced.
- **NOTE 1**: The quantum equivalence principle incorporates the Unruh effect [27].

- **NOTE 2**: We can conjecture that the *AdS*/*CFT* correspondence is at least partly a consequence of the quantum equivalence principle, combined with the weak holographic principle, because as you go to the boundary of *AdS* the radius of curvature goes to a large constant. We discuss this below.
- **NOTE 3**: The quantum equivalence principle incorporates the *equivalence of quantum and thermal fluctuations* posited by myself [17] and developed by Kolekar and Padmanabhan [33]. In that paper, [17], I raised the question of why the class of special observers that see zero temperature in the ground state are the same as the special class of inertial observers. This is suggested by the observation that they need not be the same, because the former depend on the choice of vacuum in the *QFT*, while the latter does not.
  Coming from classical physics we consider the class of inertial observers as prior to and more fundamental than, the class of zero temperature observers. In quantum gravity, we must turn this around. In the deep quantum gravity regime, governed by the quantum equivalence principle, there are no particle trajectories and no notion of inertia or acceleration. Moreover, associated to causal diamonds, there is a dimensionless notion of boost energy and boost temperature. So the fundamental notion is the dimensionless temperature, T_U, from which the dimensional temperature and corresponding acceleration emerge in the appropriate limit following the breaking of conformal invariance.
- **NOTE 4**: The quantum equivalence principle (QEP) implies the classical equivalence principle (CEP), in the limit $\hbar \rightarrow 0$ and in the absence of a cosmological constant. The dimensional reduction requires that the acceleration proceed to a boost that exceeds. for any physical length scale, L,
$$L' = \frac{L}{\gamma} < l_{Pl} \tag{7}$$
  but in the limit $\hbar \rightarrow 0$, $l_{Pl} \rightarrow 0$ so the condition is never met. Furthermore the Unruh temperature also goes to zero.
  In the presence of a positive cosmological constant, the limit to the classical equivalence principle may be modified for small accelerations, $a < a^*$, where $a^* = c^2\sqrt{\Lambda}$. This is discussed in Sect. 5, below.
- **NOTE 5**: The requirement that ρ_U be an equilibrium state implies that it is the maximal possible entropy. But that is limited by the channel capacity. Hence we have
$$S_B = -Tr\rho \ln \rho = \frac{1}{T_U} < H_B > \tag{8}$$

2.4.3 Free Fall

The classical equivalence principle has two parts, related to accelerating elevators and elevators in free fall. We started with the analogue of accelerating elevators, can we get back to a statement of a quantum equivalence principle for freely falling observers?

The key idea is that ρ_U is the maximal entropy state. A freely falling reference frame is one that will observe the minimum entropy state, which is the vacuum. This means that the notion of inertial motion, i.e. free fall, should be a consequence of the possibility of reducing entanglement entropy to a minimum by transforming from a thermal state to the vacuum, below the scale of the radius of curvature. This is essentially the hypothesis made in [17] which posits that it cannot be contingent or coincidence that observers who see minimal entanglement entropy move inertially.

To do that we have to break the conformal invariance that got us to the limit of large boosts, which made it possible to use the boost Hamiltonian and boost temperature, both of which have units of action rather than energy. The breaking of conformal invariance gives us a time scale, τ. We can then define the dimensional boost Hamiltonian and temperature,

$$\tilde{H}_B = \frac{H_{Boost}}{\tau}, \qquad \tilde{T} = \frac{\hbar}{2\pi\tau} = \frac{\hbar a}{2\pi c} \tag{9}$$

where the acceleration $a = \frac{c}{\tau}$. The thermal state is then

$$\rho_U = e^{-\frac{2\pi c}{\hbar a}\tilde{H}_B} \tag{10}$$

In the absence of a positive cosmological constant, we then take the limit of $a \rightarrow 0$ to find the minimal entropy, or ground, state,

$$\rho_0 = \lim_{a\rightarrow 0} e^{-\frac{2\pi c}{\hbar a}\tilde{H}_B} \tag{11}$$

If this state exists it will have at least approximate symmetries which generate the symmetries of spacetime, and hence lead to the recovery of the equivalence principle for freely falling observers.

The fact that the ground state has translation symmetry implies that the specification of which motions are inertial does not depend on any property of a freely falling particle. This implies the equality of gravitational and inertial mass,

$$m_I = m_g. \tag{12}$$

In Sect. 5, we will see how this is modified when the cosmological constant is positive.

3 Recovery of the Einstein Equations

We show that these principles suffice to recover the Einstein equations. We follow a strategy pioneered by Jacobson [28, 34] and Padmanabhan [4].

Suppose that we have a boosted observer inside a paired causal diamond. By the weak holographic principle, the entropy is

$$S_B = \frac{A(B)}{A_p} \tag{13}$$

By the quantum equivalence principle, the entropy is also

$$S_B = -Tr\rho \ln \rho = \frac{1}{T_U} < H_B > \tag{14}$$

This is of course the first law of thermodynamics, emerging here as a consequence of the weak holographic principle and the quantum equivalence principle.

It follows that

$$< H_B >= \frac{1}{8\pi G} A[B] \tag{15}$$

where $G = \frac{A_p}{4\hbar}$.

This has been called the first law of classical spacetimes.

Now consider three events, $A < B < C$ which generate two paired causal diamonds $\mathcal{CD}_1 = \mathcal{CD}[A, B]$ which is a subset of $\mathcal{CD}_2 = \mathcal{CD}[A, C]$. Then we can also show that

$$< \Delta H_B >=< H_{B_2} > - < H_{B_1} >= \frac{1}{8\pi G} \Delta A[B] = \frac{1}{8\pi G} [A[B_2] - A[B_1]] \tag{16}$$

We next use the correspondence principle to describe the boundary of the causal diamond in terms of a congruence of null geodesics in the emergent spacetime. This is specified by an affine parameter λ and a null tangent vector k^a. We express $\Delta A[B]$ in terms of the expansion θ.

$$\delta A = \int dA d\lambda \theta \tag{17}$$

To compute this we will use the Raychauduri equation for a congruence of null geodesics

$$\dot{\theta} = -\frac{1}{2}\theta^2 - 2\sigma^2 + 2\omega^2 - R_{ab}k^a k^b \tag{18}$$

We next note that every event e in a causal spacetime appears in the corners of many causal diamonds. We fix an arbitrary event and look for suitable causal diamonds. Suitable means that e is in the corner of the causal diamond and that the null geodesics that make up the boundary of that causal diamond satisfy that $2\sigma^2$ and ω^2 are negligible compared to $R_{ab}k^a k^b$.

$$\sigma^2, \omega^2 << R_{ab}k^a k^b \tag{19}$$

We also assume that an event e in the corner, B_1 can be chosen such that there

$$\theta_1 << R_{ab}k^a k^b \tag{20}$$

If this is not the case, but (19) are satisfied we just wait, and consider an event e' further along the light cone. This is because under our assumptions.

$$\dot{\theta} = -\frac{1}{2}\theta^2 \rightarrow \theta \approx \frac{2}{t} \tag{21}$$

so that after some time (20) will be satisfied. When it is we have

$$\delta A = \int dAd\lambda\theta = \int dAd\lambda\lambda dot\theta = -\int dAd\lambda\lambda R_{ab}k^a k^b \tag{22}$$

Now we work on the other side of (16). The assumption that the shear σ can be neglected means that there is not a lot of energy in gravitational radiation. Hence this implies that

$$\Delta H_B = \Delta Q = \int_{\mathcal{H}} T_{ab}\xi^a d\Sigma^b \tag{23}$$

where xi^a is an approximate killing field generating the boost near the corner. By the equivalence principle this must exist. We can follow Jacobson [34] in setting

$$\xi^a = -\kappa\lambda k^a \tag{24}$$

$$d\Sigma^a = k^a d\lambda dA \tag{25}$$

hence we have

$$\Delta H_B = \int_{\mathcal{H}} T_{ab}k^a k^b \lambda d\lambda dA \tag{26}$$

We now use the fact that (16) will be true for a large number of causal diamonds whose waist includes e. We then have

$$R_{ab} + g_{ab}f = T_{ab} \tag{27}$$

Making use of the Bianchi identities we have

$$R_{ab} - \frac{1}{2}g_{ab}R - g_{ab}\Lambda = 8\pi G T_{ab} \tag{28}$$

Thus we see that our principles imply that general relativity is satisfied.

4 AdS/CFT as an Example of the Quantum Equivalence Principle

The formulation of the quantum equivalence principle is inspired by early ideas of holography and the infinite momentum frame, a connection that has been emphasized by Susskind [20, 35]. However, there are interesting implications of the principle in the case of negative cosmological constant, which appear to be related to aspects of the AdS/CFT correspondence.

Notice that, as pointed out by [5, 36–39], an observer at large $r >> R$ of a Schwarzschild-*AdS* spacetime is in a situation analogous to a uniformly accelerated observer in flat spacetime, i.e. an observer in Rindler spacetime. The uniformly accelerated observer in the *IMF* limit sees all massive particles moving with respect to it in the negative direction. She sees particles initially ahead of her lose speed and then fall behind, as she passes them.

Similarly, all outgoing massive particles in asymptotically *AdS* spacetimes reach a maximal r and then fall back. The uniformly accelerating observer sees light coming from behind it to be increasingly redshifted, where that redshift goes to infinity in the *IMF* limit.

To see this we work in global coordinates [36].

$$ds^2 = -f(r)d\tau^2 + \frac{dr^2}{f(r)} + r^2 d\Omega^2 \tag{29}$$

where $f(r) = (1 - \frac{2GM}{r} + \frac{r^2}{R^2})$, with $\Lambda = -\frac{1}{R^2}$. The redshift for outgoing light is

$$\frac{\omega(r_2)}{\omega(r_1)} = \frac{\sqrt{f(r_1)}}{\sqrt{f(r_2)}} = \sqrt{\frac{1 - \frac{2GM}{r} + \frac{r_1^2}{R^2}}{1 - \frac{2GM}{r} + \frac{r_2^2}{R^2}}} \tag{30}$$

In the limit $r_2 > r_1 >> R > GM$ this is

$$\frac{\omega(r_2)}{\omega(r_1)} = \frac{r_1}{r_2} \tag{31}$$

This is the same as in Rindler spacetime where $f(r) = \frac{r^2}{R^2}$. In the limit that $r_2 \to \infty$ the redshift factor goes to zero. This has been seen as a manifestation of an IR/UV duality in which high frequency excitations in the bulk are redshifted to zero energy and infinite wavelength by the time they reach infinity.

It has also been found by Deser and Levine [36] that a uniformly accelerating observer with uniform acceleration a, in anti-deSitter spacetime observes a temperature

$$T = \frac{\sqrt{a^2 + \frac{\Lambda}{3}}}{2\pi c} = \frac{a_5}{2\pi c} = \frac{\sqrt{a^2 - \frac{1}{R^2}}}{2\pi c} \tag{32}$$

where a_5 is the acceleration of the worldline in a five dimensional Minkowski spacetime in which the AdS spacetime is embedded.

We note that for the temperature to be non-zero, the acceleration must satisfy $a > \frac{1}{R}$. This is required for the accelerated observer's worldline to have an horizon. This is *not* the case for an observer at constant r in the global coordinates (29), instead one can show that for all such observers $a < \frac{1}{R}$.

This requires modifications of the quantum equivalence principle, because there are observers with non-zero acceleration, but vanishing temperature.

Observers with $a > \frac{1}{R}$ do exist, some of them can be described as the constant ξ worldlines, in a coordinate system in which ξ replaces r and in which the AdS metric is expressed as

$$ds^2 = -\frac{\xi^2}{R^2}dt^2 + \frac{d\xi^2}{1+\frac{\xi^2}{R^2}} + \left(1+\frac{\xi^2}{R^2}\right)\left[d\chi^2 + R^2\sinh^2\left(\frac{\xi}{R}\right)d\Omega_{d-2}^2\right] \tag{33}$$

Note that $\xi = 0$ is an horizon for the accelerated observers at constant $\xi > 0$. These have accelerations

$$a^2 = \frac{1}{R^2} + \frac{1}{\xi^2} \tag{34}$$

which are all greater than $\frac{1}{R^2}$, hence by (32) they have non-zero temperatures

$$T = \frac{\hbar}{2\pi\xi} \tag{35}$$

Because $\xi = 0$ is an horizon these coordinates cover only a wedge of AdS spacetime, analogous to the way in which Rindler coordinates cover only a wedge of Minkowski spacetime. Consequently an observer who can only observe this wedge sees a thermal state.

We note that in the limit $R \to \infty$ for fixed ξ (or $\frac{\xi}{R} \to 0$), the AdS metric becomes the pure Rindler metric

$$ds^2 = -\frac{\xi^2}{R^2}dt^2 + d\xi^2 + d\chi^2 + \xi^2 d\Omega_{d-2}^2 \tag{36}$$

Thus, under these conditions the requirements of the quantum equivalence principle are satisfied. Hence we predict that for large ξ, in the presence of larger R, the physics is represented by a hot CFT on flat Minkowski spacetime.

This is indeed the case, as has been described in detail in a number of papers [5, 36–39].

5 The Origin of MOND from the Quantum Equivalence Principle

We next ask how the correspondences we described in Sect. 2.4 are modified by the presence of a positive cosmological constant, Λ. This bathes the system in a bath of low temperature horizon radiation. It is natural to hypothesize that this results in a temperature dependent renormalization of the scale τ.

$$\tau \rightarrow \tau' = \tau G^{-1}\left(\frac{T}{T^*}\right) \tag{37}$$

where G is an adjustment of the renormalization scale and T^* is the temperature associated to the cosmological constant scale.

$$T^* = \frac{\hbar c}{2\pi}\sqrt{\frac{\Lambda}{3}} \tag{38}$$

We would like to compute $G\left(\frac{T}{T^*}\right)$ from first principles, but below we will estimate it empirically.

It follows that the effective Hamiltonian, $\tilde{H}_B$ defined by (9) is renormalized

$$\tilde{H}_B \rightarrow \tilde{H}'_B = G\left(\frac{T}{T^*}\right)\tilde{H}_B \tag{39}$$

In the non relativistic limit the Hamiltonian relevant for a star in orbit in a galaxy has the form of a sum of terms

$$\tilde{H}_B \rightarrow H_{NR} = \frac{p^2}{2m_i} - m_g U_{New} \tag{40}$$

where m_i and m_g are, respectively, the inertial and passive gravitational mass. These constants can absorb the renormalization factors

$$m_i^{ren} = \frac{m_i}{G\left(\frac{T}{T^*}\right)}, \qquad m_g^{ren} = m_g G\left(\frac{T}{T^*}\right). \tag{41}$$

Hence, the ratio of gravitational and inertial mass then suffer a temperature dependent renormalization

$$\frac{m_g^{ren}}{m_i^{ren}} = G^2\left(\frac{T}{T^*}\right) \tag{42}$$

This modifies the classical equivalence principle.

But as we argued above and in [17], the quantum equivalence principle requires that temperature and acceleration be intimately related, by the equivalence of free fall observers with observers that see minimal entanglement entropy. So, this turns into an acceleration dependence

$$\frac{m_g^{ren}}{m_i^{ren}} = G^2\left(\frac{a_{obs}}{a^*}\right) \tag{43}$$

where a^* is the acceleration associated to the cosmological constant

$$a^* = c^2\sqrt{\frac{\Lambda}{3}} \tag{44}$$

and a_{obs} is the observed acceleration of a particle. Hence we conclude there must be an acceleration dependent modification of the equality of gravitational and inertial mass. This will be important for extremely small accelerations, given by the scale of the cosmological constant.

But we also know that

$$\frac{m_g^{ren}}{m_i^{ren}} = \frac{a_{obs}}{a_N} \tag{45}$$

which is the ratio of the measured radial acceleration

$$a_{obs} = \frac{v^2}{r} \tag{46}$$

to the acceleration predicted by Newtonian theory

$$a_N = \nabla^i U \tag{47}$$

Hence we have

$$a_N = a_{obs} G^{-2}\left(\frac{a_{obs}}{a^*}\right) \tag{48}$$

We can invert this to find a function $F^2\left(\frac{a_N}{a^*}\right)$ such that

$$\frac{a_{obs}}{a_N} = F^2\left(\frac{a_N}{a^*}\right) \tag{49}$$

This relation, for some function, $F(\frac{a_N}{a^*})$, is then a consequence of the quantum equivalence principle.

In a recent paper, McGaugh, Lelli and Schubert (MLS) report [21] strong confirmation of an empirical relation of this form, first proposed by Milgrom [22]. They measure $F(\frac{a_N}{a^*})$ in a survey of rotation curves of 153 galaxies in the SPARC data base [40]. They measure a_{obs}, the actual radial acceleration by (46) at 2693 radii on these rotation curves. At the same radii they estimate the Newtonian gravitational potential from baryons as observed in gas and dust, and so determine a_N. They discover that the data is well described by a simple empirical relation of the form of (49), as shown in Fig. 1. As they note, it is amazing that such a relation exists over a wide range of galaxy types, sizes and morphologies, as this represents the observed accelerations only by a function of the Newtonian accelerations due to baryons.

Furthermore MLS are able to fit a simple form for $F(a)$ to the data which is [21, 41, 42]

$$F^2(a_N) = \frac{1}{1 - e^{-\sqrt{\frac{a_N}{a_0}}}} \tag{50}$$

They fit a_0 to the data to be

$$a_0 = 1.2 \times 10^{-10} ms^{-2} \tag{51}$$

which is not far from $a^* = c^2\sqrt{\frac{\Lambda}{3}}$. Indeed we are not yet fully in a deSitter expansion.

We may note that this has the limits postulated by [22], which are clearly exhibited by the data shown in Fig. 1. For large $\frac{a_N}{a_0}$, $F \rightarrow 1$, showing that conventional Newtonian dynamics is restored. This accords with the observation that dark matter is not needed in the cores of galaxies. For small $\frac{a_N}{a_0}$, $F^2 \rightarrow \sqrt{\frac{a_0}{a_N}}$, giving the MOND formula

$$a_{obs} = \sqrt{a_N a_0} \tag{52}$$

This gives immediately the baryonic Tully- Fisher relation for the velocity of the flat rotation curves in the exterior of a galaxy, in terms of its *baryonic total mass*, M [43],

$$v^4 = GMa_0. \tag{53}$$

We may note that the baryonic Tully- Fisher relation is well confirmed [22, 44, 45].

Additionally, it must be stressed that Fig. 1 shows that the scale a_0 is clearly present in the data. Indeed, this scale characterizes the phenomenology of galaxies as it is a typical scale for spiral galaxies.

This explanation for the observed mass-discrepancy-acceleration relation (or radial acceleration relation), implies several ramifications by means of which it might be tested. One is a redshift dependence of a_0, corresponding to changes in the cosmic horizon distance and temperature. Presently there is no evidence for an evolution of a_0 in the Tully-Fisher relation out to redshift of $z = 1$ [46] or $z = 1.7$ [47].

Another is acceleration dependent modifications of the equality of inertial and passive gravitational masses, present universally in processes at very small accelerations of order a^*. These would have to be tested in space, such as at lagrangian points where the Earth's gravitational acceleration is cancelled [48].

It goes almost without saying that if this proposal has any worth, a goal of research in quantum gravity must be to predict the form of (50).[6] We note that the effect in question is based on a renormalization of coupling constants in by averaging over very large scales, and is hence a far infrared effect. Indeed, there are very good

[6]Related ideas have been suggested previously in [49].

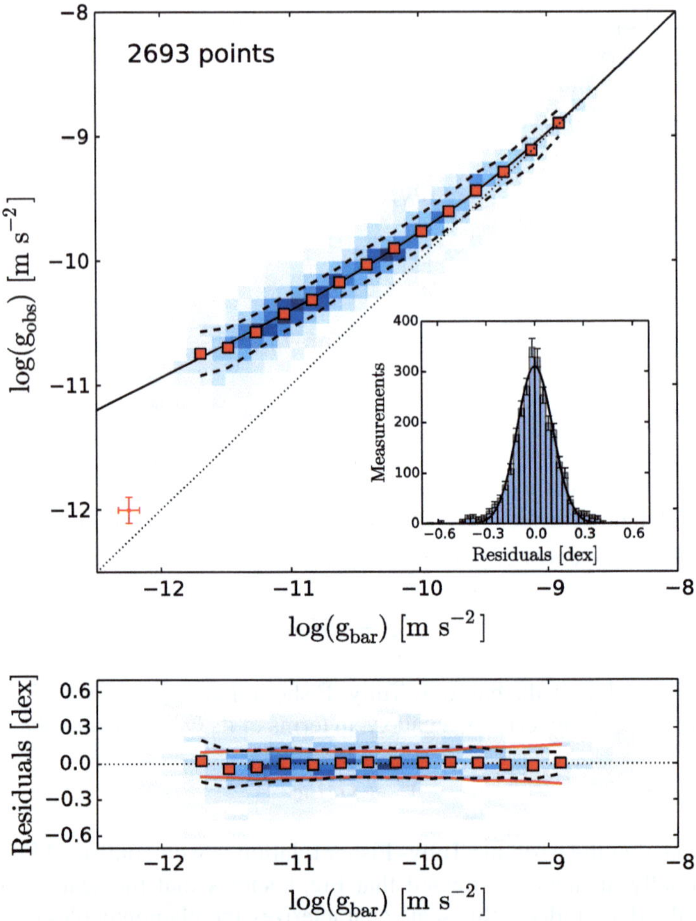

Fig. 1 The empirical radial acceleration relation, as shown in Fig. 3 of [21]. Data is taken from the SPARC data base [40]. Used with permission. From the original caption: "*The centripetal acceleration observed in rotation curves,* $a_{obs} = g_{obs} = v^2/R$, *is plotted against that predicted for the observed distribution of baryons,* $a_N = g_b$ *in the* upper panel. *Nearly 2700 individual data points for 153 SPARC galaxies are shown in grayscale. The mean uncertainty on individual points is illustrated in the* lower left corner. *Large squares show the mean of binned data.* Dashed lines *show the width of the ridge as measured by the rms in each bin. The* dotted line *is the line of unity. The* solid line *is the fit of Eq. (50), to the unbinned data using an orthogonal- distance-regression algorithm that considers errors on both variables.* The caption goes on to say, *The* inset *shows the histogram of all residuals and a Gaussian of width* $\sigma = 0.11 dex$. *The residuals are shown as a function of* g_{obs} *in the* lower panel. *The error bars on the binned data are smaller than the size of the points. The* solid lines *show the scatter expected from observational uncertainties and galaxy to galaxy variation in the stellar mass-to-light ratio. This extrinsic scatter closely follows the observed rms scatter* (dashed lines): *the data are consistent with negligible intrinsic scatter.* [21]"

reasons to think that *MOND* must be due to a new kind of non-locality in physics [49, 50]. It is intriguing to wonder if this might have something to do with relative locality [14, 15].[7]

Of course there remain the challenges to extend MOND to relativistic processes and to address large scale structure formation, the bullet cluster and others. It is possible these may be aided by the new point of view proposed here.

5.1 A Second Argument

Here is a second argument for deriving MOND from general relativity, combined with some input from the quantum equivalence principle.[8] We start with the observation of Narnhofer and Thirring [55] and Deser and Levine [36] that an observer in deSitter spacetime, with a steady four acceleration a observes a thermal spectrum with temperature,

$$T = \frac{\hbar a_5}{2\pi c} \tag{54}$$

where a_5 is the acceleration of the observer's worldline, lifted up into a flat five dimensional embedding of deSitter spacetime. This is given by (32) with positive Λ,

$$a_5 = \sqrt{a^2 + \frac{c^4 \Lambda}{3}} = \sqrt{a^2 + \frac{c^4}{R^2}} \tag{55}$$

Thus, this acceleration, a_5 is the relevant acceleration when we are taking the limit of maximally entangled thermal boosted states with decreasing acceleration to reach the minimally entangled ground state. Note that when the cosmological constant is non-zero and positive we cannot take this limit all the way, for even observers with vanishing four acceleration, a, have non-vanishing temperature, T_*. So which accelerations are relevant for expressing the equivalence principle in the Newtonian limit?

By construction, the observed kinematical acceleration, $a_k \approx \frac{v^2}{r}$, for $r << R$, is the four acceleration, a. But in what frame is the Newtonian acceleration $a_N^i = -\nabla^i \phi$ applied, and what is its value? For accelerations large compared to a_*, it doesn't matter as all candidates go to a in the limit, as accords the classical equivalence principle.

But what if we take the proposal that acceleration is tied to temperature seriously, as suggested by [17] and the quantum equivalence principle? We then might want to use a_5 for a_N in the non-relativistic limit. But this makes no sense as it has a positive lower bound, which is a_*. But we must retain that the acceleration vanishes when

[7] The idea that dark energy might be the result of non-locality in loop quantum gravity, or disordered locality [51], was suggested in [52]. The extension of this to dark matter and MOND was studied in an unpublished draft [53].

[8] A related argument was proposed in [54].

the force does. Instead we should choose for a_N, a function of a_5 (and hence of T) that goes to zero as $a \to 0$. The simplest such function is

$$\tilde{a}_5 = a_5 - a_* \tag{56}$$

Suppose we set $a_N = \tilde{a}_5$? It follows right away that

$$a = \sqrt{2a_* a_N} \tag{57}$$

which is the MOND relation for small $a_N << a_*$.

6 Conclusions

We have proposed four principles which a quantum theory of gravity should satisfy. These together imply the field equations of general relativity, they also express aspects of the AdS/CFT correspondence. We gave two tentative arguments that suggest that there are quantum gravity effects at very low acceleration, which reproduce the phenomenology of MOND.

I would like to close with an observation and a query. The observation is that all the cases we have studied situations where variances in time can be neglected, which hence involve static configurations such as uniformly accelerated observers or circular motion. In these situations there appear applications of equilibrium thermodynamics, at the classical and semiclassical level. These applications give rise to the Einstein equations, as was proposed in [34]. But what if we extend our analysis to describe strongly time dependent situations? Then we will have to extend our use of thermodynamics to non-equilibrium thermodynamics.

The query is that, given that we are working in a context in which equilibrium thermodynamics gives rise to the Einstein equations, which are symmetric under time reversal, will we now see the emergence of a time asymmetric extension of general relativity, such as are described in [56, 57]?

Acknowledgements I would like to thank Andrzej Banburski, Jacob Barnett, Linqing Chen, Marina Cortes, Bianca Dittrich, Laurent Freidel, Henriques Gomes, Andrew Liddle, Stacy McGaugh, Mordehai Milgrom, Krishnamohan Parattu, Percy Paul and Vasudev Shyam for very helpful discussions and encouragement. I am also indebted to Stacy McGaugh for permission to reproduce Fig. 1 and its caption from [21].

This research was supported in part by Perimeter Institute for Theoretical Physics. Research at Perimeter Institute is supported by the Government of Canada through Industry Canada and by the Province of Ontario through the Ministry of Research and Innovation. This research was also partly supported by grants from NSERC, FQXi and the John Templeton Foundation.

References

1. T. Padmanabhan, Limitations on the operational definition of spacetime events and quantum gravity. Class. Quantum Grav. **4**, L107–L113 (1987)
2. T. Padmanabhan, T. Roy Choudhury, *The Issue of Choosing Nothing: What Determines the Low Energy Vacuum State of Nature?*. arXiv:gr-qc/0006018
3. T. Padmanabhan, *Is gravity an intrinsically quantum phenomenon ? Dynamics of Gravity from the Entropy of Spacetime and the Principle of Equivalence*. arXiv:hep-th/0205278
4. T. Padmanabhan, Thermodynamical aspects of gravity: new insights, Rep. Prog. Phys. **73**, 046901 (2010). arXiv:0911.5004; Entropy of Static Spacetimes and Microscopic Density of States. arXiv:gr-qc/0308070, Class. Quant. Grav. **21**, 4485–4494 (2004); Equipartition of energy in the horizon degrees of freedom and the emergence of gravity. arXiv:0912.3165, Mod. Phys. Lett. A **25**, 1129–1136 (2010)
5. P. Samantray, T. Padmanabhan, *Conformal symmetry, riddler space and the AdS/CFT correspondence*. arXiv:1308.4667
6. A. Einstein, *What is the Theory of Relativity?*, The Times (London), pp. 227–232 in Ideas and Opinions. (New York: Bonanza, 1919)
7. Luca Bombelli, Joohan Lee, David Meyer, Rafael D. Sorkin, Spacetime as a causal set. Phys. Rev. Lett. **59**, 521–524 (1987)
8. M. Cortês, L. Smolin, The universe as a process of unique events. Phys. Rev. D **90**, 084007 (2014). arXiv:1307.6167 [gr-qc]
9. M. Cortês, L. Smolin, Energetic causal sets. Phys. Rev. D **90**, 044035 (2014). arXiv:1308.2206
10. M. Cortês, L. Smolin, Spin foam models as energetic causal sets. Phys. Rev. D **93**, 084039 (2016). doi:10.1103/PhysRevD.93.084039, arXiv:1407.0032
11. C. Furey, handwritten note, 2 Sept 2011, personal communication to Lee Smolin. http://www.perimeterinstitute.ca/personal/cfurey/2011notes__.pdf. See also www.perimeterinstitute.ca/personal/cfurey/essay_excerpts_2006.pdf
12. A. Criscuolo, H. Waelbroeck, Causal set dynamics: a toy model. Class. Quant. Grav. **16**(1999), 1817–1832 (1998). arXiv:gr-qc/9811088
13. F. Markopoulou, Dual formulation of spin network evolution. arXiv:gr-qc/9704013; F. Markopoulou, L. Smolin, Causal evolution of spin net- works, Nucl. Phys. B508 409; Quantum geometry with intrinsic local causality, Phys Rev D **58**, 084032
14. G. Amelino-Camelia, L. Freidel, J. Kowalski-Glikman, L. Smolin, The principle of relative locality. Phys. Rev. D **84**, 084010 (2011). arXiv:1101.0931[hep-th]
15. L. Freidel, L. Smolin, *Gamma ray burst delay times probe the geometry of momentumspace*. arXiv:1103.5626
16. F. Markopoulou, L. Smolin, Holography in a quantum spacetime. arXiv:hep-th/9910146, L. Smolin, The strong and weak holographic principles, Nucl. Phys. B **601**(2001), 209–247 (1999), arXiv:hep-th/0003056
17. L. Smolin, *On the nature of quantum fluctuations and their relation to gravitation and the principle of inertia, Classical and Quantum Gravity, 3(1986) 347–359; Stochastic mechanics, hidden variables and gravity, in Quantum Concepts in Space and*, Time edn. (Penrose Oxford University Press, C.J. Isham and R, 1985)
18. L. Freidel, R.G. Leigh, D. Minic, *Born Reciprocity in String Theory and the Nature of Spacetime*. arXiv:1307.7080; *Quantum Gravity, Dynamical Phase Space and String Theory*. arXiv:1405.3949; *Metastring Theory and Modular Space-time*. arXiv:1502.08005; *Quantum Spaces are Modular*. arXiv:1606.01829
19. G. 't Hooft, Dimensional reduction in quantum gravity, Salanfestschrift ed. by A. Alo, J. Ellis, S. Randjbar-Daemi (World Scien- tific,1993). arXiv:gr-qc/9310006.in
20. L. Susskind, The world as a hologram. J. Math. Phys. **36**, 6377 (1995). arXiv:hep-th/9409089
21. S. McGaugh, F. Lelli, J. Schombert, *The Radial Acceleration Relation in Rotationally Supported Galaxies*. arXiv:1609.05917v1

22. M. Milgrom, ApJ, 270, 365 (1983). For reviews, see B. Famaey, S. McGaugh, Liv. Rev. Rel. **15**, 10 (2012); M. Milgrom, Scale Invariance at low accelerations (aka MOND) and the dynamical anomalies in the Universe. arXiv:1605.07458v2; M. Milgrom, Scholarpedia **9**(6), 31410 (2014)
23. L. Crane, in, Categorical Physics. arXiv:hep-th/9301061, arXiv:hep-th/9308126, in, Knot Theory and Quantum Gravity ed. by J. Baez?(Oxford University Press); Clocks and Categories, is quantum gravity algebraic? J. Math. Phys. **36** 6180–6193 (1995). arXiv:gr-qc/9504038; Topological field the- ory as the key to quantum gravity
24. L. Smolin, The strong and weak holographic principles. Nucl. Phys. B **601**, 209–247 (2001). arXiv:hep-th/0003056
25. C. Rovelli, Relational quantum mechanics. Int. J. Theor. Phys. **35**, 1637–1678 (1996). arXiv:quant-ph/9609002
26. For a review see, M. Rangamani, T. Takayanagi, *Holographic Entanglement Entropy*. arXiv:1609.01287
27. W.G. Unruh, Notes on black hole evaporation. Phys. Rev. D **14**, 870 (1976)
28. T. Jacobson, *Entanglement equilibrium and the Einstein equation*. arXiv:1505.04753
29. S. Weinberg, Dynamics at Infinite Momentum, Phys. Rev. **150**, 1313 (1966); R.P. Feynman (1969) The Behavior of Hadron Collisions at Extreme Energies. High Energy Collisions: Third International Conference at Stony Brook, N.Y. Gordon & Breach. pp. 237–249. ISBN 978-0-677-13950-0; J. Bjorken, E. Paschos, (1969) Inelastic Electron-Proton and ?-Proton Scattering and the Structure of the Nucleon. Phys. Rev. **185**(5): 1975–1982. Bibcode:1969 PhRv..185.1975B. doi:10.1103/PhysRev.185.1975; J.D. Bjorken, *Asymptotic sum rules at infinite momentum*, Published in Phys. Rev. **179**, 1547–1553 (1969); J.B. Kogut*, D,E. Soper, Quantum electrodynamics in the infinite-momentum frame, Phys. Rev. D **1**, 2901–2914 (1970); J. Kogut, L. Susskind, The Parton Picture of Elementary Particles, Phys. Rep. **8**, 75 (1973)
30. D.M. Greenberger, *The Disconnect Between Quantum Mechanics and Gravity*. arXiv:1011.3719
31. L. Crane, L. Smolin, Renormalizability of general relativity on a background of spacetime foam, Nucl. Phys. B **267**, 714–757 (1986); Spacetime foam as the universal regulator, Second award, Gravity Research Foundation 1985, Gen. Relat. Gravit. **17**, 1209 (1985)
32. S. Carlip, Dimensional reduction in causal set gravity. arXiv:1506.08775, Class. Quantum Grav. **32**, 232001; L. Crane, L. Smolin, Renormalizability of general relativity on a background of spacetime foam, Nucl. Phys. B **267**, 714–757 (1986); Gen. Relat. Gravit. **17**(1985), 1209 (2015)
33. S. Kolekar, T. Padmanabhan (2015), *Indistinguishability of thermal and quantum fluctuations*, S. Kolekar, T Padmanabhan, Classical and Quantum Gravity, **32**(20). arXiv:1308.6289
34. T. Jacobson, Thermodynamics of spacetime: the Einstein equation of state. Phys. Rev. Lett. **75**, 1260–1263 (1995). arXiv:gr-qc/9504004
35. L. Susskind, Strings, black holes and Lorentz contractions, arXiv:hep-th/9308139; Phys. Rev. Lett. **71**, 2367, (1993); Phys. Rev. D **49** (1994) 6606; D50, 2700 (1994); L. Susskind, P. Griffin, Partons and black holes. arXiv:hep-th/9410306
36. S. Deser, O. Levin, Accelerated Detectors and Temperature in (Anti) de Sitter Spaces. Class. Quant. Grav. **14**, L163–L168 (1997). doi:10.1088/0264-9381/14/9/003, http://arxiv.org/abs/gr-qc/9706018
37. B. Czech, J.L. Karczmarek, F. Nogueira, M. Van Raamsdonk, *Rindler Quantum Gravity*. arXiv:1206.1323
38. V. Balasubramanian, B.D. Chowdhury, B. Czech, J. de Boer, M.P. Heller, *A hole-ographic spacetime*, Journal-ref: Phys. Rev. D **89**, 086004 (2014). arXiv:1305.085; V. Balasubramanian, B. Czech, B.D. Chowdhury, J. de Boer, *The entropy of a hole in spacetime*, JHEP **1310**, 220 (2013)
39. M. Parikh, P. Samantray, *Rindler-AdS/CFT*. arXiv:1211.7370
40. F. Lelli, S.S. McGaugh, J.M. Schombert, Astron. J. (2016). arXiv:1606.09251
41. M. Milgrom, R.H. Sanders, Rings and shells of dark matter as MOND artifacts, Astrophys. J. **678**(1), 131–143, (2008). arXiv:0709.2561
42. S.S. McGaugh, Astrophys. J. **683**, 137 (2008). arXiv:0804.1314
43. R.B. Tully, J.R. Fisher, Astron. Astrophys. **54**, 661 (1977)

44. R.H. Sanders, The published extended rotation curves of spiral galaxies: confrontation with modified dynamics. arXiv:astro-ph/9606089, Astrophys. J. **473**, 117
45. S. McGaugh, F. Lelli, J. Schombert, The small scatter of the baryonic tully-fisher relation, Astrophys. J. Lett. **816**(1), article id. L14, 6D (2016). arXiv:1512.04543
46. E.M. Di Teodoro, F. Fraternali, S.H. Miller, Flat rotation curves and low velocity dispersions in KMOS star-forming galaxies at z 1, Astrophys. Galaxies (astro-ph.GA). arXiv:1602.04942 [astro-ph.GA]
47. S.H. Miller, R.S. Ellis, M. Sullivan, K. Bundy, A.B. Newman, T. Treu, *The Assembly History of Disk Galaxies: II. Probing the Emerging Tully-Fisher Relation During* $1<z<1.7$. arXiv:1201.4386 [astro-ph.CO]
48. J Bekenstein, J. Magueijo, *MOND habitats within the solar system*. doi:10.1103/PhysRevD.73.103513, arXiv:astro-ph/0602266
49. M. Milgrom, The modified dynamics as a vacuum effect, Phys. Lett. A **253**, 273–279 (1999). doi:10.1016/S0375-9601(99)00077-8, arXiv:astro-ph/9805346; Dynamics with a nonstandard inertia-acceleration relation: an alternative to dark matter in galactic systems, Ann. Phys. **229**(2), 384–415. arXiv:astro-ph/9303012
50. R.P. Woodard, Nonlocal metric realizations of MOND, Can. J. Phys. **93**(2), 242–249. arXiv:1403.6763
51. Fotini Markopoulou, Lee Smolin, Disordered locality in loop quantum gravity states, Class. Quant. Grav. **24**, 3813–3824 (2007). arXiv:gr-qc/0702044
52. Chanda Prescod-Weinstein, Lee Smolin, Disordered locality as an explanation for the dark energy. Phys. Rev. D **80**, 063505 (2009). arXiv:0903.5303 [hep-th]
53. F. Markopoulou, I. Premont-Schwarz, L. Smolin, *Macroscopic deviations from Newtonian gravity from quantum gravity*, unpublished manuscript (2005)
54. M.H.P.M. van Putten, *Galaxy rotation curves in de Sitter space*. arXiv:1411.2665
55. H. Narnhofer, I. Peter, W. Thirring, Int. J. Mod. Phys. B **10**, 1507 (1996)
56. M. Cortes, H. Gomes, L. Smolin, Time asymmetric extensions of general relativity, Phys. Rev. D **92**, 043502 (2015). arXiv:1503.06085; M. Cortes, A.R. Liddle, L. Smolin, Cosmological signatures of time-asymmetric gravity. arXiv:1606.01256
57. L. Smolin, Dynamics of the cosmological and Newton's constant. arXiv:1507.01229. Class. Q. Gravit. **33**(2). http://dx.doi.org/10.1088/0264-9381/33/2/025011
58. L. Smolin, *What is the problem of quantum gravity?*, Introduction to Ph.D. thesis, also IAS preprint, Sept 1979
59. P. Candelas, D.W. Sciama, Irreversible thermodynamics of black holes, Phys. Rev. Lett. **38**, 1372 (1977); D.W. Sciama, *Black holes and fluctuations of quantum particles: an Einstein synthesis,* in Relativity, Quanta, and Cosmology in the Development of the Scientific Thought of Albert Einstein, vol. II, ed. by M. Pantaleo, dir., F. De Finis (Giunti Barb'era, Firenze, 1979)
60. Juan M. Maldacena, hep-th/9711200, hep-th/9803002; E. Witten, hep-th/9802150, hep-th/9803131; S.S. Gubser, I.R. Klebanov, A.M. Polyakov, Phys. Lett. B **428** (1998) 105-114, hep-th/9802109. For a review, see, O. Aharony, S. S. Gubser, J. Maldacena, H. Ooguri, Y. Oz, Large N Field Theories, String Theory and Gravity hep-th/9905111
61. H. Casini, M. Huerta, R.C. Myers, Towards a derivation of holographic entanglement entropy. JHEP **1105**, 036 (2011). doi:10.1007/JHEP05(2011)036, arXiv:1102.0440 [hep-th]
62. R. Fareghbal, A. Naseh, Rindler Contracted-CFT Correspondence, JHEP **06**, 134 (2014). doi:10.1007/JHEP06(2014)134, arXiv:1404.3937; Aspects of Flat/CCFT Correspondence. Class. Quantum Grav. **32**, 135013 (2015). doi:10.1088/0264-9381/32/13/135013, arXiv:1408.6932
63. A. Bagchi, The BMS/GCA correspondence. Phys. Rev. Lett. **105**, 171601 (2010), doi:10.1103/PhysRevLett.105.171601, arXiv:1006.3354
64. F.J. Herranz, *Non-standard quantum so(3,2) and its contractions,* J. Phys. A **30**, 6123-6129 (1997). doi:10.1088/0305-4470/30/17/019, arXiv:q-alg/9704006
65. P. Kosiski, J. Lukierski, P. Malanka, *On kappa-deformed D=4 quantum conformal group.* arXiv:hep-th/0312227; F. Falceto, K. Gawedzki, *On quantum group symmetries of conformal field theories*. arXiv:hep-th/9109023

66. G. Moore, N. Reshetikhin, *A comment on quantum group symmetry in conformal field theory*, Nucl. Phys. B **328**(3), 557–574, 25 December 1989; S. Meljanac, A. Pachol, D. Pikutic, *Twisted conformal algebra related to kappa-Minkowski space* Phys. Rev. D **92**, 105015 (2015). doi:10.1103/PhysRevD.92.105015, arXiv:1509.02115
67. Jae-Weon Lee, Hyeong-Chan Kim, Jungjai Lee, Gravity from Quantum Information. J. Korean Phys. Soc. **63**, 1094 (2013). doi:10.3938/jkps.63.1094, arXiv:1001.5445
68. S.S. McGaugh, Milky way mass models and MOND, Astrophys. J. **683**, 137 (2008). arXiv:0804.1314

What Do Detectors Detect?

L. Sriramkumar

Abstract By a detector, one has in mind a point particle with internal energy levels, which when set in motion on a generic trajectory can get excited due to its interaction with a quantum field. Detectors have often been considered as a helpful tool to understand the concept of a particle in a curved spacetime. Specifically, they have been used extensively to investigate the thermal effects that arise in the presence of horizons. In this article, I review the concept of detectors and discuss their response when they are coupled linearly as well as non-linearly to a quantum scalar field in different situations. In particular, I discuss as to how the response of detectors does not necessarily reflect the particle content of the quantum field. I also describe an interesting 'inversion of statistics' that occurs in odd spacetime dimensions for 'odd couplings', i.e. the response of a uniformly accelerating detector is characterized by a Fermi–Dirac distribution even when it is interacting with a scalar field. Moreover, by coupling the detector to a quantum field that is governed by a modified dispersion relation arising supposedly due to quantum gravitational effects, I examine the possible Planck scale modifications to the response of a rotating detector in flat spacetime. Lastly, I discuss as to why detectors that are switched on for a finite period of time need to be turned on smoothly in order to have a meaningful response.

1 Introduction

The vacuum state of a quantum field develops a non-trivial structure in the presence of a strong classical electromagnetic or gravitational background. This effect essentially manifests itself as two types of physical phenomena: polarization of the vacuum and production of pairs of particles corresponding to the quantum field. Apart from these two effects, there is another feature that one encounters in a gravitational background: the definition of the vacuum does not prove to be generally covariant. In other words, the concept of a particle turns out to be, in general, dependent on the choice of coordinates. (For a detailed discussion on these different aspects of quantum

L. Sriramkumar (✉)
Department of Physics, Indian Institute of Technology Madras, Chennai 600036, India
e-mail: sriram@physics.iitm.ac.in

J.S. Bagla and S. Engineer (eds.), *Gravity and the Quantum*,
Fundamental Theories of Physics 187, DOI 10.1007/978-3-319-51700-1_27

field theory in strong electromagnetic and gravitational fields, see the following texts [1, 2] and reviews [3].) A classic example of vacuum polarization is the Casimir effect [4]. The Schwinger effect [5, 6], *viz.* pair creation by strong electric fields, and Hawking radiation from collapsing black holes are the most famous examples of particle production [7]. The coordinate dependence of the particle concept that arises in a gravitational background is well illustrated by the flat spacetime example wherein the vacuum defined in the frame of a uniformly accelerating observer (often referred to as the Rindler vacuum) turns out to be inequivalent to the conventional Minkowski vacuum [8]. Similar issues are encountered when the behavior of quantum fields are studied in curved spacetimes. Needless to say, concepts such as vacuum and particle need to be unambiguously defined in order to determine the extent of vacuum polarization or particle production occurring in a curved spacetime.

It is in such a situation that the concept of a detector was initially introduced in the literature [9, 10]. The motivation behind the idea of detectors was to provide an operational definition for the concept of a particle in a curved spacetime. After all, 'particles are what the particle detectors detect' [11]. With this goal in mind, the response of different types of detectors have been studied in a variety of situations over the last three to four decades (in fact, there is an enormous amount of literature on the topic; for an incomplete list of early efforts in this direction, see Refs. [12–24] and, for more recent work, see, for example, Refs. [25–28]). But, what do these detectors actually detect? In particular, do their responses reflect the particle content of the field as it was originally desired? In this article, apart from attempting to address such questions with the help of a few specific examples, I shall also discuss a couple of interesting phenomena associated with detectors, including possible Planck scale effects. I should mention here that this article is essentially a review based on my earlier efforts in these directions (see Refs. [18, 21–23, 27]).

An outline of the contents of this article is as follows. In the following section, I shall discuss the response of non-inertial Unruh–DeWitt detectors (which are linearly coupled to the quantum field) in flat spacetime. Specifically, I shall focus on the response of uniformly accelerating and rotating detectors. I shall also compare the response of detectors in different situations with the results from more formal methods—such as the Bogolubov transformations and the effective Lagrangian approach—that probe the vacuum structure of the quantum field. Such an exercise helps us understand the conditions under which the detectors respond. In Sect. 3, I shall consider the response of detectors that are coupled non-linearly to a quantum scalar field. Interestingly, I shall show that, in odd spacetime dimensions, the response of the detectors exhibit an 'inversion of statistics' when they are coupled to an odd power of the quantum field. In Sect. 4, I shall consider possible Planck scale effects on the response of a rotating detector in flat spacetime. Assuming that the Planck scale effects modify the dispersion relation governing a quantum field, I shall study the response of a rotating Unruh–DeWitt detector that is coupled to such a quantum scalar field. I shall illustrate that, while super-luminal dispersion relations hardly affect the response of the detector, sub-luminal dispersion relations alter their response considerably. In Sect. 5, I shall consider Unruh–DeWitt detectors that are switched on for a finite period of time and show that divergences can arise in the response of the detector if it is turned on abruptly. Lastly, I conclude in Sect. 6 with a brief summary.

A few words on my conventions and notations are in order before I proceed. I shall adopt natural units such that $\hbar = c = 1$ and, for convenience, denote the trajectory of the detector $x^{\mu}(\tau)$ as $\tilde{x}(\tau)$, where τ is the proper time in the frame of the detector. In Sect. 3, I shall consider the response of non-linearly coupled detectors in arbitrary spacetime dimensions. In all the other sections, I shall restrict myself to working in $(3+1)$-spacetime dimensions.

2 Response of the Unruh–DeWitt Detector in Flat Spacetime

A detector is an idealized point like object whose motion is described by a classical worldline, but which nevertheless possesses internal energy levels. Such detectors are basically described by the interaction Lagrangian for the coupling between the degrees of freedom of the detector and the quantum field. The simplest of the different possible detectors is the detector due to Unruh and DeWitt [9, 10]. Consider a Unruh–DeWitt detector that is moving along a trajectory $\tilde{x}(\tau)$, where τ is the proper time in the frame of the detector. The interaction of the Unruh–DeWitt detector with a canonical, real scalar field ϕ is described by the interaction Lagrangian

$$\mathscr{L}_{\rm int}[\phi(\tilde{x})] = \bar{c}\, m(\tau)\, \phi\left[\tilde{x}(\tau)\right], \tag{1}$$

where $\bar{c}$ is a small coupling constant and m is the detector's monopole moment. Let us assume that the quantum field $\hat{\phi}$ is initially in the vacuum state $|0\rangle$ and the detector is in its ground state $|E_0\rangle$ corresponding to an energy eigen value E_0. Then, up to the first order in perturbation theory, the amplitude of transition of the Unruh–DeWitt detector to an excited state $|E_1\rangle$, corresponding to an energy eigen value E_1 $(> E_0)$, is described by the integral [2]

$$A(E) = M \int_{-\infty}^{\infty} {\rm d}\tau\, {\rm e}^{iE\tau}\, \langle\psi|\hat{\phi}[\tilde{x}(\tau)]|0\rangle, \tag{2}$$

where $M = i\,\bar{c}\,\langle E_1|m(0)|E_0\rangle$, $E = E_1 - E_0 > 0$ and $|\psi\rangle$ is the state of the quantum scalar field *after* its interaction with the detector. Note that the quantity M depends only on the internal structure of the detector, and not on its motion. Therefore, as is often done, I shall drop the quantity hereafter. The transition probability of the detector to all possible final states $|\psi\rangle$ of the quantum field is given by

$$P(E) = \sum_{|\psi\rangle} |A(E)|^2 = \int_{-\infty}^{\infty} \mathrm{d}\tau \int_{-\infty}^{\infty} \mathrm{d}\tau'\, \mathrm{e}^{-iE(\tau-\tau')}\, G^+\left[\tilde{x}(\tau), \tilde{x}(\tau')\right], \tag{3}$$

where $G^+\left[\tilde{x}(\tau), \tilde{x}(\tau')\right]$ is the Wightman function defined as

$$G^+\left[\tilde{x}(\tau), \tilde{x}(\tau')\right] = \langle 0|\hat{\phi}\left[\tilde{x}(\tau)\right]\, \hat{\phi}\left[\tilde{x}(\tau')\right]|0\rangle. \tag{4}$$

When the Wightman function is invariant under time translations in the frame of the detector—as it can occur, for example, in cases wherein the detector is moving along the integral curves of time-like Killing vector fields [12, 22]—I have

$$G^+\left[\tilde{x}(\tau), \tilde{x}(\tau')\right] = G^+(\tau - \tau'). \tag{5}$$

In such situations, the transition probability of the detector simplifies to

$$P(E) = \lim_{T\to\infty} \int_{-T}^{T} \frac{\mathrm{d}v}{2} \int_{-\infty}^{\infty} \mathrm{d}u\, \mathrm{e}^{-iE\,u}\, G^+(u), \tag{6}$$

where

$$u = \tau - \tau' \quad \text{and} \quad v = \tau + \tau'. \tag{7}$$

The above expression then allows one to define the transition probability *rate* of the detector to be [2]

$$R(E) = \lim_{T\to\infty} \frac{P(E)}{T} = \int_{-\infty}^{\infty} \mathrm{d}u\, \mathrm{e}^{-iE\,u}\, G^+(u). \tag{8}$$

For the case of the canonical, massless scalar field, in $(3+1)$-spacetime dimensions, the Wightman function $G^+\left(\tilde{x}, \tilde{x}'\right)$ in the Minkowski vacuum is given by [2]

$$G^+\left(\tilde{x}, \tilde{x}'\right) = -\frac{1}{4\pi^2}\left[\frac{1}{(t - t' - i\,\varepsilon)^2 - (\boldsymbol{x} - \boldsymbol{x}')^2}\right], \tag{9}$$

where $\varepsilon \to 0^+$ and $(t, \boldsymbol{x})$ denote the Minkowski coordinates. Given a trajectory $\tilde{x}(\tau)$, the response of the detector is obtained by substituting the trajectory in this Wightman function and evaluating the transition probability rate (8). For example, it is straightforward to show that the response of a detector that is moving on an inertial trajectory in the Minkowski vacuum vanishes identically. I had mentioned above that the quantization of a field proves to be inequivalent in the inertial and the uniformly accelerating frames in flat spacetime. Due to this reason, it seems worthwhile to examine the behavior of non-inertial detectors. In the next sub-section, I shall consider the response of uniformly accelerating as well as rotating detectors in flat spacetime.

2.1 *Response of Accelerating and Rotating Detectors*

As is commonly known, there are ten independent time-like Killing vector fields in flat spacetime. These Killing vector fields correspond to three types of symmetries, *viz.* translations, rotations and boosts. Different types of non-inertial trajectories can be generated by considering the integral curves of various linear combinations of these Killing vector fields [12, 22]. Amongst the trajectories that are possible, there exist two trajectories which have attracted considerable attention in the literature. They correspond to uniformly accelerating and rotating trajectories. In what follows, I shall consider the response of the Unruh–DeWitt detector moving along these trajectories.

2.1.1 Uniformly Accelerated Motion

The trajectory of a uniformly accelerated observer moving along the x-axis is given by

$$\tilde{x}(\tau) = g^{-1}\,[\sinh(g\,\tau),\ \cosh(g\,\tau),\ 0,\ 0]\,, \tag{10}$$

where g denotes the proper acceleration. The coordinates associated with the frame of such an observer are known as the Rindler coordinates [29]. The Wightman function in the frame of the uniformly accelerating observer is obtained by substituting the above trajectory in Eq. (9). It is given by

$$G^{+}(u) = \frac{-1}{16\,\pi^2}\,\frac{g^2}{\sinh^2[(g\,u/2) - i\,\varepsilon]} = \frac{-1}{4\,\pi^2}\sum_{n=-\infty}^{\infty}\frac{1}{(u - i\,\varepsilon + 2\,\pi\,i\,n/g)^2}, \tag{11}$$

where, recall that, $u = \tau - \tau'$. The resulting transition probability rate can be easily evaluated to be [9, 10]

$$R(E) = \frac{1}{2\,\pi}\,\frac{E}{e^{2\,\pi\,E/g} - 1}, \tag{12}$$

which is a thermal spectrum corresponding to the temperature $T = g/(2\,\pi)$. This thermal response is the famous Unruh effect (for a detailed discussion, see, for instance, Ref. [30]).

2.1.2 Rotational Motion

Let us now turn to the case of the rotating detector. The trajectory of the rotating detector can be expressed in terms of the proper time τ as follows [12, 27]:

$$\tilde{x}(\tau) = [\gamma\,\tau,\ \sigma\,\cos\,(\gamma\,\Omega\,\tau),\ \sigma\,\sin\,(\gamma\,\Omega\,\tau),\ 0]\,, \tag{13}$$

where the constants σ and Ω denote the radius of the circular path along which the detector is moving and the angular velocity of the detector, respectively. The quantity $\gamma = \left[1 - (\sigma\,\Omega)^2\right]^{-1/2}$ is the Lorentz factor that relates the Minkowski time to the proper time in the frame of the detector. The Wightman function along the rotating trajectory can be obtained to be

$$G^{+}(u) = -\frac{1}{4\,\pi^2}\left(\frac{1}{\gamma^2\,(u - i\,\varepsilon)^2 - 4\,\sigma^2\,\sin^2\,(\gamma\,\Omega\,u/2)}\right). \tag{14}$$

However, unfortunately, it does not seem to be possible to evaluate the corresponding transition probability rate $R(E)$ analytically. I have arrived at the response of the rotating detector by substituting the Wightman function (14) in the expression (8), and numerically computing the integral involved. If I define the dimensionless energy to be $\bar{E} = E/(\gamma\,\Omega)$, I find that the dimensionless transition probability rate $\bar{R}(\bar{E}) = \sigma\,R(\bar{E})$ of the detector depends only on the dimensionless quantity $\sigma\,\Omega$ that describes the linear velocity of the detector. In Fig. 1, I have plotted the transition probability rate of the detector for three different values of the quantity $\sigma\,\Omega$ [12]. I should mention here that, in order to check the accuracy of the numerical procedure that I have used to evaluate the integral (8) for the rotating trajectory, I have compared the results from the numerical code with the analytical one [*viz.* Eq. (12)] that is available for the case of the uniformly accelerated detector. This comparison clearly indicates that the numerical procedure I have adopted to evaluate the integral (8) is quite accurate [27].

In the discussion above, I had arrived at the response of the rotating detector by evaluating the Fourier transform of the Wightman function with respect to the differential proper time u in the frame of the detector. In this case, evidently, I had first summed over the normal modes (to arrive at the Wightman function) before evaluating the integral over the differential proper time. I shall now rederive the result by changing the order of these procedures. I shall express the Wightman function as a sum over the normal modes and first evaluate the integral over the differential proper time before computing the sum. This method proves to be helpful later when I shall consider the Planck scale effects on the rotating detector. As I shall illustrate, the method can be easily extended to cases wherein the scalar field is described by a modified dispersion relation.

I shall start by working in the cylindrical polar coordinates, say, (t, ρ, θ, z), instead of the cartesian coordinates, since they prove to be more convenient. In terms of the cylindrical coordinates, the trajectory (13) of the rotating detector can be written in

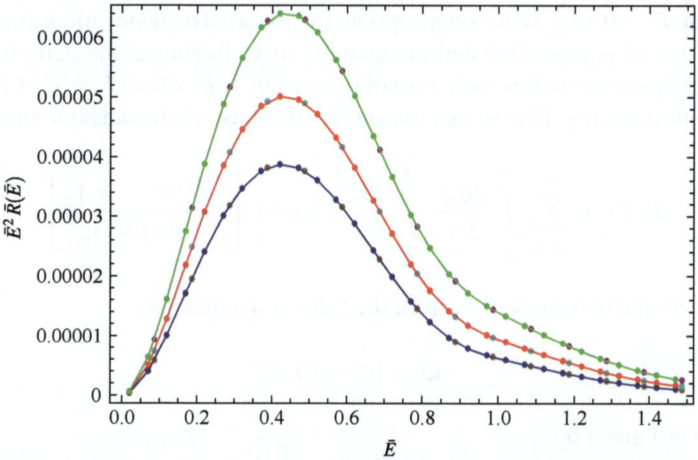

Fig. 1 The transition probability rate of the rotating Unruh–DeWitt detector that is coupled to the conventional, massless scalar field. The *dots* and the *curves* that simply link them represent numerical results arrived at from the computation of the integral (8) along the rotating trajectory. The curves correspond to the following three values of the quantity $\sigma\,\Omega = 0.325$ (in *blue*), 0.350 (in *red*) and 0.375 (in *green*). The dots of an alternate color that appear on the curves denote the numerical results that have been obtained by another method which I shall describe below [they actually correspond to the sum (24)]. Clearly, the results from the two methods match very well

terms of the proper time τ as follows:

$$\tilde{x}(\tau) = (\gamma\,\tau,\ \sigma,\ \gamma\,\Omega\,\tau,\ 0)\,. \tag{15}$$

Using well established properties of the Bessel functions, it is straightforward to show that, along the trajectory of the rotating detector, the standard Minkowski Wightman function (9) can be written as

$$G^{+}(u) = \sum_{m=-\infty}^{\infty} \int_0^{\infty} \frac{\mathrm{d}q\,q}{(2\,\pi)^2} \int_{-\infty}^{\infty} \frac{\mathrm{d}k_z}{(2\,\omega)}\, J_m^2(q\,\sigma)\; \mathrm{e}^{-i\,\gamma\,(\omega-m\,\Omega)\,u}, \tag{16}$$

where $J_m(q\,\sigma)$ denote the Bessel functions of order m, with ω being given by

$$\omega = \left(q^2 + k_z^2\right)^{1/2}. \tag{17}$$

One can then immediately express the corresponding transition probability rate of the rotating detector as [cf. Eq. (8)]

$$R(E) = \sum_{m=-\infty}^{\infty} \int_0^{\infty} \frac{\mathrm{d}q\,q}{2\,\pi} \int_{-\infty}^{\infty} \frac{\mathrm{d}k_z}{2\,\omega}\, J_m^2(q\,\sigma)\; \delta^{(1)}\left[E + \gamma\,(\omega - m\,\Omega)\right]. \tag{18}$$

Recall that, $E > 0, \omega \geq 0$ (as is appropriate for positive frequency modes), and I have assumed that Ω is a positive definite quantity as well. Hence, the delta function in the above expression will be non-zero only when $m \geq \bar{E}$, where $\bar{E} = E/(\gamma\,\Omega)$ is the dimensionless energy. Due to this reason, the response of the detector simplifies to

$$R(\bar{E}) = \sum_{m\geq\bar{E}}^{\infty} \int_0^{\infty} \frac{\mathrm{d}q\,q}{2\,\pi} \int_{-\infty}^{\infty} \frac{\mathrm{d}k_z}{2\,\omega}\, J_m^2(q\,\sigma) \left[\frac{\delta^{(1)}(k_z - \kappa_z)}{\gamma\,|(\mathrm{d}\omega/\mathrm{d}k_z)|_{\kappa_z}}\right], \tag{19}$$

where κ_z are the two roots of k_z from the following equation:

$$\omega = \left(m - \bar{E}\right)\,\Omega. \tag{20}$$

The roots are given by

$$\kappa_z = \pm\left(\lambda^2 - q^2\right)^{1/2}, \tag{21}$$

where, for convenience, I have set

$$\lambda = \bar{\lambda}\,\Omega = \left(m - \bar{E}\right)\,\Omega. \tag{22}$$

Since both the positive and negative roots of κ_z contribute equally, the dimensionless transition probability rate of the rotating detector can be obtained to be

$$\bar{R}(\bar{E}) = \sigma\,R(\bar{E}) = \frac{\sigma}{2\,\pi\,\gamma} \sum_{m\geq\bar{E}}^{\infty} \int_0^{\lambda} \mathrm{d}q\,q\,\left[\frac{J_m^2(q\,\sigma)}{\left(\lambda^2 - q^2\right)^{1/2}}\right], \tag{23}$$

where I have set the upper limit on q to be λ as κ_z is a real quantity [cf. Eq. (21)]. I find that the integral over q can be expressed in terms of the hypergeometric functions (see, for instance, Ref. [31]). Therefore, the transition probability rate of the rotating detector can be written as

$$\bar{R}(\bar{E}) = \frac{1}{2\,\pi\,\gamma} \sum_{m\geq\bar{E}}^{\infty} \frac{\left(\sigma\,\Omega\,\bar{\lambda}\right)^{(2\,m+1)}}{\Gamma\,(2\,m+2)} \times {}_1F_2\left[m + (1/2);\; m + (3/2), 2\,m + 1;\; -\left(\sigma\,\Omega\,\bar{\lambda}\right)^2\right], \tag{24}$$

where ${}_1F_2\,(a; b, c; x)$ denotes the hypergeometric function, while $\Gamma(x)$ is the usual Gamma function. Though it does not seem to be possible to arrive at a closed form expression for this sum, the sum converges very quickly, and hence proves to be easy to evaluate numerically. In Fig. 1, I have plotted the numerical results for the above sum for the same values of the linear velocity $\sigma\,\Omega$ for which I had plotted the results obtained from Fourier transforming the Wightman function (14) along the rotating trajectory. The figure clearly indicates that the results from the two different methods match each other rather well.

2.2 *Are Detectors Sensitive to the Particle Content of the Field?*

In order to clearly understand as to what detectors detect, I shall compare the response of detectors with the results from more conventional probes of the vacuum structure of the quantum fields, such as the approaches based on the Bogolubov transformations and the effective Lagrangian [22]. However, before carrying out such a comparison, let me say a few words briefly explaining these two other approaches.

Consider a quantum field that can be decomposed in terms of two complete sets of normal modes. These two sets of modes can be related to each other through the Bogolubov transformations, which are essentially characterized by two coefficients often referred to as α and β [32]. Moreover, the particle content of the field is determined by the Bogolubov coefficient β. In a gravitational background, the Bogolubov transformations can either relate the modes of a quantum field at two different times in the same coordinate system or the modes in two different coordinate systems covering the same region of spacetime. When the Bogolubov coefficient β is non-zero, in the latter context, such a result is normally interpreted as implying that the quantization in the two coordinate systems are inequivalent [8]. Whereas, in the former context, a non-zero β is attributed to the production of particles by the background gravitational field [3]. Similarly, in an electromagnetic background, a non-zero β relating the modes of a quantum field at different times (in a particular gauge) implies that the background leads to pair creation [1].

In the effective Lagrangian approach, one essentially integrates out the degrees of freedom associated with the quantum field, thereby arriving at an effective action describing the classical background [5, 6]. An imaginary part to the effective Lagrangian unambiguously suggests the decay of the quantum vacuum, i.e. the production of particles corresponding to the quantum field. The real part of the effective Lagrangian can be related to the extent of polarization of the vacuum caused by the classical background. While the effective Lagrangian approach is powerful, since it involves computing a path integral, it often proves to be technically difficult to evaluate.

In Table 1, to illustrate the conclusions I wish to draw about the response of detectors, I have tabulated the results one obtains in a handful of different situations. I have listed whether the Bogolubov coefficient β, the response of the detector [or, more precisely, the transition probability $P(E)$] and the real and the imaginary parts of the effective Lagrangian $\mathscr{L}_{\rm eff}$ are zero or non-vanishing in these contexts. Apart from the results in the non-inertial frames in flat spacetime, I have compared the results between the Casimir plates, and different types of electromagnetic backgrounds.

Table 1 A comparison of the response of a detector with the results from more formal probes of the vacuum structure of the quantum field—*viz.* the Bogolubov transformations and the effective Lagrangian approaches—in a variety of situations. Note that, in the case of the time-independent electric field background, actually, the Bogolubov coefficient β is trivially zero. I refer here to particle production that can occur in such a background due to the phenomenon called Klein paradox [33]. I should also add that, in electromagnetic backgrounds, the coupling of the detector to the quantum field (say, a complex scalar field) has to be intrinsically non-linear in order to preserve gauge-invariance [21]

	Detector response $P(E)$	Bogolubov coefficient β	Effective lagrangian	
			Re. $\mathscr{L}_{\mathrm{eff}}$	Im. $\mathscr{L}_{\mathrm{eff}}$
In inertial coordinates	0	0	0	0
In Rindler coordinates	$\neq 0$	$\neq 0$	0	0
In rotating coordinates	$\neq 0$	0	0	0
Between Casimir plates	0	0	$\neq 0$	0
In a time-dependent electric field	$\neq 0$	$\neq 0$	$\neq 0$	$\neq 0$
In a time-independent electric field	$\neq 0$	$\neq 0$	$\neq 0$	$\neq 0$
In a time-independent magnetic field	0	0	$\neq 0$	0

Let me first consider the case of the non-inertial coordinates in flat spacetime. The Bogolubov coefficient β relating the Rindler modes and the Minkowski modes turns out to be non-zero and, in fact, the expectation value of the Rindler number operator in the Minkowski vacuum yields a thermal spectrum as well [8]. In contrast, in the rotating coordinates, while the Bogolubov coefficient β turns out to be zero [12], as we have seen, the detector responds non-trivially. Also, in both these cases, one can show that the effective Lagrangian vanishes identically—in fact, this is true even in the case of the Rindler coordinates, wherein the Bogolubov coefficient β proves to be non-zero [22]. Evidently, the response of a detector can be non-zero even when the Bogolubov coefficient β and the effective Lagrangian vanish identically. Clearly, the response of a detector does not necessarily reflect the particle content of the quantum field.

Let me now turn to the response of the detector between Casimir plates and in electromagnetic backgrounds. It is well known that Casimir plates and a time-independent magnetic field lead to vacuum polarization, but not to particle production. One finds that an inertial detector does not respond in these two backgrounds. In contrast, it is found that even an inertial detector responds in an electric field background, whether time-dependent or otherwise. It is easy to argue that, in a time-dependent electric field, the evolving modes will excite the inertial detector [21, 22]. Whereas, in a time-independent electric field of sufficient strength, modes of positive norm that have negative frequencies (which lead to the so-called Klein paradox and associated pair production [22, 33], as is also reflected by the imaginary part of the effective Lagrangian [6]) are found to be responsible for a non-vanishing response of an inertial detector.[1] These clearly suggest that, irrespective of the nature of its trajectory, a detector will respond whenever particle production takes place. In that sense a detector *is* sensitive to particle production. Further, if one restricts the motion of the detector to inertial trajectories, then the effects due to non-inertial motion can be avoided and, in such cases, the detector response will be non-zero *only* when particle production takes place. However, unlike in flat spacetime or classical electromagnetic backgrounds, there exists no special frame of reference in a classical gravitational background and all coordinate systems have to be treated equivalently. This aspect of the detector proves to be a major constraint in being able to utilize it to investigate the phenomenon of particle production in a curved spacetime [11, 35].

3 'Inversion of Statistics' in Odd Dimensions

We had seen that the response of a uniformly accelerating monopole detector that is coupled to a quantized massless scalar field is characterized by a Planckian distribution when the field is assumed to be in the Minkowski vacuum [cf. Eq. (12)]. However, it has been noticed that this result is true only in even-dimensional flat spacetimes and it has been shown that a Fermi–Dirac factor (rather than a Bose–Einstein factor) appears in the response of the accelerated detector when the dimensionality of spacetime is odd [14]. Recall that the Unruh–DeWitt detector is coupled *linearly* to the quantum scalar field. Over the years, motivated by different reasons, there have also been efforts in the literature to investigate the response of detectors that are coupled *non-linearly* to the quantum field [16, 20, 21]. It will be interesting to examine whether the non-linearity of the coupling affects the result in odd-dimensional flat spacetimes that I mentioned above.

3.1 Response of Non-linearly Coupled Detectors

Consider a detector that is interacting with a real scalar field ϕ through the non-linear interaction Lagrangian [20]

[1] In fact, it is such modes—*viz.* those which have a positive norm but negative frequencies—that excite the rotating detector [22, 34].

$$\mathcal{L}_{\rm int}[\phi(\tilde{x})] = \bar{c}\, m(\tau)\, \phi^n\left[\tilde{x}(\tau)\right], \tag{25}$$

where $\bar{c}$, $m(\tau)$ and $\tilde{x}(\tau)$ are the same quantities that we had encountered earlier in the context of the Unruh–DeWitt detector. The quantity n is a positive integer that denotes the index of non-linearity of the coupling. Let me assume that the quantum field $\hat{\phi}$ is initially in the vacuum state $|0\rangle$. The transition amplitude of the non-linearly coupled detector from the ground to an excited state can be written as

$$A_n(E) = M \int_{-\infty}^{\infty} {\rm d}\tau\, {\rm e}^{i\,E\,\tau}\, \langle\psi|\, \hat{\phi}^n[\tilde{x}(\tau)]\, |0\rangle\,, \tag{26}$$

where $|\psi\rangle$ is the final state of the field, and M and E are defined in the same fashion as in the case of the Unruh–Dewitt detector.

It is important to notice that the transition amplitude $A_n(E)$ above involves products of the quantum field $\hat{\phi}$ at the *same* spacetime point. Because of this reason, one will encounter divergences when evaluating this transition amplitude. In order to avoid these divergences, I shall normal order the operators in the matrix element in the transition amplitude $A_n(E)$ with respect to the Minkowski vacuum [20]. In other words, rather than the expression (26), I shall assume that the transition amplitude is instead given by

$$\bar{A}_n(E) = \int_{-\infty}^{\infty} {\rm d}\tau\, {\rm e}^{i\,E\,\tau}\, \langle\psi|\, :\hat{\phi}^n[\tilde{x}(\tau)]:\, |0\rangle\,, \tag{27}$$

where the colons denote normal ordering with respect to the Minkowski vacuum. Then, the transition probability of the detector to all possible final states $|\psi\rangle$ of the quantum field can be written as

$$P_n(E) = \sum_{|\psi\rangle} |\bar{A}_n(E)|^2 = \int_{-\infty}^{\infty} {\rm d}\tau \int_{-\infty}^{\infty} {\rm d}\tau'\, {\rm e}^{-i\,E\,(\tau-\tau')}\, G_n\left[\tilde{x}(\tau), \tilde{x}(\tau')\right], \tag{28}$$

where $G_n\left[\tilde{x}(\tau), \tilde{x}(\tau')\right]$ is the $(2\,n)$-point function defined as

$$G_n\left[\tilde{x}(\tau), \tilde{x}(\tau')\right] = \langle 0|:\hat{\phi}^n\left[\tilde{x}(\tau)\right]::\hat{\phi}^n\left[\tilde{x}(\tau')\right]:|0\rangle\,. \tag{29}$$

In situations where the $(2\,n)$-point function $G_n\left[\tilde{x}(\tau), \tilde{x}(\tau')\right]$ is invariant under time translations in the frame of the detector, I can define a transition probability rate for the detector as follows:

$$R_n(E) = \int_{-\infty}^{\infty} \mathrm{d}u \, \mathrm{e}^{-i E u} \, G_n(u), \tag{30}$$

where, as earlier, $u = \tau - \tau'$.

3.2 *Odd Statistics in Odd Dimensions for Odd Couplings*

Let me now assume that the quantum scalar field $\hat{\phi}$ is in the Minkowski vacuum. In this case, the $(2\,n)$-point function $G_n\left(\tilde{x}, \tilde{x}'\right)$ reduces to

$$G_n\left(\tilde{x}, \tilde{x}'\right) = n! \left[G^{+}\left(\tilde{x}, \tilde{x}'\right)\right]^n, \tag{31}$$

where $G^{+}\left(\tilde{x}, \tilde{x}'\right)$ denotes the Wightman function in the Minkowski vacuum.[2] The Wightman function (9) that I had quoted earlier had corresponded to the result in $(3+1)$-spacetime dimensions. In $(D+1)$ spacetime dimensions [and for $(D+1) \geq 3$], the Wightman function for a massless scalar field in the Minkowski vacuum is given by [14]

$$G^{+}(\tilde{x}, \tilde{x}') = \frac{C_D}{\left\{(-1)\,\left[(t - t' - i\,\varepsilon)^2 - |\boldsymbol{x} - \boldsymbol{x}'|^2\right]\right\}^{(D-1)/2}}, \tag{32}$$

where it should be evident that $\boldsymbol{x} \equiv \left(x^1, x^2, \ldots, x^D\right)$, while the quantity C_D is given by

$$C_D = \Gamma\left[(D-1)/2\right] / \left[4\,\pi^{(D+1)/2}\right] \tag{33}$$

with $\Gamma\left[(D-1)/2\right]$ denoting the Gamma function.

Now, the trajectory of a detector accelerating uniformly along the x^1 direction with a proper acceleration g is given by

$$\tilde{x}(\tau) = g^{-1}\,\left[\sinh\left(g\,\tau\right),\ \cosh\left(g\,\tau\right),\ 0,\ 0, \ldots,\ 0\right], \tag{34}$$

where τ is the proper time in the frame of the detector. On substituting this trajectory in the Minkowski Wightman function (32), I obtain that [14]

---

[2] I should stress here that I would have arrived at the expression (31) for the $(2\,n)$-point function in the Minkowski vacuum even if I had started with the transition amplitude (26) [instead of the normal ordered amplitude (27)], expressed the resulting $(2\,n)$-point function in the transition probability in terms of the two-point functions using Wick's theorem and then replaced the divergent terms that arise (i.e. those two-point functions with coincident points) with the corresponding regularized expressions [20, 23].

$$G^{+}(u) = \frac{C_D\,(g/2\,i)^{(D-1)}}{\left\{\sinh\left[(g\,u/2) - i\,\varepsilon\right]\right\}^{(D-1)}}. \tag{35}$$

Therefore, along the trajectory of the uniformly accelerating detector, the $(2\,n)$-point function in the Minkowski vacuum (31) is given by

$$G_n(u) = \frac{n!\,C_D^n\,(g/2\,i)^p}{\left\{\sinh\left[(g\,u/2) - i\,\varepsilon\right]\right\}^{p}}, \tag{36}$$

where $p = (D-1)\,n$.

Upon substituting the $(2\,n)$-point function (36) in the expression (30) and carrying out the resulting integral [36], I find that the transition probability rate of the uniformly accelerated, non-linearly coupled detector can be written as [23]

$$R_n(E) = B(n,D)\begin{cases} (g^p/E)\,\underbrace{\dfrac{1}{\exp(2\,\pi\,E/g) - 1}}_{\text{Bose-Einstein factor}}\,\displaystyle\prod_{l=0}^{(p-2)/2}\left[l^2 + (E/g)^2\right] & \text{when } p \text{ is even,}\\ g^{p-1}\,\underbrace{\dfrac{1}{\exp(2\,\pi\,E/g) + 1}}_{\text{Fermi-Dirac factor}}\,\displaystyle\prod_{l=0}^{(p-3)/2}\left\{\left[(2\,l+1)/2\right]^2 + (E/g)^2\right\} & \text{when } p \text{ is odd,}\end{cases} \tag{37}$$

where the quantity $B(n,D)$ is given by

$$B(n,D) = 2\,\pi\,n!\;C_D^n/\Gamma(p). \tag{38}$$

When $(D+1)$ is even, p is even for all n and, hence, a Bose–Einstein factor will always arise in the response of the uniformly accelerated detector in an even-dimensional flat spacetime. Whereas, when $(D+1)$ is odd, evidently, p will be odd or even depending on whether n is odd or even. Therefore, in an odd-dimensional flat spacetime, a Fermi–Dirac factor will arise in the detector response when n is odd (as in the case of the Unruh–DeWitt detector), but a Bose–Einstein factor will appear when n is even!

Let me make three clarifying comments regarding the curious result I have obtained above. To begin with, the temperature associated with the Bose–Einstein and the Fermi–Dirac factors that appear in the response of the non-linearly coupled detector is the standard Unruh temperature, *viz.* $g/(2\,\pi)$. Moreover, the response of the detector is characterized *completely* by either a Bose–Einstein or a Fermi–Dirac distribution *only* in situations wherein $p < 3$. When $p \geq 3$, apart from a Bose–Einstein or a Fermi–Dirac factor, the detector response contains a term which is polynomial in E/g. Lastly, plots of the transition probability rate of the detector suggest that, though the characteristic response of the detector alternates between the Bose–Einstein and the Fermi–Dirac factors as we go from one D to another for odd n [or from one n to another when $(D+1)$ is odd], the complete spectra themselves exhibit a smooth dependence on the index of non-linearity of the coupling as well as the dimension of spacetime (in this context, see the figures in Ref. [23]).

3.3 Nature of the Odd Statistics

Despite its interesting character, the 'inversion of statistics' encountered in the response of the detector in odd dimensions for odd couplings seems to be only *apparent*. It is well known that, in the frame of the uniformly accelerating detector, the Wightman function in the Minkowski vacuum (35) is skew-periodic in imaginary proper time with a period corresponding to the inverse of the Unruh temperature [37], i.e.

$$G^{+}(u) = G^{+}\left[-u + (2\,\pi\, i/g)\right]. \tag{39}$$

This property is known as the Kubo-Martin-Schwinger (KMS) condition, as is applicable to scalar fields. Note that the above property is, in fact, satisfied by the Minkowski Wightman function in *all* dimensions [14]. Since the $(2\,n)$-point function in the Minkowski vacuum is proportional to the nth power of the Wightman function, obviously, in the frame of the accelerated detector, the $(2\,n)$-point function will also be skew-periodic in imaginary proper time for all n and D [cf. Eq. (36)]. In other words, the $(2\,n)$-point function satisfies the KMS condition (as is required for a scalar field) for *all* D and n. This implies that the appearance of the Fermi–Dirac factor (instead of the expected Bose–Einstein factor) for odd $(D+1)$ and n simply reflects a peculiar aspect of the detector rather than indicate a fundamental shift in the field theory in such situations [14, 23, 24].

4 Detecting Planck Scale Effects

Consider a typical mode that constitutes Hawking radiation at future null infinity around a collapsing black hole. As one traces such a mode back to the past null infinity where the initial conditions are imposed on the quantum field, it is found that the energy of the mode turns out to be way beyond the Planck scale [38]. (This feature seems to have been originally noticed in Ref. [39]; in this context, also see Ref. [40].) In fact, due to the rapid, virtually exponential expansion, a similar phenomenon is encountered in the context of the inflationary scenario. One finds that scales of cosmological interest can be comparable to the Planck scale at very early times when the initial conditions are imposed during inflation [41]. While the possible Planck scale corrections to Hawking radiation and the perturbations generated during inflation have cornered most of the attention [38, 41], the Planck scale effects on a variety of non-perturbative, quantum field theoretic effects in flat as well as curved spacetimes have been investigated as well (see, for example, Refs. [42–44]). In the absence of a viable quantum theory of gravity, it becomes imperative to extend such phenomenological analyses to as many physical situations as possible (in this context, see Ref. [45], and references therein).

The Unruh effect has certain similarities with Hawking radiation from black holes. Due to this reason, the Unruh effect and its variants provide another interesting domain to study the quantum gravitational effects [28]. But, due to the lack of a workable quantum theory of gravity, to investigate the Planck scale effects, one is forced to consider phenomenological models constructed by hand. These models attempt to capture one or more features expected of the actual effective theory obtained by integrating out the gravitational degrees of freedom. The approach based on modified dispersion relations has been extensively considered both in the context of black holes and inflationary cosmology. In this approach, a fundamental scale is effectively introduced into the theory by breaking local Lorentz invariance (see, for instance, Refs. [43, 46]). It should be clarified that there does not exist any experimental or observational reason to believe that Lorentz invariance could be violated at high energies. Nevertheless, theoretically, these models prove to be attractive because of the fact that they permit quantum field theories to be constructed and calculations to be carried out in a consistent fashion.

In this section, I shall adopt the approach due to the modified dispersion relations to analyze the Planck scale corrections to the response of the rotating Unruh–DeWitt detector in flat spacetime. As I shall show, the rotating trajectory turns out to be a special case wherein the transition probability rate of the rotating detector can be defined in precisely the same fashion as I had done earlier in the case of the canonical scalar field governed by the linear dispersion relation. I shall illustrate that the response of the rotating detector can be computed *exactly*, although, numerically, even when the field it is coupled to is described by a non-linear dispersion relation.

4.1 Scalar Field Governed by a Modified Dispersion Relation

I shall be interested in calculating the response of the rotating detector when it is coupled to a massless scalar field that is governed by a modified dispersion relation of the following form:

$$\omega = k\left[1 + a\,\left(k/k_{\mathrm{P}}\right)^2\right]^{1/2}. \tag{40}$$

The quantity ω is the frequency corresponding to the mode $\boldsymbol{k}$, $k = |\boldsymbol{k}|$ and k_{P} denotes the fundamental scale (that I shall assume to be of the order of the Planck scale) at which the deviations from the linear dispersion relation become important. Note that a is a dimensionless constant whose magnitude is of order unity, and the above dispersion relation is super-luminal or sub-luminal depending upon whether a is positive or negative. Clearly, if I can evaluate the Wightman function associated with the quantized scalar field described by the non-linear dispersion relation (40), I may then be able to evaluate the corresponding transition probability rate of the rotating detector as I had carried out originally. However, unlike the standard case, it turns out to be difficult to even arrive at an analytical expression for the Wightman function of such a scalar field. Therefore, I shall make use of the second method that I had adopted earlier to evaluate the response of the rotating detector—I shall first integrate over the differential proper time and then numerically sum over the normal modes to arrive at the transition probability rate.

The equation of motion of the scalar field ϕ that is described by the dispersion relation (40) is given by

$$\Box\phi + \frac{a}{k_{\rm P}^2}\,\nabla^2\left(\nabla^2\phi\right) = 0, \tag{41}$$

where $\Box$ is the d'Alembertian corresponding to the four dimensional Minkowski spacetime, while ∇^2 is the three dimensional, spatial Laplacian. Evidently, the first term in the above equation is the standard one. The non-linear term in the dispersion relation is responsible for the second term. Such terms can be generated by adding suitable terms to the original action describing the scalar field [43, 46]. While these additional terms preserve rotational invariance, they break Lorentz invariance. In fact, this property is common to all the theories that are described by a non-linear dispersion relation. It is obvious that the normal modes of such a scalar field in flat spacetime remain plane waves as in the standard case, but with the frequency and the wavenumber related by the modified dispersion relation. Moreover, the quantization of the scalar field can be carried out in the same fashion. It is straightforward to show that, in the Minkowski vacuum, the Wightman function for any such field in $(3+1)$-spacetime dimensions can be expressed as (see, for example, Ref. [43])

$$G_{\rm M}^{+}(\tilde{x}, \tilde{x}') = \int \frac{{\rm d}^3\boldsymbol{k}}{(2\,\pi)^3\,2\,\omega}\,{\rm e}^{-i\,\omega\,(t-t')}\,{\rm e}^{i\boldsymbol{k}\cdot(\boldsymbol{x}-\boldsymbol{x}')} \tag{42}$$

with ω being related to $k = |\boldsymbol{k}|$ by the given non-linear dispersion relation.

4.2 *Response of the Rotating Detector*

For a scalar field governed by a modified dispersion relation, using the expression (42) for the corresponding Wightman function, one can immediately show that, along the rotating trajectory, the function can be expressed exactly as in Eq. (16), with the frequency ω being related to the wavenumbers q and k_z by the non-linear dispersion relation. Clearly, in such a case, the transition probability rate of the detector will again be given by Eq. (19) with ω suitably defined. It is important to recognize that the result is actually applicable for *any* non-linear dispersion relation [27].

Let me now evaluate the response of the rotating detector for the dispersion relation (40). In such a case, ω is related to the wavenumbers q and k_z as follows:

$$\omega = \left(q^2 + k_z^2\right)^{1/2} \left[1 + \frac{a}{k_{\rm P}^2}\, \left(q^2 + k_z^2\right)\right]^{1/2}. \tag{43}$$

Also, one can show that the roots κ_z [from Eq. (20)] are given by

$$\kappa_z^2 = \pm \frac{k_{\rm P}^2}{2\,a} \left(1 + \frac{4\,a\,\lambda^2}{k_{\rm P}^2}\right)^{1/2} - \frac{k_{\rm P}^2}{2\,a} - q^2, \tag{44}$$

with λ defined as in Eq. (22). It ought to be noted that κ_z^2 has to be positive definite, since κ_z is a real quantity.

Let me first consider the super-luminal case when a is positive. When, say, $a = 1$, the two roots that contribute to the delta function in Eq. (19) can be written as

$$\kappa_z = \pm \left(\lambda_+^2 - q^2\right)^{1/2}, \tag{45}$$

where λ_+^2 is given by the expression

$$\begin{aligned} \lambda_+^2 &= \frac{k_{\rm P}^2}{2} \left[\left(1 + \frac{4\,\lambda^2}{k_{\rm P}^2}\right)^{1/2} - 1\right] \\ &= \frac{\bar{\lambda}_+^2}{\sigma^2} = \frac{\bar{k}_{\rm P}^2}{2\,\sigma^2} \left[\left(1 + \frac{4\,(\sigma\,\Omega\,\bar{\lambda})^2}{\bar{k}_{\rm P}^2}\right)^{1/2} - 1\right]. \end{aligned} \tag{46}$$

Note that $\bar{k}_{\rm P} = \sigma\, k_{\rm P}$ denotes the dimensionless fundamental scale and the sub-script in λ_+ refers to the fact that I am considering a super-luminal dispersion relation. Further, as κ_z is real, I require that $q \le \lambda_+$. As in the standard case, the positive and negative roots of κ_z above contribute equally. Therefore, the response of the rotating detector is given by

$$\bar{R}(\bar{E}) = \sigma\, R(\bar{E}) = \frac{\sigma}{2\,\pi\,\gamma} \sum_{m \ge \bar{E}}^{\infty} \left(1 + \frac{2\,\lambda_+^2}{k_{\rm P}^2}\right)^{-1} \int_0^{\lambda_+} {\rm d}q\, q \left[\frac{J_m^2(q\,\sigma)}{\left(\lambda_+^2 - q^2\right)^{1/2}}\right], \tag{47}$$

and the integral over q can be carried out as in the standard case to arrive at the result

$$\begin{aligned} \bar{R}(\bar{E}) = \frac{1}{2\,\pi\,\gamma} \sum_{m \ge \bar{E}}^{\infty} \frac{\bar{\lambda}_+^{(2\,m+1)}}{\Gamma\,(2\,m+2)} \left(1 + \frac{2\,\bar{\lambda}_+^2}{\bar{k}_{\rm P}^2}\right)^{-1} \\ \times {}_1F_2\left[m + (1/2);\; m + (3/2), 2\,m + 1;\; -\bar{\lambda}_+^2\right]. \end{aligned} \tag{48}$$

It should be emphasized here that this result for the transition probability rate is exact and no approximations have been made in arriving at the expression.

Since the Planck scale is expected to be orders of magnitude beyond the scales probed by experiments, the quantity $\bar{k}_{\rm P}$ is expected to be large. It is clear that, as $\bar{k}_{\rm P} \to \infty$, $\bar{\lambda}_+ \to \sigma\, \Omega\, \bar{\lambda}$ and, hence, the transition transition probability rate (48) reduces to the expression that I had arrived at earlier for the standard dispersion relation [*viz.* Eq. (24)], as required. Let me now evaluate the Planck scale corrections to the standard result by expanding the transition probability rate (48) in terms of $\lambda/k_{\rm P}$ and retaining terms upto $\mathcal{O}[(\lambda/k_{\rm P})^2]$. Note that, in such a case, λ_+ reduces to

$$\lambda_+ \simeq \lambda \left(1 - \frac{\lambda^2}{2\, k_{\rm P}^2}\right), \tag{49}$$

so that I have

$$\lambda_+^{(2\,m+1)} \simeq \lambda^{(2\,m+1)} - (2\,m+1)\, \frac{\lambda^{(2\,m+3)}}{2\, k_{\rm P}^2} \tag{50}$$

and

$$\left(1 + \frac{2\,\lambda_+^2}{k_{\rm P}^2}\right)^{-1} \simeq 1 - \frac{2\,\lambda^2}{k_{\rm P}^2}. \tag{51}$$

Moreover, in the limit of our interest, the hypergeometric function in Eq. (48) can be written as

$$\begin{aligned}
&{}_1F_2\left[m + (1/2);\; m + (3/2),\; 2\,m+1;\; -\bar{\lambda}_+^2\right] \\
&\quad \simeq {}_1F_2\left[m + (1/2);\; m + (3/2),\; 2\,m+1;\; -\left(\sigma\,\Omega\,\bar{\lambda}\right)^2\right] \\
&\qquad + \frac{(\sigma\,\Omega\,\bar{\lambda})^2}{\bar{k}_{\rm P}^2}\, \frac{[m + (1/2)]\,(\sigma\,\Omega\,\bar{\lambda})^2}{[m + (3/2)]\,(2\,m+1)} \\
&\qquad \times {}_1F_2\left[m + (3/2);\; m + (5/2),\; 2\,m+2;\; -\left(\sigma\,\Omega\,\bar{\lambda}\right)^2\right].
\end{aligned} \tag{52}$$

Upon using the above expansions, I obtain the response of the detector at $\mathcal{O}[(\lambda/k_{\rm P})^2]$ to be

$$\begin{aligned}
\bar{R}(\bar{E}) \simeq\; & \frac{1}{2\,\pi\,\gamma} \sum_{m \geq \bar{E}}^{\infty} \frac{\left(\sigma\,\Omega\,\bar{\lambda}\right)^{(2\,m+1)}}{\Gamma\,(2\,m+2)} \\
& \times {}_1F_2\left[m + (1/2);\; m + (3/2),\; (2\,m+1);\; -\left(\sigma\,\Omega\,\bar{\lambda}\right)^2\right] \\
& - \frac{1}{2\,\pi\,\gamma}\, \frac{(\sigma\,\Omega\,\bar{\lambda})^2}{\bar{k}_{\rm P}^2} \sum_{m \geq \bar{\mathscr{E}}}^{\infty} \frac{[m + (5/2)]\,\left(\sigma\,\Omega\,\bar{\lambda}\right)^{(2\,m+1)}}{\Gamma\,(2\,m+2)}
\end{aligned}$$

$$\times\, {}_1F_2\left[m+(1/2);\; m+(3/2),\, 2\,m+1;\; -\left(\sigma\,\Omega\,\bar{\lambda}\right)^2\right]$$
$$+\frac{1}{2\,\pi\,\gamma}\,\frac{(\sigma\,\Omega\,\bar{\lambda})^2}{\bar{k}_{\mathrm{P}}^2}\sum_{m\geq\bar{E}}^{\infty}\frac{[m+(1/2)]\,\left(\sigma\,\Omega\,\bar{\lambda}\right)^{(2\,m+3)}}{[m+(3/2)]\,(2\,m+1)\,\Gamma\,(2\,m+2)}$$
$$\times\, {}_1F_2\left[m+(3/2);\; m+(5/2),\, 2\,m+2;\; -\left(\sigma\,\Omega\,\bar{\lambda}\right)^2\right]. \tag{53}$$

Evidently, the first term in this expression corresponds to the conventional transition probability rate [cf. Eq. (24)], while the other two terms represent the leading corrections to the standard result.

Let me now turn to considering the sub-luminal dispersion relation. When a is negative, say, $a = -1$, the roots κ_z are given by

$$\kappa_z = \pm\left(\lambda_-^2 - q^2\right)^{1/2} \tag{54}$$

with λ_-^2 defined as

$$(\lambda_-^{\pm})^2 = \frac{k_{\mathrm{P}}^2}{2}\left[1\pm\left(1-\frac{4\,\lambda^2}{k_{\mathrm{P}}^2}\right)^{1/2}\right]$$
$$= \frac{(\bar{\lambda}_-^{\pm})^2}{\sigma^2} = \frac{\bar{k}_{\mathrm{P}}^2}{2\,\sigma^2}\left\{1\pm\left[1-\frac{4\,(\sigma\,\Omega\,\bar{\lambda})^2}{\bar{k}_{\mathrm{P}}^2}\right]^{1/2}\right\}, \tag{55}$$

where the minus sign in the sub-script represents that it corresponds to the sub-luminal case (i.e. when a is negative), while the super-scripts denote the two different possibilities of λ_-. Just as in the super-luminal case (i.e. when $a = 1$), I require $q \leq \lambda_-^{\pm}$, if κ_z is to remain real. Moreover, note that, unlike the super-luminal case, there also arises an upper limit on the sum over m. I require that $\lambda \leq k_{\mathrm{P}}/2$, in order to ensure that $\lambda_-^{\pm}$ is real. This corresponds to $m \leq \bar{E} + \bar{k}_{\mathrm{P}}/(2\,\sigma\,\Omega)$. Therefore, for the sub-luminal dispersion relation, I find that I can write the response of the rotating detector as follows:

$$\bar{R}(\bar{E}) = \frac{1}{2\,\pi\,\gamma}\sum_{m\geq\bar{E}}^{\bar{E}+\bar{k}_{\mathrm{P}}/(2\,\sigma\,\Omega)}\frac{\left(\bar{\lambda}_-^{-}\right)^{(2\,m+1)}}{\Gamma(2\,m+2)}\left(\left|1-\frac{2\,(\bar{\lambda}_-^{-})^2}{\bar{k}_{\mathrm{P}}^2}\right|\right)^{-1}$$
$$\times\, {}_1F_2\left[m+(1/2);\; m+(3/2),\, 2\,m+1;\; -\left(\bar{\lambda}_-^{-}\right)^2\right]$$
$$+\frac{1}{2\,\pi\,\gamma}\sum_{m\geq\bar{E}}^{\bar{E}+k_{\mathrm{P}}/(2\,\sigma\,\Omega)}\frac{\left(\bar{\lambda}_-^{+}\right)^{2\,m+1}}{\Gamma\,(2\,m+2)}\left(\left|1-\frac{2\,(\bar{\lambda}_-^{+})^2}{\bar{k}_{\mathrm{P}}^2}\right|\right)^{-1}$$
$$\times\, {}_1F_2\left[m+(1/2);\; m+(3/2),\, 2\,m+1;\; -\left(\bar{\lambda}_-^{+}\right)^2\right]. \tag{56}$$

The reason for the upper limit on m as well as the origin of the second term in the above expression for the response of the rotating detector can be easily understood. The quantity ω is a monotonically increasing function of q and k_z in the case of the super-luminal dispersion relation. Because of this reason, there exist only two real roots of k_z corresponding to a given ω. Moreover, ω^2 remains positive definite for all the modes. In contrast, in the sub-luminal case, after a rise, ω begins to decrease for sufficiently large values of q and k_z. Actually, ω^2 even turns negative at a suitably large value [43]. It is this feature of the sub-luminal dispersion relation which leads to the upper limit on m, and the limit ensures that we avoid complex frequencies. (Such a cut-off can be achieved if I assume that, say, the detector is not coupled to modes with m beyond a certain value, when the frequency turns complex.) There arise two additional two roots of k_z which contribute to the detector response in the sub-luminal case as a result of the decreasing ω at large q and k_z. The second term in the above transition probability rate of the rotating detector corresponds to the contributions from these two extra roots.

If one plots the result (48) for the response of the rotating detector when it is coupled to a field that is governed by a super-luminal dispersion relation, one finds

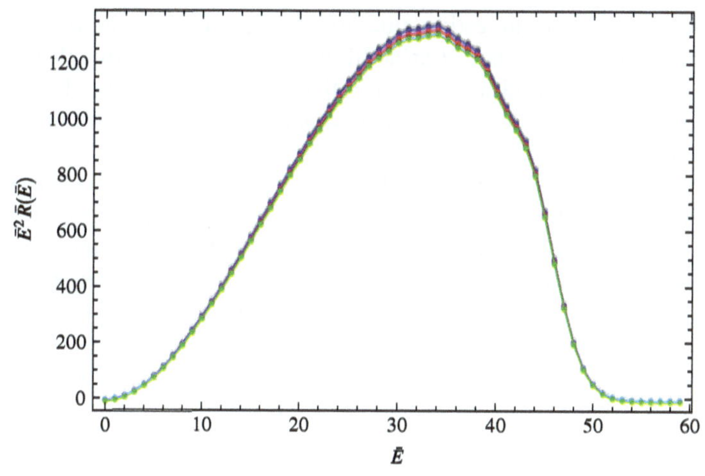

Fig. 2 The transition probability rate of the rotating Unruh–DeWitt detector that is coupled to a massless scalar field governed by the modified dispersion relation (40) with $a = -1$. The *dots* and the *curves* linking them denote the numerical results for the same set of values for the quantity $\sigma\,\Omega$ (and the same choice of colors) that I had *plotted* in the previous figure. I have set $\bar{k}_{\rm p} = 50$, which is an extremely small value for $\bar{k}_{\rm p}$. Evidently, for such a value, the modifications to the standard result (cf. Fig. 1) due to the sub-luminal dispersion relation is considerable. In fact, more realistic values of $\bar{k}_{\rm p}$ would correspond to, say, $\bar{k}_{\rm p}/\bar{E} > 10^{10}$. However, numerically, it turns out to be difficult to sum the contributions in the expression (56) up to such large values of $\bar{k}_{\rm p}$. It seems reasonable to conclude that the modifications to standard result due to the sub-luminal dispersion relation can be expected to be much smaller if one assumes $\bar{k}_{\rm p}$ to be sufficiently large. Nevertheless, my analysis unambiguously points to the fact that, as is known to occur in other situations, a sub-luminal dispersion relation modifies the standard result considerably more than a similar super-luminal dispersion relation

that it does not differ from the standard result (as plotted in Fig. 1) even for an unnaturally small value of $\bar{k}_{\rm P}$ such that, say, $\bar{k}_{\rm P}/\bar{E} \simeq 10$. This implies that super-luminal dispersion relations do not alter the conventional result to any extent. It needs to be emphasized here that similar conclusions have been arrived at earlier in the context of black holes as well as inflationary cosmology. In these contexts, it has been shown that Hawking radiation and the inflationary perturbation spectra remain unaffected due to super-luminal modifications to the conventional, linear, dispersion relation [38, 41]. In Fig. 2, I have plotted the transition probability rate (56) of the rotating Unruh–DeWitt detector corresponding to the sub-luminal dispersion relation that I have considered. I have plotted the result for a rather small value of $\bar{k}_{\rm P} = 50$. It is clear from the figure that the sub-luminal dispersion relation can lead to substantial modifications to the standard result. I believe that the modifications from the standard result will be considerably smaller (than exhibited in the figure) for much larger and more realistic values of $\bar{k}_{\rm P}$ such that, say, $\bar{k}_{\rm P}/\bar{E} > 10^{10}$.

4.3 Rotating Detector in the Presence of a Boundary

I shall now consider an interesting situation wherein I study the response of the rotating detector in the presence of an additional boundary condition that is imposed on the scalar field on a cylindrical surface in flat spacetime. Because of the symmetry of the problem, in this case too, the cylindrical coordinates turn out to be more convenient to work with.

It is well known that the time-like Killing vector associated with an observer who is rotating at an angular velocity Ω in flat spacetime becomes space-like for radii greater than $\rho_{\rm SL} = 1/\Omega$. Due to this reason, it has been argued that one needs to impose a boundary condition on the quantum field at a radius $\rho < \rho_{\rm SL}$ when evaluating the response of a rotating detector [19]. Curiously, in the presence of such a boundary, it was found that a rotating Unruh–DeWitt detector which is coupled to the standard scalar field ceases to respond. It is then interesting to examine whether this result holds true even when one assumes that the scalar field is governed by a modified dispersion relation.

In the cylindrical coordinates, along the rotating trajectory (15), the Wightman function corresponding to a scalar field that is assumed to vanish at, say, $\rho = \rho_* \ (< \rho_{\rm SL})$, can be expressed as a sum over the normal modes of the field as follows [19]:

$$G^+(u) = \sum_{m=-\infty}^{\infty} \sum_{n=1}^{\infty} \int_{-\infty}^{\infty} \frac{{\rm d}k_z}{(2\,\pi)^2\, 2\,\omega}\, [N\, J_m(\xi_{mn}\, \sigma/\rho_*)]^2\, e^{-i\,\gamma\,(\omega - m\,\Omega)\,u}, \tag{57}$$

where ξ_{mn} denotes the nth zero of the Bessel function $J_m(\xi_{mn}\, \sigma/\rho_*)$, while N is a normalization constant that is given by

$$N = \frac{\sqrt{2}}{\rho_* \, |J_{m+1}(\xi_{mn})|}. \tag{58}$$

As in the situation without a boundary, m is a real integer, whereas k_z is a continuous real number. But, due to the imposition of the boundary condition at $\rho = \rho_*$, the spectrum of the radial modes is now discrete, and is described by the positive integer n. It should be pointed out that the expression (57) is in fact valid for any dispersion relation, with ω suitably related to the quantities ξ_{mn} and k_z. For instance, in the case of the modified dispersion relation (40), the quantity ω is given by

$$\omega = \left(\frac{\xi_{mn}^2}{\rho_*^2} + k_z^2\right)^{1/2} \left[1 + \frac{a}{k_{\rm P}^2}\left(\frac{\xi_{mn}^2}{\rho_*^2} + k_z^2\right)\right]^{1/2}, \tag{59}$$

where, it is evident that, while the overall factor corresponds to the standard, linear, dispersion relation, the term involving a within the brackets arises due to the modifications to it. Since the Wightman function depends only u, the transition probability rate of the detector simplifies to

$$R(E) = \sum_{m=-\infty}^{\infty} \sum_{n=1}^{\infty} \int_{-\infty}^{\infty} \frac{{\rm d}k_z}{2\pi \; 2\omega} \; [N \, J_m(\xi_{mn}\, \sigma/\rho_*)]^2 \; \delta^{(1)} \left[E + \gamma \; (\omega - m\, \Omega)\right]. \tag{60}$$

For exactly the same reasons that I had presented in the last section, the delta function in this expression can be non-zero only when $m > 0$. In fact, the detector will respond only under the condition

$$m\, \Omega > \frac{\xi_{m1}}{\rho_*} \left(1 + \frac{a}{k_{\rm P}^2} \frac{\xi_{m1}^2}{\rho_*^2}\right)^{1/2}, \tag{61}$$

where the right hand side is the lowest possible value of ω corresponding to $n = 1$ and $k_z = 0$. However, from the properties of the Bessel function, it is known that $\xi_{mn} > m$, for all m and n (see, for instance, Ref. [47]). Therefore, *when a is positive*, $\Omega\, \rho_*$ has to be greater than unity, if the rotating detector has to respond. But, this is not possible since I have assumed that the boundary at ρ_* is located *inside* the static limit $\rho_{\rm SL} = 1/\Omega$. This is exactly the same conclusion that one arrives at in the standard case [19, 30].

Actually, it is easy to argue that the above conclusion would apply for all super-luminal dispersion relations. But, it seems that, under the same conditions, the rotating detector would be excited by a certain range of modes if I consider the scalar field to be described by a sub-luminal (such as, when $a < 0$) dispersion relation! In fact, this aspect is rather easy to understand. Consider a frequency, say, ω, associated with a mode through the linear dispersion relation. Evidently, a super-luminal dispersion relation raises the energy of all the modes, while the sub-luminal dispersion relation lowers it. Therefore, if the interaction of the detector with a standard field does not excite a particular mode of the quantum field, clearly, the mode is unlikely to be excited if its energy has been raised further, as in a super-luminal dispersion relation. However, the motion of the detector mode may be able to excite a mode of the field, if the energy of certain modes are lowered when compared to the standard case, as the sub-luminal dispersion relation does.

5 Finite Time Detectors

The response of detectors have always been studied for their entire history, *viz.* from the infinite past to the infinite future in the detector's proper time. But, in any realistic situation, the detectors can be kept switched on only for a finite period of time and due to this reason the study of the response of a detector for a finite interval in proper time becomes important. In this section, I shall illustrate that, unless the detectors are switched on smoothly, the response of the detector can contain divergent contributions [17, 18].

Consider a Unruh–DeWitt detector that has been switched on for a finite period of time with the aid of a window function, say, $W(\tau, T)$, where, as before, τ is the proper time in the frame of the detector, while T is the effective time for which the detector is turned on. The window function $W(\tau, T)$ can be expected to have the following properties:

$$W(\tau, T) \simeq \begin{cases} 1 \text{ for } |\tau| \ll T, \\ 0 \text{ for } |\tau| \gg T. \end{cases} \tag{62}$$

In such a case, instead of Eq. (6), the transition probability of the detector will be described by the integral

$$P(E, T) = \int_{-\infty}^{\infty} \mathrm{d}\tau \int_{-\infty}^{\infty} \mathrm{d}\tau'\, \mathrm{e}^{-iE\,(\tau-\tau')}\, W(\tau, T)\, W(\tau', T)\, G^{+}\left[\tilde{x}(\tau), \tilde{x}(\tau')\right]. \tag{63}$$

While abrupt switching corresponds to

$$W(\tau, T) = \Theta(T - \tau) + \Theta(T + \tau), \tag{64}$$

more gradual switching on and off can be achieved, for instance, with the aid of the window function

$$W(\tau, T) = \exp -\left(\frac{\tau^2}{2\,T^2}\right). \tag{65}$$

Consider a detector that is moving along the integral curve of a time-like Killing vector field so that $G^{+}\left[\tilde{x}(\tau), \tilde{x}(\tau')\right] = G^{+}(\tau - \tau')$. Let the detector be switched on and off with the aid of a smooth window function of the form $W(\tau/T)$. In such a situation, I can express the transition probability of the detector as

$$P(E,T) = \int_{-\infty}^{\infty} \mathrm{d}\tau \int_{-\infty}^{\infty} \mathrm{d}\tau'\, W(\tau,T)\, W(\tau',T)\, \mathrm{e}^{-iE\,(\tau-\tau')}\, G^{+}(\tau-\tau') \tag{66}$$

$$= W\left(i\,\frac{\partial}{\partial E}, T\right) W\left(-i\,\frac{\partial}{\partial E}, T\right) P(E), \tag{67}$$

where $P(E)$ is the original transition probability (6) for the case of the Unruh–DeWitt detector that has been kept on for its entire history. Let me now expand $W(\tau,T) = W(\tau/T)$ as a Taylor series around $\tau = 0$ and assume that $W(0) = 1$, $W'(0) = 0$, where the overprime denotes differentiation with respect to the argument τ/T. I can then write the window function as

$$\begin{aligned} W\left(\frac{\tau}{T}\right) &\simeq W(0) + W'(0)\left(\frac{\tau}{T}\right) + \frac{1}{2}\, W''(0)\left(\frac{\tau}{T}\right)^2 \\ &\simeq 1 + \frac{1}{2}\, W''(0)\left(\frac{\tau}{T}\right)^2, \end{aligned} \tag{68}$$

so that the transition probability becomes

$$\begin{aligned} P(E,T) &\simeq \left(1 - \frac{W''(0)}{2\,T^2}\,\frac{\partial^2}{\partial E^2}\right)^2 P(E) \\ &\simeq P(E) - \frac{W''(0)}{T^2}\,\frac{\partial^2 P(E)}{\partial E^2}. \end{aligned} \tag{69}$$

This gives the transition probability rate to be

$$R(E,T) = R(E) - \frac{W''(0)}{T^2}\,\frac{\partial^2 R(E)}{\partial E^2} + \mathcal{O}\left(\frac{1}{T^4}\right), \tag{70}$$

for any window function and trajectory. Note that the response at finite T depends on the derivatives of the window function, such as, for example, $W''(0)$. Hence, if the detector is switched on abruptly, these derivatives can diverge, thereby leading to divergent responses [17].

6 Summary

The concept of detectors was originally introduced to provide an operational definition to the concept of a particle. With this aim, the response of detectors have been studied in the literature in a wide variety of situations. In this article, I have described a few different aspects of detectors. I have highlighted the point that, while the detectors are sensitive to the phenomenon of particle production, their response do not, in general, reflect the particle content of the field. I have shown that, in odd spacetime dimensions, the response of a detector that is coupled to an odd power of the scalar field exhibits a Fermi–Dirac distribution rather than the expected Bose–Einstein distribution. I have also discussed the response of a rotating detector that is coupled to a scalar field governed by modified dispersion relations, supposedly arising due to quantum gravitational effects. I have illustrated that, as it has been encountered in other similar contexts, while super-luminal dispersion relations hardly affect the response of the detector, sub-luminal relations substantially modify the response. Finally, I have argued that detectors which are switched on abruptly can exhibit responses which contain divergences.

Acknowledgements It is a pleasure to contribute an article to the volume being put together to celebrate the sixtieth birthday of Prof. T. Padmanabhan, or Paddy as he is affectionately known. With undiminished energy and enthusiasm, Paddy continues to be an inspiration for many of us. I can not thank him adequately enough for the years of constant friendship, support and guidance.

References

1. W. Greiner, B. Müller, J. Rafelski, *Quantum Electrodynamics of Strong Fields* (Springer, Berlin, 1985); E.S. Fradkin, D.M. Gitman, S.M. Shvartsman, *Quantum Electrodynamics with Unstable Vacuum* (Springer, Berlin, 1991); V.M. Mostepanenko, N.N. Trunov, *The Casimir Effect and its Applications* (Clarendon Press, Oxford, 1997)
2. N.D. Birrell, P.C.W. Davies, *Quantum Fields in Curved Space* (Cambridge University Press, Cambridge, 1982); S.A. Fulling, *Aspects of Quantum Field Theory in Curved Spacetime* (Cambridge University Press, Cambridge, 1989); R.M. Wald, *Quantum Field Theory in Curved Spacetime and Black Hole Thermodynamics* (The University of Chicago Press, Chicago, 1994); V.F. Mukhanov, S. Winitzki, *Introduction to Quantum Effects in Gravity* (Cambridge University Press, England, 2007); L. Parker, D. J. Toms, *Quantum Field Theory in Curved Spacetime: Quantized Fields and Gravity* (Cambridge University Press, England, 2009)
3. B.S. DeWitt, Phys. Rep. **19**, 297 (1975); L. Parker, The production of elementary particles by strong gravitational fields, in *Asymptotic Structure of Spacetime*, ed. by F.P. Eposito, L. Witten (Plenum, New York, 1977); T. Padmanabhan, Pramana–J. Phys. **37**, 179 (1991); L.H. Ford, Quantum field theory in curved spacetime, in *Proceedings of the IX Jorge Andre Swieca Summer School, Campos dos Jordao, Sao Paulo, Brazil, 1997*. arXiv:gr-qc/9707062
4. H.B.G. Casimir, Proc. Kon. Ned. Akad. Wet. **51**, 793 (1948)
5. W. Heisenberg, H. Euler, Z. Phys. **98**, 714 (1936)
6. J. Schwinger, Phys. Rev. **82**, 664 (1951)
7. S.W. Hawking, Commun. Math. Phys. **43**, 199 (1975)
8. S.A. Fulling, Phys. Rev. D **7**(1973), 2850 (1973)
9. W.G. Unruh, Phys. Rev. D **14**, 870 (1976)
10. B.S. DeWitt, Quantum gravity: the new synthesis, in *General Relativity: An Einstein Centenary Survey*, ed. by S.W. Hawking, W. Israel (Cambridge University Press, Cambridge, 1979)
11. P.C.W. Davies, Particles do not exist, in *Quantum Theory of Gravity*, ed. by S.M. Christensen (Hilger, Bristol, 1984)
12. J.R. Letaw, Phys. Rev. D **23**, 1709 (1981)
13. T. Padmanabhan, Astrophys. Space Sci. **83**, 247 (1982); P.G. Grove, A.C. Ottewill, J. Phys. A: Math. Gen. **16**, 3905 (1983); J.S. Bell, J.M. Leinaas, Nucl. Phys. B **212**, 131 (1983); P.G. Grove, Class. Quantum Grav. **3**, 793 (1986); J.S. Bell, J.M. Leinaas, Nucl. Phys. B **284**, 488 (1987); J.R. Anglin, Phys. Rev. D **47**, 4525 (1993); J.I. Korsbakken, J.M. Leinaas. Phys. Rev. D **74**, 084016 (2004)

14. S. Takagi, Prog. Theor. Phys. **72**, 505 (1984); S. Takagi, Prog. Theor. Phys. **74**, 142 (1985); S. Takagi, *ibid.* **74**, 501 (1985); C.R. Stephens, *Odd statistics in odd dimensions*, University of Maryland Report, 1985 (unpublished); C.R. Stephens, *On Some Aspects of the Relationship Between Quantum Physics, Gravity and Thermodynamics*, Ph.D. thesis, University of Maryland, 1986; S. Takagi, Prog. Theor. Phys. Suppl. **88**, 1 (1986); W.G. Unruh, Phys. Rev. D **34**, 1222 (1986); H. Ooguri, Phys. Rev. D **33**, 3573 (1986); H. Terashima, Phys. Rev. D **60**, 084001 (1999)
15. G. Lifschytz, M. Ortiz, Phys. Rev. D **49**, 1929 (1994); S. Deser, O. Levin, Class. Quantum Grav. **14**, L163 (1997); T. Jacobson, ibid. **15**, 251 (1998); S. Deser, O. Levin, ibid. **15**, L85 (1998); S. Deser, O. Levin, Phys. Rev. D **59**, 064004 (1999); S. Hyun, Y.S. Song, J.H. Yee, ibid. **51**, 1787 (1995); T. Murata, K. Tsunoda, K. Yamamoto. Int. J. Mod. Phys. A **16**, 2841 (2001)
16. K.J. Hinton, J. Phys. A: Math. Gen. **16**, 1937 (1983); K.J. Hinton, Class. Quantum Grav. **1**, 27 (1984); T. Padmanabhan, T.P. Singh, Class. Quantum Grav. **4**, 1397 (1987)
17. B.F. Svaiter, N.F. Svaiter, Phys. Rev. D **46**, 5267 (1992); A. Higuchi, G.E.A. Matsas, C.B. Peres. Phys. Rev. D **48**, 3731 (1993)
18. L. Sriramkumar, T. Padmanabhan, Class. Quantum Grav. **13**, 2061 (1996)
19. P.C.W. Davies, T. Dray, C.A. Manogue, Phys. Rev. D **53**, 4382 (1996)
20. N. Suzuki, Class. Quantum Grav. **14**, 3149 (1997)
21. L. Sriramkumar, Mod. Phys. Lett. A **14**, 1869 (1999)
22. L. Sriramkumar, T. Padmanabhan, Int. J. Mod. Phys. D **11**, 1 (2002)
23. L. Sriramkumar, Mod. Phys. Lett. A **17**, 1059 (2002)
24. L. Sriramkumar, Gen. Rel. Grav. **35**, 1699 (2003)
25. J. Louko, A. Satz, Class. Quantum Grav. **23**, 6321 (2006); A. Satz, *ibid.* **24**, 1719 (2007); J. Louko, A. Satz, *ibid.* **25**, 055012 (2008); L. Hodgkinson, *Particle Detectors in Curved Spacetime Quantum Field Theory*, Ph.D. thesis, University of Nottingham, 2013; L. Hodgkinson, J. Louko, A.C. Ottewill, Phys. Rev. D **89**, 104002 (2014); K.K. Ng, L. Hodgkinson, J. Louko, R.B. Mann, E. Martin-Martinez, *ibid.* **90**, 064003 (2014); J. Louko, JHEP **1409**, 142 (2014); C.J. Fewster, B.A. Juárez-Aubry, J. Louko, Class. Quantum Grav. **33**, 165003 (2016)
26. D. Kothawala, T. Padmanabhan, Phys. Lett. B **690**, 201 (2010); L.C. Barbado, M. Visser, Phys. Rev. D **86**, 084011 (2012)
27. S. Gutti, S. Kulkarni, L. Sriramkumar, Phys. Rev. D **83**, 064011 (2011)
28. I. Agullo, J. Navarro-Salas, G.J. Olmo, L. Parker, Phys. Rev. D **77**, 104034 (2008); *ibid.* **77**, 124032 (2008); M. Rinaldi, Phys. Rev. D **77**, 124029 (2008); D. Campo, N. Obadia, arXiv:1003.0112v1 [gr-qc]; G.M. Hossain, G. Sardar, arXiv:1411.1935 [gr-qc]; V. Husain, J. Louko, Phys. Rev. Lett. **116**, 061301 (2016); N. Kajuri, Class. Quantum Grav. **33**, 055007 (2016); G.M. Hossain, G. Sardar, arXiv:1606.01663 [gr-qc]
29. W. Rindler, Am. J. Phys. **34**, 1174 (1966)
30. L.C.B. Crispino, A. Higuchi, G.E.A. Matsas, Rev. Mod. Phys. **80**, 787 (2008)
31. A.P. Prudnikov, YuA Brychkov, O.I. Marichev, *Integrals and Series*, vol. 2 (Gordon and Breach Science Publishers, New York, 1986), p. 212
32. N.N. Bogolubov, Sov. Phys. JETP **7**, 51 (1958)
33. C.A. Manogue, Ann. Phys. (N.Y.) **181**, 261 (1988); A. Calogeracos, N. Dombey, Int. J. Mod. Phys. A **14**, 631 (1999)
34. J.R. Letaw, J.D. Pfautsch, Phys. Rev. D **24**, 1491 (1981)
35. P. Candelas, D.J. Raine, J. Math. Phys. **17**, 2101 (1976); S.A. Fulling, J. Phys. A: Math. Gen. **10**, 917 (1977); U.H. Gerlach, Phys. Rev. D **40**, 1037 (1989); S. Winters-Hilt, I.H. Redmount, L. Parker, Phys. Rev. D **60**, 124017 (1999)
36. I.S. Gradshteyn, I.M. Ryzhik, *Table of Integrals, Series and Products* (Academic Press, New York, 1980)
37. W. Troost, H. Van Dam, Phys. Lett. B **71**, 149 (1977); S.M. Christensen, M.J. Duff, Nucl. Phys. B **146**, 11 (1978); W. Troost, H. Van Dam, *ibid.* **152**, 442 (1979)

38. T. Jacobson, Phys. Rev. D **48**, 728 (1993); *ibid.* **53**, 7082 (1996); W.G. Unruh, *ibid.* **51**, 2827 (1995); R. Brout, S. Massar, R. Parentani, Ph. Spindel, *ibid.* **52**, 4559 (1995); N. Hambli, C.P. Burgess, *ibid.* **53**, 5717 (1996); S. Corley, T. Jacobson, *ibid.* **54**, 1568 (1996); T. Jacobson, Prog. Theor. Phys. Suppl. **136**, 1 (1999); R. Brout, C. Gabriel, M. Lubo, Ph. Spindel, Phys. Rev. D **59**, 044005 (1999); C. Barrabes, V. Frolov, R. Parentani, *ibid.* **59**, 124010 (1999); *ibid.* **62**, 044020 (2000); R. Parentani, *ibid.* **63** 041503 (2001); R. Casadio, P.H. Cox, B. Harms, O. Micu, *ibid.* **73**, 044019 (2006); I. Agullo, J. Navarro-Salas, G.J. Olmo, Phys. Rev. Lett. **97**, 041302 (2006); I. Agullo, J. Navarro-Salas, G.J. Olmo, L. Parker, Phys. Rev. D **76**, 044018 (2007); R. Schutzhold, W.G. Unruh, *ibid.* **78**, 041504 (2008); D.A. Kothawala, S. Shankaranarayanan, L. Sriramkumar, JHEP **0809**, 095 (2008)
39. R. Wald, Phys. Rev. D **13**, 3176 (1976)
40. R. Wald, *General Relativity* (University of Chicago Press, Chicago, 1984). Footnote on p. 406
41. R. Brandenberger, J. Martin, Mod. Phys. Lett. A **16**, 999 (2001); C.S. Chu, B.R. Greene, G. Shiu, *ibid.* **16**, 2231 (2001); J. Martin, R. Brandenberger, Phys. Rev. D **63**, 123501 (2001); J.C. Niemeyer, *ibid.* **63**, 123502 (2001); A. Kempf, J.C. Niemeyer, *ibid.* **64**, 103501 (2001); J.C. Niemeyer, R. Parentani, *ibid.* **64**, 101301 (2001); R. Easther, B.R. Greene, W.H. Kinney, G. Shiu, *ibid.* **64**, 103502 (2001); A.A. Starobinsky, Pisma Zh. Eksp. Teor. Fiz. **73**, 415 (2001); M. Lemoine, M. Lubo, J. Martin, J.P. Uzan, Phys. Rev. D **65**, 023510 (2002); J. Martin, R. Brandenberger, *ibid.* **65**, 103514 (2002); U.H. Danielsson, *ibid.* **66**, 023511 (2002); R. Brandenberger, P.M. Ho, *ibid.* **66**, 023517 (2002); R. Easther, B.R. Greene, W.H. Kinney, G. Shiu, *ibid.* **66**, 023518 (2002); N. Kaloper, M. Kleban, A.E. Lawrence, S. Shenker, *ibid.* **66**, 123510 (2002); F. Lizzi, G. Mangano, G. Miele, M. Peloso, JHEP **0206**, 049 (2002); U.H. Danielsson, *ibid.* **0212**, 025 (2002); L. Bergstrom, U.H. Danielsson, *ibid.* **0212**, 038 (2002); R. Brandenberger, J. Martin, Int. J. Mod. Phys. A **17**, 3663 (2002); J. Martin, R. Brandenberger, Phys. Rev. D **68**, 063513 (2003); S. Shankaranaryanan, Class. Quantum Grav. **20**, 75 (2003); C.P. Burgess, J.M. Cline, F. Lemieux, R. Holman, JHEP **0302**, 048 (2003); S.F. Hassan, M.S. Sloth, Nucl. Phys. B **674**, 434 (2003); J. Martin, C. Ringeval, Phys. Rev. D **69**, 083515 (2004); S. Shankaranarayanan, L. Sriramkumar, *ibid.* **70**, 123520 (2004); L. Sriramkumar, T. Padmanabhan, *ibid.* **71**, 103512 (2005); R. Easther, W.H. Kinney, H. Peiris, JCAP **0505**, 009 (2005); *ibid.* **0508**, 001 (2005); L. Sriramkumar, S. Shankaranarayanan, JHEP **0612**, 050 (2006)
42. K. Srinivasan, L. Sriramkumar, T. Padmanabhan, Phys. Rev. D **58**, 044009 (1998)
43. T. Jacobson, D. Mattingly, Phys. Rev. D **63**, 041502 (2001); *ibid.* **64**, 024028 (2001)
44. U. Harbach, S. Hossenfelder, Phys. Lett. B **632**, 379 (2006); S. Hossenfelder, Phys. Rev. D **73**, 105013 (2006); Class. Quant. Grav. **25**, 038003 (2008); D.A. Kothawala, L. Sriramkumar, S. Shankaranarayanan, T. Padmanabhan. Phys. Rev. D **80**, 044005 (2009)
45. S. Hossenfelder, L. Smolin, Phys. Canada **66**, 99 (2010)
46. D. Mattingly, Liv. Rev. Rel. **8**, 5 (2005); T. Jacobson, S. Liberati, D. Mattingly. Ann. Phys. **321**, 150 (2006)
47. M. Abramowitz, I.A. Stegun, *Handbook of Mathematical Functions* (Dover, New York, 1965)

Stability Longevity and All That: False Vacua and Topological Defects

Urjit A. Yajnik

1 Cuisine and a Canine

I got to know Paddy during the last year of MSc at IIT Bombay as my senior Kandaswamy had joined PhD in Prof. Jayant Narlikar's group, the same group as Paddy. Paddy already had a reputation of sorts as a very sharp student and a friendly guy. I ended up going to Texas Austin for PhD, and as luck would have it, he showed up there in the last year of my PhD. He had been invited for a visit by my advisor Prof. E.C. George Sudarshan, who was then busy reinvigorating Institute of Mathematical Sciences, Chennai known as Matscience in those days, as its new Director.

Paddy took interest in the work I had just then wound up for my thesis. One of the papers dealt with phase transition induced by cosmic strings [1]. I had identified the mechanism qualitatively and set up an example to demonstrate it, and was able to simulate the false vacuum decay. These were very interesting possibilities to explore as the dynamics of inflationary Universe was then believed to rely on the availability of a metastable or false vacuum. However there was a need to make the detailed mechanism explicit. Elegant arguments such as demonstrating the existence of an instanton were not working, that is, proving too difficult. Paddy who was not himself working in any of those topics listened carefully and made a pragmatic suggestion for my open problem. That was to was to analyse the small oscillations and see the emergence of a zero frequency mode as the temperature was lowered. This worked out nicely and appeared as Ref. [2].

Out of general friendship, and because he was a visitor to my Centre, and because he was also a senior, but perhaps also because he was making such fruitful suggestions for research, it was only appropriate that I invited him home once in a while. His apartment was nearby but then some neighbours used to leave their dog free to roam at night, causing understandable concerns for Paddy especially on a short visit in a

U.A. Yajnik (✉)
Indian Institute of Technology Bombay, Powai, Mumbai 400076, India
e-mail: yajnik@iitb.ac.in

J.S. Bagla and S. Engineer (eds.), *Gravity and the Quantum*,
Fundamental Theories of Physics 187, DOI 10.1007/978-3-319-51700-1_28

foreign country. So I also used to drop him off by car. It was during these sessions that he ended up discovering that I was reasonably good at making pasta sauce and for that reason, I assume among others, he came to decide that I could be a good addition as a visiting fellow to his group in TIFR. And so the following January, 1987, I joined Theoretical Astrophysics Group as a Visiting Fellow.

I knew him much better over the next few years in TIFR where he was a patient and amused listener (he still is I believe) of my daily observations about the way the world worked and also continued providing useful comments on the work I was doing. He was during that time as prolific as he has ever been. I remember particularly hearing from him about the Dark Matter problem which had not till then pricked the conscience of Particle Physicists sufficiently, and a "white paper" on inflationary density fluctuations, while some ideas that occupied him then included emergence of a fundamental length scale at the Planck scale. It was a very pleasant and fruitful time and a time to make some bold calculations on particle production during inflation [3]. Subsequently some of the undergraduate students from IIT Bombay went and worked with him at IUCAA during summers and we were in touch in that way.

Paddy has been a wellspring of innumerable new ideas and insights, has moved into all the emerging areas of General Relativity and Cosmology that seemed to bear on the fundamental issues, written many instructional and pedagogical articles and books, and has marked out a space for India in the world in those topics. Here I choose to contribute on the problem that was our first and only joint work, presenting some interesting further evolution of the ideas originated then, with the help of other colleagues as appears in Refs. [4, 5].

2 Introduction

In this contribution I cover two works, one in which an otherwise unstable topological configuration becomes stabilised and the other, in which a topological object can hasten the decay of a false vacuum. While an infinite cosmic string can be topologically stable, the same when made into a closed loop becomes unstable to decay by shrinking. However such strings can bind zero modes of fermions, which in some cases requires assigning fractional fermion number to the string. Such fermions can be Dirac or Majorana. Majorana fermions do not have a conserved fermion number. Yet it is interesting to prove that when a Majorana fermion is bound to a cosmic string loop, both configurations by themselves unstable, the combination becomes stable to spontaneous decay. This is due to the unusual assignment of fermion number of the configuration, whose quantum numbers cannot be matched to those of the vacuum. This is shown in Sects. 3 and 4. An interesting aspect of this analysis is that instead of an unfathomable *Dirac sea* we have a small *Majorana pond* at the energy threshold. The occurrence of a finite number of gapless states degenerate with the vacuum has been argued in [6] to be a signature of spontaneous symmetry breaking with a fermionic order parameter.

In a second study which takes us back to the problem on which I collaborated with Paddy, cosmic strings can induce the decay of a false vacuum. In our attempts then we showed that there were reasonable theoretical grounds for the phenomenon and proved it by numerical simulations. But in more recent work, thanks to the collaborators cited, we were able to doctor the model sufficiently that there was more theoretical control. In this case, it is possible to extend the well known theoretical tool of an "instanton bounce" to this case. Unlike in a translation invariant false vacuum, quite a bit of more ingenuity is required for deducing a bounce in the presence of a topological object, which breaks the translation invariance. However when everything is put together one gets an explicit dependence of the decay rate of the false vacuum on the parameters of the model. We can then see the enhancement in the rate, as also a possible regime of instability implied only due to the presence of the topological object. While the analytical answer could be obtained in the $2 + 1$ dimensional case of a vortex [7], here we take up the $3 + 1$ dimensional cosmic string which presents interesting theoretical challenges. In this case numerical calculations assist substantially in visualising and verifying the bounce. This is shown in Sects. 5 and 6.

3 Topological Solutions and Fractionalised Fermion Number

Cosmic strings and vortices are examples of solitons, extended objects occurring as stable states [8, 9] within Quantum Field Theory present the curious possibility of fermionic zero-energy modes trapped on such configurations. Their presence requires, according to well known arguments [10, 11], an assignment of half-integer fermion number to the solitonic states. Dynamical stability of such objects was pointed out in [12], in cosmological context in [13, 14] and later studied also in [15–17]. Fractional fermion number phenomenon also occurs in condensed matter systems and its wide ranging implications call for a systematic understanding of the phenomenon.

The impossibility of connecting half-integer valued states to integer valued states suggests that a superselection rule [18, 19] is operative. In a theory with a conserved charge (global or local), a superselection rule operates among sectors of distinct charge values because the conservation of charge is associated with the inobservability of rescaling operation $\Psi \rightarrow e^{iQ}\Psi$. In the case at hand, half-integer values of fermion number occur, preventing such states from decaying in isolation to the trivial ground state [12, 13].

Here we construct an example in which the topological object of a low energy theory is metastable due to the embedding of the low energy symmetry group in a larger symmetry group at higher energy. Examples of this kind were considered in [20]. Borrowing the strategies for bosonic sector from there, we include appropriate fermionic content to obtain the required zero-modes. Consider a theory with local $SU(3)$ symmetry broken to $U(1)$ by two scalars, Φ an octet acquiring a VEV $\eta_1 \lambda_3$ (λ_3

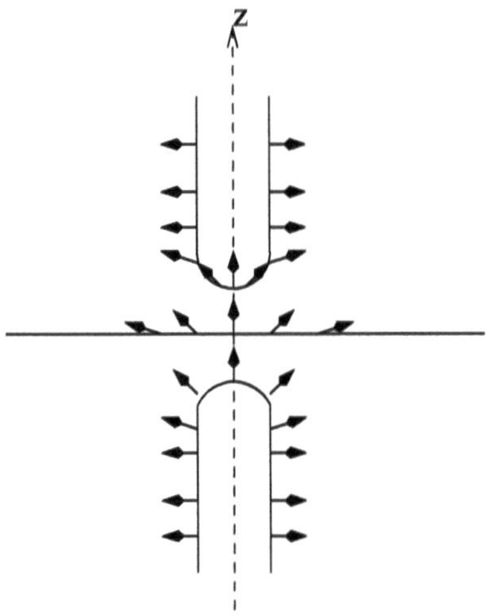

Fig. 1 Schematic configuration depicting break of a string into monopole anti-monopole pair. The diagram shows isospin vectors after the rupture of the string. Internal orientations are mapped to external space. They are shown just outside the core of the two resulting pieces and on the mid-plane symmetrically separating the two

here being the third Gell-Mann matrix) and ϕ, a $\bar{3}$, acquiring the VEV $\langle\phi^k\rangle = \eta_2\delta^{k2}$, with $\eta_2 \ll \eta_1$. Thus

$$SU(3)\xrightarrow{8}U(1)_3 \otimes U(1)_8\xrightarrow{\bar{3}}U(1)_+ \tag{1}$$

Here $U(1)_3$ and $U(1)_8$ are generated by λ_3 and λ_8 respectively, and $U(1)_+$ is generated by $(\sqrt{3}\lambda_8 + \lambda_3)/2$ and likewise $U(1)_-$ to be used below. It can be checked that this pattern of VEVs can be generically obtained from the quartic scalar potential of the above Higgses. The effective theory at the second breaking $U(1)_- \to \mathbb{Z}$ gives rise to cosmic strings. However the $\mathbb{Z}$ lifts to identity in the $SU(3)$ so that the string can break with the formation of monopole–antimonopole pair. See Fig. 1.

Now add a multiplet of left-handed fermions belonging to $\overline{15}$. Its mass terms arise from the following coupling to the $\overline{3}$

$$\mathscr{L}_{\text{Majorana}} = h_M \overline{\psi^C}_k^{\{ij\}} \psi_n^{\{lm\}} \phi^r (\varepsilon_{ilr}\delta_j^n \delta_m^k) \tag{2}$$

The indices symmetric under exchange have been indicated by curly brackets. No mass terms result from the 8 because it cannot provide a singlet from tensor product with $\overline{15} \otimes \overline{15}$ [21]. On substituting the vacuum expectation value (vev) of ϕ we get the mass matrix M of the fermions. A systematic enumeration shows that all but the two components $\psi_1^{\{22\}}$ and $\psi_3^{\{22\}}$ acquire majorana masses at the second stage of the breaking. Specifically we find the majorana mass matrix to be indeed rank 13. In the cosmic string sector, the vev of ϕ and therefore the mass matrix M becomes space dependent, $e^{in\theta}$, where θ is angle in a plane perpendicular to the string, and n is the winding number which has to be integer. The lowest energy bound states resulting from this coupling are characterized by a topological index, [22] $\mathscr{I} \equiv n_L - n_R$ where n_L and n_R are the zero modes of the left handed and the right handed fermions respectively. This index can be computed using the formula [22, 23]

$$\mathscr{I} = \frac{1}{2\pi i}(\ln \det M)|_{\phi=0}^{2\pi} \tag{3}$$

where M is the position dependent effective mass matrix for the fermions.

Thus, using either of the results [24] or [23] i.e., Eq. (3) we can see that there will be 13 zero modes present in the lowest winding sector of the cosmic string. Thus the induced fermion number differs from that of the vacuum by half-integer as required. According to well known reasoning [10] to be recapitulated below, this requires the assignment of either of the values $\pm 1/2$ to the fermion number of this configuration.

4 Assignment of Fermion Number

We now recapitulate the reasoning behind the assignment of fractional fermion number. We focus on the Majorana fermion case, which is more nettlesome, while the treatment of the Dirac case is standard [10, 11]. In the prime example in $3+1$ dimensions of a single left-handed fermion species Ψ_L coupled to an abelian Higgs model according to

$$\mathscr{L}_\psi = i\overline{\Psi_L}\gamma^\mu D_\mu \Psi_L - \frac{1}{2}\left(h\phi\overline{\Psi_L^C}\Psi_L + h.c.\right) \tag{4}$$

the following result has been obtained [24]. For a vortex oriented along the z-axis, and in the winding number sector n, the fermion zero-modes are of the form

$$\psi_0(\mathbf{x}) = \begin{pmatrix}1\\0\end{pmatrix}\left[U(r)e^{il\phi} + V^*(r)e^{i(n-1-l)\phi}\right] g_l(z+t) \tag{5}$$

In the presence of the vortex, τ^3 (here representing Lorentz transformations on spinors) acts as the matrix which exchanges solutions of positive frequency with those of negative frequency. It is therefore identified as the "particle conjugation" operator. In the above ansatz, the ψ in the zero-frequency sector are charge self-conjugates, $\tau^3\psi = \psi$, and have an associated left moving zero mode along the vortex. The functions satisfying $\tau^3\psi = -\psi$ are not normalizable. The situation is reversed when the winding sense of the scalar field is reversed, i.e., for $\sigma_u \sim e^{-in\phi}$. In the winding number sector n, regular normalizable solutions [24] exist for for $0 \le l \le n-1$. The lowest energy sector of the vortex is now 2^n-fold degenerate, and each zero-energy mode needs to be interpreted as contributing a value $\pm 1/2$ to the total fermion number of the individual states [10]. This conclusion is difficult to circumvent if the particle spectrum is to reflect the charge conjugation symmetry of the theory [25]. The lowest possible value of the induced number in this sector is $-n/2$. Any general state of the system is built from one of these states by additional integer number of fermions. All the states in the system therefore possess half-integral values for the fermion number if n is odd.

One puzzle immediately arises, what is the meaning of negative values for the fermion number operator for *Majorana* fermions? In the trivial vacuum, we can identify the Majorana basis as

$$\psi = \frac{1}{2}(\Psi_L + \Psi_L^C) \tag{6}$$

This leads to the Majorana condition which results in identification of particles with anti-particles according to

$$\mathscr{C}\psi\mathscr{C}^\dagger = \psi \tag{7}$$

making negative values for the number meaningless. Here $\mathscr{C}$ is the charge conjugation operator. We shall first verify that in the zero-mode sector we must indeed assign negative values to the number operator. It is sufficient to treat the case of a single zero-mode, which generalizes easily to any larger number of zero-modes. The number operator possesses the properties

$$[N, \psi] = -\psi \quad \text{and} \quad [N, \psi^\dagger] = \psi^\dagger \tag{8}$$

$$\mathscr{C}N\mathscr{C}^\dagger = N \tag{9}$$

Had it been the Dirac case, there should be a minus sign on the right hand side of Eq. (9). This is absent due to the Majorana condition. The fermion field operator for the lowest winding sector is now expanded as

$$\psi = c\psi_0 + \left\{ \sum_{\kappa,s} a_{\kappa,s}\chi_{\kappa,s}(x) + \sum_{\mathbf{k},s} b_{\mathbf{k},s}u_{\mathbf{k},s}(x) + h.c. \right\} \tag{10}$$

where the first summation is over all the possible bound states of non-zero frequency with real space-dependence of the form $\sim e^{-\kappa\cdot\mathbf{x}_\perp}$ in the transverse space directions $\mathbf{x}_\perp$, and the second summation is over all unbound states, which are asymptotically plane waves. These summations are suggestive and their exact connection to the Weyl basis mode functions [26] are not essential for the present purpose. Note however that no "$h.c.$" is needed for the zero energy mode which is self-conjugate. Then the Majorana condition (7) requires that we demand

$$\mathscr{C}\, c\, \mathscr{C}^\dagger = c \quad \text{and} \quad \mathscr{C}\, c^\dagger\, \mathscr{C}^\dagger = c^\dagger \tag{11}$$

Unlike the Dirac case, the c and $c^\dagger$ are not exchanged under charge conjugation. The only non-trivial irreducible realization of this algebra is to require the existence of a doubly degenerate ground state with states $|-\rangle$ and $|+\rangle$ satisfying

$$c|-\rangle = |+\rangle \quad \text{and} \quad c^\dagger|+\rangle = |-\rangle \tag{12}$$

with the simplest choice of phases. Now we find

$$\mathscr{C}\, c\, \mathscr{C}^\dagger \mathscr{C}|-\rangle \;=\; \mathscr{C}|+\rangle \tag{13}$$

$$\Rightarrow\;\; c(\mathscr{C}|-\rangle) = (\mathscr{C}|+\rangle) \tag{14}$$

This relation has the simplest non-trivial solution

$$\mathscr{C}|-\rangle \;=\; \eta_C^-|-\rangle \qquad \text{and} \qquad \mathscr{C}|+\rangle \;=\; \eta_C^+|+\rangle \tag{15}$$

where, for the consistency of (12) and (14) η_C^- and η_C^+ must satisfy

$$(\eta_C^-)^{-1}\eta_C^+ \;=\; 1 \tag{16}$$

Finally we verify that we indeed get values $\pm 1/2$ for N. The standard fermion number operator which in the Weyl basis is

$$N_F = \frac{1}{2}[\Psi_L^\dagger \Psi_L - \Psi_L \Psi_L^\dagger] \tag{17}$$

acting on these two states gives,

$$\frac{1}{2}(c\, c^\dagger \;-\; c^\dagger\, c)\;|\pm\rangle \;=\; \pm\frac{1}{2}|\pm\rangle \tag{18}$$

The number operator indeed lifts the degeneracy of the two states. For s number of zero modes, the ground state becomes 2^s-fold degenerate, and the fermion number takes values in integer steps ranging from $-s/2$ to $+s/2$. For s odd the values are therefore half-integral. Although uncanny, these conclusions accord with some known facts. They can be understood as spontaneous symmetry breaking for fermions [6]. The negative values of the number thus implied occur only in the zero-energy sector and do not continue indefinitely to $-\infty$. Instead of an unfathomable *Dirac sea* we have a small *Majorana pond* at the threshold.

5 Energetics and Dynamics of the Thin, False String

In the introductory section I mentioned the need for an instanton type description for tunneling in the string induced decay of the false vacuum. It was only many years later, in collaboration with Montréal and Seoul colleagues that this goal finally got taken up. While the insights were certainly new, there was a strong impetus for completing the argument because of the tremendous growth and ease of use of computing powers.

The work presented is a continuation of the earlier joint work [7, 27]. Here we consider the case of cosmic strings in a spontaneously broken $U(1)$ gauge theory, a generalized Abelian Higgs model. The potential for the complex scalar field has a local minimum at a nonzero value and the true minimum is at vanishing scalar field. We assume the energy density splitting between the false vacuum and the true vacuum is very small. The spontaneously broken vacuum is the false vacuum. In the scenario that we have described, the true vacuum lies at the regions of vanishing scalar field, thus the interior of the cosmic string is in the true vacuum while the exterior is in the false vacuum.

Related work of similar nature can be found in [28, 29]. More recently this phenomenon has drawn attention in other field theoretic contexts [30, 31], and in superstring theory, similar results are obtained regarding brane induced vacuum decay. See for instance [32–39].

5.1 Set-Up

We consider the abelian Higgs model (spontaneously-broken scalar electrodynamics) with a modified scalar potential corresponding to our previous work [7] but now generalized to 3 + 1 dimensions. The Lagrangian density of the model has the form

$$\mathscr{L} = -\frac{1}{4}F_{\mu\nu}F^{\mu\nu} + (D_\mu\phi)^*(D^\mu\phi) - V(\phi^*\phi), \tag{19}$$

where $F_{\mu\nu} = \partial_\mu A_\nu - \partial_\nu A_\mu$ and $D_\mu\phi = (\partial_\mu - ieA_\mu)\phi$. The potential is a sixth-order polynomial in ϕ [27, 40], written

$$V(\phi^*\phi) = \lambda(|\phi|^2 - \varepsilon v^2)(|\phi|^2 - v^2)^2. \tag{20}$$

Note that the Lagrangian is no longer renormalizable in 3 + 1 dimensions, however the understanding is that it is an effective theory obtained from a well defined renormalizable fundamental Lagrangian. The fields ϕ and A_μ, the vacuum expectation value v have mass dimension 1, the charge e is dimensionless and λ has mass dimension 2 since it is the coupling constant of the sixth order scalar potential. The potential energy density of the false vacuum $|\phi| = v$ vanishes, while that of the true vacuum has $V(0) = -\lambda v^6\varepsilon$. We rescale analogous to [7]

$$\phi \to v\phi \quad A_\mu \to vA_\mu \quad e \to \lambda^{1/2}ve \quad x \to x/(v^2\lambda^{1/2}) \tag{21}$$

so that all fields, constants and the spacetime coordinates become dimensionless, then the Lagrangian density is still given by Eq. (19) where now the potential is

$$V(\phi^*\phi) = (|\phi|^2 - \varepsilon)(|\phi|^2 - 1)^2. \tag{22}$$

and there is an overall factor of $1/(\lambda v^2)$ in the action.

Initially, the cosmic string will be independent of z the coordinate along its length and will correspond to a tube of radius R with a trapped magnetic flux in the true vacuum inside, separated by a thin wall from the false vacuum outside. R will vary in Euclidean time τ and in z to yield an instanton solution. Thus we promote R to a field $R \to R(z, \tau)$. Hence we will look for axially-symmetric solutions for ϕ and

A_μ in cylindrical coordinates (r, θ, z, τ). We use the following ansatz for a vortex of winding number n:

$$\phi(r,\theta,z,\tau) = f(r, R(z,\tau))e^{in\theta}, \qquad A_i(r,\theta,z,\tau) = -\frac{n}{e}\frac{\varepsilon^{ij} r_j}{r^2} a(r, R(z,\tau)), \tag{23}$$

where ε^{ij} is the two-dimensional Levi-Civita symbol. This ansatz is somewhat simplistic, it is clear that if the radius of the cosmic string swells out at some range of z, the magnetic flux will dilute and hence through the (Euclidean) Maxwell's equations some "electric" fields will be generated. In 3 dimensional, source free, Euclidean electrodynamics, there is no distinct electric field, the Maxwell equations simply say that the 3 dimensional magnetic field is divergence free and rotation free vector field that satisfies superconductor boundary conditions at the location of the wall. It is clear that the correct form of the electromagnetic fields will not simply be a diluted magnetic field that always points along the length of the cosmic string as with our ansatz, however the correction will not give a major contribution, and we will neglect it. Indeed, the induced fields will always be smaller by a power of $1/c^2$ when the usual units are used.

In the thin wall limit, the Euclidean action can be evaluated essentially analytically, up to corrections which are smaller by at least one power of $1/R$. The method of evaluation is identical to that in [7], we shall not repeat the details, we find

$$S_E = \frac{1}{\lambda v^2}\int d^2x \frac{1}{2} M(R(z,\tau))(\dot{R}^2 + R'^2) + E(R(z,\tau)) - E(R_0) \tag{24}$$

where

$$M(R) = \left[\frac{2\pi n^2}{e^2 R^2} + \pi R\right] \tag{25}$$

$$E(R) = \frac{n^2 \Phi^2}{2\pi R^2} + \pi R - \varepsilon\pi R^2 \tag{26}$$

and R_0 is the classically stable thin tube string radius.

6 Instantons and the Bulge

6.1 Tunnelling Instanton

We look for an instanton solution that is $O(2)$ symmetric, the appropriate ansatz is

$$R(z,\tau) = R(\sqrt{z^2+\tau^2}) = R(\rho) \tag{27}$$

with the imposed boundary condition that $R(\infty) = R_0$.

Such a solution will describe the transition from a string of radius R_0 at $\tau = -\infty$, to a point in $\tau = \rho_0$ say at $z = 0$ when a soliton anti-soliton pair is started to be created. The configuration then develops a bulge which forms when the pair separates to a radius which has to be again ρ_0 because of $O(2)$ invariance and which is the bounce point of the instanton along the z axis at $\tau = 0$. Finally the subsequent Euclidean time evolution continues in a manner which is just the (Euclidean) time reversal of evolution leading up to the bounce point configuration until a simple cosmic string of radius R_0 is re-established for $\tau \geq \rho_0$ and all z, i.e. $\rho \geq \rho_0$. The action functional is given by

$$S_E = \frac{2\pi}{\lambda v^2} \int d\rho\ \rho \left[\frac{1}{2} M(R(\rho)) \left(\frac{\partial R(\rho)}{\partial \rho} \right)^2 + E(R(\rho)) - E(R_0) \right]. \tag{28}$$

The instanton equation of motion is

$$\frac{d}{d\rho}\left(\rho M(R) \frac{dR}{d\rho} \right) - \frac{1}{2} \rho M'(R) \left(\frac{dR}{d\rho} \right)^2 - \rho E'(R) = 0 \tag{29}$$

with the boundary condition that $R(\infty) = R_0$, and we look for a solution that has $R \approx R_1$ near $\rho = 0$. The solution necessarily "bounces" at $\tau = 0$ since $\partial R(\rho)/\partial \tau|_{\tau=0} = R'(\rho)(\tau/\rho)|_{\tau=0} = 0$. (The potential singularity at $\rho = 0$ is not there since a smooth configuration requires $R'(\rho)|_{\rho=0} = 0$.) The equation of motion is better cast as an essentially conservative dynamical system with a "time" dependent mass and the potential given by the inversion of the energy function of Eq. (26), but in the presence of a "time" dependent friction where ρ plays the role of time:

$$\frac{d}{d\rho}\left(M(R) \frac{dR}{d\rho} \right) - \frac{1}{2} M'(R) \left(\frac{dR}{d\rho} \right)^2 - E'(R) = -\frac{1}{\rho} \left(M(R) \frac{dR}{d\rho} \right). \tag{30}$$

As the equation is "time" dependent, there is no analytic trick to evaluating the bounce configuration and the corresponding action. However, we can be reasonably sure of the existence of a solution which starts with a given $R \approx R_1$ at $\rho = 0$ and achieves $R = R_0$ for $\rho > \rho_0$, by showing the existence of an initial condition that gives an overshoot and another initial condition that gives an undershoot, in the same manner of proof as in [41]. Actually, numerically integrating to $\rho \approx 80{,}000$ the function falls back to the minimum of the inverted energy functional Eq. 26. On the other hand, we increase the starting point by 0.0001, the numerical solution overshoots the maximum at $R = R_0$. Thus we have numerically implemented the overshoot/undershoot criterion of [41].

The cosmic string emerges with a bulge described by the function numerically evaluated and represented in Fig. 2 which corresponds to $R(z, \tau = 0)$. A cross section of the bounce is visualised as the symmetrised figure obtained by reflecting the graph of Fig. 2 in the R axis, with $-\infty < \rho < \infty$. A 3-dimensional depiction of the bounce point is given in Fig. 3.

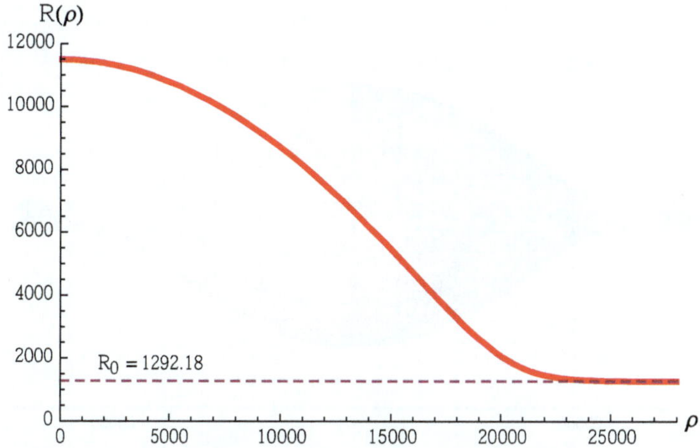

Fig. 2 The collective variable R signifying the radius of the string in the thin wall approximation, as a function of 2-dimensional radius ρ in the Euclidean $\tau - z$ plane

This radius function has argument $\rho = \sqrt{z^2 + \tau^2}$. Due to the Lorentz invariance of the original action, the subsequent Minkowski time evolution is given by $R(\rho) \rightarrow R(\sqrt{z^2 - t^2})$, which is of course only valid for $z^2 - t^2 \geq 0$. Fixed $\rho^2 = z^2 - t^2$ describes a space-like hyperbola that asymptotes to the light cone. The value of the function $R(\rho)$ therefore remains constant along this hyperbola. This means that the point at which the string has attained the large radius moves away from $z \approx 0$ to $z \rightarrow \infty$ at essentially the speed of light. The other side of course moves towards $z \rightarrow -\infty$. Thus the soliton anti-soliton pair separates quickly moving at essentially the speed of light, leaving behind a fat cosmic string, which is subsequently, classically unstable to expand and fill all space.

6.2 Tunnelling Amplitude

In the $3 + 1$ dimensional example we have presented the bounce action needs to be found numerically. It is reasonable to expect that as $\varepsilon \rightarrow 0$ the tunnelling barrier will get progressively bigger and at some point the tunnelling amplitude will vanish. On the other hand, there should exist a limiting value, call it ε_c, where the tunnelling barrier disappears at the so-called dissociation point [1, 28, 42], such that as $\varepsilon \rightarrow \varepsilon_c$, the action of the instanton will vanish, analogous to what was found in [7]. This is not possible to demonstrate in this case. However, for the bounce found above, we compare its implications to the other situations. Given the bounce action, the decay rate per unit length of the cosmic string will be of the form

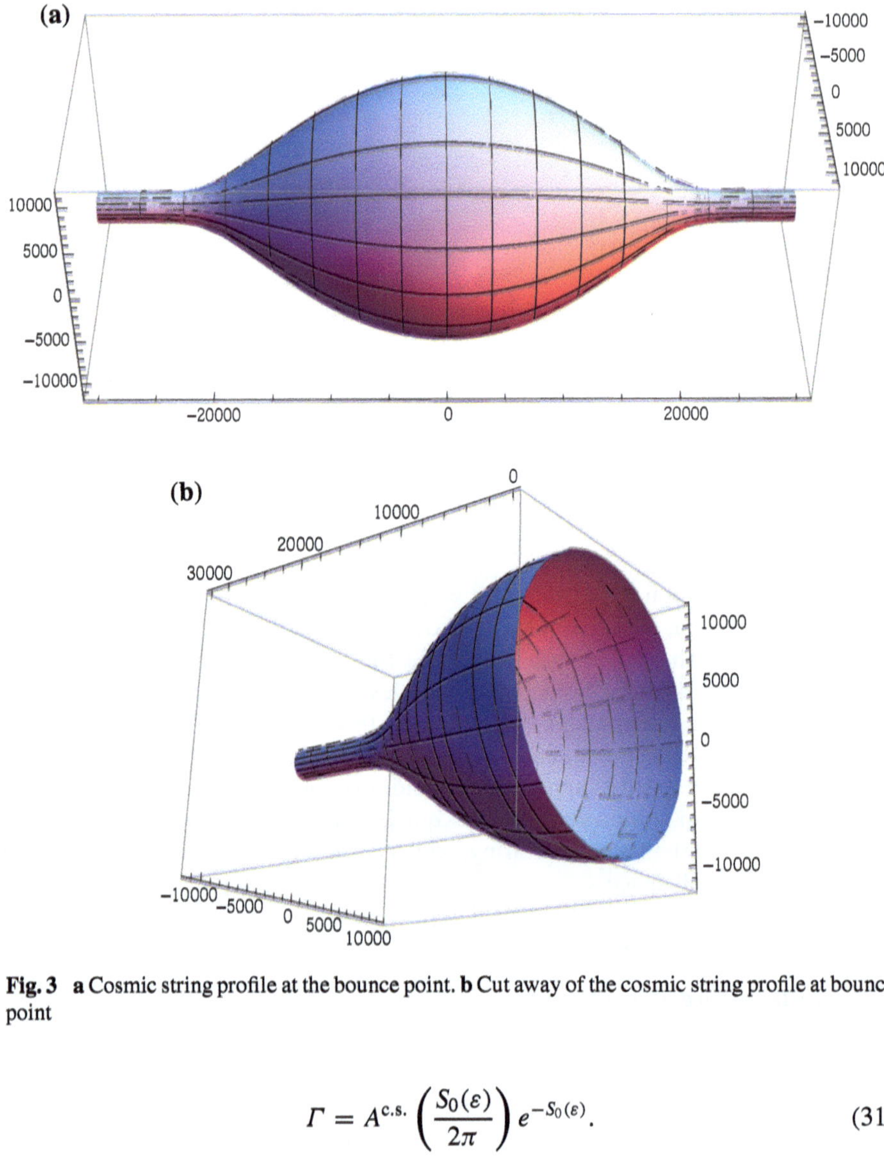

Fig. 3 **a** Cosmic string profile at the bounce point. **b** Cut away of the cosmic string profile at bounce point

$$\Gamma = A^{\text{c.s.}} \left(\frac{S_0(\varepsilon)}{2\pi} \right) e^{-S_0(\varepsilon)}. \tag{31}$$

where $A^{\text{c.s.}}$ is the determinantal factor excluding the zero modes and $\left(\frac{S_0(\varepsilon)}{2\pi}\right)$ is the correction obtained after taking into account the two zero modes of the bulge instanton. These correspond to invariance under Euclidean time translation and spatial translation along the cosmic string [41]. In general, there will be a length L of cosmic string per volume L^3. For a second order phase transition to the metastable vacuum, L is the correlation length at the temperature of the transition which satisfies $L^{-1} \approx \lambda v^2 T_c$ [43]. For first order transitions, it is not clear what the density of cosmic strings will be. We will keep L as a parameter but we do expect that it is microscopic. Then in a large volume Ω, we will have a total length NL of cosmic string, where $N = \Omega/L^3$. Thus the decay rate for the volume Ω will be

$$\Gamma \times (NL) = \Gamma\left(\frac{\Omega}{L^2}\right) = A^{\text{c.s.}}\left(\frac{S_0(\varepsilon)}{2\pi}\right)e^{-S_0(\varepsilon)}\frac{\Omega}{L^2} \tag{32}$$

or the decay rate per unit volume will be

$$\frac{\Gamma}{L^2} = \frac{A^{\text{c.s.}}\left(\frac{S_0(\varepsilon)}{2\pi}\right)e^{-S_0(\varepsilon)}}{L^2}. \tag{33}$$

A comparable calculation with point-like defects [7] would give a decay rate per unit volume of the form

$$\frac{\Gamma^{\text{point like}}}{L^3} = \frac{A^{\text{point like}}\left(\frac{S_0^{\text{point like}}(\varepsilon)}{2\pi}\right)^{3/2}e^{-S_0^{\text{point like}}(\varepsilon)}}{L^3} \tag{34}$$

and the corresponding decay rate from vacuum bubbles (without topological defects) [41] would be

$$\Gamma^{\text{vac. bubble}} = A^{\text{vac. bubble}}\left(\frac{S_0^{\text{vac. bubble}}(\varepsilon)}{2\pi}\right)^2 e^{-S_0^{\text{vac. bubble}}(\varepsilon)}. \tag{35}$$

Since the length scale L is expected to be microscopic, we would then find that the number of defects in a macroscopic volume (i.e. universe) could be incredibly large, suggesting that the decay rate from topological defects would dominate over the decay rate obtained from simple vacuum bubbles la Coleman [41]. Of course the details do depend on the actual values of the Euclidean action and the determinantal factor that is obtained in each case.

7 Conclusion

Metastable classical lumps, also referred to as embedded defects can be found in several theories. The conditions on the geometry of the vacuum manifold that give rise to such defects were spelt out in [20]. We have studied the related question of fermion zero-energy modes on such objects. It is possible to construct examples of cosmic strings in which the presence of zero-modes signals a fractional fermion number both for Dirac and Majorana masses. It then follows that such a cosmic string cannot decay in isolation because it belongs to a distinct superselected Quantum Mechanical sector. Thus a potentially metastable object can enjoy induced stability due to its bound state with fermions.

Although decay is not permitted in isolation, it certainly becomes possible when more than one such objects come together in appropriate numbers. In the early Universe such objects could have formed depending on the unifying group and its breaking pattern. Their disappearance would be slow because it can only proceed through encounters between objects with complementary fermion numbers adding up to an integer. Another mode of decay is permitted by change in the ground state in the course of a phase transition. When additional Higgs fields acquire a vacuum expectation value, in turn altering the boundary conditions for the Dirac or Majorana equation, the number of induced zero modes may change from being odd to even thus imparting the strings an integer fermion number. The decay can then proceed at the rates calculated in [20]. Such a possibility, for the case of topologically stable string can be found for realistic unification models in [14, 15, 44, 45].

There are many instances where the vacuum can be meta-stable. The symmetry broken vacuum can be metastable. Such solutions for the vacuum can be important for cosmology and for the case of supersymmetry breaking see [46] and the many references therein. In string cosmology, the inflationary scenario that has been obtained in [47], also gives rise to a vacuum that is meta-stable, and it must necessarily be long-lived to have cosmological relevance.

In a condensed matter context symmetry breaking ground states are also of great importance. For example, there are two types of superconductors [48]. The cosmic string is called a vortex line solution in this context, and it is relevant to type II superconductors. The vortex line contains an unbroken symmetry region that carries a net magnetic flux, surrounded by a region of broken symmetry. If the temperature is raised, the true vacuum becomes the unbroken vacuum, and it is possible that the system exists in a superheated state where the false vacuum is meta-stable [49]. This technique has actually been used to construct detectors for particle physics [50]. Our analysis might even describe the decay of vortex lines in superfluid liquid 3Helium [51].

The decay of all of these metastable states could be described through the tunnelling transition mediated by instantons in the manner that we have computed in this article. For appropriate limiting values of the parameters, for example when $\varepsilon \to \varepsilon_c$, the suppression of tunnelling is absent, and the existence of vortex lines or cosmic strings could cause the decay of the meta-stable vacuum without bound. Experimental observation of this situation would be interesting.

Acknowledgements I thank the editors of this volume for giving me the opportunity to contribute. This article is patched together from [4, 5], and it is a pleasant duty to acknowledge the partnership of the coauthors of both the papers where the work was originally reported. I also acknowledge the support of the Department of Science and Technology, India, and the Ministère des relations internationales du Québec.

References

1. U.A. Yajnik, Phys. Rev. D **34**, 1237 (1986)
2. U.A. Yajnik, T. Padmanabhan, Phys. Rev. D **35**, 3100 (1987)

3. U.A. Yajnik, Phys. Lett. B **234**, 271 (1990)
4. N. Sahu, U.A. Yajnik, Phys. Lett. B **596**, 1 (2004). arXiv:hep-th/0405140
5. B.H. Lee, W. Lee, R. MacKenzie, M.B. Paranjape, U.A. Yajnik, D.h. Yeom, Phys. Rev. D **88**(10), 105008 (2013). arXiv:1310.3005 [hep-th]
6. R. MacKenzie, F. Wilczek, Phys. Rev. D **30**, 2194 (1984)
7. B.H. Lee, W. Lee, R. MacKenzie, M.B. Paranjape, U.A. Yajnik, D.h Yeom, Phys. Rev. D **88**, 085031 (2013)
8. H.B. Nielsen, P. Olesen, Nucl. Phys. B **61**, 45 (1973)
9. A. Abrikozov, JETP (Sov. Phys.) **5**, 1173 (1957)
10. R. Jackiw, C. Rebbi, Phys Rev. D **13**, 3398 (1976)
11. R. Jackiw, Rev. Mod. Phys. **49**, 681 (1977)
12. H. de Vega, Phys. Rev. D **18**, 2932 (1978)
13. A. Stern, Phys. Rev. Lett. **52**, 2118 (1983)
14. A. Stern, U.A. Yajnik, Nucl. Phys. B **267**, 158 (1986)
15. S.C. Davis, A.C. Davis, W.B. Perkins, Phys. Lett. B **408**, 81 (1997)
16. A.C. Davis, S.C. Davis, W.B. Perkins, in *Proceedings of COSMO99*. arXiv:hep-ph/0005091
17. A.C. Davis, T.W.B. Kibble, M. Pickles, D.A. Steer, Phys. Rev. D **62**, 083516 (2000)
18. G.C. Wick, A.S. Wightman, E.P. Wigner, Phys. Rev. **88**, 101 (1952)
19. S. Weinberg, *The Quantum Theory of Fields*, vol. I, sec. 3.3 (Cambridge University Press, Cambridge, 1996)
20. J. Preskill, A. Vilenkin, Phys. Rev. D **47**, 2324 (1993)
21. R. Slansky, Phys. Rep. **79**, 1 (1981)
22. E.J. Weinberg, Phys. Rev. D **24**, 2669 (1981)
23. N. Ganoulis, G. Lazarides, Phys. Rev. D **38**, 547 (1988)
24. R. Jackiw, P. Rossi, Nucl. Phys. B **190**, 681 (1981)
25. E.C.G. Sudarshan, U.A. Yajnik, Phys. Rev. D **33**, 1830 (1986)
26. See for instance M. Fukugita, T. Yanagida, in *Physics and Astrophysics of Neutrinos*, ed. by M. Fukugita, A. Suzuki (Springer, Berlin, 1994), pp. 1–248
27. B. Kumar, M.B. Paranjape, U.A. Yajnik, Phys. Rev. D **82**, 025022 (2010)
28. B. Kumar, U.A. Yajnik, Nucl. Phys. B **831**, 162 (2010)
29. A. Aguirre, M.C. Johnson, M. Larfors, Phys. Rev. D **81**, 043527 (2010)
30. M. Haberichter, R. MacKenzie, M.B. Paranjape, Y. Ung, J. Math. Phys. **57**(4), 042303 (2016)
31. É. Dupuis, Y. Gobeil, R. MacKenzie, L. Marleau, M.B. Paranjape, Y. Ung, Phys. Rev. D **92**(2), 025031 (2015)
32. M. Eto, Y. Hamada, K. Kamada, T. Kobayashi, K. Ohashi, Y. Ookouchi, JHEP **1303**, 159 (2013)
33. K. Kamada, T. Kobayashi, K. Ohashi, Y. Ookouchi, JHEP **1305**, 091 (2013)
34. A. Kasai, Y. Nakai, Y. Ookouchi, JHEP **1606**, 029 (2016)
35. A. Kasai, Y. Ookouchi, JHEP **1506**, 098 (2015)
36. A. Kasai, Y. Ookouchi, Phys. Rev. D **91**(12), 126002 (2015)
37. B.H. Lee, W. Lee, D.h. Yeom. Phys. Rev. D **92**(2), 024027 (2015)
38. B. Garbrecht, P. Millington, Phys. Rev. D **92**, 125022 (2015)
39. Y. Nakai, Y. Ookouchi, arXiv:1608.01232 [hep-th]
40. P.J. Steinhardt, Phys. Rev. D **24**, 842 (1981)
41. S.R. Coleman, Phys. Rev. D **15**, 2929 (1977); C.A. Callan, S. Coleman, Phys. Rev. D **16**, 1762 (1977)
42. B. Kumar, U.A. Yajnik, Phys. Rev. D **79**, 065001 (2009)
43. T.W.B. Kibble, J. Phys. A **9**, 1387 (1976); T.W.B. Kibble, Phys. Rep. **67**, 183 (1980); W.H. Zurek, Nature (London) **317**, 505 (1985); W.H. Zurek, Acta Phys. Pol. B **24**, 1301 (1993)
44. U.A. Yajnik, H. Widyan, D. Choudhari, S. Mahajan, A. Mukherjee, Phys. Rev. D **59**, 103508 (1999)
45. S.C. Davis, W.B. Perkins, A.C. Davis, Phys. Rev. D **62**, 043503 (2000)
46. S.A. Abel, C.-S. Chu, J. Jaeckel, V.V. Khoze, JHEP **0701**, 089 (2007). arXiv:hep-th/0610334; W. Fischler, V. Kaplunovsky, L. Mannelli, M. Torres, C. Krishnan, JHEP **03**, 107 (2007)

47. S. Kachru, R. Kallosh, A.D. Linde, S.P. Trivedi, Phys. Rev. D **68**, 046005 (2003). arXiv:hep-th/0301240
48. N.W. Ashcroft, N.D. Mermin, Solid State Physics (Harcourt College Publishers, 1976)
49. A.J. Dolgert, S.J. Di Bartolo, A.T. Dorsey, Phys. Rev. B **53**, 5650 (1996)
50. Superconductive Particle Detectors, ed. by A. Barone (World Scientific, Singapore, 1987); K. Pretzl, J. Low Temp. Phys. **93**, 439 (1993)
51. A.J. Leggett, A theoretical description of the new phases of liquid 3He. Rev. Mod. Phys. **47**, 331 (1975)

MIX
Papier aus verantwortungsvollen Quellen
Paper from responsible sources
FSC® C105338

If you have any concerns about our products,
you can contact us on
ProductSafety@springernature.com

In case Publisher is established outside the EU,
the EU authorized representative is:
Springer Nature Customer Service Center GmbH
Europaplatz 3, 69115 Heidelberg, Germany

Printed by Libri Plureos GmbH
in Hamburg, Germany